Biomaterials in Orthopedics

Biomaterials in Orthopedics

edited by

Michael J. Yaszemski
Mayo Clinic
Rochester, Minnesota, U.S.A.

Debra J. Trantolo
Cambridge Scientific, Inc.
Cambridge, Massachusetts, U.S.A.

Kai-Uwe Lewandrowski
Cleveland Clinic
Cleveland, Ohio, U.S.A.

Vasif Hasirci
Middle East Technical University
Ankara, Turkey

David E. Altobelli
DEKA Research and Development Corporation
Manchester, New Hampshire, U.S.A.

Donald L. Wise
Northeastern University
Boston, and Cambridge Scientific, Inc.
Cambridge, Massachusetts, U.S.A.

MARCEL DEKKER, INC. NEW YORK • BASEL

Library of Congress Cataloging-in-Publication Data
A catalog record for this book is available from the Library of Congress.

ISBN: 0-8247-4294-X

This book is printed on acid-free paper.

Headquarters
Marcel Dekker, Inc., 270 Madison Avenue, New York, NY 10016, U.S.A.
tel: 212-696-9000; fax: 212-685-4540

Distribution and Customer Service
Marcel Dekker, Inc., Cimarron Road, Monticello, New York 12701, U.S.A.
tel: 800-228-1160; fax: 845-796-1772

Eastern Hemisphere Distribution
Marcel Dekker AG, Hutgasse 4, Postfach 812, CH-4001 Basel, Switzerland
tel: 41-61-260-6300; fax: 41-61-260-6333

World Wide Web
http://www.dekker.com

The publisher offers discounts on this book when ordered in bulk quantities. For more information, write to Special Sales/Professional Marketing at the headquarters address above.

Current printing (last digit):

10 9 8 7 6 5 4 3 2 1

PRINTED IN THE UNITED STATES OF AMERICA

Preface

This reference text *Biomaterials in Orthopedics* contains cutting-edge presentations by leading authorities dealing with the critical issues surrounding materials for bone repair and reconstruction. The chapters cover the use of a wide range of biomaterials from bioabsorbables to ceramics and metals. With input from scientific, engineering, and clinical professionals the text highlights the multi-disciplinary nature of biomaterial development and application.

Beginning with discussions of the fundamental aspects of biocompatibility and interfacial phenomena, the text then moves on to discussions of emerging bioabsorbable materials and novel advancements in time-honored ceramic and metallic bone repair biomaterials. Applications to traditional orthopedic sectors are considered along with those to oral and maxillofacial reconstruction and ever-challenging spinal applications. Despite the long history of orthopedic biomaterials, it is amazingly clear that the field begs for new solutions to clinical demands. Active lifestyles and aging populations drive a market that showcases the need for a strengthening in the battery of biomaterials in the surgeon's toolcase.

This text offers a wealth of valuable data and experience that will be of use to all bioengineers, materials scientists, and clinicians concerned with the properties, performance, and use of bone repair biomaterials—from research engineers faced with designing materials to surgeons interested in material biocompatibility and performance. The chapters, some of which include case studies, provide rich insights into our experiences today with a broad spectrum of contributing authors. The book focuses on discussion of the following:

Issues of biomaterial performance and biocompatibility
The rationale for designing bioabsorbable biomaterials for bone repair
Techniques for enhancing the surface properties of biomaterials
Developments in mechanical optimization of orthopedic biomaterials, and
Advances in fillers, cements and devices.

The orthopedic industry is currently one of the strongest market performers, and biomaterials are a key ingredient to this dynamic growth. Optimization of orthopedic biomaterials is in a constant state of activity, as old materials fail to withstand the tests of time and modern techniques and procedures drive the demand for new materials and devices. This text highlights the aggressive approaches necessary to address this demand.

Michael J. Yaszemski
Debra J. Trantolo
Kai-Uwe Lewandrowski
Vasif Hasirci
David E. Altobelli
Donald L. Wise

Contents

Biomaterials in Orthopedics

1
Hard Tissue–Biomaterial Interactions

Petek Korkusuz
Hacettepe University Faculty of Medicine
Ankara, Turkey

Feza Korkusuz
Middle East Technical University
Ankara, Turkey

I. INTRODUCTION: BONE AS A FUNCTIONAL ORGAN

Bone and its several associated elements—cartilage, connective tissue, vascular elements, and nervous components—act as a functional organ. They provide support and protection for soft tissues and act together with skeletal muscles to make body movements possible. Bones are relatively rigid structures and their shapes are closely related to their functions. Bone metabolism is mainly controlled by the endocrine, immune, and neurovascular systems, and its metabolism and response to internal and external stimulations are still under assessment.

Long bones of the skeletal system are prone to injury, and internal or external fixation is a part of their treatment. Joint replacement is another major intervention where the bone is expected to host biomaterials. Response of the bone to biomaterial intervenes with the regeneration process. Materials implanted into the bone will, nevertheless, cause local and systemic biological responses even if they are known to be inert. Host responses with joint replacement and fixation materials will initiate an adaptive and reactive process [1].

The objective of this article is to review the tissue response to biomaterials implanted into the bone for a better understanding of interactions of the hard tissue and the implant. Metals, ceramics, and polymers and/or their composites and coatings are evaluated for their tissue response. The spectrum of response with metals lies between aseptic loosening and carcinogenesis. Ceramics, on the other hand, may cause a nonspecific inflammation and bone marrow depletion. Hydroxyapatite and calcium phosphate particles are shown to be capable of stimulating the expression and secretion of cytokines and proteases that enhance bone resorption. Polymethylmethacrylate and polylactide and/or polyglycolide materials are frequently used polymers in hard tissues. Extensive research on improving the biocompatibility of these polymers used in clinical applications is going on. Various factors such as the type, structure, origin, and composition define the foreign body reaction toward the polymer. Polyhydroxybutyrate (PHBV) seems to cause a milder tissue response when compared with other polymers. Implants of metal should be of low profile, and their properties should be improved to overcome wear debris. Less use of metals for bone and joint replacement in the future is expected.

II. METALS

A. Biocompatibility

Metals have been used successfully for decades in fracture fixation and joint replacement. Mechanisms of implant failure were recently the target of intensive research as longevity and expectations from such implants are increasing [2,3]. An estimated 11 million people in the United States reported having at least one medical device in 1988 [3]. Fixation devices and artificial joints comprise 44% of all medical devices. The percentage of usage of fixation devices and artificial joints with one or more problem were 33.2 and 31.6%, respectively [3]. The demand for such medical device implants is expected to increase in the coming years.

Currently used metal implants are expected to be inert when implanted into the human bone. They are supposed to be bioactive as their surfaces are porous or coated. Metallic fixation devices are usually used alone, whereas artificial joints can comprise several parts other than metal including polymer and ceramic. If only metal has been used as in the case of uncemented endoprostheses, in a young and active patient, the head of the prosthesis may be bipolar. Cemented prostheses once again became popular using the third generation cementing techniques (i.e., medullary plug, centralizers, viscous cement, pressurising). It is obvious that the rate of complication will increase as the number of materials used in an artificial joint increases. The type of metal, manufacturer and its standards, alloy, composition, processing conditions, and mechanical properties influence the interaction of metal and the bone. Stainless steel, cobalt, titanium, and their alloys are widely used in the production of artificial joints and fixation devices. The advantages of titanium over cobalt alloys are lower modulus of elasticity and higher biocompatibility [4]. The rate of reaction toward metals is more severe in artificial joint surgery than fracture fixation as motion in the prior and immobilization in the latter are the ultimate aims.

Long-term stability is closely related to bone–implant integration. Bone cells mediate initial response to the implant. The interaction between osteoblasts and biomaterial surfaces was evaluated extensively. Response of osteoblastic cells toward commonly used titanium and cobalt alloys revealed cellular extension on both alloys during the first 12 h [5]. Osteoblasts spread relatively less on rough titanium alloy than cobalt alloy. Vinculin immunostaining at focal adhesion contacts distributed throughout the cells adhering to titanium alloy, but were relatively sparse and localized to cellular processes on cobalt alloy [5]. Cell attachment was directly to implant materials through integrins [6]. Thus, the initial interaction between the implant and surrounding bone might differ to the origin of osteoblastic cells [7]. Both titanium and cobalt alloys demonstrate good biocompatibility [8]. Osseointegration was less on cobalt alloy surfaces though cartilage, and osteoid tissue was observed more frequently on the cobalt alloy than on the titanium alloy surface [8]. Cobalt alloys were also presented to release large amounts of metal ions, which could mediate cytokine release and hypersensitivity reaction [9]. Osseointegration established extensively when titanium was implanted into bone marrow [10]. Thus, some bone marrow cells formed an incomplete layer in contact with the titanium implant and presented morphologic characteristics of macrophages and multinucleated giant cells [10].

Implant wear is identified as the most important cause of aseptic loosening in artificial joint surgery [11–16]. Generation of wear debris and the subsequent tissue reaction to it are the major concerns of this type of surgery. Particles of wear debris of bone cement, polyethylene, and metal itself initiate an inflammatory reaction that induces bone resorption and implant loosening [17,18]. Metal debris is produced as a result of adhesive, abrasive, or fatigue (also known as delamination) wear. Corrosion is another mechanism that can generate debris. Wear and corrosion may couple their effects. Debris is most commonly produced at the articular surface, modular implant junction, and various interfaces such as the implant–bone, implant–cement, and cement–bone. The amount, chemical composition, and physical aspect of wear debris identify the type and feature of tissue reaction [19]. Debris particles elicit a cell-mediated inflammatory response that results in either a foreign body giant cell granuloma or a massive release of osteolytic factors affecting bone biology and metabolism [20]. Release of chemokines by macrophages in response to wear particles may contribute to chronic inflammation at the bone–implant interface [21]. A study with x-ray scanning analytical microscopy (XSAM) revealed severe tissue damage around Ni and Cu implants, while fibrous connective tissue was formed around the Fe implant [22]. Wear particles induce endotoxins responsible of adverse tissue response that can be controlled prior to implantation [23].

Clinical features of aseptic loosening in artificial joints are pain and loss of range of motion. Radiography reveals osteolysis at the bone–implant interface. Osteolysis can be recognized with cemented and uncemented implants. Osteolysis may be asymptomatic in some patients with uncemented implants, demonstrating that osteolysis alone may not be of clinical importance and a sign of loosening. Osteolysis is known to increase with years of follow-up in cemented [24] and uncemented implants [12,25]. In cemented implants, osteolysis may vary according to the type of cement and application procedure. Effect of bone cement on bone will be discussed in coming sections. It was found that most of the debris belonged to the ultra high molecular weight polyethylene (mean size, approximately 0.5μm) of the acetabular cup in loose, uncemented artificial hip joints [26]. In cemented artificial hip joints, wear particles arise from the bone cement itself, acetabular cup polyethylene, and metal, respectively [24]. Metal and polymer particles initiate the complex, biomaterial-initiated osteolytic and/or adaptive cascade (Fig. 1) in a size- and dose-dependent manner [15]. Metal particles are also defined to cause apoptosis in cells of tissue around the implant [27]. Numerous macrophages, foreign body giant cells, and fibroblasts generally surround abundant particle debris [16]. Phagocytosis of debris by macrophages may serve as a stimulus for cellular activation with synthesis and secretion of bone-resorbing factors. Such factors include proinflammatory mediators interleukin-1 (IL-1) [28,29], interleukin-4 (IL-4) [30], interleukin-6 (IL-6) [28,29,31], interleukin-8 (IL-8) [32], granulocyte macrophage colony stimulating factor (GM-CSF) [30], tumor necrosis factor-α (TNF.-α), and prostaglandin E_2 (PGE_2) [28,29,33,34]. mRNA levels of inducible nitric oxide synthetase (iNOS) and cytosolic phospholipase A_2 ($cPLA_2$) together with TNF-α were up-regulated in uncemented implants [35]. Interleukin-4 was found to down-regulate particle-induced activation of macrophages [30], whereas titanium particles up-regulated the expression of matrix metalloproteinases stromelysin and collagenase in fibroblasts [28]. Nitric oxide [36] and cyclooxygenase 2 (COX-2) [13] play important roles in wear debris. Thus, nitric oxide production at titanium surface was not detected in one study [37]. Release of selected chemokines (MCP-1, MIP-1α, and RANTES) was found to initiate macrophage accumulation around wear debris [21]. Macrophage subgroups interact differently to polyethylene and titanium implants. Macrophages positive for ED-1 are involved in the tissue response of polyethylene and titanium [38]. Chemokines and cytokines mediate inflammation [39]. Mononuclear osteoclast precursors, stimulated by monocyte colony stimulating factor (M-CSF), initiate osteoclastic activity, and bone resorption begins. One article also demonstrated that even osteoblasts that become positive for macrophage marker CD68 might play a role in periprosthetic bone resorption [40]. Osteoblasts present phenotypic differentiation depending on the chemical composition of the debris particles [15]. Particles are usually found in the cytosol of the cells following phagocytosis. Osteoblasts present extensive ruffled cell membranes, less developed endoplasmic reticulum, swollen mitochondria, and vacuolar inclusions [15].

Metallic particles and their side effects are not only limited to the peri-implant site; they are also found in other organs, such as the peripheral blood, liver, spleen, and lymph nodes [41]. Metallic particles in the liver or spleen were more prevalent in patients who had had a

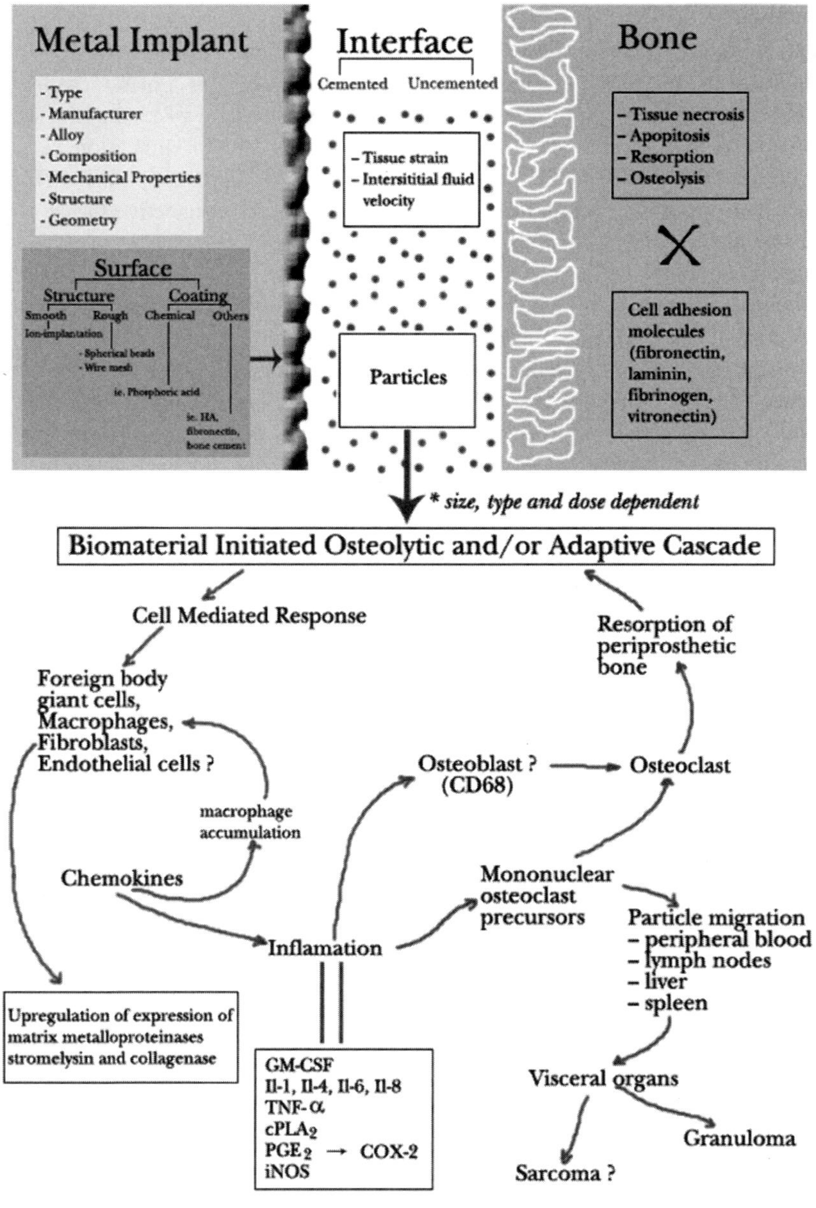

Figure 1 Metal implant–hard tissue interface and the biomaterial-initiated osteolytic and/or adaptive cascade.

failed arthroplasty. In one living patient, dissemination of titanium particles from a hip prosthesis with mechanical failure was associated with visceral granulomatosis reaction and hepatosplenomegaly, which required operative and medical treatment [41,42]. Even in well-functioning prostheses the serum and urine concentrations of titanium and chromium were found to be higher than in the normal population [12]. Serum levels of bone-resorbing cytokine GM-CSF level increased significantly in patients with aseptic loosening of hip prostheses [43]. Patients having revision arthroplasty of the hip presented increased chromosome translocations and aneuploidy in their peripheral blood [44]. Although intraarticular testing of titanium and chromium alloys in rats revealed no local tumor development [45] a study of 12 cases on orthopedic implant–related sarcoma revealed using metallic implants as artificial joints might lead to severe end results [46]. Two of the high-grade sarcoma of Keel's study were located in the soft tissue and 10 in bone [46]. Seven patients were reported to develop osteosarcoma, four malignant fibrous histiocytoma, and one a malignant peripheral nerve sheath tumor. Alloys that contain nickel had higher carcinogenic and toxic potencies [47]. One important aspect of sarcoma arising from artificial joints is the differential diagnosis of infection. Chronic and long-lasting infections may trigger sarcoma. Aggressiveness, high-grade, and metastasis of sarcoma arising from artificial joints need precaution and awareness of the symptoms. Further studies related with this severe complication are essential. It is recommended that surgeons should (1) select prostheses with minimal susceptibility to metal corrosion and wear, (2) replace implanted prostheses when there is evidence of corrosion and mechanical failure, (3) carry out epidemiological studies to quantify cancer risk in patients with various types of metal implants, and (4) improve in vitro assays for carcinogenicity of alloys intended for use in bone tissue [48].

B. Effectiveness of Metal Coatings

Coatings or ion implantation [49–51] are usually used to improve the biocompatibility of implants and decrease metallic wear and corrosion. Rough [52] or porous [53] surfaces allow cell attachment. One simple method to allow tissue ingrowth into the implant is to modify its surface by implanting spherical beads [54] or wire mesh. Though manufacturers' manuals indicate these surface modifications allow bone cells to grow into the implants and increase their mechanical strength and biocompatibility, longitudinal, randomized, prospective clinical studies with long-term follow-up are lacking. A case report concerning bone ingrowth in a porous-coated knee arthroplasty revealed that the prosthesis was held in situ by collagenous tissue, and calcified bone did not appear to interact with the metallic coating [55]. One in vitro experimental study, on the other hand, revealed that rough Ni-Ti surface promoted transforming growth factor beta (TGF-β) expression, a mediator of bone healing and differentiation [56]. Another autopsy study of five femurs indicated that circumferential porous coating of uncemented femoral components could prevent distal migration of polyethylene wear debris. [57].

An alternative method is the use of biocompatible chemicals [58] and materials such as ceramics for coating. Titanium surfaces were modified using phosphoric acid in an in vitro study to improve the biocompatibilty of dental implants. Results indicated that pretreatment of the implant with phosphoric acid caused no cytotoxicity to the osteoblasts [59]. Micro arc oxidation method in phosphoric acid on titanium implants provided chemical bonding sites for calcium ions during mineralization [60]. Hydroxyapatite (HA) coating is a proven method to improve the implants' mechanical bonding [61,62] and biocompatibility [63–66]. It is demonstrated that when the gap between the coating and bone is 1.0 mm or less, mechanical attachment strength and bone ingrowth increase significantly at all time periods [63]. Alkaline phosphatase activity, a marker of osteogenic activity, increases significantly with respect to the uncoated titanium in hydroxyapatite-coated implants [65]. The quality and thickness of coating may vary between manufacturers, and thick coatings on metal surfaces are prone to delamination [67]. Bone ingrowth and attachment mainly take place on the distal and medial parts of the HA-coated surface of femoral implants [64].

Hydroxyapatite coating may lead to the attachment of other cells than osteoblasts. Hydoxyapatite coating increases susceptibility to contamination of bacteria [68]. Hydroxyapatite coating may also increase the risk of heterotropic bone formation [69]. One recent study thus indicated that HA coating of implants could prevent distal migration of polyethylene wear debris as the tightly bonded bone on the surface of the implant will form a seal and inhibit peri-implant migration of polyethylene particles [70]. On the other hand, when particles arise from the HA coating and migrate into the joint space, the risk of polyethylene wear might increase [71]. Careful follow-up of patients with HA coating is therefore recommended [72]. Fibronectin [73] or type I collagen [74] coating of titanium alloys increased cell binding and osseointegration.

III. CERAMICS

Ceramics used in orthopedic surgery and traumatology as bone tissue substitutes are mainly of hydroxyapatite, tricalcium phosphate (TCP), or glass ionomer origin [75,76]. Ceramics can be categorized as (1) fast-resorbing, (2) slow-resorbing, and (3) injectable ones [77]. Ceramic composites have found their place in promoting healing of bone in clinical practice alone or in combination with other materials with their osteogenic, osteoconductive, and/or osteoinductive properties [78–81]. These ceramics can also be used as carriers of bone cells, growth factors [82–85], or drugs [86] such as antibiotics [87,88] and anticancer medicine [89]. Advantages of ceramics over metals are their favorable bioactivity and interaction with the host tissue. Bioactivity of ceramics is mainly limited to osteoconduction as long as they do not carry cells and/or growth factors. Thus, clinical and basic research results lack a detailed understanding of these materials' exact biological effects [90].

The ultimate aim of porous degradable ceramics implanted into bone is natural organ replacement at load-bearing or void-filling sites [91,92]. Normal tissue interacting with these ceramics is supposed to replace the implant in time. Tricalcium phosphate is known to degrade more rapidly than HA and is used in non-weight-bearing sites. The degradation rate of HA and TCP may change depending on the manufacturer, pore size, porosity, composition, and sintering temperature. The rate of degradation per year of TCP and HA is about 35 and 1–3%, respectively

A

Figure 2 From left to right: control, allogenic bone chips, natural apatite ceramic, synthetic hydroxyapatite, and calcium carbonate implantation into the mandible of mongrel dogs. (A) cavities opened in the mandible; (B) biomaterial implantation; and (C) macroscopy at 4-week follow-up. Also note periosteal reaction at sites where biomaterials were in contact with the implants.

[93]. One recent study, however, indicates that TCP degradation does not occur even after 6 months and a thin fibrous layer surrounds the nonloaded ceramic at all times [94]. Mechanical properties of hydroxyapatites in general were superior compared to TCP. However, bending and torsional stresses may fracture HA easily [95].

Apatite ceramics of natural and synthetic origin, allogenic bone chips, and calcium carbonate are also frequently used in dentistry. One study [96] compared the effects of these ceramics in defects created in the mandible of mongrel dogs (Fig. 2). The results of that study indicate in 1 week natural apatite of coral origin established loose connective tissue with some osteoblasts adjacent to it (Fig. 3). Natural apatite resorbed in 4 weeks leaving its place to bone trabecules. Active osteoclasts were observed in the newly establishing Haversian system. Foreign body reaction and inflammation was not observed with natural apatite. Only granules detached from the coral elucidated fibrous encapsulation and osteoclastic activity. In 1 week, calcium carbonate disappeared totally leaving a cavity of granulation tissue. Some osteoblasts were observed at

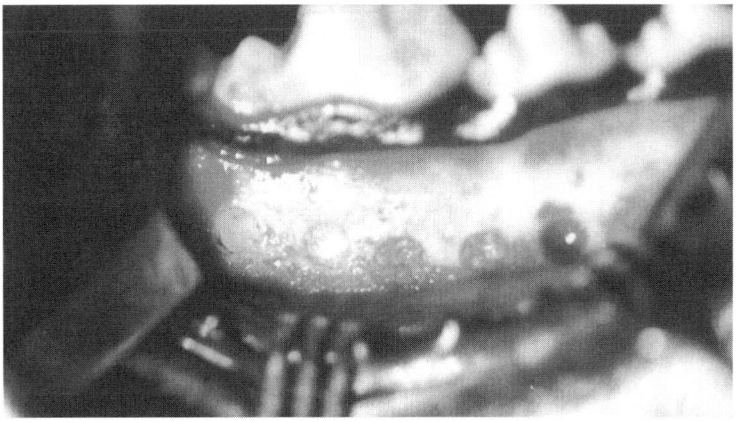

B

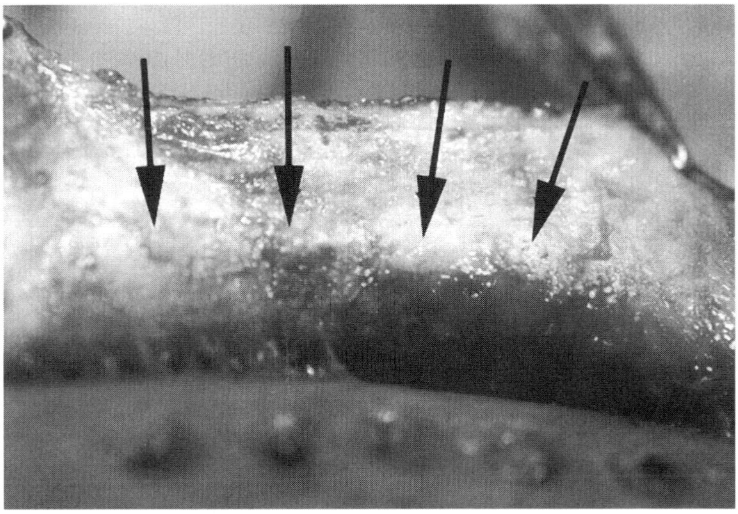

C

Figure 2 Continued.

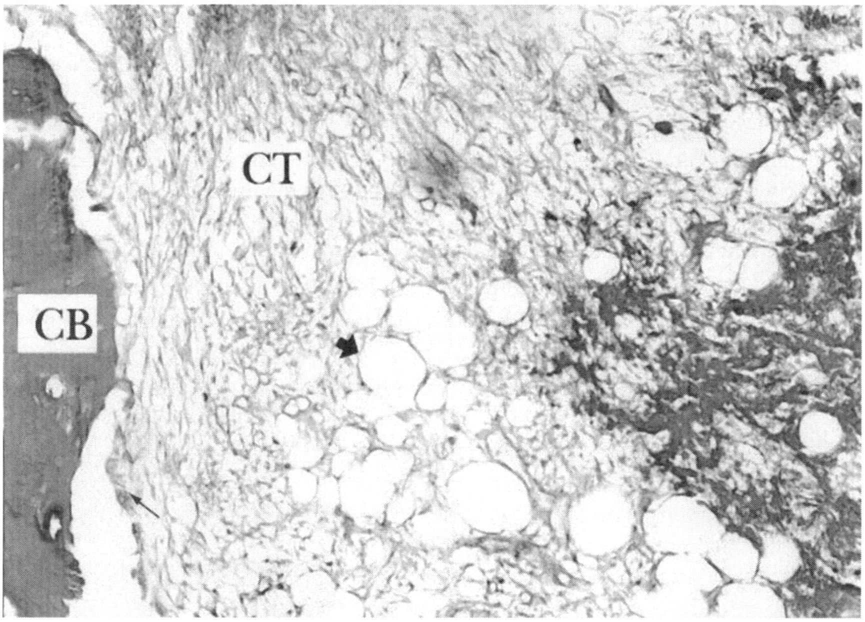

A

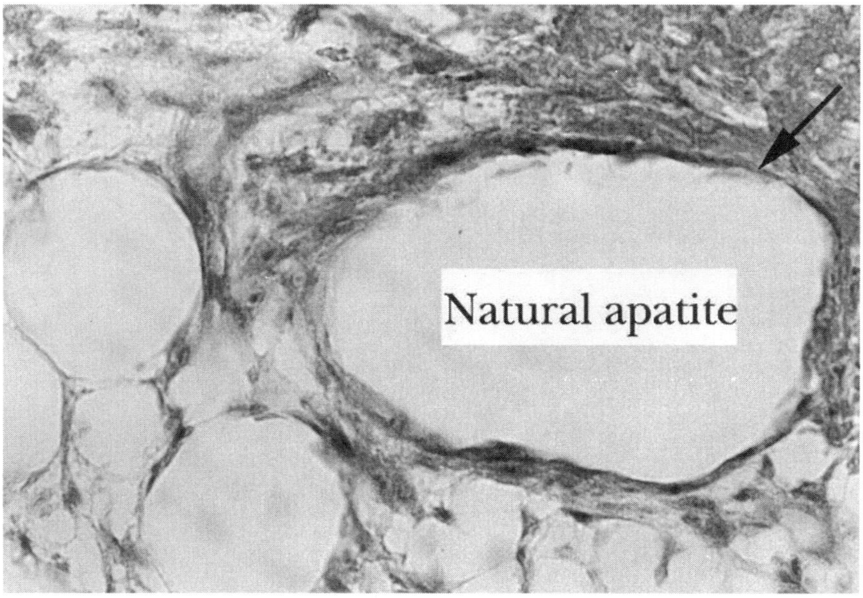

B

Figure 3 Natural apatite of coral origin. (A) Cellular connective tissue (CT) in between cortical bone (CB) and implant containing minimal osteoblasts at week 1. Arrow indicates voids of cavities belonging to the implant. Massons Trichrome 40×. (B) Voids of implant surrounded by fibrous connective tissue. HE 40×.

the bone–cavity border. In 4 weeks, the granulation tissue was replaced by dense connective tissue. Findings were inferior with calcium carbonate than coral apatite. Dense connective tissue also established with synthetic apatites; however, osteoblastic activity with these ceramics at the implant–bone interface was better than that of calcium carbonate. Thin new bone trabecules were surrounding the synthetic HA in some locations. Synthetic HA presented a favorable bone-healing sequence, with no foreign body reaction and osteoclasts at 1 week when compared to the other materials (Fig. 4). New bone did not grow well in cavities where allogenic bone chips were implanted. Bone healing was always from the peripheral to the central part of the implant. All implants presented an osteoconductive property. Reaction to these implants by bone was limited probably due to the dense cortical structure of the mandible. Best results were attained with natural apatite followed by synthetic apatite (Fig. 5). Allogenic bone chips and calcium carbonate followed (Fig. 6) these two materials in effectiveness means of bone healing. Hydroxy-apatite particles in the periosteum elaborated a significant osteoclastic activity (Fig. 7) [96]. Thus, bone healing of the mandible is known to be significantly better than of the femur of rabbits [97].

Ceramics, so far, have been identified as compatible and biologically active materials. They are not toxic and do not cause cell death at the surrounding tissue. Biological response to these ceramics follows a similar cascade observed in fracture healing. This cascade includes (1) hematoma formation, (2) inflammation, (3) neovascularization, (4) osteoclastic resorption, and (5) new bone formation. Surrounding tissue is supposed to replace these ceramics as they degrade (Fig. 8A and B). A fibrous tissue capsule rarely occurs, and an interfacial bond between the ceramic and the bone is established [98]. Particle size is one important factor in ascertaining the ostogen-esis with ceramics [99]. Recent research, however, demonstrates that these materials can also induce an early and nonspecific inflammatory reaction (Fig. 9) followed by cellular depletion (Fig. 10) when implanted into the bone marrow [100,101]. This early response was found to subside in about 14 weeks [102]. It can be concluded that the marrow [103] and soft tissues [104] are more sensitive to ceramic implantation than the cancellous and cortical bone sites.

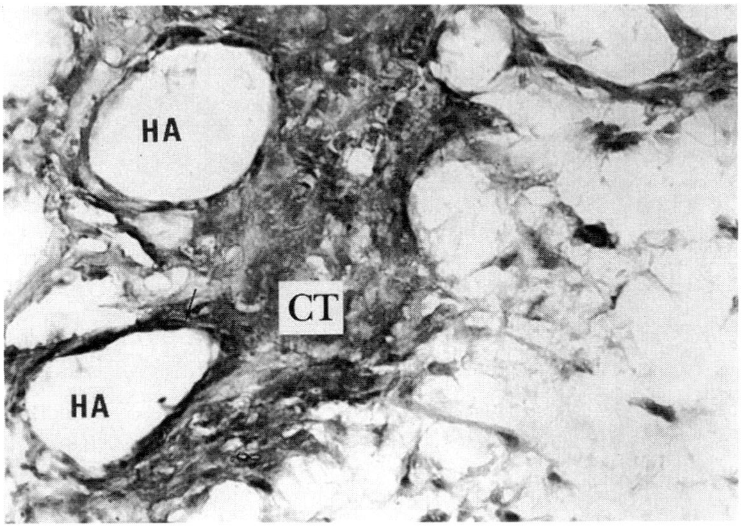

Figure 4 Synthetic hydroxyapatite (HA). Favorable healing sequence without osteoclasts at week 1. Hydroxyapatite granules are surrounded by dense connective tissue (CT). HE 400×.

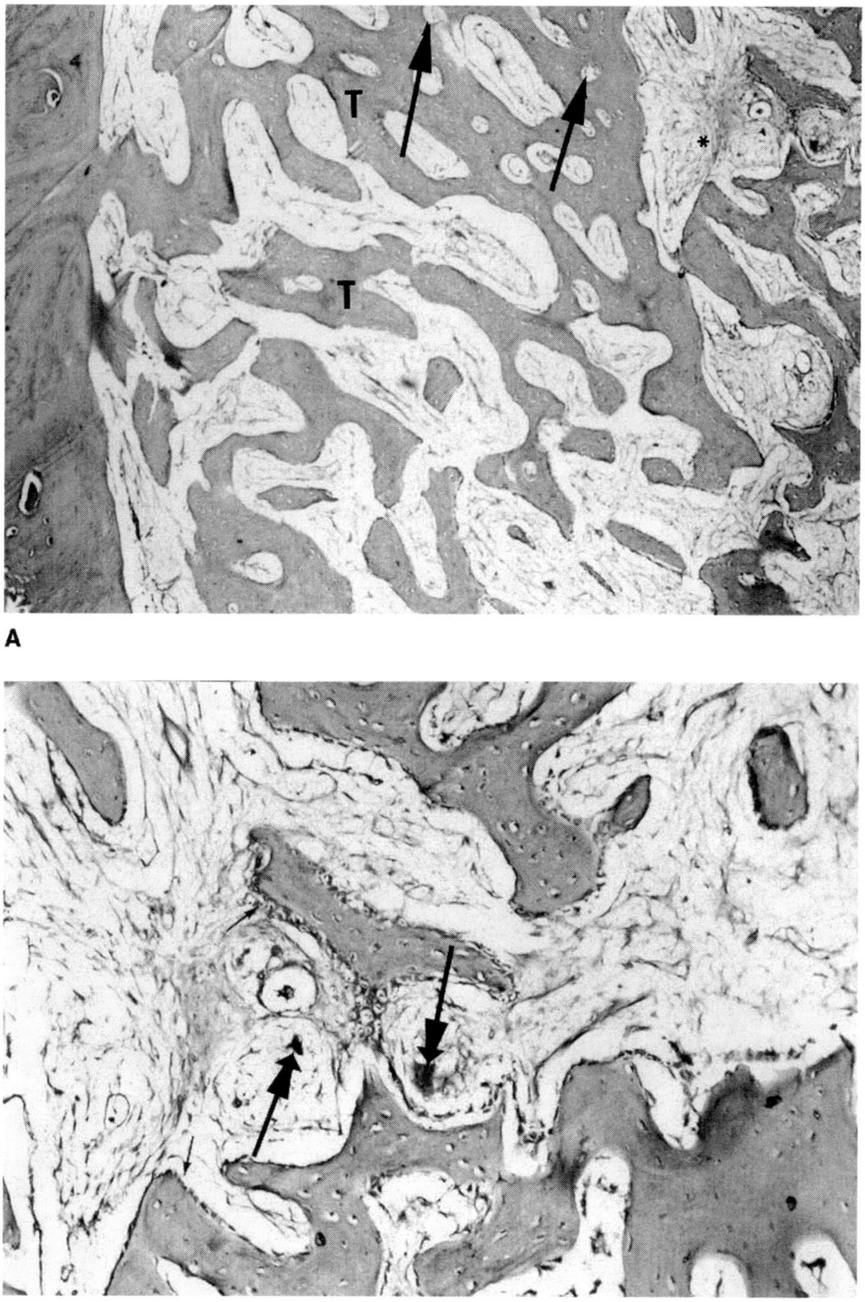

A

B

Figure 5 Synthetic hydroxyapatite. (A) New trabecular bone (arrows) healing at week 4. Arrows indicate the new establishing Haversian canals. HE 40×. (B) Osteoblasts and osteoclasts (arrows) can be seen around the new bone trabecules. HE 10×.

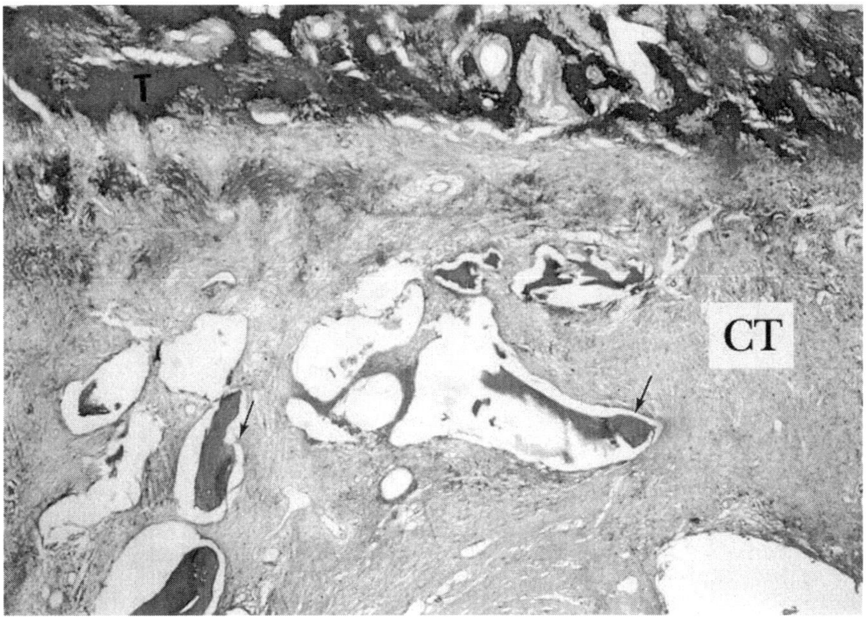

Figure 6 Allogenic bone chips. Dense connective tissue (CT) can be observed in close contact with the new bone trabecules. Trichrome 40×.

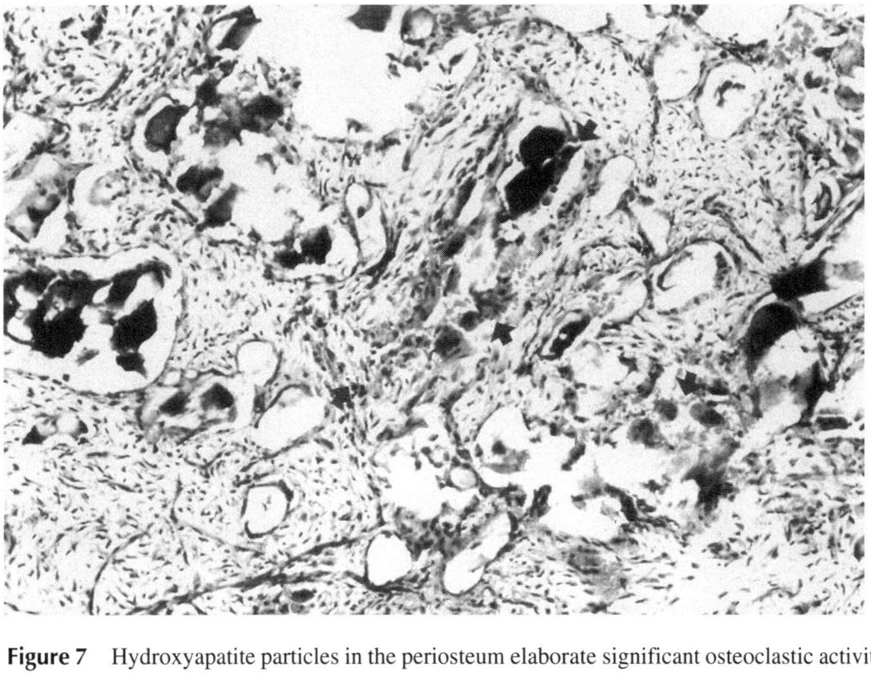

Figure 7 Hydroxyapatite particles in the periosteum elaborate significant osteoclastic activity. HE 100×.

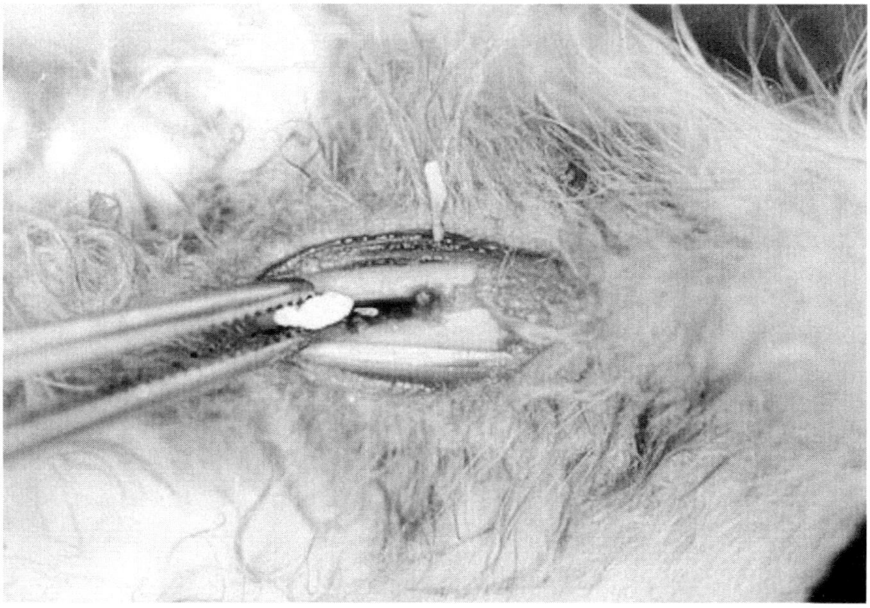

A

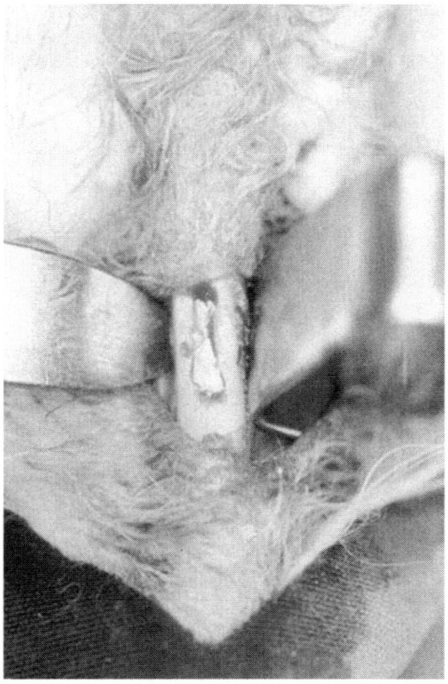

B

Figure 8 Implantation of porous HA particles into the bone marrow site of rabbit tibia. (A) Establishment of the unicortical aperture and (B) porous HA particles placed into the bone marrow and cortical defect site.

Blood cells and osteoblasts are among the first cells to react to the implanted ceramic [105]. Ceramic particles do also interact with monocytes [106,107], and they are capable of stimulating the expression and secretion of cytokines and proteases that enhance bone formation and/or resorption [108]. Macrophages are presented to be the major infiltrating cells when HA and TCP are implanted [109]. These cells secrete H+ and degrade the ceramics [109]. Aluminum-containing ceramics, furthermore, significantly increase catalase activity and lipid peroxides in

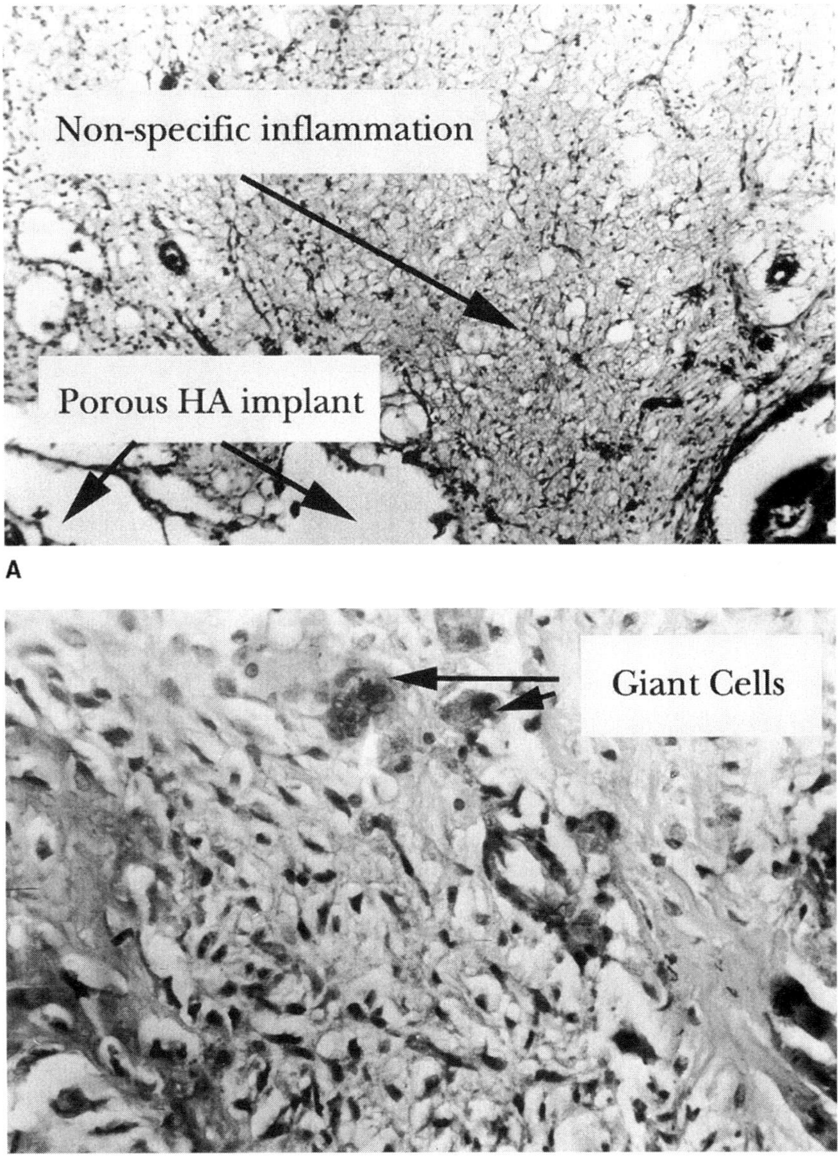

B

Figure 9 (A) Nonspecific inflammation after 1 week following implantation of porous HA particles into the bone marrow. HE 10×. (B) A few giant cells were present in the inflammation area. HE 400×.

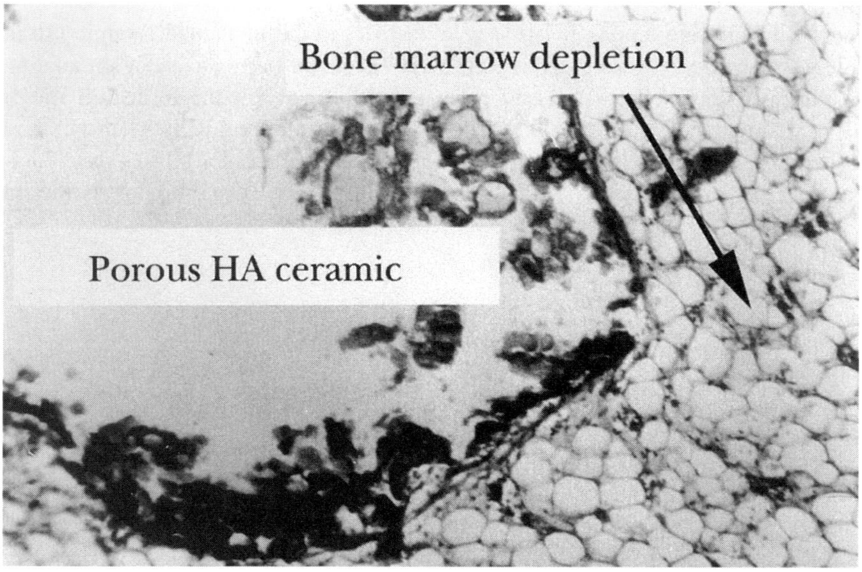

Figure 10 Bone marrow depletion at the porous HA particle implantation area at week 2. Note the fattylike appearance of the marrow. HE 100×.

macrophages and may elaborate free radicals. Interleukin-1, IL-6, and TNF-α secretion increases in human fibroblasts with HA particles. The action of HA particles was found to be at the level of transcription of such mediators. Gelatinolytic activity of the fibroblasts also increased with HA particles [108]. Hydroxyapatite sintered at low temperatures was found to cause more toxicity than that of HA sintered at higher temperatures [107]. Thus, new production methods of bioactive ceramics are under investigation [110] to overcome the toxicity problems. Toxicity of ceramics also depends on their solubility [111]. Natural ceramics are presented to be more effective in attracting cells and favoring their proliferation than synthetic ones [112]. Large amounts of mineral powder may also down-regulate osteogenic markers such as alkaline phosphatase activity and osteocalcin release (Fig. 11) [113].

Bone mineral density gradually increases after HA implantation [114,115]. Thus, the mechanical properties of HA-implanted bone differ from those of a normal bone. Stiffness properties of HA-implanted bone in compression do not change significantly throughout the healing process. The pattern of fracture in a defect-created control and HA-implanted bone differ from each other. The HA-implanted bone fractures from its end plates revealing a stiffer area at the implantation site that prevents the propagation of a longitudinal fracture. Control bone with the defect created, however, fractures longitudinally as the defect itself creates a weak point that allows the fracture to pass through it. A gradual increase in stiffness is also observed with HA-implanted bones in three-point bending [115]. One other study [116] also indicated that HA implantation increased the torsional stiffness of bone. Healing progress of HA and HA/TCP composite–implanted bones in load-bearing segmental defect sites was evaluated by modal analysis [95]. Resonant frequencies yield comparable results with the three-point bending tests at the early stages of healing. As the flexural resonant frequency is proportional with the square root of the stiffness of the structure and square of the length of the structure, it is recommended not to use this method in later stages of healing due to callus formation. Hydroxyapatite and HA/TCP-implanted bones gained 30% of impact strength of normal bone at 18 weeks. Mechani-

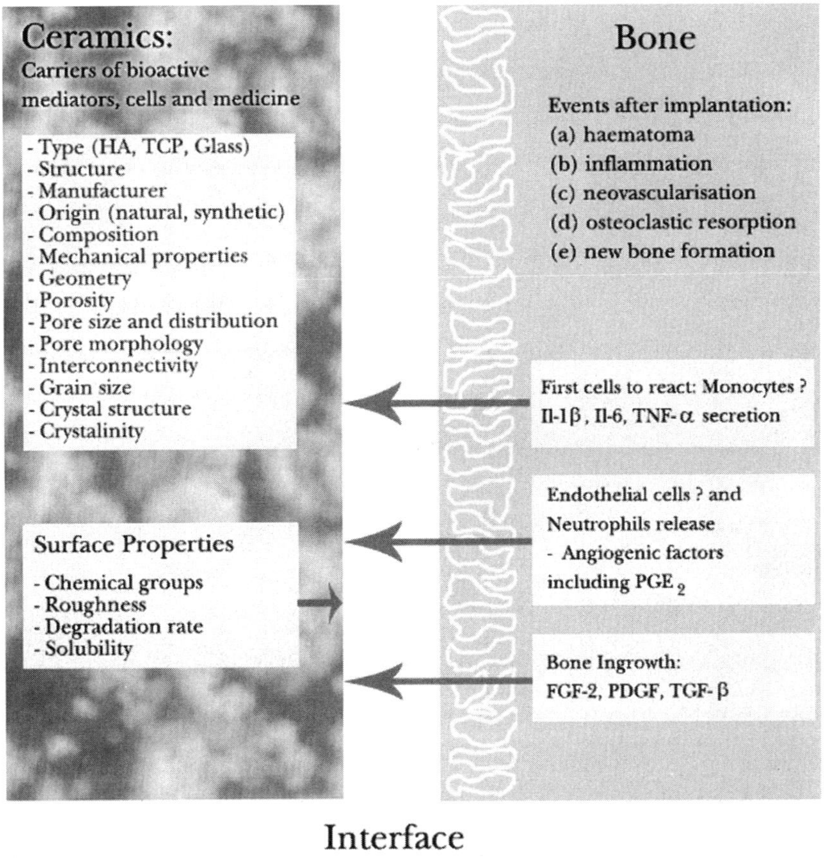

Figure 11 Ceramic implant–hard tissue interface.

cal vibration analysis results were in accordance with the results of bending tests. Results of mechanical testing of ceramics implanted into load-bearing sites indicate the need for advancement of mechanical properties of such implants [95].

The osteoconductive performance of ceramics in vivo mainly depends on the contact area of the implant and the living bone (Fig. 12). Mineralization is presented to directly start on the implant surface of macro- and micropores [117]. Needlelike new microcrystals form at the micropores of the ceramics [118]. There might be a delay in the calcification process due to the initial bone marrow depletion and mineralization-related increase in cellular matrix vesicles that is observed after 6 days of ceramic implantation [100]. A gap more than 50 μm between the ceramic and bone may cause fibrous encapsulation. An ultrastructural study demonstrated an organized network of collagen fibers between the bone and ceramic. These fibers mineralized subsequently. A 50 to 600-nm-wide collagen-free granular deposition was also observed on the ceramics [119]. This unmineralized zone was measured as at least 600 μm in another SEM study [120]. New bone between the implant and ceramic is of normal lamellar type [121]. A light and laser scanning microscopy study revealed mineralized bone apposition directly on HA [122]. In unmineralized parts osteoid interposition was observed. A thin layer of fluorescent material was also observed at the interface [122].

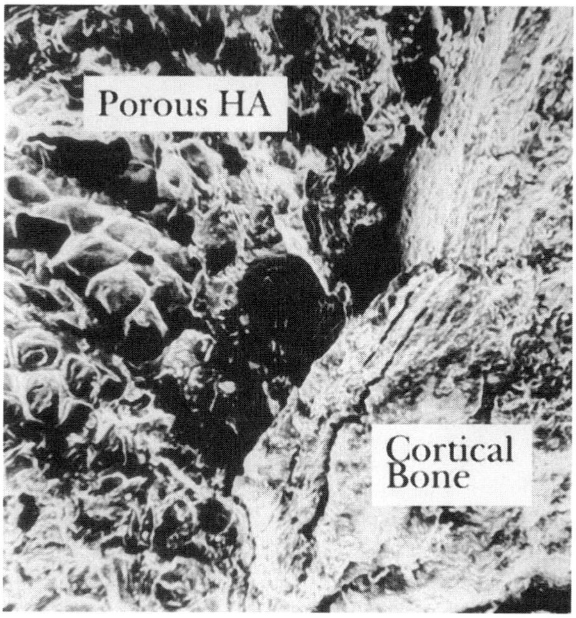

Figure 12 Favorable bonding of porous HA ceramic and cortical bone. SEM 1400×.

Porous structure and pore sizes ranging from 100 to 400 μm enhance bone ingrowth rate [123,124]. Mechanical properties of the ceramic decrease as the pore size increases. When osteoblastic cells were cultured with HA these cells spread quickly on the ceramic [125], then stopped spreading 12 h after cell seeding [126]. Results of in vitro studies may differ from those of in vivo studies where the osteoblastic activity increases at 2 weeks [127]. Other factors, such as surface roughness and reactivity of the ceramic, are determinants of cell adhesion, proliferation, and differentiation [128]. Adhesion and detachment strength of cells increase as surface roughness increases [129]. One should keep in mind that the reactive and adaptive response (Fig. 13) of the host may differ between humans and other species [130]. The blood circulation of the ceramic implantation site and interventions such as irradiation may also alter these responses extensively [131].

Adding cells [132–137] or bioactive materials, such as BMP, to the ceramics can enhance bone integration. The origin of cells found in the ceramics needs to be clarified when cell-containing ceramics are implanted into bone. It is not known whether the cells implanted with the ceramic can survive and advance osseointegration or if they go through an apoptotic cascade. Cells in ceramics may also migrate from the hosting bone tissue. Thus, studies reveal that adding osteogenic cells advanced the osseointegration of porous ceramics [134,138]. Although organic matrix and bone mineral deposition are presented in osteoblasts in direct apposition to HA [139], another study revealed that ceramics had an inhibitory effect on growth of these cells [140]. These studies elaborate the necessity of further research on cell and ceramic interactions. Interconnections between the host responses to the biomaterial and transplanted cells are stated to determine the biocompatibility of the implant [141].

Glass ionomers improved the mechanical properties and biocompatibility of ceramics used in the medical field. These special types of ceramics were also presented to enhance osteogenic activity compared to HA ceramics alone [142]. It was found that even small amounts of glass ionomer can effectively bind apatite particles (unpublished data) (Fig. 14). Glassceramics lead

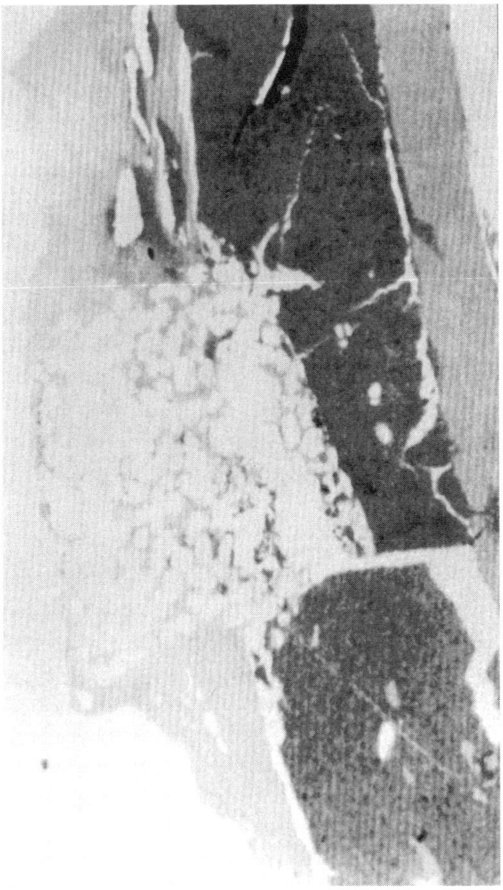

Figure 13 Cortical and periosteal adaptation of rabbit bone to porous HA implant. HE 40×.

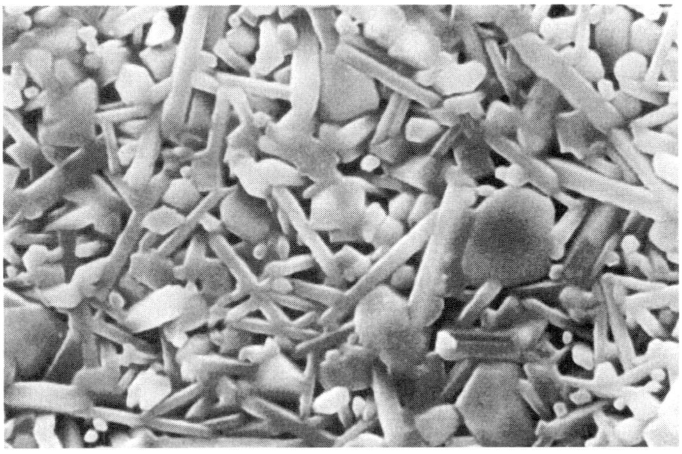

Figure 14 Apatite-Wollastonite glass ceramic. Rods are apatite and hexagonal constituents are glass particles. SEM 3000×. (Courtesy of M. Timuçin, Middle East Technical University.)

to a more rapid bone proliferation than HA [143,144]. There was a direct bond between osteoblast cells and mineralized layers established on the surface of the glass ceramics [145]. Glass ionomers, on the other hand, may have adverse effects on neural tissue [146]. These implants should therefore be used with caution when implanted nearby neural tissues as in spinal surgery. Another study with peritoneal macrophages, monocytes, and bioactive glasses revealed that these biomaterials had a moderate toxic effect on these cells [147]. Bioglass particles led to an increased release of TNF-α and expression of TNF-α mRNA. In vivo, induced rapid bone growth appeared to activate an autocrinelike process [147].

Ceramics, in general, are publicized as bioactive materials (Fig. 15) with minimal side effects when implanted into the bone. They elaborate inflammation followed by depletion when implanted into bone marrow. Giant cells and macrophages are rarely seen around the implant, and this nonspecific early reaction does not lead to chronic inflammation. Osseointegration and tissue ingrowth occur in porous ceramics [148]. Tissue ingrowth into the pores of the ceramic is usually limited (Fig. 16) to its surface, and new bone formation comes to a halt in bulky and slow-resorbing ceramics.

IV. BONE CEMENTS AND NOVEL BIOACTIVE CEMENTS

Thermal reaction during cement curing is a critical factor in the determination of cement biocompatibility. Cements with higher curing temperature may cause tissue necrosis. A synovium-like membrane formation containing macrophages and foreign body giant cells around the cement is usually observed following the acute inflammatory stage. Particles of polyethylene can be found in the cytoplasm of the macrophages. Interleukin-1 and PGE_2 levels increase enormously at the bone–cement interface. It is speculated that titanium and polymethylmethacrylate (PMMA) particles smaller than 10μm are able to stimulate IL-1 and PGE_2 secretion and initiate the osteolytic process [149]. Free radicals are produced in fibroblasts by PMMA [150]. High levels

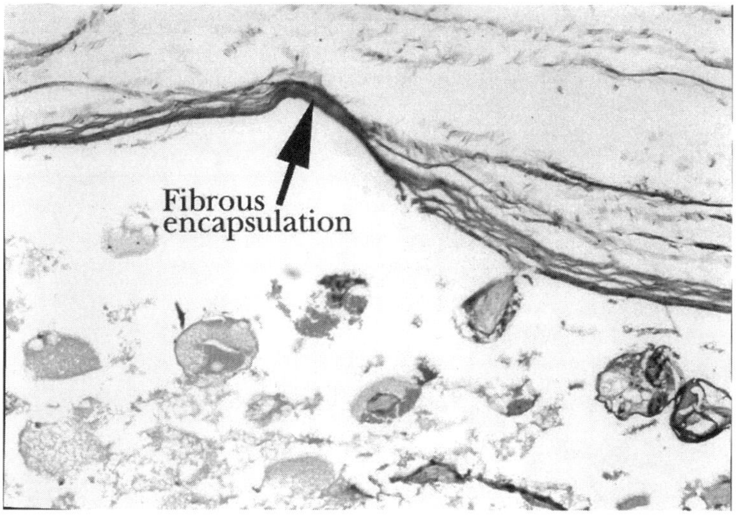

Figure 15 Porous HA-TCP ceramic implanted into a 1-cm-long critical size defect area of the weight-bearing rabbit tibia. (A) A very limited fibrous encapsulation can occasionally be seen at week 1. HE 10×. (B) Endochondral bone formation can follow the fibrous encapsulation stage at 1 month. HE 100×. (C) In 2 months, new bone formation can be observed at the margin of the implant. HE 40×.

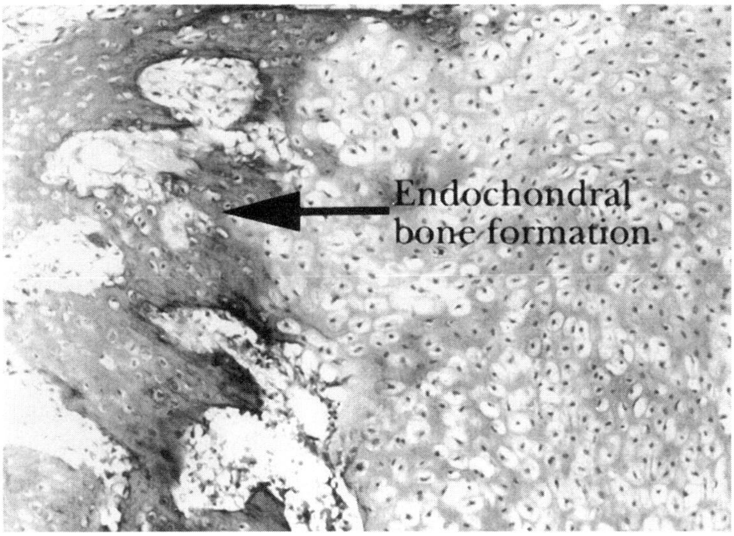

B

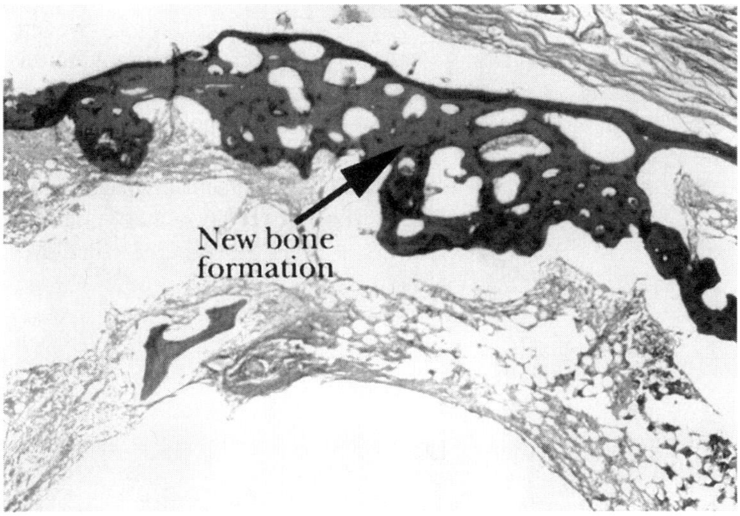

C

Figure 15 Continued.

of apoptosis of osteoblasts were observed with PMMA directly after polymerization [151]. Polymethyl methacrylate particles also suppressed osteoblast differentiation [152]. Additives such as barium sulfate may cause an increase in the inflammatory response to PMMA [153]. Changing the activator in bone cements may improve their biocompatibility [154].

Attempts to decrease the curing temperature and increase biocompatibility without changing biomechanical properties of bone cements have been common in recent years [155–157]. Very low viscosity cement compositions were prepared by mixing PMMA particles with two different molecular weights in order to achieve a proper and homogeneous distribution of HA particles in the polymer matrix in a recent study [158]. Addition of HA into the cement in that

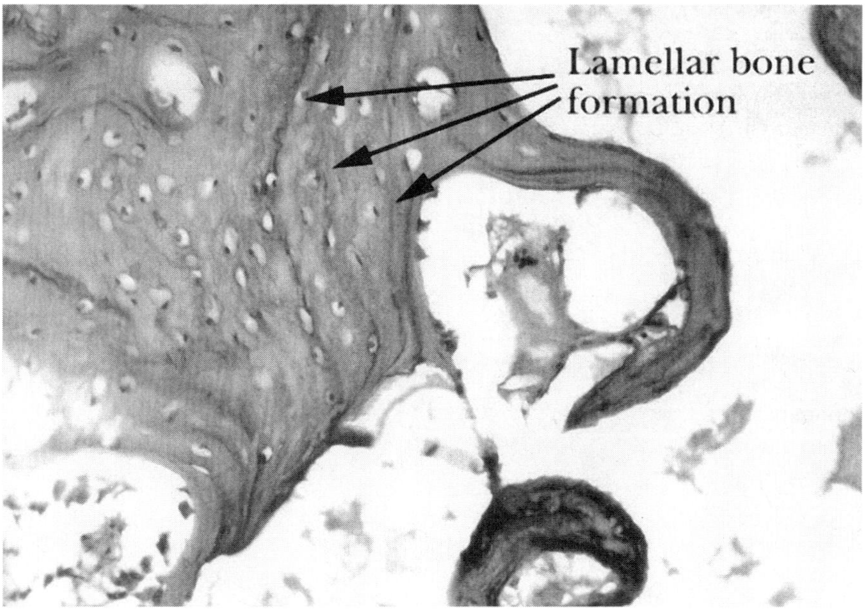

A

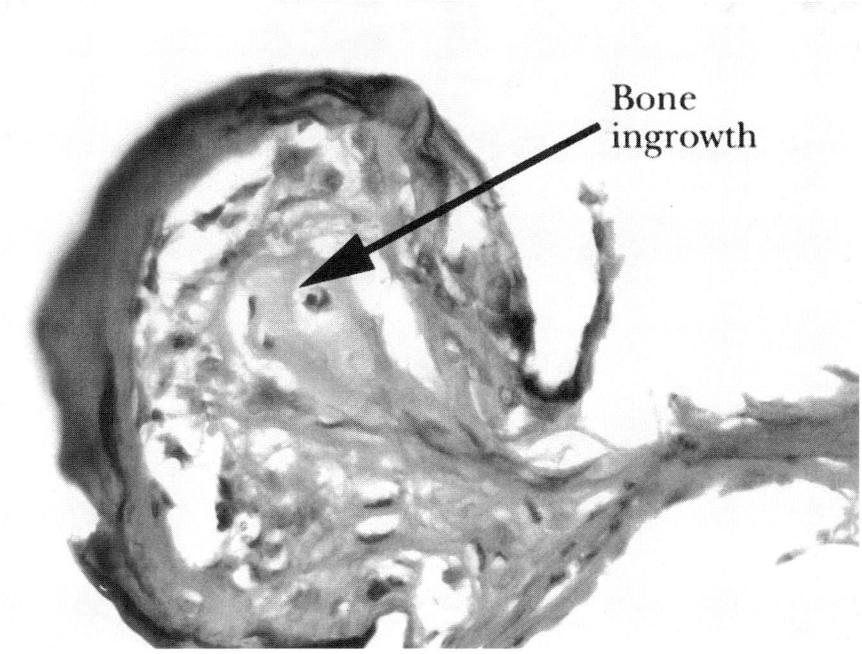

B

Figure 16 Limited bone ingrowth on the surface of porous HA-TCP ceramic. (A) Lamellar bone formation can be seen at the adjacent site of the ceramic in 4 months. HE 200×. (B) Bone ingrowth into the pores of the composite at 4 months. HE 400×.

study made workability easier by increasing viscosity. Polymerization temperature decreased from 111 to 87°C. Compressive strength increased from 96 to 122 MPa, and sufficient compressive fatigue strength of 77 MPa at 10^6 cycles was obtained with this novel formulation. Hydroxyapatite-containing acrylic bone cements demonstrated higher mechanical strength than the reference cement [158]. In-vivo application of HA-containing cement revealed better biological compatibility compared to currently used cements [159]. It is concluded that HA-containing acrylic bone cement may effectively be used in the clinical field in the future. Other in vitro studies also reveal that osteoblast-like cells proliferated and differentiated significantly better on HA containing PMMA than PMMA alone [155]. Phenotype of cells of that study retained up to 21 days.

Self-curing calcium phosphate cements (CPC) are novel materials of injectable form [160–164]. These cements revealed extensive bone formation immediately after implantation without any inflammatory tissue response [161]. Bone colonization occurred much earlier and faster in CPC than calcium phosphate ceramics [165]. Polymers [166], gelatin [167], and collagen [168] can be added to improve the biocompatibility of calcium phosphate cements. Chitosan and citric acid were added to improve their biocompatibility and decrease the initial inflammatory response of these self-curing cements [169]. Addition of 20% citric acid decreased the initial inflammatory response, and good bone bonding was observed in that study. Preosteoblastic cells can also be stimulated in vitro by adding TGF-α into calcium phosphate cement [170]. Thus, one study found that CPC particles could adversely affect osteoblast function related with particle size [171]. The concern on the biomechanical effectiveness and slow degradation due to the nonporous structure of CPC still remains.

V. POLYMERS

Polymers are mainly used in fracture fixation, bone replacement, cartilage repair, fixation of ligaments, and drug delivery. Polylactides (PLA), polyglycolides (PGA), and polyhydroxybutyrates (PHBV) are the most common types of polymers used in hard tissue engineering [172]. Material properties including type, composition, surface geometry, chemistry, porosity, and degradation rate of polymers define their interaction with bone tissue. A larger surface area and the addition of quinone dye are documented to increase the risk of adverse reaction. Hydrophilic polymers need surface modification for cell adhesion and growth [173]. Degradation rate of polymers in vivo is slower than in vitro. The first cells that interact with polymer are generally of mesenchymal origin (Fig. 17). Mesenchymal cells use fibronectin to anchor to collagen in the extracellular matrix. Tissue transglutaminase (tTG) that binds with high affinity to fibronectin has recently been used as a surface coating to enhance biocompatibility of polymers [174]. Integrins, small proteins of the extracellular matrix, will act on these cells to initiate the biological response. Protooncogenes c-fos, c-jun, and zif/268, on the other hand, will activate osteoblasts following mechanical stimuli [175]. The following events of cellular and humoral mechanisms in hard tissue–implant interaction are identical with other materials. The first cells that interact with polymers of lactic acid in culture were granulocytes, monocytes, and lymphocytes [176]. Fibroblasts were the ancestors of the first cells forming fibrous encapsulation. Macrophages predominated around polymers between 12 weeks [177] and 4 years [178] depending on the degradation profile of the implant. The severity of tissue reaction may increase when polymers are implanted into or close to immunologically more active sites such as the synovium of a joint [179]. Implantation depth in intra-articular applications is a critical point [180]. In PGA devices, monocytes and lymphocytes dominated the inflammatory response. DNA synthesis was not induced, but major histocompatibility complex II (MHC-II) antigen and IL-2R activation

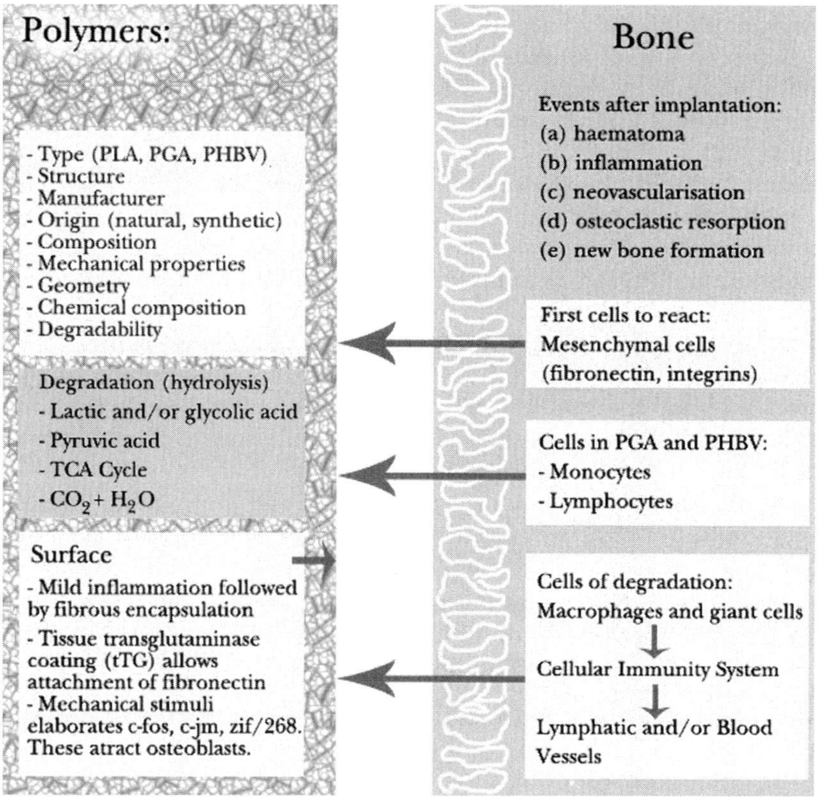

Figure 17 Polymer–hard tissue interface.

marker expression were seen when monocytes were cultured on these implants. These findings suggested that PGA in cell culture is relatively inert; however, it can still induce inflammatory mononuclear cell migration and adhesion leading to a slight nonspecific lymphocyte activation [181].

Interaction of polymers with bone cells is usually studied in culture. Screening implant toxicity using osteoblastic cells became an integral part of biocompatibility testing. In vitro studies with rat marrow–derived stromal osteoblasts revealed decreases of number of cells, mineralization, and [^{3}H]-thymidine incorporation with an increase of the concentration of polymer particles [182]. That study also suggested those polymeric particles may affect the bone remodeling process.

Foreign body reaction toward polymers is well documented in experimental and clinical studies [183,184]. The predominating signs of reaction after polymeric implant application are sterile drainage and osteolysis around the implant. One study [184] of patients operated on with bioabsorbable PLA fixation devices revealed a reaction in 107 out of 2528 patients. Average time from surgery to reaction was 11 weeks. Polyglycolic acid elicited a reaction 4.3 years after surgery in a single case. The severity of the reaction varied from a painful erythematous papule to sinus discharge with extensive osteolytic lesions. The histological picture was that of a nonspecific tissue reaction [184].

An other approach in dentistry also revealed foreign body granuloma with polymers following replacement of the hard tissue [185]. Two out of 27 patients required surgical drainage and debridement following a destructive and nonspecific foreign body reaction encountered with absorbable rods used in foot surgery [186]. Recovery rate and time were inferior with PGA than metallic rods used for the fixation of distal radial fractures [187]. Eighty-three patients with malleolar fractures were treated with polymeric screws, on the other hand, and none of them presented any early or late drainage due to the polymeric implant [188]. Concerns on biodegradable rod fixation include the demanding technique and postoperative instability [189]. Reaction toward the polymer increases when the size of the implant increases. Fibrous encapsulation, lymphocyte infiltration, and giant cells and plasma cells surrounding the intramedullary-inserted polymeric rod were predominating the histological picture at 30 months in an experimental study where 21 adult female sheep were used [190].

Poly(caprolactone) (PCL) is another type of polymer used for bone tissue engineering. Studies with PCL revealed bone marrow cells on this implant did not show alkaline phosphatase activity [191]. Foreign body reaction toward PHBV implants was also not persistent (Fig. 18) [192]. Connective tissue cells surrounded PHBV. On week 1, a thick and highly cellular layer consisting of fibroblasts and mononuclear phagocytic cells surrounded the polymers (Fig. 19). Giant cells were observed in limited areas. On weeks 3 and 6, the fibrous layer became thinner and allowed interaction between the implant and neighboring bone in some locations (Fig. 20). The number of macrophages decreased on week 6. Osteoblasts replaced the fibroblasts and macrophages. One other implant, the calcium phosphate/gelatine composite, eroded tissue response when implanted into bone [193]. Connective tissue was replaced by new bone trabecules. Stages of tissue reaction to polymers can be summarized as (1) early stages: thin connective tissue encapsulation with few lymphocytes (Fig. 21), (2) middle stages: fibroblasts within denser connective tissue (Fig. 22), and (3) late stages: giant cells and macrophages clearing material particles (Fig. 23).

Adding alkaline salts or antibodies of inflammatory mediators into implants was recommended to decrease the side effects of polylactides [175]. Calcium acetate as soluble salt filler [194] or cross-linking of polypropylene fumarate with ethyleneglycol dimethacrylate [195] was also proposed to increase the in vivo biocompatibility of bioresorbable polymers. One other alternative proposed to lessen the inflammatory response is the addition of bone morphogenic protein into the polymer [196]. Synthesis of polymers, processing them into implants, and sterilization modalities need to be improved to overcome the tissue reaction problem [197,198]. Long shelf-life may also alter biomechanical and chemical properties of polymers.

The rate of failure of loss of fixation and bacterial wound infection with polymeric rods is given as 4% each [199]. The number of sites where polymeric implants are used in the human body is increasing as their biocompatibility is increasing. Anterior cruciate ligament fixation, cartilage and meniscal repair, rotator cuff repair with anchors, and capsule repair of the shoulder with biodegradable polymeric implants besides fracture fixation are increasing [183,200]. Degradation rate of polymers can be adjusted according to the requirements of the bone tissue. Tissue response to degradable polymers will frequently cause fibrous encapsulation. This encapsulation can be less with comparatively more biocompatible polymers such as PHBV. The degradation process will be mediated through macrophages and giant cells. Improving mechanical properties of polymers without altering the acceptable host response is a novel research field. Changing the chemical composition of the implants from acidic to neutral pH will improve their biocompatibility.

VI. CERAMIC/POLYMER COMPOSITES: THE FUTURE

Advantages of ceramics and polymers are combined in recent studies. Ceramic polymer composites are used as bone graft substitutdes in some cases [178,201–205]. They are also used in a

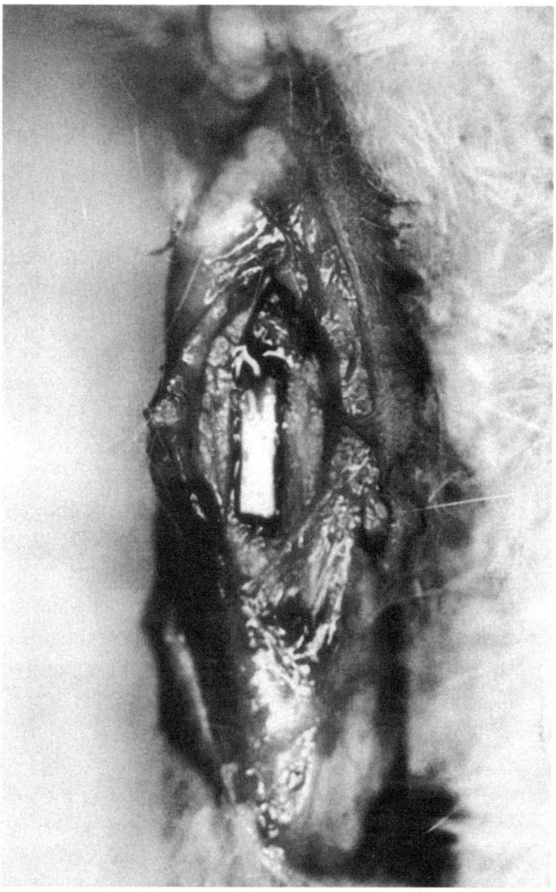

Figure 18 Polyhydroxybutyrate (PHBV) implantation into rabbit bone.

gene therapy approach to bone regeneration [206]. In these composites, polymer and ceramic are supposed to mimic the bone collagen and mineral, respectively. Cells seeded on ceramicpolymer matrices are presented to retain their characteristic morphology and grew in a multilayer fashion [207]. Hydroxyapatite particles in polymer appeared to provide an anchor for the attachment of cells [202]. Apatite crystals, furthermore, kept the pH of the environment within the physiological range. Acid reaction around the implantation site with PLA and PGA implants can be prevented when polymers are used together with apatites [203]. Thus, a strong inflammatory response was seen according to the degradation of the polymer at 24 months even when they are integrated into the composites [178]. It is concluded that the balance between the polymer and ceramic is delicate and chemical events and cellular reaction during polymer degradation may counteract complementary bone ingrowth [208].

The future of hard tissue engineering lies between the appropriate composition of a fascilitating matrix, mediators, and osteogenic cells [209,210]. The need to create a tissue close to the original tissue is essential. Tissue engineers should keep the elastic and rigid properties of bone in mind and seek for a better matrix with equal biomechanical properties of the original tissue. Cortical and cancelleous replacement of bone can be evaluated separately. Growth factors act in a dose- and time-dependent manner. Appropriate growth factors during each phase of

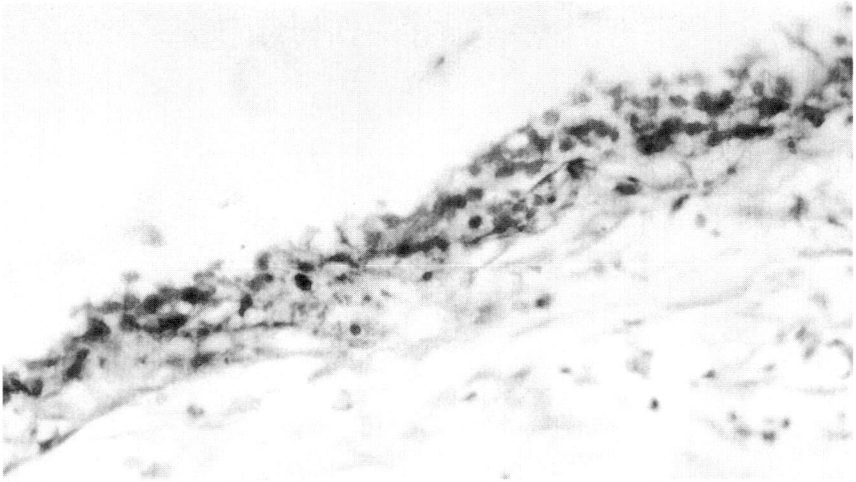

A

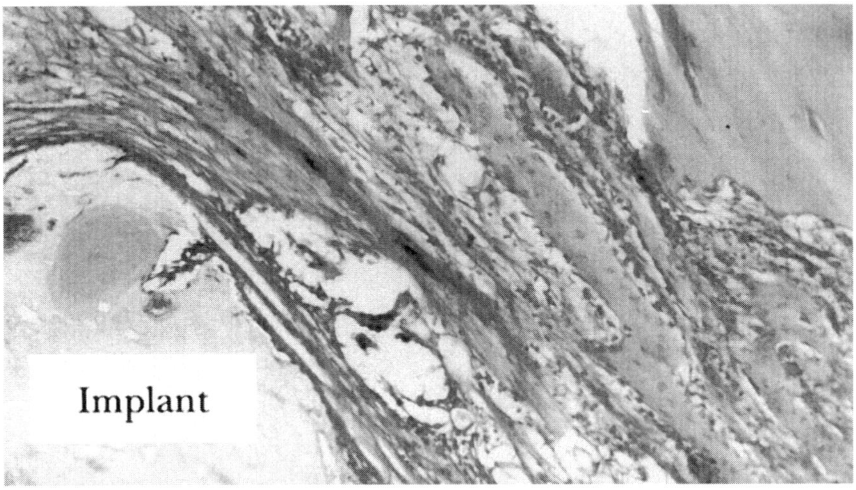

B

Figure 19 Fibrous encapsulation of gentamicin containing PHBV implant. (A) Thick and cellular connective tissue layer surrounds the implant in week 1. Massons Trichrome 400× (B) Few macrophages are still present on week 3 (I = implant). HE 40×. (C) Inflammatory cells invading the cavities of PHBV. Massons Trichrome 400×.

healing should be delivered to the tissue. Cells capable of proliferating and differentiating at the implanted site should be identified. In vivo implantation effects will overlap with the ongoing response of the host indicating that all delivered material, cells, and growth factors may be inhibited. Phenotypic and genotypic characteristics of progenitor cells and growth factors that affect bone formation and remodeling need to be clarified before success. One other important aspect is the monitoring of tissue–implant interactions on site. Radiography is the conventional method of metal monitoring. Other techniques including bone scintigraphy (Fig. 24), bone mineral density measurements, quantitative computerized tomography, and magnetic resonance im-

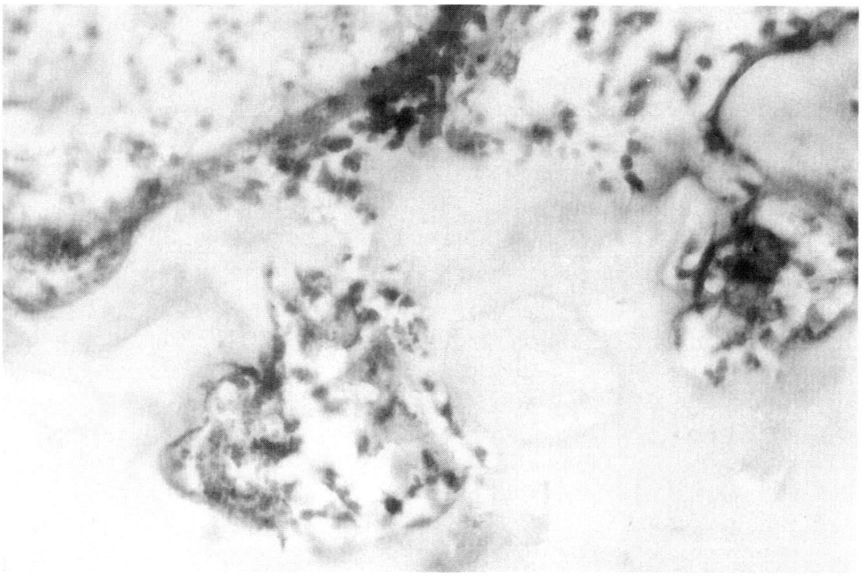

C

Figure 19 Continued.

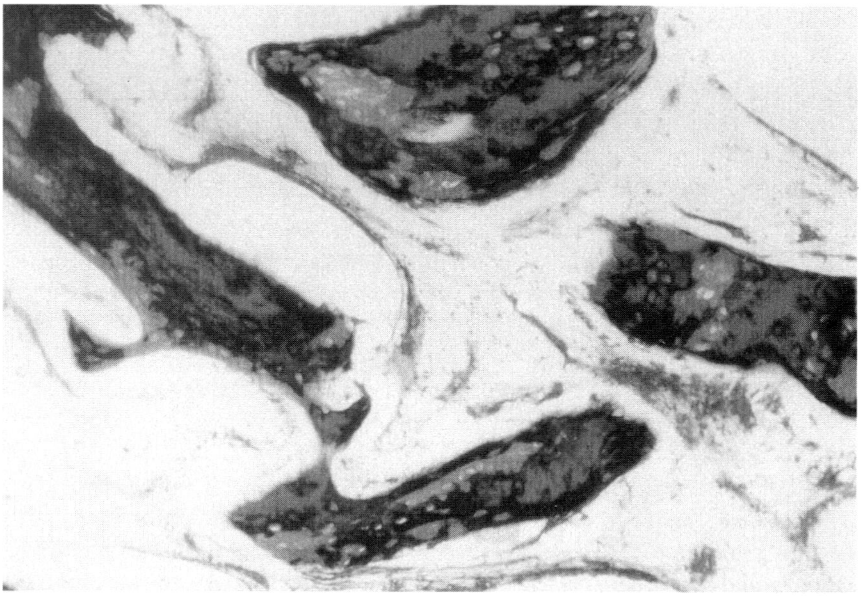

Figure 20 New bone formation within the anastomosing bony trabecules is demonstrated in the PHBV polymer. Massons Trichrome 400×.

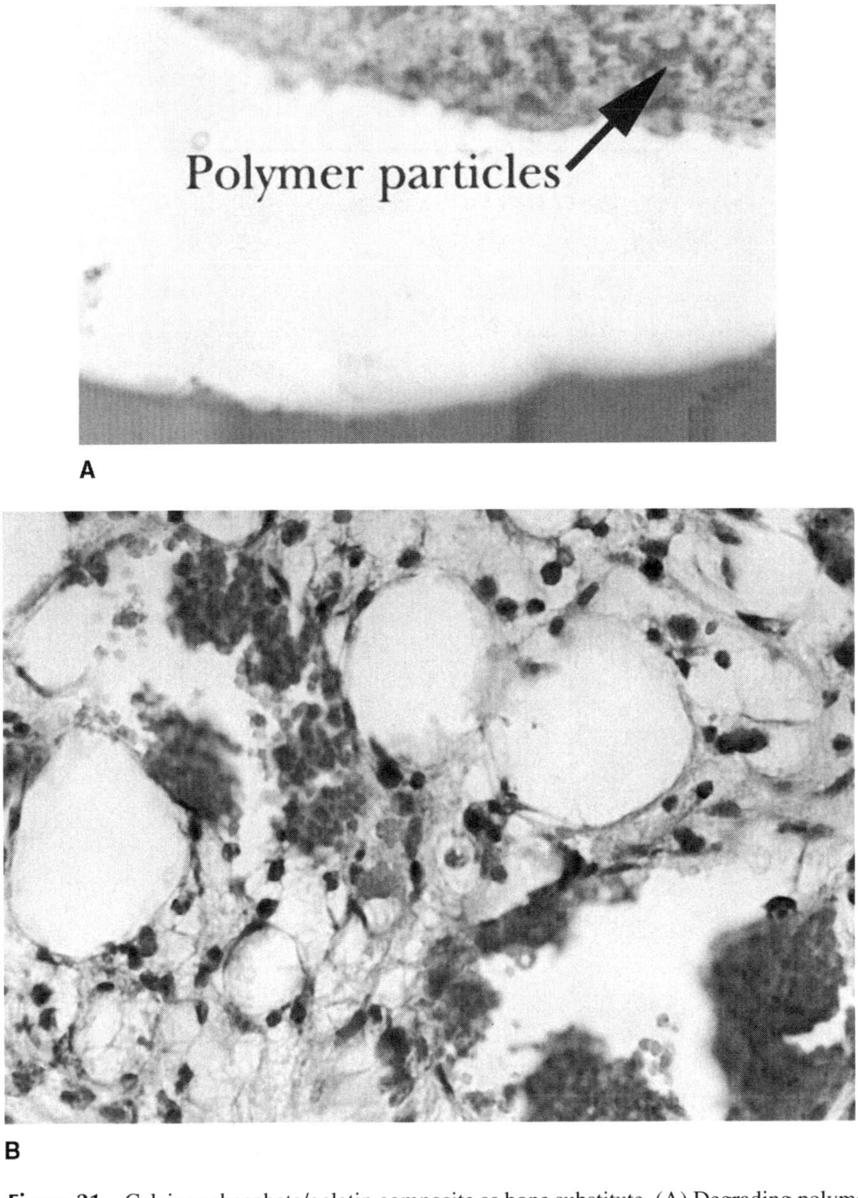

Figure 21 Calcium phosphate/gelatin composite as bone substitute. (A) Degrading polymer particles into the surrounding soft tissue at week 1. HE 100×. (B) Foreign body reaction toward the polymer is minimal. HE 100×.

aging methods are advancing. In vivo physiological changes are now prone to monitoring using magnetic resonance. Advanced magnetic resonance techniques may be used to monitor degradable or non degradable implants in vivo in the near future.

VII. CONCLUSION

The number of surgeries using implants of biological origin is expected to increase in the near future. Implants of metal should be of low profile, and their properties should be improved to

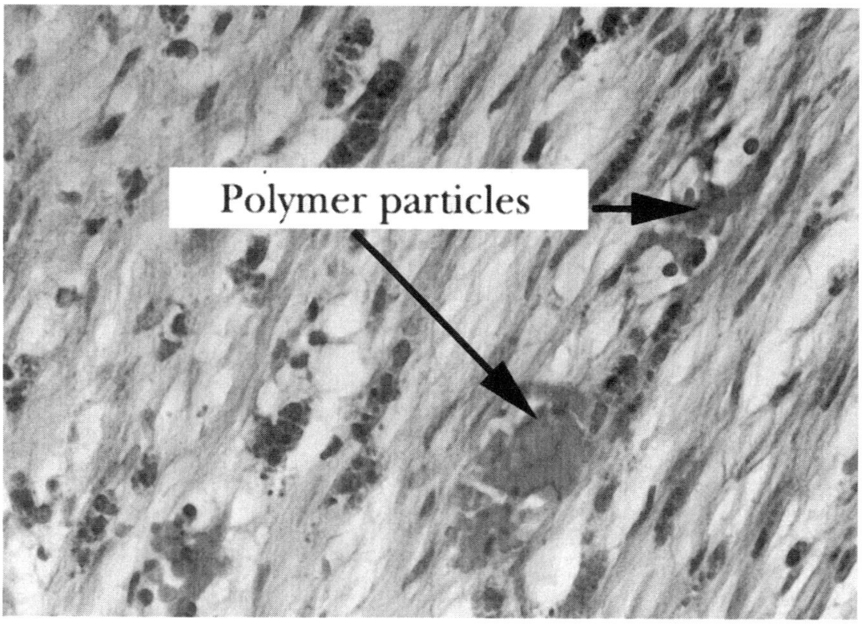

A

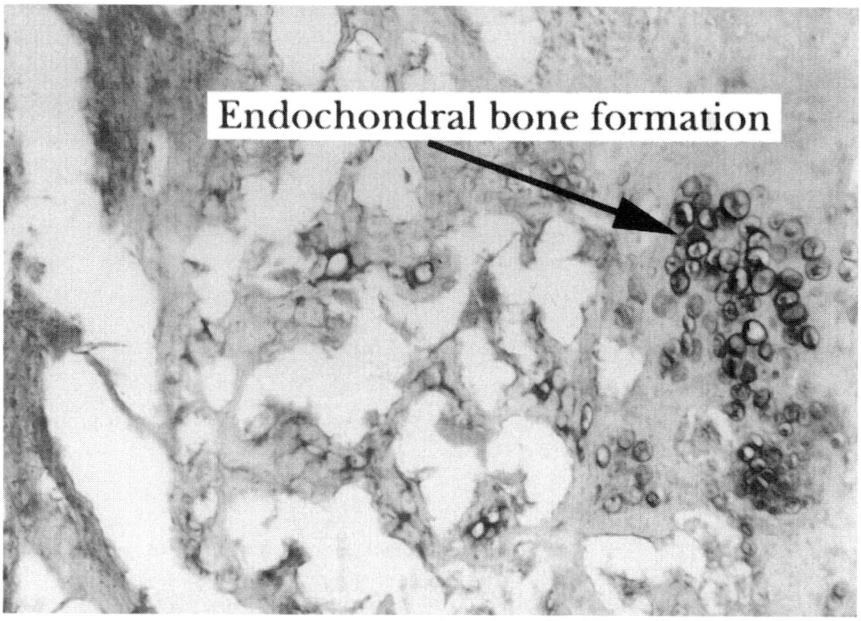

B

Figure 22 Calcium phosphate/gelatin composite as bone substitute. (A) Connective tissue surrounding the implant became denser in week 2. HE 400×. (B) Endochondral bone formation can be observed at week 3. HE, 100×.

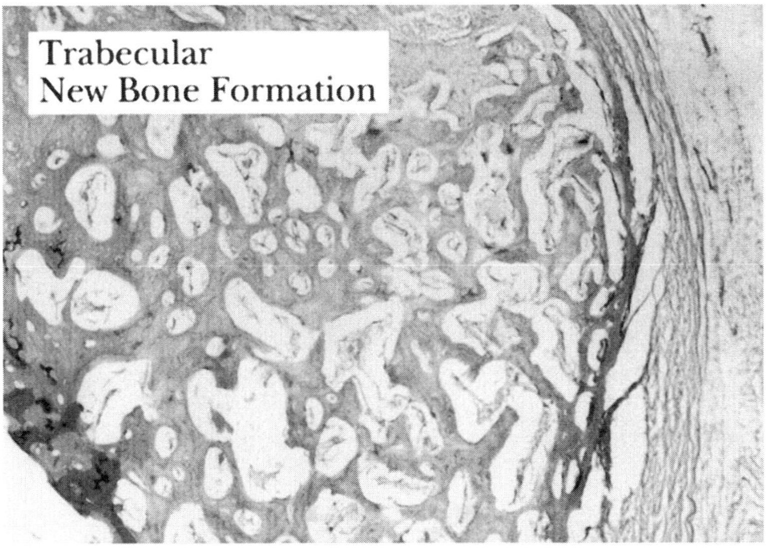

A

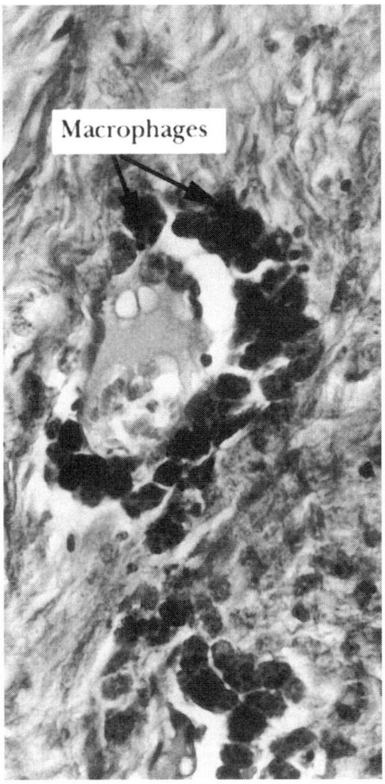

B

Figure 23 Calciumphosphate gelatin composite as bone substitute. (A) New bone formation is observed in week 4. HE 40×. (B) Calcium phosphate particles belonging to the implant are surrounded by numerous macrophages. Massons Trichrome 400×.

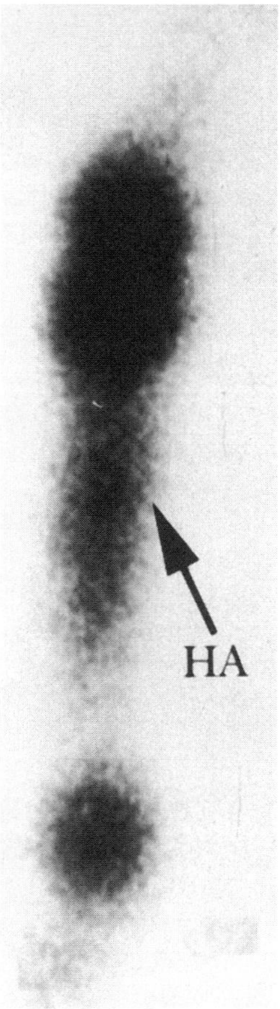

Figure 24 Tc-99m bone scintigraphy in HA-implanted rabbit tibia. Quantitative analysis was possible in monitoring tissue integration and bone healing.

overcome wear debris. It is obvious that less metal will be used in the future. Improving methods of production, processing, and sterilization should decrease the adverse response to ceramics and polymers. New materials should undergo a set of well-defined immunological tests to better understand their biological response before they are introduced into clinical use.

REFERENCES

1. Santavirta S, Gristina A, Konttinen YT. Cemented versus cementless hip arthroplasty: a review of prosthetic biocompatibility. Acta Orthop. Scand 1992; 63:225–232.
2. Total hip replacement. NIH Consensus Statement; Sep 12–14, 1994; 12(5):1–31.
3. Praemer A, Furner S, Rice DP. Musculoskeletal conditions in the United States. Illinois: American Academy of Orthopaedic Surgeons, 1992:127–141.
4. Head WC, Bauk DJ, Emerson RH. Titanium as the material of choice for cementless femoral components in total hip arthroplasty. Clin. Orthop. Rel. Res 1995; 311:85–90.

5. Shah AK, Sinha RK, Hickok NJ, Tuan RS. High-resolution morphometric analysis of human osteoblastic cell adhesion on clinically relevant orthopedic allyos. Bone 1999; 24:499–506.
6. Gronowicz G, McCarthy MB. Response of human osteoblasts to implant materials: integrin-mediated adhesion. J. Orthop. Res 1996; 14:878–887.
7. Kurachi T, Nagao H, Nagura H, Enomoto S. Effect of titanium surface on bone marrow–derived osteoblastic cells in vitro. Arch. Oral Biol 1997; 42:465–468.
8. Jinno T, Goldberg VM, Davy D, Stevenson S. Osseointegration of surface-blasted implants made of titanium alloy and cobalt–chromium alloy in a rabbit intramedullary model. J. Biomed. Mater. Res 1998; 42:20–29.
9. Granchi D, Ciapetti G, Stea S, Savarino L, Filippini F, Sudanse A, Zinghi G, Montanaro L. Cytokine release in mononuclear cells of patients with Co-Cr hip prosthesis. Biomaterials 1999; 20: 1079–1086.
10. Rahal MD, Branemark PI, Osmond DG. Response of bone marrow to titanium implants. Int. J. Oral Maxillofacial Implants 1993; 8:573–579.
11. Davidson JA, Poggie RA, Mishra AK. Abrasive wear of ceramic, metal and UHMPE bearing surfaces from third-body bone, PMMA bone cement and titanium debris. Biomed. Mater. Eng 1994; 4:213–229.
12. Jacobs JJ, Skipor AK, Patterson LM, Hallab NJ, Paprosky WG, Black J, Galante JO. Metal release in patients who have had a primary total hip arthroplasty. J. Bone Joint Surg 1998; 80-A:1447–1458.
13. Zhang X, Morham SG, Langenbach R, Young DA, Xing L, Boyce BF, Puzas EJ, Rosier RN, O'Keefe RJ, Schwarz EM. Evidence for a direct role of cyclo-oxygenase 2 in implant wear debris–induced osteolysis. J. Bone Miner. Res 2001; 16:660–670.
14. Bi Y, van de Motter RR, Ragab AA, Goldberg VM, Anderson JM, Greenfield EM. Titanium particles stimulate bone resorption by inducing differentiation of murine osteoclasts. J. Bone Joint Surg 2001; 83-A:501–508.
15. Lohmann CH, Schwartz Z, Koster G, Jahn U, Buchhorn GH, MacDougall MJ, Casasola D, Liu Y, Sylvia VL, Dean DD, Boyan BD. Phagocytosis of wear debris by osteoblasts affects differentiation and local factor production in a manner dependent on particle composition. Biomaterials 2000; 21: 551–561.
16. Neale SD, Athanasou NA. Cytokine receptor profile of arthroplasty macrophages, foreign body giant cells and mature osteoclasts. Acta Orthop. Scand 1999; 70:452–458.
17. Bauer TW, Schils J. The pathology of total joint arthroplasty. Skeletal Radiol 1999; 28:483–497.
18. Jacobs JJ, Shanbhag A, Glant TT, Black J, Galante JO. Wear debris in total joint replacements. J. Am. Acad. Orthop. Surg 1994; 2:212–220.
19. Pizzoferrato A, Stea S, Sundase A, Toni A, Nigrisoli M, Gualtieri G, Squarzoni S. Morphometric and microanalytical analyses of alumina wear particles in hip prostheses. Biomaterials 1993; 14: 583–587.
20. Pipino F. The bone–prosthesis interaction. J. Orthopaed. Traumatol 2000; 1:3–9.
21. Nakashima Y, Sun DH, Trindade MCD, Chun LE, Song Y, Goodman SB, Schurman DJ, Maloney WJ, Smith RL. Induction of macrophage C-C chemokine expression by titanium alloy and bone cement particles. J. Bone Joint Surg 1999; 81-B:155–162.
22. Uo M, Watari F, Yokoyama A, Matsuno H, Kawasaki T. Tissue reaction around metal implants observed by x-ray scanning analytical microscopy. Biomaterials 2000; 22:677–685.
23. Ragab AA, van de Motter R, Lavish SA, Goldberg VM, Ninomiya JT, Carlin CR, Greenfield EM. Measurement and removal of adherent endotoxin from titanium particles and implant surfaces. J. Orthop. Res 1999; 17:803–809.
24. Bos I. Gewebereaktionen um gelockerte hüftgelenkendoprothesen. Eine histologische studie an sekundärkapseln und interfacemembranen. Orthopäde 2001; 30:881–889.
25. Bonelli A. Current concepts in orthopaedic biomaterials and implant fixation. J. Bone Joint Surg 1994; 76:1897–1898.
26. Shanbhag AS, Jacobs JJ, Glant TT, Gilbert JL, Black J, Galante JO. Composition and morphology of wear debris in failed uncemented total hip arthroplasty. J. Bone Joint Surg 1994; 76-A:60–67.
27. Stea S, Visentin M, Granchi D, Cenni E, Ciapetti G, Sudanese A, Toni A. Apoptosis in peri-implant tissue. Biomaterials 2000; 21:1393–1398.

28. Glant TT, Jacobs JJ, Mikecz K, Yao J, Chubinskaja S, Williams JM, Urban RL, Shanbhag AS, Lee SH, Sumner DR. Particulate-induced, prostaglandin- and cytokine-mediated bone resorption in an experimental system and in failed joint replacements. Am. J. Ther 1996; 3:27–41.
29. Robinson TM, Manley PA, Sims PA, Albrecht R, Darien BJ. Cytokine and eicosanoid production by cultured human monocytes exposed to titanium particulate debris. J. Arthroplasty 1999; 14: 781–787.
30. Trindade MC, Nakashima Y, Lind M, Sun DH, Goodman SB, Maloney WJ, Schurman DJ, Smith RL. Interleukin-4 inhibits granulocyte-macrophage colony-stimulating factor, interleukin-6, and tumour necrosis factor-alpha expression by human monocytes in response to polymethylmethacrylate particle challenge in vitro. J. Orthop. Res 1999; 17:797–802.
31. Shida J, Trindade MC, Goodman SB, Schurman DJ, Smith RL. Induction of interleukin-6 release in human osteoblast-like cells exposed to titanium particles in vitro. Calcif. Tissue Int 2000; 67: 151–155.
32. Lassus J, Waris V, Xu JW, Li TF, Hao J, Nietosvaara Y, Santavirta S, Konttinen YT. Increased interleukin-8 (IL-8) expression is related to aseptic loosening of total hip replacement. Arch. Orthop. Trauma. Surg 2000; 120:328–332.
33. Horowitz SM, Rapuano BP, Lane JM, Burstein AH. The interaction of the macrophage and the osteoblast in the pathophysiology of aseptic loosening of the joint replacements. Calcif. Tissue Int 1994; 54:320–324.
34. Stea S, Visentin M, Granchi D, Ciapetti G, Donati ME, Sudanese A, Zanotti C, Toni A. Cytokines and osteolysis around total hip prostheses. Cytokine 2000; 12:1575–1579.
35. Suh KT, Chang JW, Jung JS. The role of inducible nitric oxide synthetase in aseptic loosening after total hip arthroplasty. J. Bone Joint Surg 2002; 84-B:753–757.
36. Shanbhag AS, Macaulay W, Stefanovic-Racic M, Rubash HE. Nitric oxide release by macrophages in response to particulate wear debris. J. Biomed. Mater. Res 1998; 41:497–503.
37. Tengvall P, Elwing H, Sjoqvist L, Lundstrom I, Bujersten LM. Interaction between hydrogen peroxide and titanium: a possible role in the biocompatibility of titanium. Biomaterials 1989; 10:118–120.
38. Overgaard L, Danielsen N, Bujersten LM. Anti-inflammatory properties of titanium in the joint environment. J. Bone Joint Surg 1998; 80-B:888–893.
39. Rader CP, Sterner T, Jakob F, Schutze N, Eulert J. Cytokine response of human macrophage-like cells after contact with polyethylene and pure titanium particles. J. Arthoplasty 1999; 14:840–488.
40. Heinemann DE, Lohmann C, Siggelkow H, Alves F, Engel I, Koster G. Human osteoblast-like cells phagocytose metal particles and express the macrophage marker CD68 in vitro. J. Bone Joint Surg 2000; 82-B:283–289.
41. Urban RM, Jacobs JJ, Tomlinson MJ, Gavrilovic J, Black J, Peoch M. Dissemination of wear particles to the liver, speen and the abdominal lymph nodes of patients with hip or knee replacement. J. Bone Joint Surg 2000; 82-A:457–477.
42. Mak KH, Wong TK, Poddar NC. Wear debris from total hip arthroplasty presenting as an intrapelvic mass. J. Arthroplasty 2001; 16:674–676.
43. Granchi D, Verri E, Ciapetti G, Stea S, Savarino L, Sudanse A, Mieti M, Rotini R, Dallari D, Zinghi G, Montanaro L. Bone-resorbing cytokines in serum of patients with aseptic loosening of hip prostheses. J. Bone Joint Surg 1998; 80-B:912–917.
44. Doherty AT, Howell RT, Ellis LA, Bisbinas I, Learmonth ID, Newson R, Case CP. Increased chromosome translocations and aneuploidy in peripheral blood lymphocytes of patients having revision arthroplasty of the hip. J. Bone Joint Surg 2001; 83-B:1075–1081.
45. Lewis CG, Belniak RM, Plowman MC, Hopfer SM, Knight JA, Sunderman FW. Intraarticular carcionogenesis bioassay of CoCrMo and TiAlV alloys in rats. J. Arthroplasty 1995; 10:75–82.
46. Keel SB, Jaffe KA, Petur Nielsen G, Rosenberg AE. Orthopaedic implant–related sarcoma: a study of twelve cases. Modern Pathology 2001; 14:969–977.
47. Takamura K, Hayashi K, Ishinishi N, Yamada T, Sugioka Y. Evaluation of carcinogenicity and chronic toxicity associated with orthopedic implants in mice. J. Biomech. Mater. Res 1994; 28: 583–589.
48. Sunderman FW. Carciongenity of metal alloys in orthopedic prostheses. Fundam. Appl. Toxicol 1989; 13:205–216.

49. Sovak G, Weiss A, Gotman I. Osseointegration of Ti6Al4V alloy implants coated with titanium nitride by a new method. J. Bone Joint Surg 2000; 82-B:290–296.
50. Sawase T, Wennerberg A, Baba K, Tsuboi Y, Sennerby L, Johansson CB, Albrektsson T. Application of oxygen ion implantation to titanium surfaces: effects on surface characteristics, corrosion resistance and bone response. Clin. Implant Dent. Rel. Res 2001; 3:221–229.
51. Krupa D, Baszkiewicz J, Kozubowski JA, Barcz A, Sobczak JW, Bilinski A, Lewandrowska-Szumiel M, Rajchel B. Effect of phosphorus-ion implantation on the corrosion resistance and biocompatibility of titanium. Biomaterials 2002; 23:3329–3340.
52. Li D, Liu B, Wu J, Chen J. Bone interface of dental implants cytologically influenced by a modified sandblasted surface. Implant Dent 2001; 10:132–138.
53. Rhalmi S, Odin M, Assad M, Tabrizian M, Rivard CH, Yahia LH. Hard, soft tissue and in vitro cell response to porous nickel–titanium: a biocompatibility evaluation. Biomed. Mater. Eng 1999; 9:151–162.
54. Hofmann AA, Bloebaum RD, Bachus KN. Progression of human bone ingrowth into porous-coated implants. Acta Orthop. Scand 1997; 68:161–166.
55. Hainau B, Reimann I, Dorph S, Rechnagel K, Henschel A, Kragh F. Porous-coated knee arthroplasty: a case report concerning bone ingrowth. Clin. Orthop. Rel. Res 1989; 239:178–184.
56. Kapanen A, Kinnunen A, Ryhanen J, Tuukkanen J. TGF-beta secretion of ROS-17/2.8 cultures on NiTi implant material. Biomaterials 2002; 23:3341–3346.
57. von Knoch M, Engh CA, Sychterz CJ, Engh CA, Willert HG. Migration of polyethylene wear debris in one type of uncemented femoral component with circumferential porous coating. J. Arthroplasty 2000; 15:72–78.
58. Nanci A, Wuest JD, Peru L, Brunet P, Sharma V, Zalzal S, McKee MD. Chemical modification of titanium surfaces for covalent attachment of biological molecules. J. Biomed. Mater. Res 1998; 40:324–335.
59. Viorney C, Guenther HL, Aronsson BO, Pechy P, Descouts P, Gratzel M. Osteoblast culture on polished titanium disks modified with phosphonic acids. J. Biomed. Mater. Res 2002; 62:149–155.
60. Sul YT, Johansson CB, Kang Y, Jeon DG, Kang Y, Jeong DG, Albrektsson T. Bone reaction to oxidized titanium implants with electrochemical anion sulphuric acid and phosphoric acid incorporation. Clin. Implant Dent. Relat. Res 2002; 4:78–87.
61. Cook SD, Thomas KA, Kay JF, Jarcho M. Hydroxyapatite-coated porous titanium for use as an orthopedic biologic attachment system. Clin. Orthop. Rel. Res 1988; 230:303–312.
62. Rashmir-Raven AM, Richardson DC, Aberman HM, DeYoung DJ. The response of cancellous and cortical canine bone to hydroxyapatite-coated and uncoated titanium rods. J. App. Biomater 1995; 6:237–242.
63. Dalton JE, Cook SD, Thomas KA, Kay JF. The effect of operative fit and hydroxyapatite coating on the mechanical and biological response to porous implants. J. Bone Joint Surg 1995; 77-A: 97–110.
64. Coathulp MJ, Blunn GW, Flynn N, Williams C, Thomas NP. A comparison of bone remodelling around hydroxyapatite-coated, porous-coated and grit-blasted hip replacements retrieved at postmortem. J. Bone Joint Surg 2001; 83-B:118–123.
65. Montanaro L, Arciola CR, Campoccia D, Cervellati M. In vitro effects of MG63 osteoblast-like cells following contact with two roughness-differing fluorohydroxyapatite-coated titanium alloys. Biomaterials 2002; 23:3651–3659.
66. Simske SJ, Sachdeva R. Cranial bone apposition and ingrowth in a porous nickel–titanium implant. J. Biomed. Mater. Res 1995; 29:527–533.
67. Nimb L, Gotfredsen K, Steen-Jensen J. Mechanical failure of hydroxyapatite-coated titanium and cobalt–chromium–molybdenum alloy implants. Acta Orthop. Belg 1993; 59:333–338.
68. Manero JM, Salsench J, Nogueras J, Aparicio C, Padros A, Balcells M, Gil FJ, Planell JA. Growth of bioactive surfaces on dental implants. Implant Dent 2002; 11:170–175.
69. Huracek J, Spirig P. The effect of hydroxyapatite coating on the fixation of hip prostheses. Arch. Orthop. Trauma. Surg 1994; 113:72–77.
70. Rahbek O, Overgaard S, Lind M, Bendix K, Bunger C, Soballe K. Sealing effect of hydroxyapatite coating on peri-implant migration of particles. J. Bone Joint Surg 2001; 83:441–447.

71. Bauer TW, Taylor SK, Jiang M, Medendorp SV. An indirect comparison of third-body wear in retrieved hydroxyapatite-coated, porous and cemented femoral components. Clin. Orthop. Rel. Res 1994; 298:11–18.
72. Bloebaum RD, Beeks D, Dorr LD, Savory CG, Dupont JA, Hofmann AA. Complications with hydroxyapatite particulate separation in total hip arthroplasty. Clin. Orthop. Rel. Res 1994; 298: 19–26.
73. Kornu R, Maloney WJ, Kelly MA, Smith RL. Osteoblast adhesion to orthopaedic implant alloys: effects of cell adhesion molecules and diamond-like carbon coating. J. Orthop. Res 1996; 14: 871–877.
74. Roehlecke C, Witt M, Kasper M, Schulze E, Wolf C, Hofer A, Funk RW. Synergistic effect of titanium alloy and collagen type I on cell adhesion, proliferation and differentiation of osteoblast-like cells. Cells Tissues Organs 2001; 168:178–187.
75. Lemons JE. Ceramics: past, present and future. Bone 1996; 19:121S–128S.
76. Koç N, Timuçin M. Medical ceramics. Arthroplasty Arthroscopic Surgery 1999; 10:104–109.
77. Fleming JE, Cornell CN, Muschler GF. Bone cells and matrices in orthopedic tissue engineering. Orthop. Clin. North Am 2000; 31:357–374.
78. Uchida A, Nade SML, McCartney ER, Ching W. The use of ceramics for bone replacement. J. Bone Joint Surg 1984; 66-B:269–275.
79. Uchida A, Nade S, McCartney E, Ching W. Bone ingrowth into three different porous ceramics implanted into the tibia of rats and rabbits. J. Orthop. Res 1985; 3:65–77.
80. Korkusuz F, Korkusuz P. Calcium hydroxyapatite ceramics in orthopedic surgery. Acta Orthopaedica et Traumatologica Turcica 1997; 31:63–67.
81. Khan SN, Tomin E, Lane JM. Clinical applications of bone graft substitutes. Orthop. Clin. North Am 2000; 31:389–398.
82. Takaoka K, Nakahara H, Yoshikawa H, Masuhara K, Tsuda T, Ono K. Ectopic bone induction on and in porous hydroxyapatite combined with collagen and bone morphogenic protein. Clin. Orthop. Rel. Res 1988; 234:250–254.
83. Young SW, Andrews A, Muller H, Constanz B. Induction of fracture healing using fibrous calcium phosphate composite spherulites. Invest. Radiol 1991; 26:470–473.
84. Frayssinet P, Trouillet JL, Rouquet N, Azimus E, Autefage A. Osseointegration of macroporous calcium phosphate ceramics having a different chemical composition. Biomaterials 1993; 14: 423–429.
85. Noshi T, Yoshikawa T, Dohi Y, Ikeuchi M, Horiuchi K, Ichijima K, Sugimura M, Yonemasu K, Ohgushi H. Recombinant human bone morphogenic protein-2 potentiates the in vivo osteogenic ability of marrow/hydroxyapatite composites. Artificial Organs 2001; 25:201–208.
86. Krajewski A, Ravaglioli A, Roncari E, Pinasco P, Montanari L. Porous ceramic bodies for drug delivery. J. Mat. Sci. Mat. Med 2000; 12:763–771.
87. Shinto Y, Uchida A, Korkusuz F, Araki N, Ono K. Calcium hydroxyapatite ceramic used as a delivery system for antibiotics. J. Bone Joint Surg 1992; 74-B:600–604.
88. Korkusuz F, Uchida A, Shinto Y, Araki N, Inoue K, Ono K. Experimental implant-related osteomyelitis treated by antibiotic–calcium hydroxyapatite ceramic composites. J. Bone Joint Surg 1993; 75-B:111–114.
89. Uchida A, Araki N, Shinto Y, Yoshikawa H, Ono K, Kurosaki E. The use of hydroxyapatite ceramic in bone tumour surgery. J. Bone Joint Surg 1990; 72-B:298–302.
90. Bauer TW, Muschler GF. Bone graft materials. Clin. Orthop. Rel. Res 2000; 371:10–27.
91. Block JE, Thorn MR. Clinical indications of calcium-phosphate biomaterials and related composites for orthopedic procedures. Calcif. Tissue. Int 2000; 66:234–238.
92. Holmes RE, Bucholz RW, Mooney V. Porous hydroxyapatite as a bone-graft substitute in metaphyseal defects. J. Bone Joint Surg 1986; 68-A:904–911.
93. Yamaguchi K, Hirano T, Yoshida G, Iwasaki K. Degradation-resistant character of synthetic hydroxyapatite blocks filled in bone defects. Biomaterials 1995; 16:983–985.
94. Handschel J, Wiesmann HP, Stratmann U, Kleinheiz J, Meyer U, Joos U. TCP is hardly resorbed and not osteoconductive in a non-loading calvarial model. Biomaterials 2002; 23:1689–1695.

95. Balçık C. Biomechnaics of porous ceramic implants used in long bone segmental defect healing. PhD Thesis, Middle East Technical University, Ankara, Turkey, 2002.
96. Gürmeriç A. Experimental evaluation of new bone formation using hydroxyapatite, allogenic bone chips, natural coral and calcium carbonate. PhD Thesis, Hacettepe University, Ankara, Turkey, 1995.
97. Tassery H, Pertot WJ, Camps J, Proust JP, Dejoy J. Comparison of two implantation sites for testing intraosseous biocompatibility. J. Endod 1999; 25:615–618.
98. Hench LL, Wilson J. Surface active biomaterials. Science 1984; 226:630–636.
99. Boyan BD, Hummert TW, Den DD, Schwartz Z. Role of material surface in regulating bone and cartilage cell response. Biomaterials 1996; 17:137–146.
100. Schwartz Z, Braun G, Kohavi D, Brooks B, Amir D, Sela J, Boyan B. Effects of hydroxyapatite implants on primary mineralization during rat tibial healing. J. Biomed. Mater. Res 1993; 27: 1029–1038.
101. Korkusuz F, Uluoğlu Ö. Non-specific inflammation and bone marrow depletion due to intramedullary porous hydroxyapatite application. Bull. Hosp. Joint Diseases 1999; 58:86–91.
102. Klinge B, Alberius P, Isaksson S, Jonsson J. Osseous response to implanted natural bone mineral and synthetic hydroxyapatite ceramic in the repair of experimental skull bone defects. J. Oral Maxillofac. Surg 1992; 50:241–249.
103. Lu JX, Gallur A, Flautre B, Anselme K, Descamps M, Thierry B, Hardouin P. Comparative study of tissue reactions to calcium phosphate ceramcis among cancellous, cortical and medullary bone sites in rabbits. J. Biomed. Mater. Res 1998; 42:357–367.
104. Misiek DJ, Kent JN, Carr RF. Soft tissue responses to hydroxyapatite particles of different shapes. J. Oral Maxillofac. Surg 1984; 42:150–160.
105. Lin TC, Su CY, Chang CS. Stereomorphologic observation of bone tissue response to hydroxyapatite using SEM with the EDTA-KOH method. J. Biomed. Mater. Res 1997; 36:91–97.
106. Goodman SB, Davidson JA, Fornasier VL. Histological reaction to titanium alloy and hydroxyapatite particles in the rabbit tibia. Biomaterials 1993; 14:723–727.
107. Laquerriere P, Kilian L, Bouchot A, Jallot E, Grandjean A, Guenounou M, Balossier G, Frayssinet P, Bonhomme P. Effect of hydroxyapatite sintering temperature on intracellular ionic concentration of monocytes. J. Biomed. Mater. Res 2001; 58:238–246.
108. Ninomiya JT, Struve JA, Stelloh CT, Toth JM, Crosby KE. Effects of hydroxyapatite particulate debris on the production of cytokines and proteases in human fibroblasts. J. Orthop. Res 2001; 19: 621–628.
109. Zhi-Dao X, Tong Z, Jing-Yuan D, Qi-Xing Z, Lan W, Shi-Pu L, Chao-Ying C, Shi-Yuan F. Macrophages in degradation of collagen/hydroxylapatite (CHA), beta-tricalcium phosphate ceramcis (TCP) artificial bone graft. Chinese Med. J 1994; 107:845–849.
110. Taş AC, Korkusuz F, Timuçin M, Akkaş N. An investigation of the chemical synthesis and high-temperature sintering behaviour of calcium hydroxyapatite and tricalcium phosphate bioceramcis. J. Mat. Sci. Mat. Med 1997; 8:91–96.
111. Ignatus AA, Schmidt C, Kaspar D, Claes LE. In vitro biocompatibility of resorbable experimental glass ceramics for bone substitutes. J. Biomed. Mater. Res 2001; 55:285–294.
112. Guizzardi S, Montanari C, Migliaccio S, Strocchi R, Solmi R, Martini D, Ruggeri A. Quality assessment of natural apatite in vitro and in vivo. J. Biomed. Mater. Res 2000; 53:227–234.
113. Midy V, Dard M, Hollande E. Evaluation of the effect of three calcium phosphate powders on osteoblast cells. J. Mat. Sci. Mater. Med 2001:259–265.
114. Mosheiff R, Klein BY, Leichter I, Chaimsky G, Nyska A, Peyser A, Segal D. Use of dual energy x-ray absorptiometry (DEXA) to follow mineral content changes in small ceramic implants in rats. Biomaterials 1992; 13:462–466.
115. Korkusuz F, Karamete K, Irfanoğlu B, Yetkin H, Hastings GW, Akkaş N. Do porous calcium hydroxyapatite ceramics cause porosis in bone? A bone densitometry and biomechanical study on cortical bones of rabbits. Biomaterials 1995; 16:537–543.
116. Parsons JR, Ricci JL, Alexander H, Bajpai PK. Osteoconductive composition grouts for orthopedic use. Ann. NY Acad. Sci 1988; 523:190–207.

117. Jianguo Z, Xingdong Z, Müller-Mai C, Gross U. The early host and material response of hydroxyapatite/beta-tricalciumphosphate porous ceramics after implantation into the femur of rats. J. Mat. Sci. Mater. Med 1994; 5:243–251.
118. Daculsi G, LeGeros RZ, Heughebaert M, Barbieux I. Formation of carbonate–apatite crystals after implantation of calcium phosphate ceramics. Calcif. Tissue Int 1990; 46:20–27.
119. de Bruijn JD, van Blitterswijk CA, Davies JE. Initial bone matrix formation at the hydroxyapatite interface in vivo. J. Biomed. Mater. Res 1995; 29:89–99.
120. Krajewski A, Ravaglioli A, Mongiorgi R, Moroni A. Mineralization and calcium fixation within a porous apatitic ceramic material after implantation in the femur rabbits. J. Biomed. Mater. Res 1988; 22:445–457.
121. de Lange GL, de Putter C, de Wijs FLJA. Histological and ultrastructural appearance of the hydroxy-apatite–bone interface. J. Biomed. Mater. Res 1990; 24:829–845.
122. Piattelli A, Trisi P. A light and laser scanning microscopy study of bone/hydroxyapatite-coated titanium implants interface. J. Biomed. Mater. Res 1994; 28:529–536.
123. Eggli PS, Müller W, Schenk RK. Porous hydroxyapatite and tricalcium phosphate cylinders with two different pore size ranges implanted in the cancellous bone of rabbits. Clin. Orthop. Rel. Res 1988; 232:127–138.
124. Tampieri A, Celotti G, Spiro S, Delcogliano A, Franzese S. Porosity-graded hydroxyapatite ceramics to replace natural bone. Biomaterials 2001; 22:1365–1370.
125. Yoshikawa T, Ohgushi H, Tamai S. Immediate bone forming capability of prefabricated osteogenic hydroxyapatite. J. Biomed. Mater. Res 1996; 32:481–492.
126. Okamura A, Goto M, Goto T, Yoshinari M, Matsuko S, Katsuki T, Tanaka T. Substrate effects to initial attachment and subsequent behaviour of human osteoblastic cells (Saon-2). Biomaterials 2001; 22:2263–2271.
127. Daculsi G, Passuti N, Martin S, Deudon C, Legeros RZ, Raher S. Macroporous calcium phosphate ceramic for long bone surgery in humans and dogs. Clinical and histological study. J. Biomed. Mater. Res 1990; 24:379–396.
128. Ducheyne P, Qui Q. Bioactive ceramics: the effect of surface reactivity on bone formation and bone cell function. Biomaterials 1999; 20:2287–2303.
129. Deligianni DD, Katsala ND, Koutsoukos PG, Missirlis YF. Effect of surface roughness of hydroxyapatite on human bone marrow cells adhesion, proliferation, differentiation and detachment strength. Biomaterials 2001; 22:87–96.
130. Yang Z, Yuan H, Tong W, Zou P, Chen W, Zhang X. Osteogenesis in extraskeletally implanted porous calcium phosphate ceramics: variability among different kinds of animals. Biomaterials 1996; 17:2131–2137.
131. Kudo M, Matsui Y, Ohno K, Michi K. A histomorphometric study of the tissue reaction around hydroxyapatite implants irradiated after placement. J. Oral Maxillofac. Surg 2001; 59:293–300.
132. Ohgushi H, Goldberg VM, Caplan AI. Repair of bone defects with marrow cells and porous ceramics. Acta Orthop. Scand 1989; 60:334–339.
133. Okamura M, Ohgushi H, Tamai S. Bonding osteogenesis in coraline hydroxyapatite combined with bone marrow cells. Biomaterials 1991; 12:411–416.
134. Frayssinet P, Primout I, Rouquet N, Autefage A, Guilhem A, Bonnevialle P. Bone cell grafts in bioreactor. J. Mat. Sci. Mat. Med 1991; 2:217–221.
135. Sempuku T, Ohgushi H, Okumura M, Tamai S. Osteogenic potential of allogenic rat marrow cells in porous hydroxyapatite ceramics. J. Orthop Res 1996; 14:907–913.
136. Bareille R, Lafage-Proust MH, Faucheux C, Laroche N, Wenz R, Dard M, Amedee J. Various evaluation techniques of newly formed bone in porous hydroxyapatite loaded with human bone marrow cells implanted in an extra-osseous site. Biomaterials 2000; 21:1345–1352.
137. Petite H, Viateau V, Bensaid W, Meunier A, de Pollak C, Bourguignon M, Oudina K, Sedel L, Guillemin G. Tissue-engineered bone regeneration. Nature Biotech 2000; 18:959–963.
138. Goshima J, Goldberg VM, Caplan AI. The origin of bone formed in composite grafts of porous calcium phosphate ceramic loaded with marrow cells. Clin. Orthop. Rel. Res 1991; 269:274–283.
139. Bagambisa FB, Joos U, Schilli W. Interaction of osteogenic cells with hydroxyapatite implant materials in vitro and in vivo. Int. J. Maxillofac. Implants 1990; 5:217–226.

140. Sun JS, Tsuang YH, Liao CJ, Liu HC, Hang YS, Lin FH. The effects of calcium phosphate particles on the growth of osteoblasts. J. Biomed. Mater. Res 1997; 37:324–334.
141. Mikos AG, McIntire LV, Anderson JM, Babensee JE. Host response to tissue engineered devices. Adv. Drug. Deliv. Rev 1998; 33:111–139.
142. Salih V, Georgiou G, Knowles JC, Olsen I. Glass reinforced hydroxyapatite for hard tissue surgery. Biomaterials 2001; 22:2817–2824.
143. Wolfe LA, Boyde A. Biocompatibility tests on a novel glass–ceramic system. J. Appl. Biomater 1992; 3:217–224.
144. Oonishi H, Hench LL, Wilson J, Sugihara F, Tsuji E, Kushitani S, Iwaki H. Comparative bone ingrowth behaviour in granules of bioceramic materials of various sizes. J. Biomed. Mater. Res 1999; 44:31–43.
145. Sautier JM, Kokubo T, Ohtsuki T, Nefussi JR, Boulekbache H, Oboeuf M, Loty S, Loty C, Forest N. Bioactive glass–ceramic containing crystalline apatite and wollastonite initiates biomineralization in bone cell cultures. Calcif. Tissue Int 1994; 55:458–466.
146. Brook IM, Hatton PV. Glass ionomers: bioactive implant materials. Biomaterials 1998; 19:565–571.
147. Bosetti M, Hench L, Cannas M. Interaction of bioactive glasses with peritoneal macrophages and monocytes in vitro. J. Biomed. Mater. Res 2002; 60:79–85.
148. Gao TJ, Lindholm TS, Kommonen B, Ragni P, Paronzini A, Lindholm TC. Microscopic evaluation of bone–implant contact between hydroxyapatite, bioactive glass and tricalcium phosphate implanted in sheep diaphyseal defects. Biomaterials 1995; 16:1175–1179.
149. Glant TT, Jacobs JJ, Molnar G, Shanbhag AS, Valyon M, Galante JO. Bone resorption activity of particulate-stimulated macrophages. J. Bone Miner. Res 1993; 8:1071–1079.
150. Vale FM, Castro M, Monteiro J, Couto FS, Pinto R, Giao-Toscano-Rico JM. Acrylic bone cement induces the production of free radicals by cultured human fibroblasts. Biomaterials 1997; 18: 1133–1135.
151. Gough JE, Downes S. Osteoblast cell death on methacrylate polymers involves apoptosis. J. Biomed. Mater. Res 2001; 57:497–505.
152. Ohsawa K, Neo M, Matsuoka H, Akiyama H, Ito H, Nakamura T. Tissue response around polymethylmethacrylate particles implanted into bone. J. Biomed. Mater. Res 2001; 54:501–508.
153. Lazarus MD, Cuckler JM, Schumacher HR, Ducheyne P, Baker DG. Comparison of the inflammatory response to particulate polymethylmethacrylate debris with and without barium sulfate. J. Orthop. Res 1994; 12:532–541.
154. Deb S, Silvio LD, Vazquez B, San Roman J. Water absorption characteristics and cytotoxic and biological evaluation of bone cements formulated with a novel activator. J. Biomed. Mater. Res 1999; 48:719–725.
155. Dalby MJ, diSilvio L, Harper EJ, Bonfield W. Initial interaction of osteoblasts with the surface of a hydroxyapatite–poly(methacrylate) cement. Biomaterials 2001; 22:1739–1747.
156. Şerbetçi K, Korkusuz F, Hasırcı N. Mechanical and thermal properties of hydroxyapatite-impregnated bone cement. Turk. J. Med. Sci 2000; 30:543–549.
157. Yap AU, Pek YS, Kumar RA, Cheang P, Khor KA. Experimental studies on a new bioactive material: HA ionomer cements. Biomaterials 2002; 23:955–962.
158. Şerbetçi K. Formulation and biomechanical evaluation of bone cements. PhD Thesis, Middle East Technical University, Ankara, Turkey, 2001.
159. Şerbetçi K, Orhun S, Korkusuz F, Hasırcı N. In vivo biocompatibility of hydroxyapatite containing bone cement. Arthroplasty Arthroscopic Surgery 2002; 13:259–263.
160. Saito M, Maruoka A, Mori T, Sugano N, Hino K. Experimental studies on a new bioactive bone cement: hydroxyapatite composite resin. Biomaterials 1994; 15:156–160.
161. Sarkar MR, Wachter N, Patka P, Kinzl L. First histological observations on the incorporation of a novel calcium phosphate bone substitute material in human cancellous bone. J. Biomed. Mater. Res 2001; 58:329–334.
162. Stelnicki EJ, Ousternout DK. Hydroxyapatite paste used as an onlay implant for supraorbital and malar augmentation. J. Craniofac. Surg 1997; 8:367–372.
163. Ginebra MP, Fernandez E, de Maeyer EA, Verbeeck RM, Boltong MG, Ginebra J, Driessens FC, Planell JA. Setting reaction and hardening of an apatitic calcium phosphate cement. J. Dent. Res 1997; 76:905–912.

164. Grimandi G, Weiss P, Millot F, Daculsi G. In vitro evaluation of a new injectable calcium phosphate material. J. Biomed. Mater. Res 1998; 39:660–666.
165. Gauthier O, Goyenvalle E, Bouler JM, Guicheux J, Pilet P, Weiss P, Daculsi G. Macroporous biphasic calcium phosphate ceramic versus injectable bone substitute. J. Mat. Sci. Mat. Med 2001; 12:385–390.
166. Simon CG, Khatri CA, Wight SA, Wang FW. Preliminary report on the biocompatibility of a moldable, resorbable, composite bone graft considering of calcium phosphate cement and poly(lactide-co-glycolide) microspheres. J. Orthop. Res 2002; 20:473–482.
167. Fujishiro Y, Takahashi K, Sato T. Preparation and compressive strength of alpha-tricalcium phosphate/gelatin gel composite cement. J. Biomed. Mater. Res 2001; 54:525–530.
168. Miyamoto Y, Ishikawa K, Takechi M, Toh T, Yuasa T, Nagayama M, Suzuki K. Basic properties of calcium phosphate cement containing atelocollagen in its liquid or powder phase. Biomaterials 1998; 19:707–715.
169. Yokoyama A, Yamamoto S, Kawasaki T, Kohgo T, Nakasu M. Development of calcium phosphate cement using chitosan and citric acid for bone substitute materials. Biomaterials 2002; 23: 1091–1101.
170. Blom EJ, Klein-Nulend J, Klein CP, Kurashina K, van Wass MA, Burger EH. Transforming growth factor-beta1 incorporated during setting in calcium phosphate cement stimulates bone cell differentiation in vitro. J. Biomed. Mater. Res 2000; 50:67–74.
171. Pioletti DP, Takei H, Lin T, vanLanduyt P, Ma QJ, Kwon SY, Sung KL. The effects of calcium phosphate cement particles on osteoblast function. Biomaterials 2000; 21:1103–1114.
172. Hasirci V, Lewandrowski K, Gresser JD, Wise DL, Trantolo DJ. Versatility of biodegradable biopolymers: degradability and an in vivo application. J. Biotech 2001; 86:135–150.
173. Deschamps AA, Claase MB, Sleijster WJ, deBruijn JD, Grijpma DW, Feijen J. Design of segmented poly(ether ester) materials and structures for the tissue engineering of bone. J. Control Release 2002; 78:175–186.
174. Heath DJ, Christian P, Griffin M. Involvement of tissue transglutaminase in the stabilization of biomaterial/tissue interfaces important in medical devices. Biomaterials 2002; 23:1519–1526.
175. Böstman O, Pihlajmaki H. Clinical biocompatibility of biodegradable orthopaedic implants for internal fixation. Biomaterials 2000; 21:2615–2621.
176. Yağmurlu MF, Korkusuz F, Gürsel I, Korkusuz P, Örs Ü, Hasırcı V. Sulbactamcetoperazone PHBV local antibiotic delivery system. In vivo effectivity and biocompatibility in treatment of implant related experimental osteomyelitis. J. Biomed. Mater. Res. 1999; 46:494–503.
177. Nordstorm P, Pihlajamaki H, Toivonen T, Tormala P, Rokkanen P. Tissue response to polyglycolide and polylevolactide pins in osteotomized cancellous bone. Clin. Orthop. Rel. Res 2001; 382: 247–257.
178. Ignatus AA, Betz O, Augat P, Claes LE. In vivo investigation on composites made of resorbable ceramics and poly(lactide) used as bone graft substitutes. J. Biomed. Mater. Res 2001; 58:701–709.
179. Tegnander A, Engebretsen L, Bergh K, Eide E, Holen KJ, Iversen OJ. Activation of the complement system and adverse effects of biodegradable pins of polylactic acid in osteochondrritis dissecans. Acta Orthop. Scand 1995; 66:573–574.
180. Koskikare K, Toivonen T, Rokkanen P. Tissue response to bioabsorbable self-reinforced polylevolactide and polyglycolide pins implanted intra-articularly and directly into the bone on different levels. Arch. Orthop. Trauma Surg 1998; 118:149–155.
181. Santavirta S, Konttinen YT, Saito T, GrÖnblad M, Partio E, Kemppinen P, Rokkanen P. Immune response to polyglycolic acid implants. J. Bone Joint Surg 1990; 72-B:597–600.
182. Wake MC, Gerecht PD, Lu L, Mikos AG. Effects of biodegradable polymer particles on rat marrow–derived stromal osteoblasts in vitro. Biomaterials 1998; 19:1255–1268.
183. Burkhart SS. The evolution of clinical applications of biodegradable implants in arthroscopic surgery. Biomaterials 2000; 21:2631–2634.
184. Böstman O, Pihlajmaki H. Adverse tissue reactions to bioabsorbable fixation devices. Clin. Orthop. Rel. Res 2000; 371:216–227.
185. Girard B, Baker G, Mock D. Foreign body granuloma following placement of hard tissue replacement material. A case report. J. Periodont 2000; 71:517–520.

186. Miketa JP, Prigoff MM. Foreign body reactions to absorbable implant fixation of ostetomies. J. Foot Ankle Surg 1994; 33:623–627.
187. Casteleyn PP, Handelberg F, Haentjens P. Biodegradable rods versus kirschner wire fixation of wrist fractures. J. Bone Joint Surg 1992; 74-B:858–861.
188. Bucholz RW, Henry S, Henley MB. Fixation with bioabsorbable screws for the treatment of fractures of the ankle. J. Bone Joint Surg 1994; 76-A:319–324.
189. Ahl T, Dalen N, Lundberg A, Wykman A. Biodegradable fixation of ankle fractures. Acta Orthop. Scand 1994; 65:166–170.
190. van der Elst M, Klein CPAT, de Blieck-Hogervorst JM, Patka P, Haarman HJTM. Bone tissue response to biodegradable polymers used for intramedullary fracture fixation: a long term in vivo study in sheep femora. Biomaterials 1999; 20:121–128.
191. Calvert JW, Marra KG, Cook L, Kumta PN, diMilla PA, Weiss LE. Characterization of osteoblast-like behaviour of cultured bone marrow stromal cells on various polymer surfaces. J. Biomed. Mater. Res 2000; 52:279–284.
192. Korkusuz F, Korkusuz P, Ekşioğlu F, Gürsel I, Hasırcı V. In vivo response to biodegradable controlled antibiotic release systems. J. Biomed. Mater. Res 1999; 55:217–228.
193. Yaylaoğlu MB, Korkusuz P, Örs Ü, Korkusuz F, Hasırcı V. Development of a calcium phosphate–gelatin composite as a bone substitute and its use in drug release. Biomaterials 1999; 20: 711–719.
194. Lewandrowski KU, Gresser JD, Wise DL, Trantolo DJ. Bioresorbable bone graft substitutes of different osteoconductivities: a histologic evaluation of osteointegration of poly(propylene glycol-co-fumaric acid)–based cement implants in rats. Biomaterials 2000; 21:757–764.
195. Lewandrowski KU, Gresser JD, Wise DL, Trantolo DJ, Hasırcı V. Tissue response to molecularly reinforced polylactide-co-glycolide implants. J. Biomater. Sci. Polym. Ed 2000; 11:401–414.
196. Andriano KP, Chandrashekar B, McEnery K, Dunn RL, Moyer K, Balliu CM, Holland KM, Garrett S, Huffer WE. Preliminary in vivo studies on the osteogenic potential of bone morphogenic proteins delivered from an absorbable putty-like polymer matrix. J. Biomed. Mater. Res 2000; 53:36–43.
197. Athanasiou KA, Niederauer GG, Agrawal CM. Sterilization, toxicity, biocompatibility and clinical applications of polylactic acid/polyglycolic acid copolymers. Biomaterials 1996; 17:93–102.
198. Gogolewski S. Bioresorbable polymers in trauma and bone surgery. Injury 2000; 31:S-D28-32.
199. Rokkanen PU, Böstman O, Hirvensalo E, Antero-Makela E, Partio EK, Patiala H, Vainionpaa S, Vihtonen K, Tormala P. Bioabsorbable fixation in orthopaedic surgery and traumatology. Biomaterials 2000; 21:2607–2613.
200. An YH, Woolf SK, Friedman RJ. Pre-clinical evaluation of orthopaedic bioabsorbable devices. Biomaterials 2000; 21:2635–2652.
201. Daculsi G, Weiss P, Bouler JM, Gauthier O, Millot F, Aguado E. Biphasic calcium phosphate/hydrosoluble polymer composites: a new concept for bone and dental substitution Biomaterials. Bone 1999; 25:59S–61S.
202. Rizzi SC, Heath DJ, Coombes AG, Bock N, Textor M, Downes S. Biodegradable polymer/hydroxyapatite composites: surface analysis and initial attachment of human osteoblasts. J. Biomed. Mater. Res 2001; 55:475–486.
203. Linhart W, Peters F, Lehman W, Schwartz K, Shilling AF, Amling M, Rueger JM, Epple M. Biologically and chemically optimized composites of carbonated apatite and polyglycolide as bone substitution materials. J. Biomed. Mater. Res 2001; 54:162–171.
204. Zhang K, Ma Y, Francis LF. Porous polymer/bioactive glass composites for soft-to-hard tissue interfaces. J. Biomed. Mater. Res 2002; 61:551–563.
205. Matsumoto M, Chosa E, Nabeshima K, Shikinami Y, Tajima N. Influence of bioabsorbable, unsintered hydroxyapatite/poly-l-lactide composite films on spinal cord, nerve root and epidural space. J. Biomed. Mater. Res 2002; 60:101–109.
206. Laurencin CT, Attawia MA, Lu LQ, Borden MD, Lu HH, Gorum WJ, Lieberman JR. Poly(lactide-co-glycolide)/hydroxyapatite delivery of BMP-2-producing cells. A regional gene therapy approach to bone regeneration. Biomaterials 2001; 32:1271–1277.

207. Laurencin CT, Attawia MA, Elgendry HE, Herbert KM. Tissue engineered bone-regeneration using degradable polymers: the formation of mineralized matrices. Bone 1996; 19:93S–99S.
208. Schliephake H, Kage T. Enhancement of bone regeneration using resorbable ceramics and a polymer–ceramic composite material. J. Bone Miner. Res 2001; 56:128–136.
209. Niklason LE. Engineering of bone grafts. Nature Biotechnology 2000; 18:929–930.
210. Ripamonti U, Tasker JR. Advances in biotechnology for tissue engineering of bone. Curr. Pharma. Biotech 2000; 1:47–55.

2

Material Characteristics and Biocompatibility of Low Rigidity Titanium Alloys for Biomedical Applications

Mitsuo Niinomi
Toyohashi University of Technology
Toyohashi, Japan

Tomokazu Hattori
Meijo University
Nagoya, Japan

Shigeo Niwa
Aichi Medical University
Nagakute, Japan

I. INTRODUCTION

Titanium and its alloys are the most suitable materials for structural biomaterials among metallic biomaterials. Nowadays, Pure Ti and Ti-6Al-4V ELI are the most widely used in the biomedical fields. However, for replacing failed hard tissue, the strength of pure Ti is not sufficient. For Ti-6Al-4V ELI, V has been pointed out to be toxic [1]. Therefore, V-free titanium alloys such as Ti-6Al-7Nb, Ti-5Al-2.5Fe, etc., have been developed for biomedical applications [2,3]. These alloys are certainly applicable for replacing failed hard tissue.

Very recently, stress transmission between hard tissue and biomaterials is getting attention from the point of view of bone absorption. That is, when the stress transmission between bone and biomaterials is insufficient, bone absorption occurs. In general, the rigidity of metallic biomaterials is much greater than that of bone, as shown in Fig. 1 [4] Among the metallic biomaterials, the rigidity of Ti-6Al-4V ELI is much lower than that of stainless steels or Co-Cr type alloys for biomedical applications. However, the rigidity of Ti-6Al-4V ELI is still much greater than that of bone. Therefore, the research and development of low rigidity titanium alloys were started in the United States.

Demand for low rigidity titanium alloys is increasing based on these backgrounds. Therefore, the research and development of low rigidity titanium alloys for biomedical applications will be described in the present chapter.

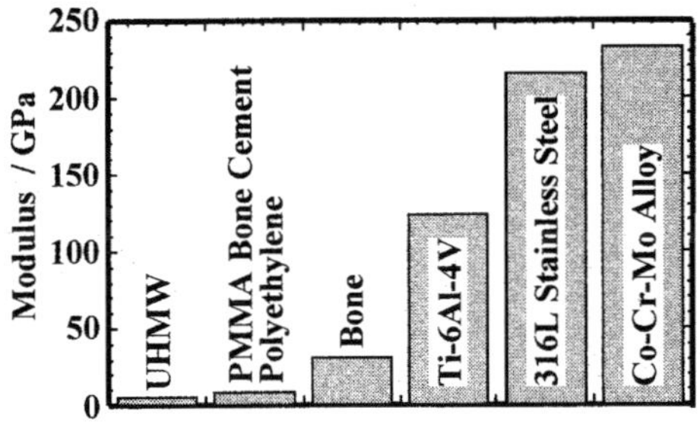

Figure 1 Elastic moduli of UHMW polyethylene, PMMA bone cement, bone, Ti-6Al-4V, 316L stainless steel and Co-Cr-Mo alloy.

II. REPORTED REPRESENTATIVE TITANIUM ALLOYS FOR BIOMEDICAL APPLICATIONS

Titanium alloys for biomedical applications which have so far been introduced are listed in Table 1 [4]. This table includes both those alloys that are already licensed and implemented for use in living bodies and those at the research and development stage, which have not yet been licensed or implemented. The alloys are tabulated approximately in the order of their introduction. The alloys marked by a circle were developed for biomedical applications from the beginning.

Table 1 Titanium Alloys for Biomedical Applications

	1.	Pure Ti (ASTM F67–89) Grade 1, 2, 3, and 4 Purity → Decrease (add N, Fe, and O) Strength → Increase Ductility → Decrease
	2.	Ti-6Al-4V ELI (ASTM F136-84, F620-87): $\alpha+\beta$ type
	3.	Ti-6Al-4V (ASTM F1108-88): $\alpha+\beta$ type
○	4.	Ti-6A1-7Nb (ASTM F1295-92, ISO5832-11): $\alpha+\beta$ type (Swiss)
○	5.	Ti-5Al-2.5Fe (ISO5832-10): $\alpha+\beta$ type (Germany)
○	6.	Ti-5Al-3Mo-4Zr: $\alpha+\beta$ type (Japan)
○	7.	Ti-15Sn-4Nb-2Ta-0.2Pd: $\alpha+\beta$ type (Japan)
○	8.	Ti-15Zr-4Nb-2Ta-0.2Pd: $\alpha+\beta$ type (Japan)
○	9.	Ti-13Nb-13Zr (ASTM F1713-96): near β type (United States), low modulus
○	10.	Ti-12Mo-6Zr-2Fe (ASTM F1813-97): β type (United States), low modulus
○	11.	Ti-15Mo (ASTM2066-01): β type (United States), low modulus
	12.	Ti-16Nb-10Hf: β type (United States), low modulus
	13.	Ti-15Mo-5Zr-3Al: β type (Japan), low modulus
	14.	Ti-15Mo-2.8Nb-0.2Si-0.26O: β type (United States), low modulus
○	15.	Ti-35Nb-7Zr-5Ta: β type (United States), low modulus
○	16.	Ti-29Nb-13Ta-4.6Zr: β type (Japan), low modulus
○	17.	Ti-40Ta, Ti-50Ta: β type (United States), high corrosion resistance

○: Developed for biomedical applications.

Pure titanium and $\alpha + \beta$ types Ti-6Al-4V ELI and Ti-6Al-4V were originally designed for use as general structural materials, especially for aerospace structures and were only later adopted for biomedical applications. However, the toxicity of the β stabilizing element of V was later pointed out [1]. Therefore, V in the Ti-6Al-4V has been replaced by other β stabilizing elements such as Fe or Nb, both of which are considered to be safer for biomedical use compared to V. Subsequently, $\alpha + \beta$ types Ti-5Al-2.5Fe [5] and Ti-6Al-7Nb [6] have been developed. Ti-5Al-2.5Fe has been registered only in ISO standardization [7], while Ti-6Al-7Nb has been registered in both ASTM and ISO standardizations [8,9]. Based on the same concept, other $\alpha + \beta$ type biomedical titanium alloys like Ti-6Al-6Nb-1Ta and Ti-6Al-2Nb-1Ta, have been developed [10]. The latter one is said to be basically developed for the structural titanium alloy in military use. Subsequently, $\alpha + \beta$ type titanium alloys that do not contain V or Al, such as the Ti-15Zr–based and Ti-15Sn–based alloys, were developed [11].

The elastic moduli of biomaterials have been said to be desirable to be equal to that of cortical bone because if the elastic moduli of biomaterials are much greater than that of cortical bone, bone resorption occurs. The elastic modulus of $\alpha + \beta$ type titanium alloy, Ti-6Al-4V, is much lower than those of stainless steel and Co–based alloys. However, the elastic modulus of Ti-6Al-4V is still much greater than that of cortical bone. The elastic moduli of β type titanium alloys are known to be smaller than those of α or $\alpha + \beta$ type titanium alloys. Therefore, mainly β type titanium alloys have been developed[12–18], and most of them are aimed at a low modulus of elasticity. For the β type titanium alloys for biomedical applications, Ti-13Nb-13Zr [12], Ti-12Mo-6Zr-2Fe [14], and Ti-15Mo [13] have been registered, and Ti-35Nb-7Zr-5Ta [16] will be registered in the ASTM standardization. The low rigidity titanium alloys contain a large amount of Nb, Mo, Ta, or Hf and a small amount of Zr, Fe, or Al.

III. DEVELOPMENT OF LOW RIGIDITY TYPE TITANIUM ALLOYS FOR BIOMEDICAL APPLICATIONS

A. Alloy Design

Nontoxic metallic elements can be selected based on the reported data of cytotoxicity in pure metals [19] and the reported data on corrosion resistance (polarization resistance) and biocompatibility in pure metals and representative metallic biomaterials [20]. It is important to select metallic alloying elements that do not cause metal allergy [3]. Subsequently, Nb, Ta, and Zr were selected as nontoxic and nonallergic alloying elements for Ti. Therefore, a low rigidity Ti-Nb-Ta-Zr system β type alloy with a good balance of strength and ductility is determined to be desirable.

The alloy can be designed more efficiently by using a rationally configured alloy design method instead of an empirical method. The d electron alloy design method [21] is suitable. It uses the value of bond order, B_o, as the index of the bonding strength between Ti and added elements and the value of d electron energy level, M_d, as the index of the chemical stability. Using the values of $\bar{B}_o$ and $\bar{M}_d$ in average composition, a $\bar{B}_o$-$\bar{M}_d$ diagram can be prepared for the various conventional titanium alloys. Each area of α, $\alpha + \beta$, and β type titanium alloys is clearly defined on the $\bar{B}_o$-$\bar{M}_d$ diagram as shown in Fig. 2 [17] The compositional direction of $\bar{B}_o$ and $\bar{M}_d$ by adding Nb, Ta, or Zr to Ti is also shown in Figure 2. Entering the modulus of elasticity at the position of each alloy in that diagram will clarify the direction of the values of $\bar{B}_o$ and $\bar{M}_d$. From the point of view of mechanical properties and processability, the position of $\bar{B}_o$-$\bar{M}_d$ has been found to be better to situate in the β area near the boundary between the $\alpha + \beta$ area and β area. According to these concepts, the target area for $\bar{B}_o$ and $\bar{M}_d$ values for getting

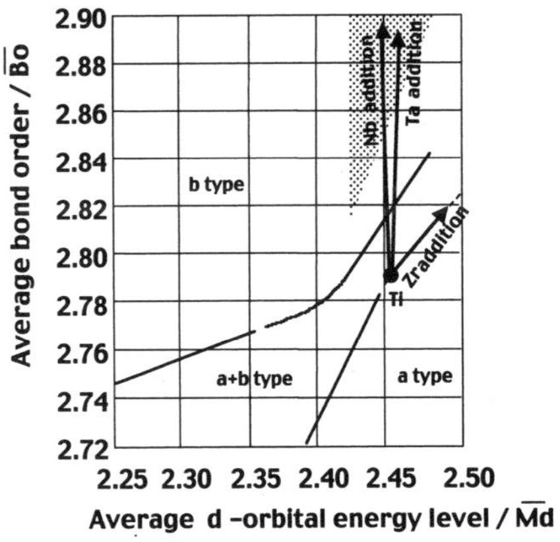

Figure 2 Schematic explanation of d electron alloy design method.

the desired β type alloy can be indicated as the dotted area on the $\bar{B}_o$-$\bar{M}_d$ diagram in Fig. 2. Subsequently, Ti-29Nb-13Ta-4.6Zr has been determined preferable.

B. Fabrication

Ti-29Nb-13Ta-4.6Zr contains a large amount of high melting elements such as Ta and Nb. Their specific gravity is much different from that of Ti. Therefore, the possibility of segregation of alloying elements in ingot is high. In order to solve this problem, the levitation double melting method is effective. The levitation double melting method using a Levicast [22] furnace is schematically shown in Fig. 3. In this method, melted metal does not contact with a Cu mold wall, and is stirred strongly. After the first melting, solidified ingot is cut into several pieces and then melted again. After the second melting, homogeneous solidified ingot can be obtained. The distribution of each element in the ingot of Ti-29Nb-13Ta-4.6Zr made by levitation double melting is shown in Fig. 4 [23]. Each element is homogeneously distributed from the top through the bottom of the ingot in the range of target composition.

C. Rigidity of Elasticity

Young's moduli as rigidity of Ti-29Nb-13Ta-4.6Zr conducted with representative heat treatment or thermomechanical treatment, and conventional β type titanium alloy Ti-13Nb-13Zr and α + β type titanium alloy Ti-6Al-4V, are shown in Fig. 5 [23]. Young's moduli of Ti-29Nb-13Ta-4.6Zr that are solution treated (ST), cold-rolled, or aged. are lower that those of Ti-6Al-4V ELI that are solution-treated or aged (STA), and are nearly equal to those of Ti-13Nb-13Zr that are solution-treated or aged The Young's modulus increases as aging advances.

D. Cytotoxicity

The cytotoxicity of Ti-29Nb-13Ta-4.6Zr and conventional biocompatible pure titanium and Ti-6Al-4V ELI were evaluated using the method as schematically shown in Fig. 6 [3]. The compo-

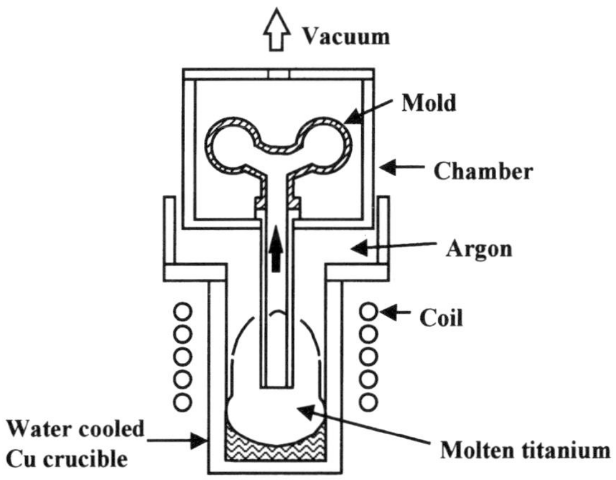

Figure 3 Schematic drawing of Levicast process.

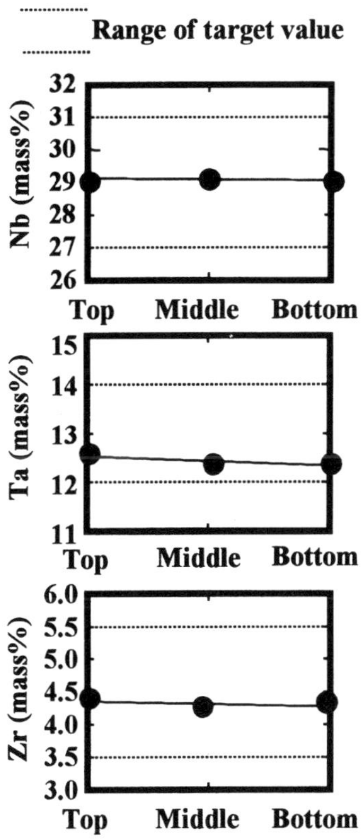

Figure 4 Distribution of Nb, Ta, or Zr as a function of position in ingot of Ti-29Nb-13Ta-4.6Zr (● is a measured value).

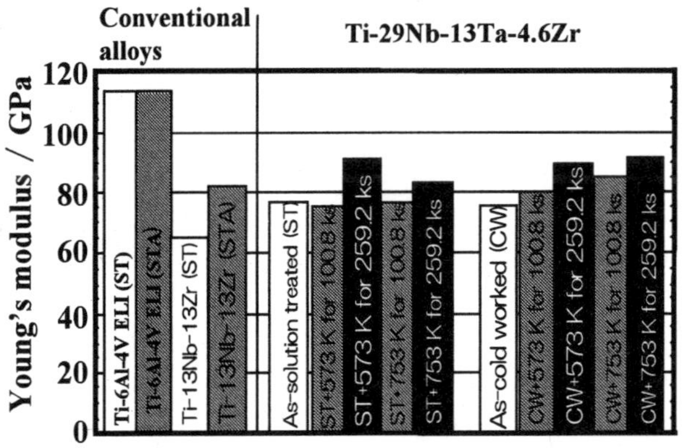

Figure 5 Comparison of Young's moduli of conventional titanium alloys (Ti-6Al-4V ELI and Ti-13Nb-13Zr) and Ti-29Nb-13Ta-4.6Zr conducted with various aging treatments after solution treatment. ST, solution treated; STA, aged after solution treatment; CW, cold rolled.

nents of the alloy were extracted by putting the plate shape sample of the alloy on zirconia balls in a chamber with culture solution and then rotating the vessel. A rotating speed was 240 rpm. The extract periods were 7 days and 14 days. The raw extract and filtrated extract were then prepared. The cell viability of L929 cells derived from mice was evaluated using the NR assay and MTT assay. The results by the NR assay and MTT assay were nearly the same. Therefore, the results of evaluation by the MTT assay are shown in Fig. 7 [23]. Ti-29Nb-13Ta-4.6Zr and pure Ti show similar cell viability both in raw (nonfiltrated) and filtrated extracts. On the other hand, Ti-6Al-4V ELI shows lesser cell viability comparing with Ti-29Nb-13Ta-4.6Zr and pure Ti. In consequence, the developed alloy, Ti-29Nb-13Ta-4.6Zr, has a high bicompatibility. As can be expected, the cell viability of extract for 14 days is smaller than that of extract for 7

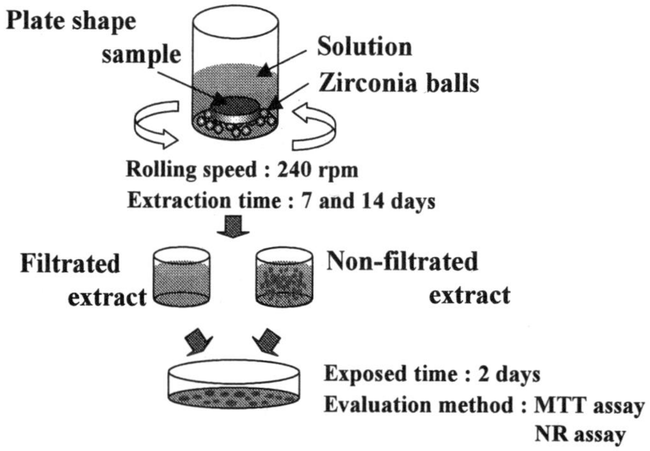

Figure 6 Schematic explanation of evaluation method of cytotoxicity.

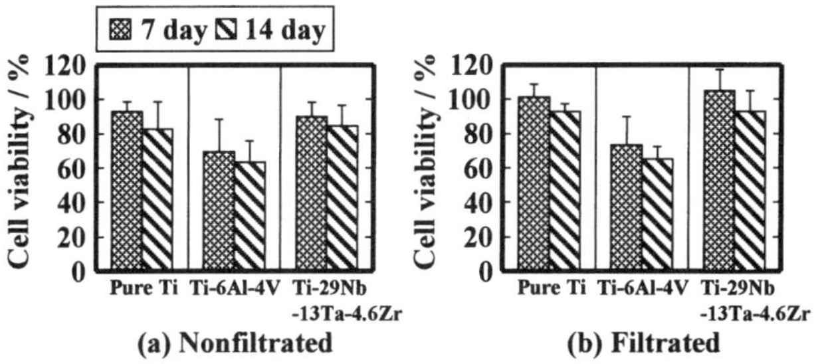

Figure 7 Cell viability of L-929 in (a) nonfiltrated and (b) filtrated cultivate solution evaluated by MTT method for pure titanium, Ti-6Al-4V, and Ti-29Nb-13Ta-4.6Zr.

days. Since the cell viability is higher in filtrated extract than in nonfiltrated extract, the wear debris is apparent to reduce the cell viability.

E. Tensile Properties

The balance of tensile strength and elongation of Ti-29Nb-13Ta-4.6Zr conducted with aging after solution treatment or aging directly after cold rolling is shown in Fig. 8 [23] with the range of strength and elongation of conventional biocompatible Ti-6Al-4V ELI. By conducting aging after solution treatment or thermomechanical treatment, the balance of strength and elongation of Ti-29Nb-13Ta-4.6Zr is improved greater compared with that of Ti-6Al-4V ELI. The balance of strength and elongation of Ti-29Nb-13Ta-4.6Zr can be controlled through heat treatment of thermomechanical treatment, which leads to microstructural control.

F. Fatigue Strength in Air and Simulated Body Environment

S-N curves of Ti-29Nb-13Ta-4.6Zr in as-solutionized and aged conditions and Ti-6Al-4V ELI in aged conditions are shown in Fig. 9 [23]. In the case of Ti-29Nb-13Ta-4.6Zr, S-N curves

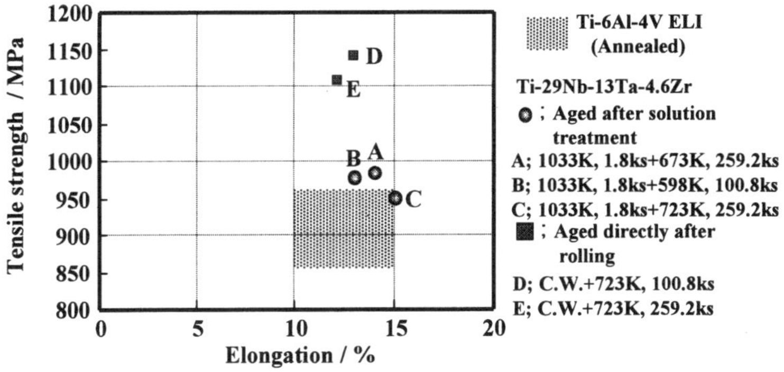

Figure 8 Balance of tensile strength and elongation of annealed Ti-6Al-4V ELI and Ti-29Nb-13-Ta-4.6Zr conducted with various aging treatments after solution treatment or various aging treatments directly after cold rolling.

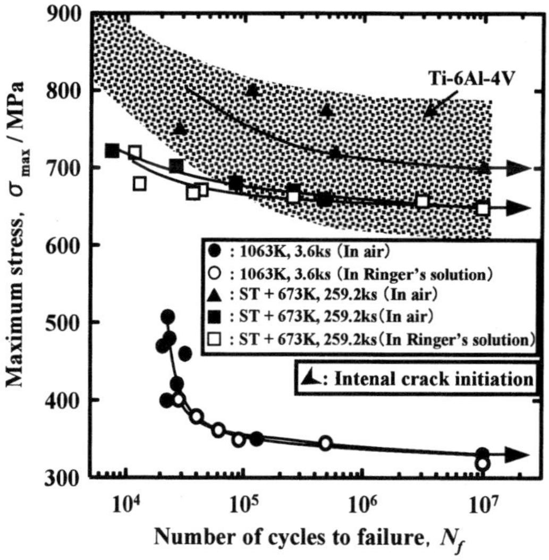

Figure 9 S-N curve of Ti-29Nb-13Ta-4.6Zr conducted with each heat treatment in air and Ringer's solution and range of S-N data of Ti-6Al-4V in air. Data indicated by ▲: data obtained from Ti-29Nb-13Ta-4.6Zr where Nb, Ta, Zr, and O are 31.5, 11.6, 4.7, and 0.14 mass %, respectively. ■, □, ●, and ○: data obtained from Ti-29Nb-13Ta-4.6Zr where Nb, Ta, Zr, and O are 29.2, 12.3, 4.4, and 0.10 mass %, respectively.

obtained in both air and simulated body environment, that is, in Ringer's solution are shown in Fig. 9. The fatigue strength of Ti-29Nb-13Ta-4.6Zr is equal to that of Ti-6Al 4V ELI, especially in the high cycle fatigue life region by conducting aging after solution treatment.

The fatigue strength of Ti-29Nb-13Ta-4.6Zr is not degraded in Ringer's solution as compared with the data in air. Therefore, the corrosion fatigue resistance in the living body environment will be sufficient.

G. Fretting Fatigue Strength in Air and Simulated Body Environment

Fretting fatigue sometimes occurrs for example between bone plate and screw or bone and stem. Fretting fatigue is also a very important issue for biomaterials. Fretting fatigue and plane fatigue strength data of Ti-29Nb-13Ta-4.6Zr in as-solutionized (ST) and aged (STA) conditions, and Ti-15Mo-5Zr-3Al in annealed conditions as a comparison, are shown in Fig. 10 [24]. The fretting fatigue data of Ti-29Nb-13Ta-4.6Zr obtained in Ringer's solution are shown in Fig. 11 [24] in addition to the data obtained in air. Plane fatigue limit, P_f, is the greatest in Ti-15Mo-5Zr-3Al and the smallest in as-solutionized Ti-29Nb-13Ta-4.6Zr. However, fretting fatigue limit, F_f, is the greatest in Ti-29Nb-13Ta-4.6Zr conducted with aging. The fatigue damage is more accelerated by fretting in Ti-15Mo-5Zr-3Al. According to the observation of the fretting area, the slipping distance is the greatest in Ti-15Mo-5Zr-3Al and the smallest in as-solutionized Ti-29Nb-13Ta-4.6Zr. These phenomena can be considered to be caused by the degree of sink of pad into specimen, which is related to the Young's modulus of the specimen. Therefore, The fretting damage on fatigue strength, P_f/F_p, can be related with Young's modulus as shown in

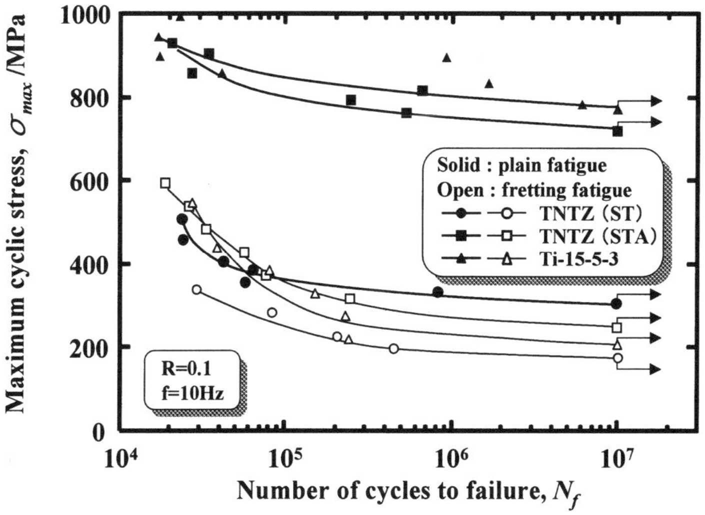

Figure 10 S-N curves of Ti-29Nb-13Ta-4.6Zr (TNTZ) and Ti-15Mo-5Zr-3Al (T_i15-5-3) conducted with each heat treatment obtained from plain fatigue and fretting fatigue tests in air.

Fig. 12 [24]. The low rigidity β type titanium alloy is less sensitive to fretting fatigue comparing with the high modulus titanium alloy.

The fretting fatigue strength of as-solutionized Ti-29Nb-13Ta-4.6Zr is a little greater in Ringer's solution than in air in the low cycle fretting fatigue life region, but a little smaller in Ringer's solution than in air in the high cycle fatigue life region. The friction force is always greater in air than in Ringer's solution, as shown in Fig. 13 [24]. On the other hand, the pit formation can be observed in the contact area in both low and high cycle fatigue life regions.

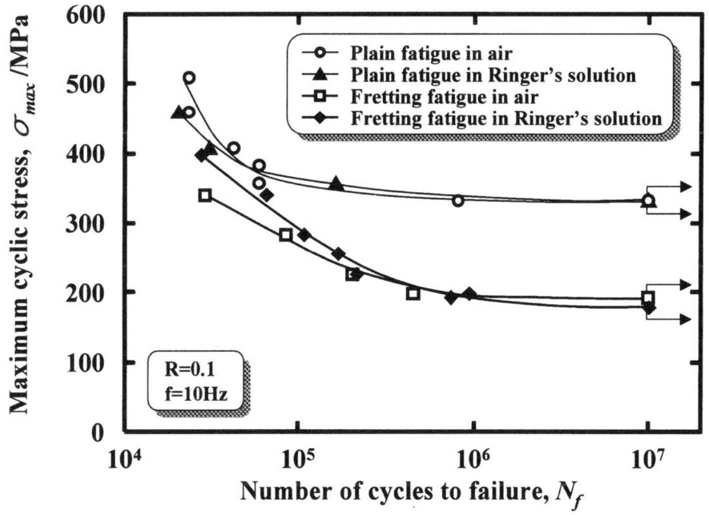

Figure 11 S-N curves of Ti-29Nb-13Ta-4.6Zr conducted with solution treatment obtained from plain fatigue and fretting fatigue tests in air and Ringer's solution.

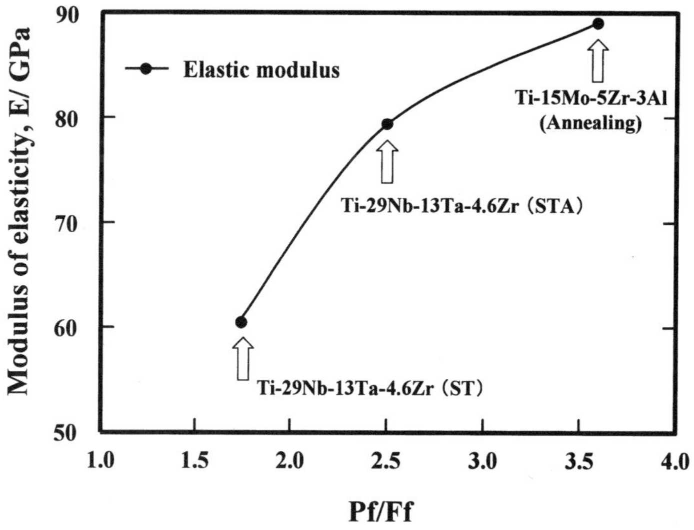

Figure 12 Relationships between fretting damage ratio, P_f/F_f, and modulus of elasticity. P_f, plain fatigue limit; F_f, fretting fatigue limit.

The fretting fatigue crack does not initiate from pit in the low cycle fretting fatigue life region, but initiates from pit in the high cycle fretting fatigue life region. Therefore, in the low cycle fretting fatigue life region, the effect of lubrication by Ringer's solution is relatively greater, but the effect of corrosion is relatively greater in the high cycle fretting fatigue region.

H. Wear Characteristics

The wear resistance of titanium alloys is in general poor. However, when compared with Ti-6Al-4V ELI, the wear resistance of Ti-29Nb-13Ta-4.6Zr is better than that of Ti-6Al-4V ELI when the mating material is a zirconia ball, as shown in Fig. 14 [25]. However, when the mating material is an alumina ball, the wear resistance of Ti-29Nb-13Ta-4.6Zr is inferior to that of Ti-6Al-4V ELI.

The wear resistance of Ti-29Nb-13Ta-4.6Zr can be improved by simple oxidation treatment in air at lower temperature compared with Ti-6Al-4V ELI, as shown in Fig. 15 [26].

I. Bioactive Surface Modification

A method that is expected to be effective in further improving the biocompatibility of Ti-29Nb-13Ta-4.6Zr is to coat the surface of the alloy with calcium phosphate recrystalized invert glass. In this method, glass composed of $60CaO \cdot (90 - x)P_2O_5 \cdot 3TiO_2 \cdot (10\text{-}y)Na_2O$ is prepared, from which glass paste is made using a ball mill. After drying, it is heat treated. The results of x-ray diffraction analysis of glass of this composition after being subjected to 1-hour heat treatment at 1023 K is shown in Fig. 16 [27]. β-TCP ($Ca_3(PO_4)_2$) crystals and β-CPP($CaPO_7$) crystals, both of which have a high biocompatibility, precipitate, and the precipitation of β-TCP crystals prevails at this heat treatment temperature. If heat treatment is performed in a vacuum, cracks occur between the coating layer and the alloy, but no cracks occur when heat treatment

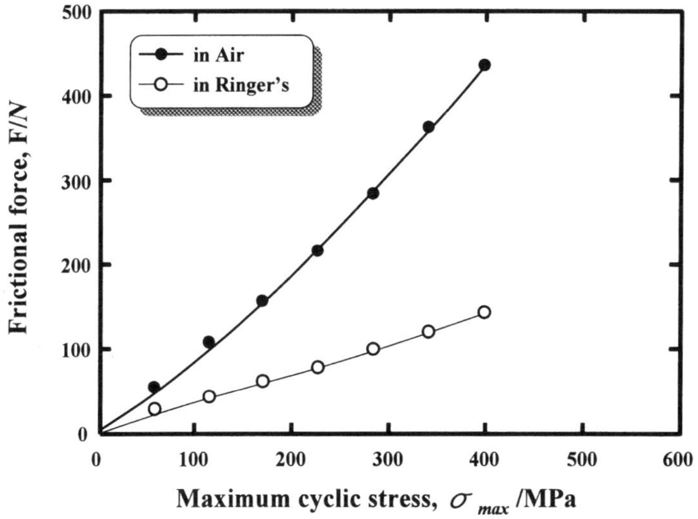

Figure 13 Relationships between frictional force and maximum cyclic stress of Ti-29Nb-13Ta-4.6Zr conducted with solution treatment in air and in Ringer's solution.

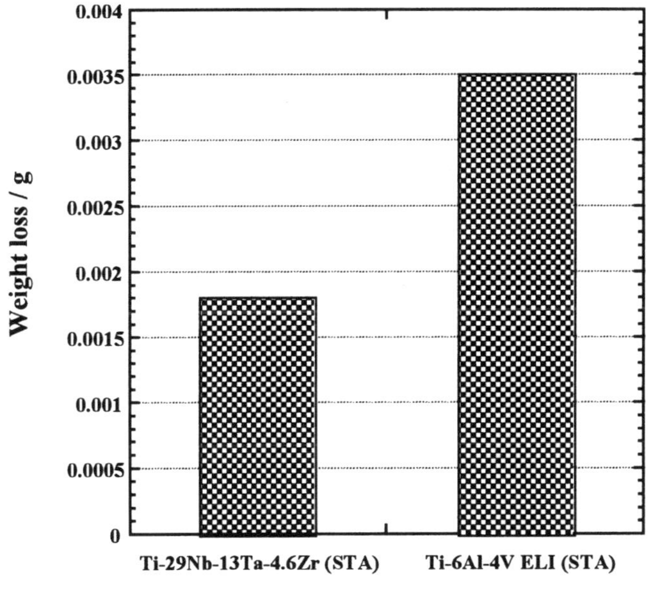

Figure 14 Weight loss of Ti-29Nb-13Ta-4.6Zr and conventional alloy, Ti-6Al-4V ELI, against a zircon ball.

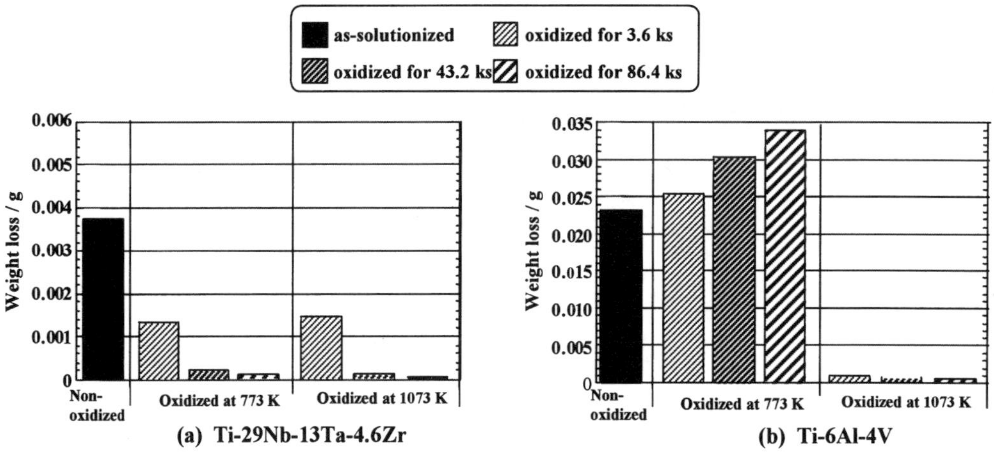

Figure 15 Weight loss of (a) Ti-29Nb-13Ta-4.6Zr and (b) Ti-6Al-4V oxidized at 773 K and 1073 K for various times after solutionizing in friction wear tests in Ringer's solution.

is carried out in atmosphere, resulting in a sound coating layer as shown in Fig. 17 [27]. This coating can be more easily performed in Ti-29Nb-13Ta-4.6Zr than in pure Ti or Ti-6Al-4V ELI.

Ti-29Nb-13Ta-4.6Zr coated with calcium phosphate crystallized glass is dipped into the simulated body liquid, SBF solution, and then the hydroxyapatite, which has high biocompatibility, precipitates on the surface of the coating layer as shown in Fig. 18 [27].

J. Fabrication of Implants by Machining

Dental implant parts can be machined from Ti-29Nb-13Ta-4.6Zr using a CNC (computer numerical control) automatic precision lathe as shown in Fig. 19. Stem for artificial hip joints can also be machined from Ti-29Nb-13Ta-4.6Zr using vertical machining center as shown in Fig. 20.

K. Dental Casting

Ti-29Nb-13Ta-4.6Zr can also be applicable for dental precision casting products like attachment post, denture, porcelain glazed metal crown, crown, inlay, etc., using alumina-based investment

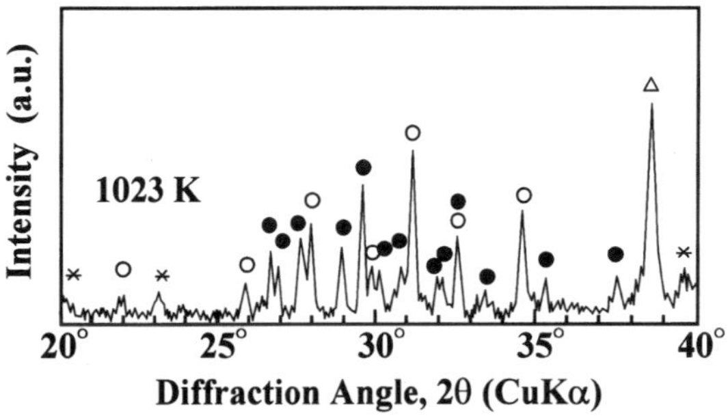

Figure 16 X-ray diffraction patterns of coating layers on the specimens prepared by heating at 1023 K for 3.6 ks in air. ○: TCP; ●: CPP; △: β-Ti; *: unknown phase.

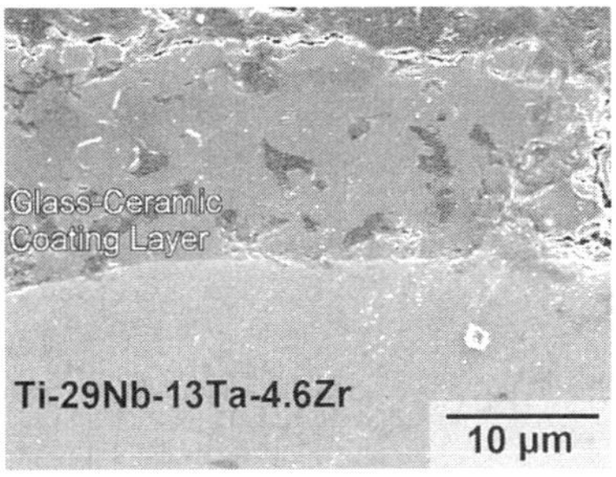

Figure 17 SEM micrograph of the cross section of the coating layer obtained by heating at 1073 K for 3.6 ks in air.

material. These dental prostheses made of Ti-29Nb-13Ta-4.6Zr using alumina based investment material are shown in Fig. 21. In future, much more suitable investment materials, for example, magnesia-based or calcia-based investment materials for precision castings of Ti-29Nb-13Ta-4.6Zr should be developed because the melting point of the alloy is relatively high due to high melting point alloying elements like Nb and Ta compared with conventional pure Ti or Ti-6Al-4V for dental precision casting.

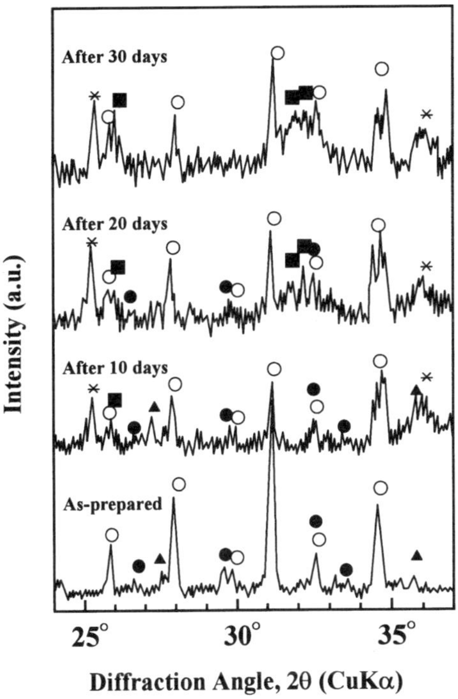

Figure 18 X-ray diffraction pattern of the surface of the glassceramic layer coated on the alloy before or after soaking in SBF for 10, 20, or 30 days. ■: HA; ○: TCP; ●: CPP; ▲: TiO_2, and * unknown phase.

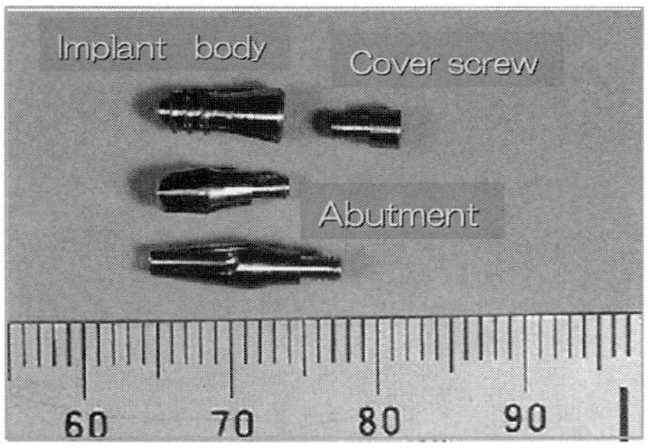

Figure 19 Parts for dental implant made of Ti-29Nb-13Ta-4.6Zr by machining. (From Aktoment Ltd. Co.)

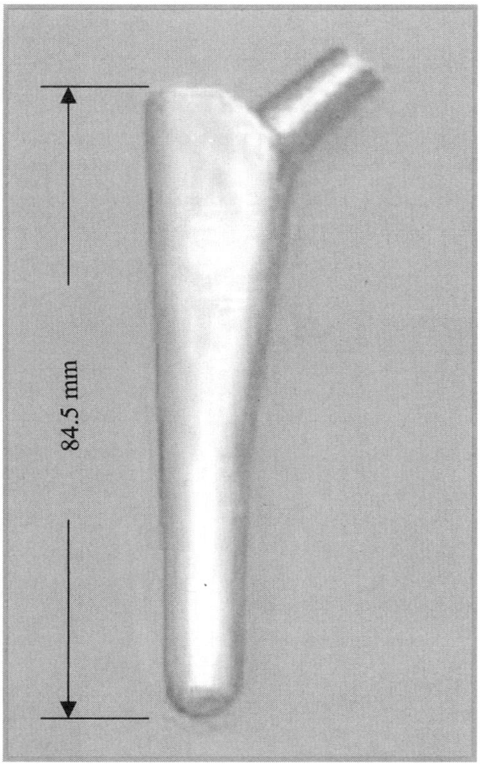

Figure 20 Stem for artificial joint made of Ti-29Nb-13Ta-4.6Zr by machining. (From Aktoment Ltd. Co.)

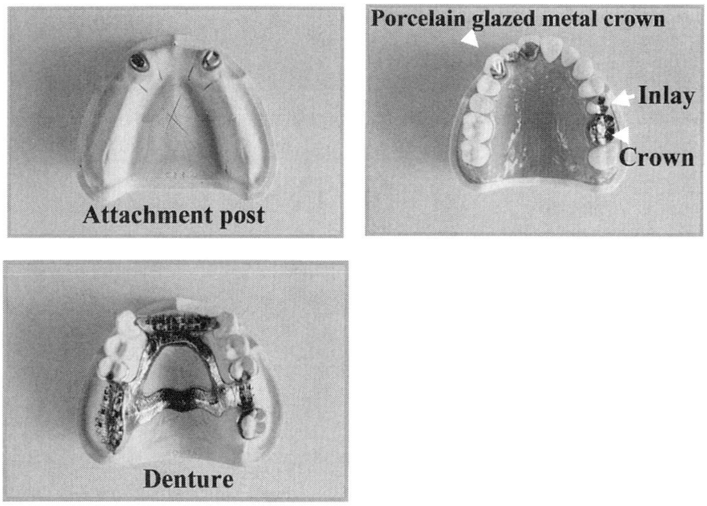

Figure 21 Examples of dental prostheses made of Ti-29Nb-13Ta-4.6Zr. (From Fukui, Aichi-Gakuin University.)

L. Biocompatibility

To be accepted as biomaterial a new material's biocompatibility must be evaluated by in vivo animal experiment. Regarding solid materials for orthopedic implants, a small specimen is implanted into femoral or tibial bone of small animals. After a certain period, histological and radiological observations are performed in terms of tissue reaction around the implanted material. When the implanted material is recognized as the foreign body due to the release of chemical substances or stimulative surface, although tiny particles or small decomposition products in small amount can be eliminated by phagocytic activity of foreign body giant cells and macrophages, the implanted material will be surrounded by interstitial fibrous tissue to isolate from the living body, which is called encapsulation known as one of typical foreign body reactions in which the interstitial fibrous tissue confines the foreign body, and minimize the harmful influence. And the thickness of the fibrous tissue increases depending on the degree of the harmfulness. However, in case of the material with a good chemical stability and nonstimulative surface, the implanted material will exist without the interstitial fibrous tissue, and newly formed bone tissue can directly contact to the material surface. Furthermore, by using the material made of calcium phosphate, which is the main substance of bone mineral, the bone tissue can chemically or biologically bond to the implanted material surface. Therefore, either the appearance or thicknesses of the interstitial fibrous tissue or the bone formation around the implanted material demonstrate the degree of biocompatibility.

According to the tissue reaction phenomena, the biocompatibility of orthopedic implant materials was classified into three categories by Heimke. [28] such as ''biotolerant,'' showing distant osteogenesis (bone formation with indirect contact to the material); ''bioinert,'' showing contact osteogenesis (bone formation with direct contact to the material), and ''bioactive,'' showing bonding osteogenesis (bone formation with chemical or biological bonding to the material). And this classification and category names have been widely used in biomaterial research and development.

In order to investigate the biocompatibility of the developed titanium alloy of Ti-29Nb-13Ta-4.6Zr, columnar specimens of $\phi 5 \times 10$ mm were implanted into lateral condyles of rabbit

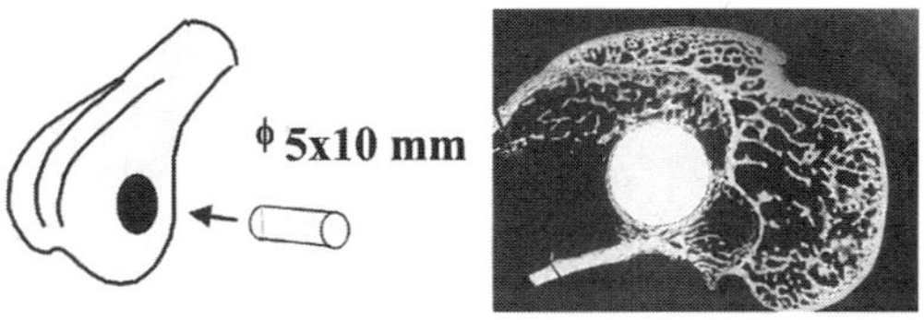

Figure 22 Implantation of columnar specimen.

femurs (Fig. 22). As control materials, columnar specimens made of SUS316L stainless steel and Ti-6Al-4V were also implanted in the same manner. At 4, 8 and 24 weeks and 1 year after the implantation, histological observation was performed with Fuchsin staining and contact microradiogram (CMR). At 4 weeks after implantation, all the implanted materials were surrounded by newly formed bone tissue, and the direct contact was partially observed in the material surface. Even in SUS316L stainless steel, which is the biotolerable material showing distant osteogenesis, bone tissue directly contacted to the material surface because bone formation in repairing phase is very active as a reaction to the implantation (Fig. 23). At 8 weeks, the implanted materials were surrounded by mature bone tissue, and the direct contact was observed at in both titanium alloys of Ti-29Nb-13Ta-4.6Zr and Ti-6Al-4V. However, in SUS316 stainless, a thin radiolucent line was observed at the interface between the bone tissue and the material surface, in which the radiolucent line represents the interstitial fibrous tissue. At 24 weeks, most of the circumference was closely covered by mature bone tissue, and the direct contact was often observed in both titanium alloys. However, in SUS316L stainless steel mature bone tissue was observed but away from the material surface in which the wide radiolucent line existed. At 1 year after implantation, the same findings of bone formation and the direct contact were obtained in both titanium alloys, but the implanted material of SUS316L stainless steel fell off from the femoral condyle during harvesting operation.

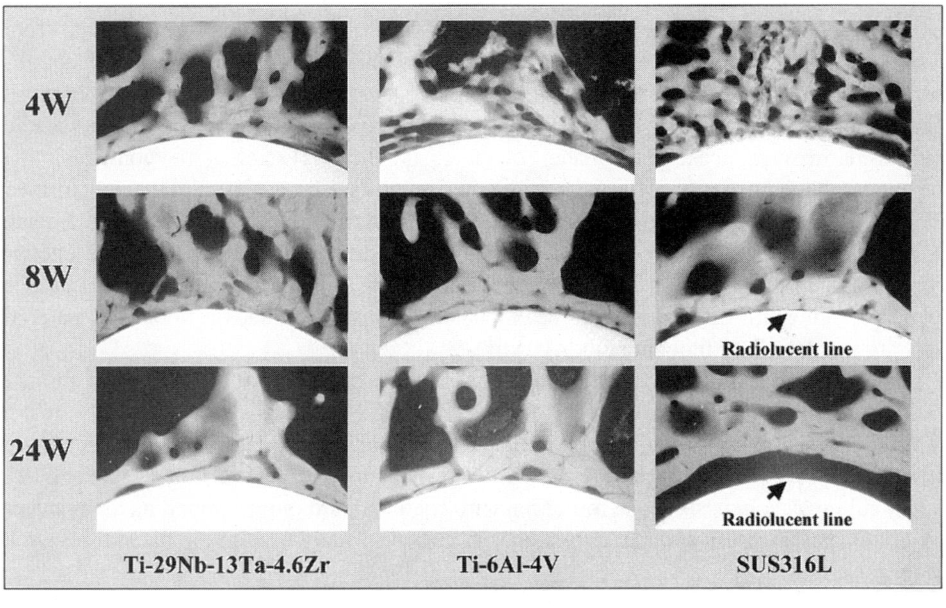

Figure 23 Bone formation around the implanted materials.

From these results, the developed titanium alloy Ti-29Nb-13Ta-4.6Zr is classified into the bioinert material showing contact osteogenesis at the same as Ti-6Al-4V titanium alloy. And the biocompatibility of the developed titanium alloy is far superior to SUS316L stainless steel, and equal to Ti-6Al-4V, or possibly greater than Ti-6Al-4V in the long term because Ti-29Nb-13Ta-4.6Zr consists of nontoxic elements but Ti-6Al-4V contains aluminum and vanadium, known as harmful elements [29–31].

M. Bone Tissue Reaction to Material Rigidity

It is well known as Wolff's law of functional restoration that bone tissue is necessarily remodeled under the influence of the mechanical stress, in which the decrease or absence of mechanical stress can cause atrophy or absorption of bone tissue [32]. Also in fracture healing, the mechanical stress greatyly influences callus formation and following bone remodeling [33]. Conventional implant alloys with high mechanical strength typically have a high rigidity of 150–200 GPa in the elastic modulus, which is approximately ten times higher than 18–20 GPa of human living cortical bone. The big difference in the material rigidity causes a load transmission failure with the reduction of the mechanical stress, in particular part of the surrounding bone tissue, and induces the bone atrophy or absorption and the further component loosening in implant surgery [34,35]. This is the so-called stress-shielding phenomenon. Recently, Ti-6Al-4V has become popular as an implant metal with good biocompatibility and relatively lower rigidity, around 100 GPa; however, the rigidity is still five times higher than the living cortical bone. The developed low rigidity titanium alloy is expected to improve the reduction of the mechanical stress, and promote bone formation and bone remodeling in the implant surgery such as internal fracture fixation and replacement arthroplasty.

In order to investigate bone tissue reaction to the material rigidity, and to consider biomechanical significance of the low rigidity, an animal study of intramedullary fixation was performed in experimental tibia fracture of rabbit using metal rods made of Ti-29Nb-13Ta-4.6Zr with a rigidity of 58 GPa in the elastic modulus, Ti-6Al-4V with 108 GPa, and SUS316L stainless steel with 161 GPa (Fig. 24). In terms of bone healing, remodeling and bone atrophy, bone tissue reaction was observed for 24 weeks by x-ray pictures taken at every 2 weeks (Fig. 25). And then histological observation was performed with Fuchsin staining and contact microradiogram. At 4 weeks after the fixation, callus formation was clearly observed at the posterior part of the fracture gap in Ti-29Nb-13Ta-4.6Zr and Ti-6Al-4V, but large callus formation was observed at the anterior part in SUS316L stainless (Fig. 26). At 6 weeks, the fracture was almost

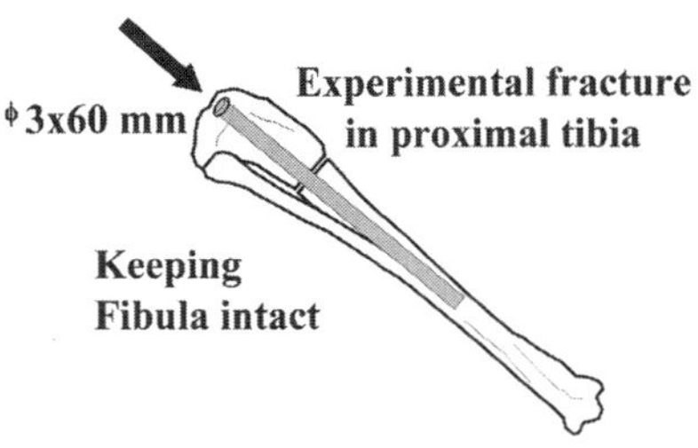

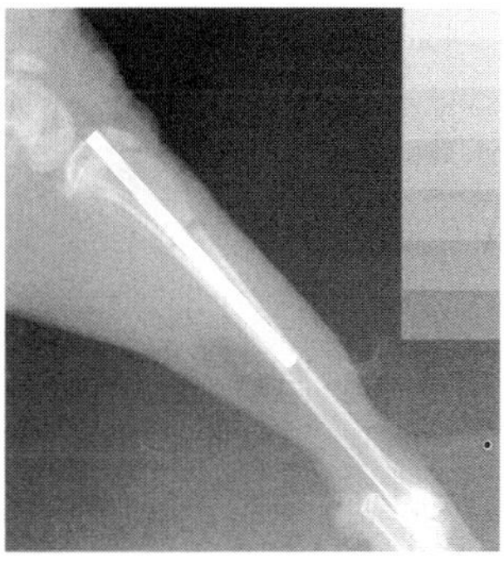

Figure 24 Intramedullary fixation of tibial fracture model.

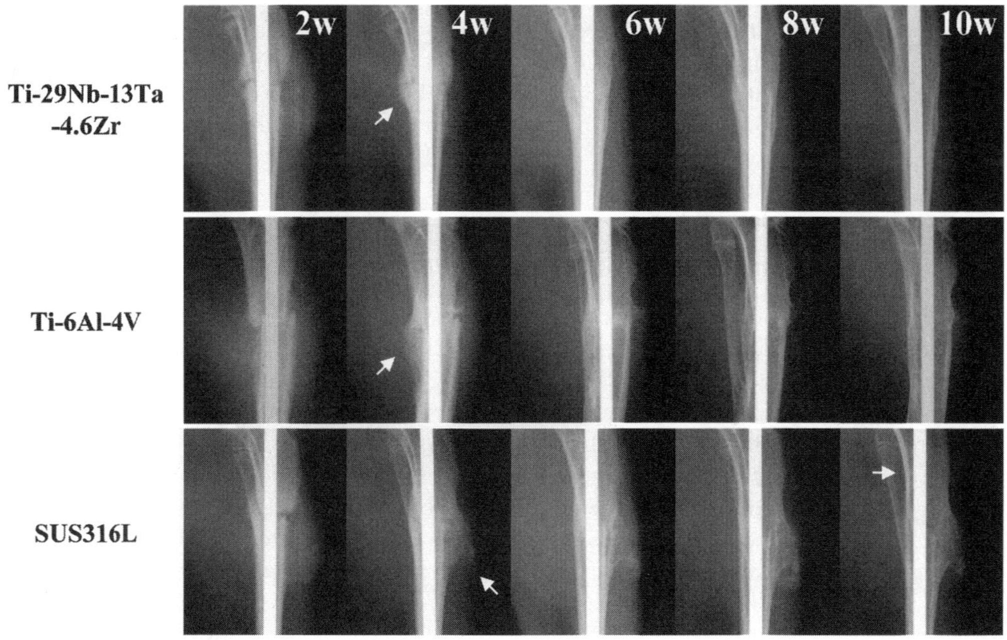

Figure 25 Fracture healing and following remodeling up to 10 weeks.

healed with the smooth outline of the callus and no fracture gap in Ti-29Nb-13Ta-4.6Zr, while similar findings were observed at 8 weeks in Ti-6Al-4V and SUS316L stainless steel. However, the anterior callus formation was remained up to 24 weeks as ectopic bone formation in SUS316L stainless steel. Additionally, regardless of material, bone swelling was often observed at the anterior part of the tibia shaft.

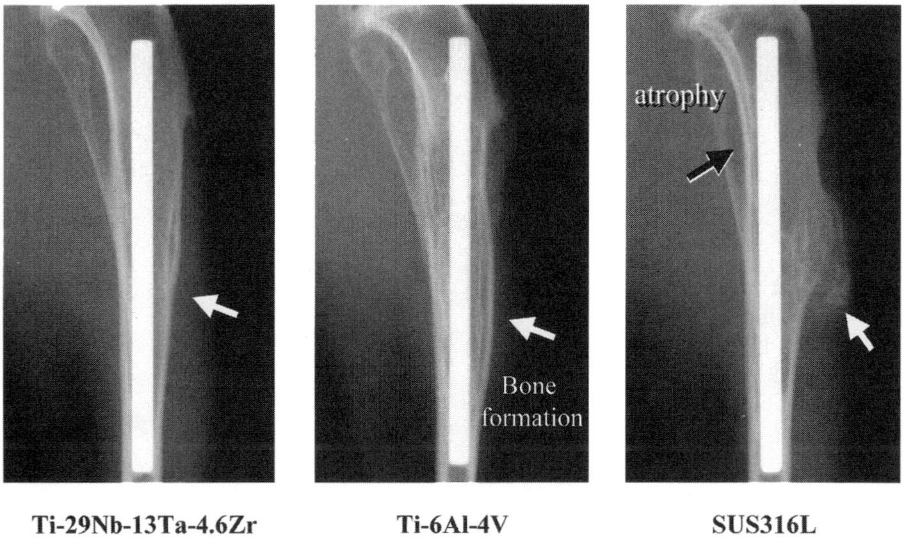

Figure 26 Remodeling at 24 weeks.

As for histological findings at 24 weeks after the fixation, the bone atrophy in SUS316L stainless steel at the posterior part of proximal tibia was confirmed as lamination of cortical bone with the absence of osteons. On the contrary, a number of osteons, the result of internal remodeling of cortical bone, were observed in thick cortical bone in Ti-29Nb-13Ta-4.6Zr and Ti-6Al-4V (Fig. 27). At the tibia shaft, the anterior swelling in Ti-29Nb-13Ta-4.6Zr and Ti-6Al-4V, and the remained anterior bone formation in SUS316L stainless steel, seemed to be the reorganization of cortical bone to adapt to a new loading condition with the intramedullary fixation. However, the new cortical bone was thick with smooth shape in Ti-29Nb-13Ta-4.6Zr as compared with Ti-6Al-4V and SUS316L stainless steel (Fig. 28).

From these results, it is confirmed that the mechanical stress greatly influences fracture healing and bone remodeling corresponding to Wolff's law of functional restoration and the stress-shielding phenomenon. They also suggest that Ti-29Nb-13Ta-4.6Zr as a low rigidity titanium alloy can improve load transmission failure, providing the moderate mechanical stress to the surrounding bone tissue, which promotes faster fracture healing and bone remodeling as compared with conventional implant metals with high rigidity.

IV. FUTURE PROSPECTS OF IMPLANT METAL ALLOY

Recent technology in metallurgy allows the development of new titanium alloys with different properties, such as nontoxic elements, high mechanical strength, and superelasticity. Also advanced technology in surface modification allows the control of chemical properties and nanostracture of the material surface to obtain the best biocompatibility. The future implant metal developed by these technologies will overthrow previous empirical common knowledge on implant surgery because the implant design and clinical applications have been restricted by

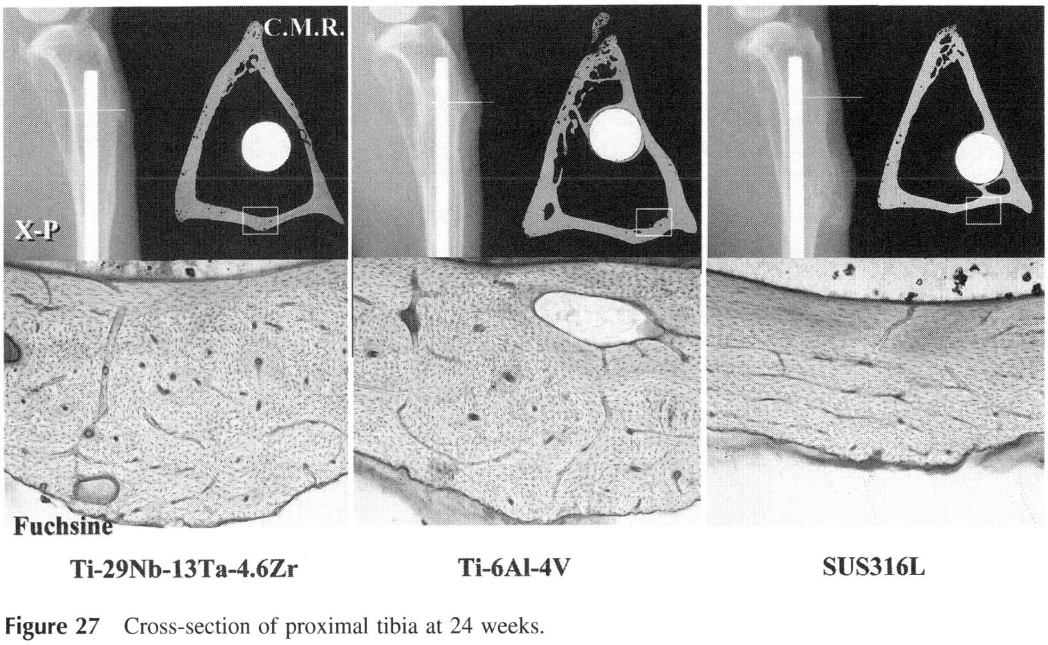

Figure 27 Cross-section of proximal tibia at 24 weeks.

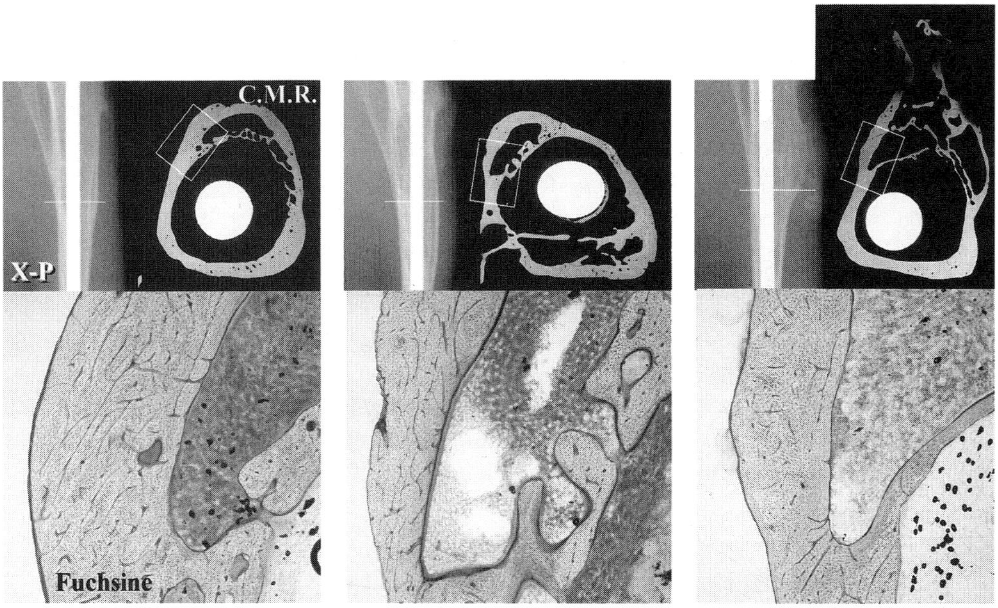

Figure 28 Cross-section of tibia shaft at 24 weeks.

material properties, and they changed with every new material. It is expected that technical innovation will produce new biomaterials and improve the evolution of implant surgery as well.

V. CONCLUSIONS

Newly developed low rigidity β type titanium alloy Ti-29Nb-13Ta-4.6Zr can be successfully fabricated by the Levicast method. The rigidity of Ti-29Nb-13Ta-4.6Zr is much lower than that of conventional Ti-6Al-4V ELI. The cytotoxicity of Ti-29Nb-13Ta-4.6Zr is equivalent to that of pure Ti, and is lower than that of Ti-6Al-4V ELI. Ti-29Nb-13Ta-4.6Zr shows a good balance of strength and ductility. The fatigue strength of Ti-29Nb-13Ta-4.6Zr is equivalent to that of conventional α + β type Ti-6Al-4V ELI. The fatigue strength of Ti-29Nb-13Ta-4.6Zr is not degraded in simulated body environments. The decreasing ratio in fatigue strength of Ti-29Nb-13Ta-4.6Zr by fretting is relatively lower comparing with that of conventional β type Ti-15Mo-5Zr-3Al. Wear resistance of Ti-29Nb-13Ta-4.6Zr is greater than that of conventional Ti-6Al-4V ELI. Bioactive coating of calcium phosphate crystallized glass can be easily performed on Ti-29Nb-132Ta-4.6Zr. The bioactive hydroxyapatite is formed on the surface of coating layer of calcium phosphate crystallized glass after soaking the calcium phosphate crystallized glass coated Ti-29Nb-13Ta-4.6Zr in SBF. Dental implant parts, stem for artificial hip joints, etc., can be made from Ti-29Nb-13Ta-4.6Zr by machining. Ti-29Nb-13Ta-4.6Zr can be applicable for dental precision casting products. Ti-29Nb-13Ta-4.6Zr is classified into the bioinert material showing contact osteogenesis as the same as Ti-6Al-4V titanium alloy. The biocompatibility of Ti-29Nb-13Ta-4.6Zr is far superior to SUS316L, and equal to Ti-6Al-4V, or possibly greater than Ti-6Al-4V in the long term because Ti-29Nb-13Ta-4.6Zr consists of nontoxic element but Ti-6Al-4V contains aluminium and vanadium, known as harmful elements. Ti-29Nb-13Ta-4.6Zr of the low rigidity titanium alloy can improve the load transmission failure providing the moderate mechanical stress to the surrounding bone tissue, which promotes faster fracture healing and bone remodeling as compared with conventional implant metals with high rigidity.

ACKNOWLEDGMENTS

The authors gratefully acknowledge Prof. H. Fukui, School of Dental Materials Science, Aichi-Gakuin University, Nagoya, Japan, Assoc. Prof. T. Kasuga, with Department of Materials Science and Engineering, Nagoya Institute of Technology, Nagoya, Japan, Dr. K. Kyo, Actoment Co. Ltd., Kasukabe, Japan, Mr. A. Suzuki, Daido Steel Co., Ltd., Nagoya, Japan, and Mr. S. Katsura, Yamahachi Dental Mfg Co., Gamagori, Japan for their contributions. Some parts of this study are supported by NED (New Energy and Industrial Technology Development Organization, Tokyo, Japan), Grant-in-Aid for Promoting Scientific Frontier Research from the Ministry Education, Science and Culture (Tokyo, Japan), Grant-in-Aid for Scientific Research from Japan Society for Promotion of Science (Tokyo, Japan), the Mitsubishi Foundation (Tokyo, Japan), the Tokai Foundation (Toyohashi, Japan), the Iron and Steel Institute of Japan (Tokyo, Japan), the Light Metal Education Foundation (Osaka, Japan), and the Suzuki Foundation (Hamamatsu, Japan).

REFERENCES

1. Zwicker R, Buehler K, Beck H, Schmid HJ. Mechanical properties and tissue reaction of a titanium alloy for implant material. In: Kimura H, Izumi O, eds. Titanium'80. Vol. 2. Warrendale, PA: TMS-AIME, 1980:505–514.
2. Niinomi M. Mechanical properties of biomedical titanium alloys. Mater Sci & Eng A 1998; A243: 231–236.
3. Niinomi M. Research and development of low modulus titanium alloys composed of non-toxic elements for biomedical applications. Rec. Res Devel Metallurg & Mater Sci 2001; 5:1–9.
4. Davidson JA, Gergette FS. State of the art materials for orthopedic prosthetic devices. In:. Proc. Implant Manufacturing and Material Technology, December 2–4. Itaca, IL, 1986:EM87-122-1–EM87-122-26.
5. Browy K-H, Kramer K-H. On the properties of a new titanium alloy (TiAl5Fe2.5) as implant material. In: Lütjering G, Zwicker U, Bunk W, eds. Titanium Science and Technology Vol. 2 Deutsche Gesellshaft fur Metallkunde E. V., 1985:1381–1386.
6. Semlitsch M. Classic and new titanium alloys for production of artificial hip joints. Titan 1986; 2: 721–740.
7. Implant for surgery—Metallic materials—Part 10: wrought titanium 5-aluminium 2.5-iron alloy, ISO 5832-10, Genève, Switzerland, 1996.
8. Standard specification for wrought titanium 6Al-7Nb alloy for surgical implant. ASTM Designation F1295-92, ASTM, Philadelphia, PA, USA, 1994.
9. Implant for surgery–Metallic materials—Part II: Wrought titanium 6-aluminum 7-niobium alloy, ISO 5832-11, Genève, Switzerland, 1994.
10. Sasaki Y, Doi K, Matsushita T. New titanium alloys for artificial hip joints. Kinzoku 1996; 66: 812–817.
11. Okazaki Y, Ito Y, Tateishi T, Ito A. Effect of heat treatment on microstructure and mechanical properties of new titanium alloys for surgical implantation. J. Jpn Inst. Metals 1995; 59:108–115.
12. Mishra AK, Davidson JA, Poggie RA, Kovacs P, Fitzgerald TJ. Mechanical and tribological properties and biocompatibility of diffusion hardened Ti-13Nb-13Zr—a new titanium alloy for surgical im-

plants. In: Brown SA, Lemons JE, eds. Medical Applications of Titanium and Its Alloys ASTM STP 1272. West Conshohocken, PA: ASTM, 1996:96–116.

13. Zardiackas LD, Mitchell DW, Disegi JA. Characterization of Ti-15Mo beta titanium alloy for orthopaedic implant. In: Brown SA, Lemons JE, eds. Medical Applications of Titanium and Its Alloys ASTM STP 1272. West Conshohocken, PA: ASTM, 1996:60–75.
14. Wang KK, Gustavson LJ, Dumbleton JH. Microstructure and properties of a new beta titanium alloy, Ti-12Mo-6Zr-2Fe, developed for surgical implants. In: Brown SA, Lemons JE, eds. Medical Applications of Titanium and Its Alloys ASTM STP 1272. West Conshohocken, PA: ASTM, 1996:76–87.
15. Bhambri SK, Shetty RH, Gilberton LN. Optimization of properties of Ti-15Mo-2.8Nb-3Al-0.2Si and Ti-15Mo-2.8Nb-0.2Si-0.26O beta titanium alloys for application in prosthetic implants. In: Brown SA, Lemons JE, eds. Medical Applications of Titanium and Its Alloys ASTM STP 1272. West Conshohocken, PA: ASTM, 1996:88–95.
16. Ahmed T, Long M, Silvestri J, Ruiz C, Rack HJ. A new low modulus, biocompatible titanium alloy. In: Blenkinsop PA, Evans WJ, Flower HM, eds. Titanium '95. Vol. II. London, UK: Institute of Metals, 1996:1760–1767.
17. Kuroda D, Niinomi M, Morinaga M, Kato Y, Yashiro T. Design and mechanical properties of new β type titanium alloys for implant materials. Mater Sci Eng A 1998; A243:244–249.
18. Niinomi M. Recent metallic materials for biomedical applications. Metall Mater Trans A 2002; 33A: 477–486.
19. Kawahara H, Ochi S, Tanetani K, Kato K, Isogai M, Mizuno Y, Yamamoto H, Yamaguchi A. A biological test of dental materials, Effect of pure metals upon the mouse subcutaneous fibroblast, strain L cell in tissue culture. J Jpn Soc Dent Apparat Mater 1963; 4:65–67.
20. S. G. Steinemann. Corrosion of surgical implants—in vitro and in vitro tests. In: Winter GD, Leray JL, de Groot K, eds. Evaluation of Biomaterials: John Wiley and Sons, 1980:1–34.
21. Morinaga M, Yukawa N, Adachi H. Electronic structure and phase stability of titanium alloys. Tetsu-to-Hagane 1986; 72:555–562.
22. Demukai N. Development of a new type titanium casting technology/LEVICAST process. In: Hanada S, Zhong Z, Nam SW, Wright RN, eds. Proc. Fourth International Conference on Advanced Materials and Processing (PRICM 4): Japan Institute of Metals, 2001:369–372.
23. Niinomi M, Hattori T, Morikawa K, Kasuga T, Suzuki A, Fukui H, Niwa S. Development of low rigidity β-type titanium alloy for biomedical applications. Mater Trans 2002; 43:2970–2977.
24. Niinomi M, Akahori T, Yabunaka T, Fukui H, Suzuki A. Fretting fatigue characteristics of new biomedical β type titanium alloy. Tetsu-to-Hagane. 2002; 88:553–560.
25. Niinomi M, Kuroda D, Fukunaga K, Morinaga M, Kato Y, Yashiro T, Suzuzki A. Corrosion wear fracture of new β-type biomedical titanium alloys. Mater Sci Eng A 1999; A263:193–199.
26. Niinomi M, Akahori T, Nakamura S, Fukui H, Suzuki A. Wear characteristics of surface oxidation treated new biomedical β-type titanium alloy in simulated body environment. Tetsu-to-Hagane 2002; 88:567–574.
27. Kasuga T, Watanabe M, Nogami M, Niinomi M. Calcium phosphate glass-ceramic joined by self-development of compositionally gradient layers on a titanium alloy. Biomaterials 2001; 22:577–582.
28. Heimke G, Stock D. Clinical application of ceramic osseo– or soft tissue–integrated implant. Orthoped Ceram Implants 1984; 4:1–19.
29. Merritt K, Brown SA. Metal sensitivity reactions to orthopedic implant. J. Dermat. 1981; 20:89–94.
30. Woodman SL. Metal ion release from titanium-based prosthetic segmental replacement of long bones in baboons: a long term study. J Orthoped Red 1984; 1:421–430.
31. Goyer RA. Toxic effects of metals. Chapter 19. Macmillian. New York: Toxicology—The Basic Science of Poisons, 1986.
32. Wolff J. The Law of Bone Remodeling. (Translated by Maquet, P. Furlong, R.). Berlin: Springer-Verlag, 1986.
33. Rockwood CA. Fractures in Adults. 4th ed:. Philadelphia: Lippincott Williams & Wilkins, 1996: 269–271.
34. Burstein AH, Wright TM. Fundamentals of Orthopaedic Biomechanics, Williams & Wilkins. 1994: 200–201.
35. Black G. Orthopaedic Biomaterials in Research and Practice, Churchill Livingstone. 1988.

3

Corrosion and Biocompatibility of Orthopedic Implants

Nadim James Hallab, Robert M. Urban, and Joshua J. Jacobs
Rush Medical College
Chicago, Illinois, U.S.A.

I. INTRODUCTION

All metallic implants electrochemically corrode to some extent. This is disadvantageous for two main reasons: (1) the process of degradation reduces the structural integrity and (2) degradation products may react unfavorably with the host. Metallic implant degradation results from both electrochemical dissolution and wear, but most frequently occurs through a synergistic combination of the two [1,2]. Electrochemical corrosion processes include both generalized dissolution uniformly affecting an entire surface and localized areas of a component. Locally these areas tend to be at both identifiable areas relatively shielded from the environment (e.g., crevice corrosion) and at random sites on the surface (e.g., pitting corrosion). In the past, these electrochemical and other mechanical processes have interacted to cause premature structural failure and accelerated metal release (e.g., stress corrosion cracking, corrosion fatigue, and fretting corrosion). Current clinical questions persist regarding the degree to which elevated local and systemic metal concentrations and particulate corrosion products persist in peri-implant tissues.

This chapter will overview implant corrosion basic science, in vitro corrosion testing techniques, corrosion properties of orthopedic alloys, and biocompatibility issues such as metal release, local tissue response to implant corrosion products, and implications for potential systemic effects.

II. GENERAL CORROSION CONCEPTS

Metal implant corrosion is controlled by (1) the extent of the thermodynamic driving forces which cause corrosion (oxidation/reduction reactions) and (2) physical barriers which limit the kinetics of corrosion. In practice these two parameters that mediate the corrosion of orthopedic biomaterials can be broken down into a number of variables: geometric variables (e.g., taper geometry in modular component hip prostheses), metallurgical variables (e.g., surface microstructure, oxide structure, and composition), mechanical variables (e.g., stress and/or relative motion), and solution variables (e.g., pH, solution proteins, enzymes).

A. Thermodynamic Considerations: How and Why Metals Corrode

The basic reaction that occurs during corrosion is the increase of the valence state (i.e., loss of electrons) of the metal atom.

$$M \rightarrow M^{z+} + ze^{-}(\text{oxidation}) \quad (1)$$

This oxidation event (loss of electrons and increase in valence) may result in the release of free ions from the metal surface into solution (which can then migrate away from the metal surface), or may result in many other reactions such as the formation of metal oxides, metal chlorides, organometallic compounds, or other species. These ''end'' products may also be soluble in solution or may precipitate to form solid phases. Solid oxidation products may be subdivided into those that form adherent compact oxide films, or those that form nonadherent oxide (or other) particles that can migrate away from the metal surface.

For corrosion to occur at all there must be a thermodynamic driving force for the oxidation of metal atoms. This driving force can be quantified thermodynamically using the Gibbs function, or free energy equation (the Gibbs function incorporates both the entropy and enthalpy changes of the above chemical reaction, or the total work to reach equilibrium).

$$\Delta G_{red} = \Delta G^{o} + RT\ \text{In} \frac{[M]}{[M^{z+}][e^{-}]^{z}} \quad (2)$$

where

ΔG_{red} is the free energy change for the reduction reaction (i.e., the reverse reaction from above)
ΔG^{o} is the free energy of the reaction in some quantifiable standard state (typically where the ions are at unit activity or fully saturated)
$[M]$, $[M^{z+}]$, and $[e^{-}]^{z}$ are approximate activities (or concentrations) of the species involved in the reaction. This assumption is only really true in infinitely dilute solutions where released ions do not interact and molality equals activity, but it remains a good approximation for dilute solutions as well.

By convention, if $\Delta G > 0$, then the process requires energy; or if $\Delta G < 0$, then the oxidation process releases energy and will spontaneously occur.

There are two interrelated sources of energy to be considered in corrosion processes: chemical and electrical (charge separation). The chemical driving force (ΔG) determines whether or not corrosion will take place under the conditions of interest. When the free energy for oxidation is less than zero, oxidation is energetically favorable and will take place spontaneously. The second energy force relates to how the positive and negative charges (metal ions and electrons, respectively) are separated from one another during corrosion. This charge separation contributes to what is known as the electrical double layer and creates an electrical potential across the metal–solution interface (similar to that of a capacitor), which can be quantified by the expression:

$$\Delta G = -zF\Delta E \quad (3)$$

where

ΔG is the free energy change
z is the valence of the ion
F is known as the Faraday constant (i.e., 96,500 coulombs/mol electrons)
ΔE is the voltage across the metal–solution interface.

This potential is also a measure of the reactivity of the metals, or the driving force for metal oxidation. It shows that the more negative the potential of a metal in solution, the more reactive it will tend to be (i.e., the greater is ΔG for reduction).

At equilibrium, the chemical energy balances with the electrical energy, which can be quantified using the Nernst equation, which defines the electrical potential across an idealized metal–solution interface when in a solution.

$$\Delta E = \Delta E^{o} + \frac{RT}{zF} \ln \frac{[M^{z+}]}{[M]} \tag{4}$$

where

ΔE is the potential of the metal
ΔE^{o} is the standard electrode potential of the metal
R is the gas constant (8.314 J/mol)
T is temperature in Kelvin (25°C is 298 K)
z is the valency
F is the Faraday constant (96,490 coulombs/mol).

From this equation, a theoretical scale of metal reactivity can be established, known as the electrochemical series, which is a ranking of the equilibrium potential from most positive (i.e., least reactive or most noble) to most negative (most reactive or most base). Be aware that this ranking is based only on thermodynamic equilibrium. That is, it is only true if we assume that there are no barriers (i.e., no surface oxide formation) to the oxidation (loss of electrons/corrosion) of the metal; then these potentials would be the ones that would exist across the metal–solution interface. Table 1 shows some selected idealized reactions and their electrochemical potential (using a standard hydrogen electrode). Certain metals owe their corrosion resistance to the fact that their equilibrium potentials are very positive. Gold and platinum are examples of metals that have little or no driving force for oxidation in aqueous solutions, and thus they tend to corrode very little in the human body. However, most orthopedic metals have very negative potentials, indicating that from a chemical driving force perspective they are much more likely to corrode. For example, titanium has a very large negative potential, − 1.6 V (SHE), indicating that there is a large chemical driving force for corrosion (oxidation). If surface oxide formation (or passivation) did not intervene, pure titanium would react with its surroundings (typically oxygen, water, or other oxidizing species) and corrode vigorously. But it doesn't, thanks to the formation of metal oxides.

B. Kinetic Barriers to Corrosion: Oxide Film Formation

The second primary factor that governs the corrosion process of metallic biomaterials is the formation of stable surface barriers or limitations to the kinetics of corrosion. These barriers prevent corrosion by physically limiting the rate at which oxidation or reduction processes can take place. The formation of a metal–oxide passive film on a metal surface is one example of a kinetic limitation to corrosion. The general reaction that governs this formation is as follows:

$$M^{z+} + \frac{Z}{2} H_2O \rightarrow \frac{z}{2} MO + zH + ze^{-} \tag{5}$$

In general, kinetic barriers to corrosion prevent either the migration of metallic ions from the metal to the solution, the migration of anions from solution to metal, or the migration of electrons across the metal–solution interface. Passive oxide films are the most well known forms of kinetic barriers in corrosion, but other kinetic barriers exist including manufactured polymeric coatings.

Table 1 Standard Electrochemical Series for Selected Metals

	Reaction	Potential (V)
Noble (corrosion resistant)		
	$Au^{3+} + 3e^- \Leftrightarrow Au$	1.42
	$Pt^{2+} + 2e^- \Leftrightarrow Pt$	1.20
	$Ag^+ + e^- \Leftrightarrow Ag$	0.80
	$O_2 + 2H_2O + 4e^- \Leftrightarrow OH^-$	0.40
	$Ti(OH)^{3+} + H^+ + e^- \Leftrightarrow Ti^{3+} + H_2O$	0.06
	$H^+ + e^- \Leftrightarrow 1/2\ H_2$	0.00
	$Fe^{3+} + 3e^- \Leftrightarrow Fe$	−0.04
	$Co^{2+} + 2e^- \Leftrightarrow Co$	−0.28
	$Fe^{2+} + 2e^- \Leftrightarrow Fe$	−0.41
	$Cr^{2+} + 2e^- \Leftrightarrow Cr$	−0.56
	$Cr^{3+} + 3e^- \Leftrightarrow Cr$	−0.74
	$2H_2O + 2e^- \Leftrightarrow 2OH^-$	−0.83
	$TiO_2 + 4H^+ + 4e^- \Leftrightarrow Ti + 2H_2O$	−0.86
	$Ti^{2+} + 2e^- \Leftrightarrow Ti$	−1.60
	$Mg^+ + e^- \Leftrightarrow Mg$	−2.37
Active (corrosion prone)	$Na^+ + e^- \Leftrightarrow Na$	−2.71

These values are based on the standard hydrogen electrode scale. The more noble metals at the top of the list are less reactive, while the more active metals toward the bottom are more reactive and have a higher driving force for oxidation (corrosion).

Orthopedic alloys rely almost entirely on the formation of passive films to prevent significant oxidation (corrosion) from taking place. These films consist of metal oxides (ceramic films) which form spontaneously on the surface of the metal in such a way that they prevent further transport of metallic ions and/or electrons across the film. To be effective barriers, the films must be compact and fully cover the metal surface; they must have an atomic structure that limits the migration of ions and/or electrons across the metal oxide–solution interface; and they must be able to remain on the surface of these alloys even with mechanical stressing or abrasion, expected with orthopedic devices.

Passivating oxide films spontaneously grow on the surface of metals. These oxide films are very thin (on the order of 5 to 70 Å) and may be amorphous or crystalline, which depends on the potential across the interface as well as solution variables like pH [3,4]. Since the potential across the metal solution interface for these reactive metals is typically 1 to 2 V and the distances are so small, the electric field across the oxide is very high, on the order of 10^6–10^7 V/cm. One of the more widely accepted models, by Mott and Cabrera [5], states that oxide film growth depends on the electric field across the oxide. If the potential across the metal oxide–solution interface is decreased (i.e., made closer to the electrochemical series potential; see Table 1), then the film thickness will decrease by reductive dissolution processes at the oxide to keep the electric field strength constant. Increasing the voltage will correspondingly increase the thickness of the film. In fact, oxide thickness is often determined by an anodization rate which is given as oxide thickness per volt [4]. The film will change its thickness by growth or dissolution until the rates of both are equal, giving rise to a film thickness that is dependent on metal oxide–solution potential. If the interfacial potential is made sufficiently negative or the pH of the solution is made low enough, then these oxide films will no longer be thermodynamically stable and will undergo reductive dissolution, or there will be no driving force for the formation of the oxide, and the metal surface will become unprotected.

Oxide films are not flat smooth continuous sheets of adherent oxide covering the metal. Transmission electron microscopy (TEM) and atomic force microscopy (AFM) techniques have shown that oxides of titanium, for instance, consist of needle [6] or dome [7] shapes. Also, mechanical factors such as fretting, micromotion, or applied stresses may be such that the oxide films are abraded or fractured. When an oxide film is detached from the metal substrate, unoxidized metal is exposed to solution. These films tend to reform or repassivate, and the magnitude of the repassivation currents may be large. This is due to the normally large driving forces that are present for the oxidation process which, when the kinetic barrier is removed, can operate to cause oxidation. However, the extent and duration of the oxidation currents will depend on the repassivation kinetics for oxide film formation. Hence, the mechanical stability of the oxide films, as well as the driving force associated with their formation, are central to the performance of oxide films in orthopedic applications.

C. The Solution–Metal Interface: Metal, Oxide Layer, Biofilm, and Solution

The interface between a passivating metal and body fluid can then be summed up with the following description (see Fig. 1). Prior to implantation, the metal surface spontaneously reacts with its surroundings to form a passive metal oxide film which may be nonuniform in cross-section (domed or needle shaped) and, at least initially, amorphous. An oxide film nucleates and grows on the metal surface and contains within it defects which allow for electronic and ionic transport of charged species across the film. There are species such as oxygen, phosphates [8], hydroxides, or proteins [9] adsorbed from solution onto the surface of the oxide film which may change the properties of the film. There also exists a large electric field that is the driving force for the movement of these ions across the film. If the electric field strength is changed (by changing the applied potential, for instance), then the oxide film will grow or shrink to maintain a constant field strength (at least at low temperatures). Also, depending on conditions, the oxide film will change crystal structure, size, and thickness. Several treatments have been investigated to see if improvements in the barrier effect of the oxide films can be achieved. These treatments include a hot, concentrated nitric acid bath treatment, boiling in distilled water [10], and anodization. However, optimizing changes in oxide film structure with these treatments remains incomplete.

Oxidation and reduction reactions upon metal surfaces are typically separated spatially from each other (i.e., the rate of oxidation of a species may be heterogeneously distributed about a metal surface). That is, regions where oxidation is occurring may be well separated spatially from where the corresponding reduction process is operating. These variations can be due to local microstructural heterogeneities or differences between grain boundaries and grain interiors in the alloy, differences in strain energy, or geometry (as occurs in crevice corrosion or pitting corrosion). During crevice corrosion the region inside a crevice develops into a deaerated environment and may have a lower pH than the bulk solution. This tends to accelerate the release of metal ions in the crevice (oxidation), while away from the crevice the reduction reactions involving oxygen can take place. In this case oxidation is localized inside the crevice and reduction takes place outside of the crevice.

III. IN VITRO CORROSION TESTING METHODS

There are a wide variety of corrosion test methodologies used to assess the corrosion properties of orthopedic alloys. These include immersion tests such as anodic polarization, linear polariza-

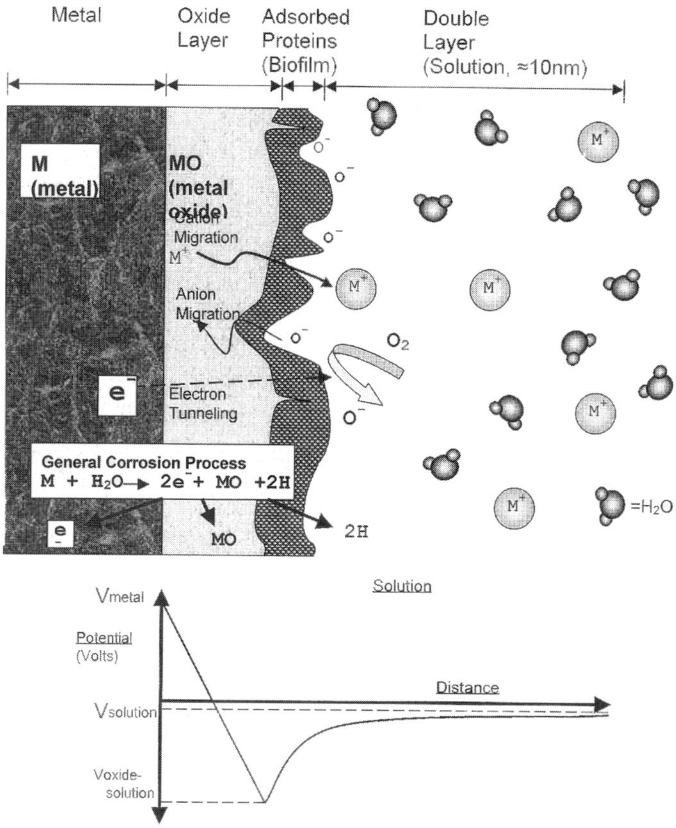

Figure 1 The interface of a metal surface in contact with a biological environment where the metal is covered by an oxide layer followed by the adsorbed proteins (biofilm) through which metal ions can pass into solution as shown.

tion, and specialized tests such as impedance spectroscopy [11], stripping analysis, and newer techniques such as scanning electrochemical microscopy [12]. All these tests are used to investigate the rate of ion release, the electrochemical conditions which cause oxidation and reduction processes, and the electrical nature of the interface. There are other test methods employed in orthopedic laboratories that combine mechanical-electrochemical processes, such as fretting corrosion, stress corrosion, and corrosion fatigue. These tests typically evaluate how an alloy resists corrosion during physiologically relevant loading.

A. Open Circuit Potential

One of the more basic electrochemical tests which is used to assess the corrosion properties of orthopedic alloys simply measures the electrical current (amps) and potential (volts) of the metal–solution interface. Metal–solution interfaces have an associated potential that is determined by the bulk concentration of species present at the interface (i.e., the oxide films and other species that may adsorb to the surface; see Fig. 1) and the electrolyte. In the complicated environment of the human body the presence of multiple metals in an alloy, the potential of the interface (known as the open circuit potential, OCP) will represent the sum of all reactions that are ongoing at the interface. At equilibrium for this ''multielectrode'' system the net external current that flows is zero. This means that any and all oxidation processes ongoing at the surface must be balanced by reduction processes that are also occurring at the same surface.

B. Polarization Testing—Current/Potential Response

A standard electrochemical test used to evaluate the corrosion resistance of implant alloys is the polarization test. This test assesses the current/potential characteristics of the metal solution interface by varying the potential of the interface and measuring the resulting current. Varying the potential across the metal–oxide–solution interface in a controlled fashion forces the oxidation and reduction reactions, which are at a dynamic equilibrium at the open circuit potential to deviate from equilibrium. For instance, for metals which do not form passivating films (not implant alloys), the current/potential relationship is represented with the Butler-Volmer equation [13]:

$$i_{corr} = i_o \left[\exp^{\eta/b_c} - \exp^{\eta/b_a}\right] \tag{6}$$

where

i_{corr} is the corrosion current at the applied potential
η is the overpotential (the potential difference between open circuit and the applied potential)
b_a and b_c are the Tafel constants
i_o is the exchange current density.

From these polarization tests, information characteristic of the corrosion currents and potentials can be obtained, as well as the nature of the electrochemical behavior when the potential of the interface is different from the resting OCP.

The typical instrumentation used for polarization tests is a potentiostat. Polarization tests are performed by applying a potential (voltage) to a metal sample (working electrode) relative to a standardized reference electrode. A counterelectrode is used to complete the circuit and to provide a current sink so that the reference electrode is not affected by the current that flows. A typical curve for a polarization test of a passivating metal is shown schematically in Fig. 2.

There are four regions of this plot that are the result of different behavior of the electrode surface. At sufficiently negative potentials (i.e., below the equilibrium potential of the metal in question), there is no driving force for oxidation of the metal to form either ions in solution or oxide on the metal surface. At this potential the metal is immune. At potentials positive of the equilibrium potential, the metal has a thermodynamic driving force for oxidation and is in the active region, i.e., it actively corrodes and releases ions into solution. A passive film has not formed to the extent that it can limit the rate of corrosion, although incomplete oxide film formation may be occurring. This continues until the potential is increased to a sufficiently more positive potential where the metal will spontaneously begin to form an oxide film on its surface. When this film fully covers the metal substrate, the metal is said to become passive, and the potential at which this occurs is known as the passivating potential, E_{pp}. This kinetic barrier prevents further oxidation, and the current that flows drops dramatically. As long as the oxide layer remains intact on the surface, further increases in electrode potential will not significantly increase the current. However as the potential is increased further in the passive range there is an increase in the thickness of the oxide film. This process is known as anodization and can be used to thicken the oxide film. For titanium, the anodization rate has been reported to be in the range of 20 Å/V [4].

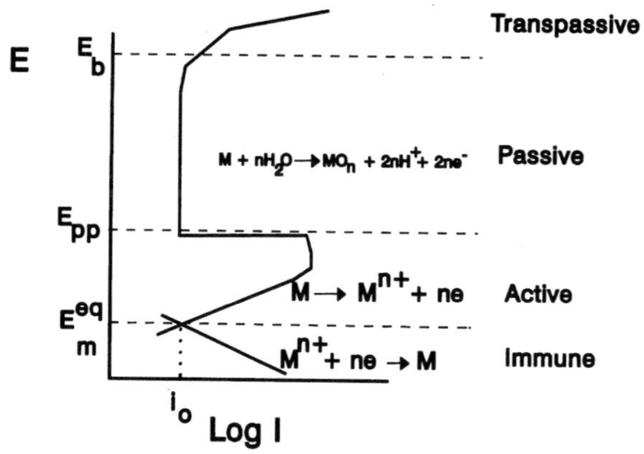

Figure 2 Schematic for a polarization test of a pure metal in water with no other electrochemical reactions taking place. The y axis represents the voltage of the metal surface relative to a standard electrode of some sort, and the x axis represents the log of the current produced by the corrosion. The areas of different corrosion behavior are indicated by the dotted lines. Below the equilibrium potential (E_m) metal ions tend to come out of solution onto the metal surface. Between the equilibrium potential and the passivation potential (E_{pp}) metal ion release from the surface is thermodynamically favorable, without inhibition by surface oxide formation. However, between the passivation potential and the breakdown potential (E_b) the surface of the metal is protected by the formation of an oxide layer (i.e., passive region). Above E_b the driving force for ionization overcomes the protective surface oxide, resulting in corrosion within the transpassive region.

Once the potential is increased further to a value where physical limitations to oxide layer growth reduce its ability to kinetically limit the high thermodynamic driving force for oxidation, the corrosion currents can increase and the electrode is said to enter transpassive behavior. This transition potential is referred to as the breakdown potential, E_b. Changes in the barrier effect of the oxide film may be the result of changes in the oxide structure or composition, valence of the metal ions in the oxide, or fracture of the oxide layer. For titanium, the breakdown potential is in the tens of volts, generally well outside of any potential capable of being induced. However, Co–Cr and stainless steel alloys have a breakdown potential of about +550 mV due to the Cr_2O_3 oxide layer breakdown.

Please note that Fig. 2 is an idealized plot of passive behavior and does not show any other oxidation or reduction reactions which may be ongoing concurrently (i.e., no oxide layer formation, precipitates, and/or ion interactions in solution). If a second electrode reaction, typically the reduction of oxygen, is present, the resultant polarization curve will be the sum of the two reactions. The summation effect can be seen in Fig. 3 and represents more real world conditions. The corrosion potential (or open circuit potential) for the combined reactions will be where the O_2 reduction reaction curve intersects with the oxidation reaction of the metal. This more complex graph more accurately represents what happens when performing this type of testing on implant alloys using physiologically relevant solutions where there are hundreds to thousands of reactions occurring.

C. Electrochemical Impedance Spectroscopy

This technique is based on the fact that metal–oxide interfaces have characteristics which are related to electrical circuits. For instance the transfer of metal ions across the interface can be

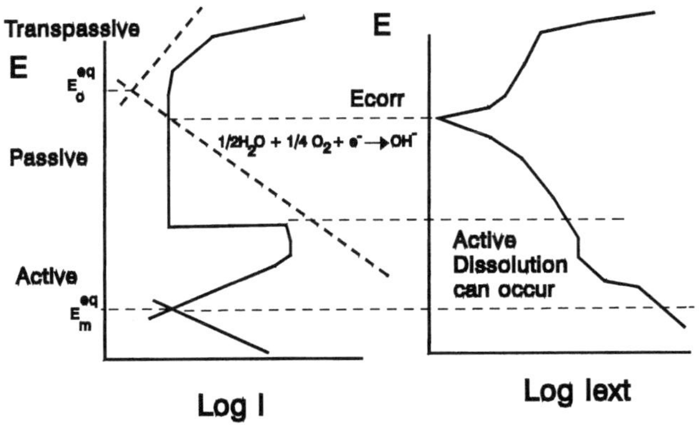

Figure 3 Schematic showing a polarization test in which there are two electrode reactions. One is a passivating metal and the other a reduction reaction (i.e., oxygen reduction). This more realistic schematic of an actual metal implant surface demonstrates the difficulty in ascertaining distinct electrochemical characteristics from real world samples (e.g., Ti-based alloy in bovine serum electrolyte).

thought of as a current whose driving force is the potential drop across the interface and the resistance due to the kinetic barriers to the movement of these ions. Also, at the interface there are positive and negative charges separated from one another, known as the electrical double layer, which creates an equivalent capacitor at the interface. Thus, the interface can be analogous to a resistor in parallel with a capacitor. Impedance spectroscopy uses alternating current techniques to determine the resistive and capacitive nature of the interface. From these experimentally derived R and C values one can determine how difficult or easy it is to transport charge across the interface and also to determine the nature of the electrical double layer. Additional information can be obtained about the growth and structure of the oxide layer as well [11]. One of the results of these types of experiments is the determination of the polarization resistance. This is a term that describes the ease of ion transport across the interface. Higher polarization resistance implies lower corrosion rates. When this technique was used to assess the polarization resistance of Ti-6Al-4V in Ringer's solution, Ringer's with serum, and Ringer's at pH $= 1$ [14]. It was found that the polarization resistance of this alloy decreased with the addition of bovine serum and with a decrease in pH, implying that the corrosion rate increased. This underscores the importance of using appropriate electrolyte fluid when conducting any corrosion testing

D. Scanning Electrochemical Microscopy

This is a relatively new technique that can be used to analyze and image the local microscopic heterogeneous corrosion behavior of metal–solution interfaces [12]. Scanning electrochemical microscopy uses a solid microelectrode probe to investigate the release of ions from a metal surface on the microscopic scale. It has the ability to obtain images of the corrosion reactions at a metallic surface under a wide variety of conditions. These include assessment of the ease and distribution of oxidation and reduction processes on metal surfaces. While this technique is relatively new to orthopedic biomaterials analysis, it may have significant application to the study of electrochemical processes at implant surfaces.

E. Surface Analytical Techniques

These techniques are used to evaluate the surface of metal alloys after they have been exposed to body simulating environments. Surface sensitive techniques include x-ray photoelectron spectroscopy (XPS), auger electron spectroscopy (AES), secondary ion mass spectroscopy (SIMS), and others. These techniques are very sensitive and are used to evaluate the outermost surfaces of alloys. These techniques rely on photon–surface interactions and electron–surface interactions to provide chemical information about the oxide layer. They are restricted to the outermost surface because the signal generated comes only from the outer 5 nm or so of the surface. One limitation to many of these techniques involves the use of instruments that require very high vacuums and may alter or affect the nature of the surface.

IV. CORROSION-RESISTANT ORTHOPEDIC ALLOYS

There are three principal metal alloys used in orthopedics and particularly in total joint replacement: titanium based alloys, cobalt based alloys, and stainless steel alloys. The elemental composition of these three alloys is shown in Table 2. Alloy-specific differences in strength, ductility, and hardness generally determine which of these three alloys is used for a particular application or implant component. However, it is primarily the high corrosion resistance of all metal alloys that has led to their widespread use as implant materials. Implant alloys were originally developed for maritime and aviation uses where mechanical properties such as corrosion resistance and high strength are paramount.

A. Stainless Steel Alloys

The form of stainless steel most commonly used in orthopedic practice is designated 316LV (American Society for Testing and Materials F138, ASTM F138). The designation 316 classifies the material as austenitic, the L denotes the low carbon content, and V the vacuum under which it is formed. The carbon content must be kept at a low level to prevent carbide (chromium–carbon) accumulation at the grain boundaries. This carbide formation weakens the material by allowing a combination of corrosion and stress to degrade the material at its grain boundaries. In the past, elevated levels of carbon have been associated with the fracture of some orthopedic implants in vivo. Molybdenum is added to enhance the corrosion resistance of the grain boundaries, while chromium dissipated evenly within the microstructure allows the formation of chromium oxide (Cr_2O_3) on the surface of the metal. The ionic bonds associated with this coating protect the surface from electrochemical degradation. Stainless steels are surface treated (e.g., in nitric acid) to promote the growth and thickening of this passive oxide layer [2,15–17].

1. New Stainless Steels

The relatively poor corrosion resistance and biocompatibility of stainless steels when compared to Ti and Co–Cr–Mo alloys provides incentive for development of improved stainless steels. New alloys such as BioDur 108 (Carpenter Technology Corp.) attempt to solve the problem of corrosion with an essentially nickel-free austenitic stainless alloy. This steel contains a high nitrogen content to maintain its austenitic structure and boasts improved levels of tensile yield strength, fatigue strength, and improved resistance to pitting corrosion and crevice corrosion as compared to nickel-containing alloys such as Type 316L (ASTM F138).

B. Cobalt–Chromium Alloys

Cobalt–chromium implant alloys fall into one of two categories, those with nickel and other alloying elements and those without. Of the many Co–Cr alloys available, the two most com-

Table 2 Approximate Weight Percent of Different Metals and Other Elements Within Major Orthopedic Alloys

Alloy	Ni	N	Co	Cr	Ti	Mo	Al	Fe	Mn	Cu	W	C	Si	V
Stainless steel														
ASTM F138	10–15.5	<0.5	*	17–19	*	2–4	*	61–68	*	<0.5	<2.0	<0.06	<1.0	*
CoCrMo alloys														
ASTM F75	<2.0	*	61–66	27–30	*	4.5–7.0	*	<1.5	<1.0	*	*	<0.35	<1.0	*
ASTM F90	9–11	*	46–51	19–20	*	*	*	<3.0	<2.5	*	14–16	<0.15	<1.0	*
ASTM F562	33–37	*	35	19–21	<1	9.0–11	*	<1	<0.15	*	*	*	<0.15	*
Ti alloys														
CPTi (ASTM F67)	*	*	*	*	99	*	*	0.2–0.5	*	*	*	<0.1	*	*
Ti-6Al-4V (ASTM F136)	*	*	*	*	89–91	*	5.5–6.5	*	*	*	*	<0.08	*	3.5–4.5
Ti-45Ni	55	*	*	*	45	*	*	*	*	*	*	*	*	*

* Indicates less than 0.05%.

Note: Alloy compositions are standardized by the American Society for Testing and Materials (ASTM Vol. 13.01).

monly used as implant alloys (Table 2) are (1) cobalt–chromium–molybdenum (Co–Cr–Mo), which is designated ASTM F-75 and F-76, and (2) cobalt–nickel–chromium–molybdenum (Co–Ni–Cr–Mo), designated as ASTM F-562. Others approved for implant use include one that incorporates tungsten (Co–Cr–Ni–W, ASTM F-90) and another with iron (Co–Ni–Cr–Mo–W-Fe, ASTM F-563). Co–Ni–Cr–Mo alloys that contain large percentages of Ni (25–37%) promise increased corrosion resistance yet raise concerns of possible toxicity and/or immunogenic reactivity (discussed later) from released Ni. The biologic reactivity of released Ni from Co–Ni–Cr alloys is cause for concern under static conditions. Due to their poor frictional (wear) properties, Co–Ni–Cr alloys are also inappropriate for use in articulating components. Therefore the dominant implant alloy used for total joint components remains Co–Cr–Mo (ASTM F-75).

C. Titanium Alloys

While CPTi is most commonly used in dental applications, the stability of the oxide layer formed on CPTi (and consequently its high corrosion resistance) and its relatively higher ductility (i.e., the ability to be cold worked) compared to Ti-6Al-4V have led to its use in porous coatings (e.g., fiber metal) of TJA components. Generally, Ti-6Al-4V (ASTM F-136) is used for joint replacement components because of its superior mechanical properties in comparison to CPTi (Table 3). The Ti-6Al-4V alloy (also known as Ti-6-4) is composed of grains of two phases: an HCP phase and a BCC phase, referred to as the alpha and beta phases, respectively. Aluminum (5.5–6.5 % by weight) stabilizes the HCP phase and vanadium (3.5–4.5 % by weight) stabilizes the BCC phase. The microstructure and mechanical properties of this alloy are highly dependent on the thermomechanical processing treatments. The Ti-6Al-4V alloy microstructure is generally composed of a fine-grained HCP phase with a sparse distribution of the BCC phase. If the material is cooled too slowly the BCC phase becomes more prominent and lowers the strength and corrosion resistance of the alloy.

Titanium alloys are particularly good implant materials because of their high corrosion resistance compared with stainless steel and Co–Cr–Mo alloys. A passive oxide film (primarily of TiO_2) protects both Ti-6Al-4V and CPTi. This stable and adherent passive oxide film protects Ti alloys from pitting corrosion, intergranular corrosion, and crevice corrosion attack and in large part is responsible for the excellent biocompatibility of Ti alloys. Generally the strength of Ti-6Al-4V exceeds that of stainless steel, with a flexural rigidity roughly half of stainless steel and Co–Cr–Mo alloys. The torsional and the axial stiffness (moduli) of Ti alloys are therefore closer to bone and theoretically provide less stress shielding than do Co alloys and stainless steel. This attribute, along with excellent biocompatibility and corrosion resistance, is primarily responsible for the popularity of titanium alloys in fracture fixation devices (plates, screws), spinal fixation devices, and total hip replacement femoral components. Ti-6Al-4V alloy is an example of a material which can be approximately 15% softer than Co–Cr–Mo alloys, yet when used in bearing applications results in significantly more (15% greater) wear than Co–Cr–Mo, e.g., TKA or THA femoral heads. Thus, Ti alloys are seldom used as materials where resistance to wear is a primary concern [1,16–20].

D. Zirconium and Tantulum Alloys

Zirconium (Zr) and tantalum (Ta) are characterized as refractory metals (others include molybdenum and tungsten) because of their relative chemical stability (passive oxide layer) and high melting points. Zr and Ta alloys are currently in use and may be gaining popularity as orthopedic metals. Because of the surface oxide layer stability, Zr and Ta (like Ti) are highly corrosion

Table 3 Mechanical Properties of Implant Alloys

Implant alloy	ASTM designation	Trade name and company	Elastic modulus (GPa)	Yield strength (MPa)	Ultimate strength (MPa)	Fatigue strength (Endurance Limit) (MPa)	Hardness (HVN)[a]	Elongation at fracture (%)
Stainless steel								
	ASTM F138	Protusul S30, Sulzer	190	792	930	241–820	130–180	43–45
CoCrMo alloys								
	ASTM F75	Alivium, Biomet	210–253	448–841	655–1277	207–950	300–400	4–14
		CoCrMo, Biomet						
		Endocast SIL, Krupp						
		Francobal, Benoist Girard						
		Orthochrome, DePuy						
		Protosul 2, Sultzer						
		Vinertia, Deloro						
		Vitallium Howmedica						
		Zimaloy, Zimmer						
		Cobalt, Chromium, Smith and Nephew						
	ASTM F90	Vitallium W, Howdmedica	210	448–1606	1896	586–1220	300–400	10–22
	ASTM F562	HS251, Haynes Stellite	200–230	300–2000	800–2068	340–520	8–50 RC[b]	10–40
		MP35N, Steel Corp.						
	ASTM 1537	TJA 1537, Allevac	200–300	960	1,300	200–300	41 RC[b]	20
		Metasul, Sulzer						
Ti alloys								
cpTi	ASTM F67	CSTi, Sulzer	110	485	760	300	120–200	14–18
		Pure Titanium, Smith and Nephew						
Ti-6Al-4V	ASTM 136	Isotan, Aesculap Werke	116	897–1034	965–1103	620–689	310	8
		Protosul 64WF, Sulzer						
		Tilastan, Waldemar Link						
		Tivaloy 12, Biomet						
		Tivanium, Zimmer						
		Titanium alloy, Smith and Nephew						
		Nephew						
Ti-5 Al-2.5Fe	*		100–110	780	860	300–725	310	7–13
Ti-45Ni	*	Nitinol, Nitinol Medical Technologies	28–110	621–793	827–1172	<200	40–62 RC[b]	1–60

HVN = Vickers hardness number, kg/mm.
[b] RC = Rockwell hardness scale.
* No current ASTM standard.

resistant. Corrosion resistance generally correlates with biocompatibility (although not always) because more stable metal alloys tend to be less chemically active and less participatory in biologic reactions. Additionally, these refractory metals generally possess high levels of hardness (12 Gpa) and wear resistance (approximately ten fold that of Co and Ti alloys, using abrasion testing), which makes them well suited for bearing surface applications. The thickness of the surface oxide layer (approximately 5 μm) and ability to extend ceramic-like material properties (i.e., hardness) into the material through techniques such as oxygen enrichment have resulted in the production of TJA components using these alloys (e.g., oxidized zirconium TKA femoral components, Smith and Nephew). As difficulties associated with forming and machining these metals are overcome the use of these materials is expected to grow [2,16,18–20].

V. PRIMARY MECHANISMS OF IMPLANT CORROSION

There are significant clinical problems relating to the corrosion of implant alloys in the current state-of-the-art implants which will likely continue to be a potential hazard for the near future, one of which is corrosion observed in the taper connections of retrieved modular joint replacement components. With the large and growing number of total joint designs that use modular connections (i.e., metal-to-metal press-fit conical tapered connections), the effects of crevices, stress, and motion take on increasing importance. Retrieval studies [21,22] have shown that severe corrosion attack can take place in the crevices formed by these tapers in vivo (see Fig. 4). It has been postulated that this corrosion process is the result of a combination of stress and motion at the taper connection and the crevice geometry of the taper. The stresses resulting from use cause fracturing and abrasion of the oxide film covering these passive metal surfaces. This, in turn, causes significant changes in the metal surface potential and in the crevice solution chemistry as the oxides are continuously fractured and repassivate. These changes may result in deaeration (loss of O_2) of the crevice solution and a lowering of the pH in the crevice [23] as is expected in crevice corrosion attack. The ultimate result of this process is a loss of the oxide film and its kinetic barrier effect and an increase in the rate of corrosive attack in the taper region.

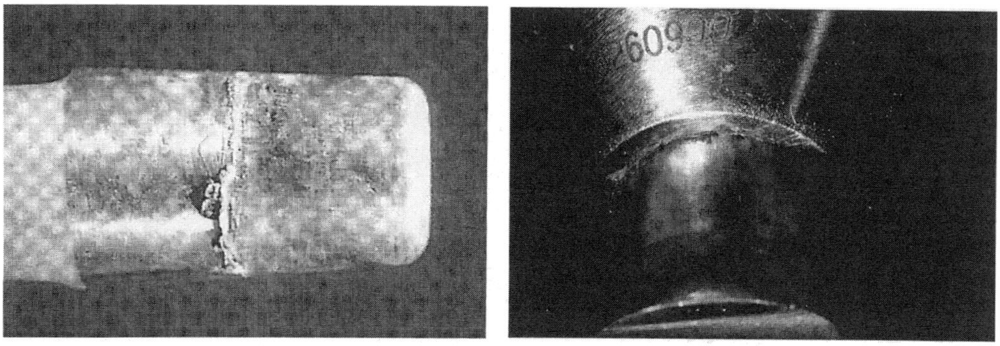

(a) (b)

Figure 4 Retrieved joint replacement components showing corrosion around the rims of early model conical taper connections. (a) Cobalt-based alloy stem showing evidence of corrosion precipitates. (b) Deposits of $CrPO_4$ corrosion products on the rim a modular cobalt–chrome femoral component. Both fretting and crevice corrosion are responsible for generating this type of implant degradation.

Severe corrosion attack has been associated with Co–Cr alloy modular taper connections. While less common, corrosion attack of titanium alloy stems can also occur. In general, Co–Cr alloys undergo intergranular corrosion, etching, selective dissolution of cobalt, and the formation of Cr-rich particles that are most likely oxides or oxychlorides. The corrosion products generated at the taper connections can migrate into periprosthetic tissues and in between articulating polymeric surfaces. In the past there have been instances where retrieved implants have corroded to such an extent that intergranular corrosion resulted in fatigue failure in the neck of Co–Cr stems. It is to be emphasized that it is the mechanical integrity of the oxide films that form on these alloys that determines long-term stability and performance of metallic components. Relatively little is known about the mechanical stability of oxide films and the electrochemical reactions which occur when an oxide film is fractured. What is known is that when the oxide films of these orthopedic alloys are abraded or removed from the surface by fretting the open circuit potential can decrease to values as low as −500 mV (vs. SCE) [24].

These potential excursions may be significant enough and prolonged enough to cause changes in the oxide structure and stability by bringing the interface potential into the active range of the alloy, thereby dramatically accelerating the corrosion rate. Known corrosion properties of popular implant alloys are listed in Table 4 and discussed in the following sections.

VI. METAL ION/SOLUBLE METAL LEVELS

Normal human serum levels of prominent implant metals are approximately as follows: 1–10 ng/mL aluminum, 0.15 ng/mL chromium, <0.01 ng/mL vanadium, 0.1–0.2 ng/mL cobalt, and

Table 4 Electrochemical Properties of Implant Metals (Corrosion Resistance) in 0.1 M NaCl at pH 7 (except where indicated)

Alloy	ASTM standard	Density (g/cm^3)	Corrosion potential (vs. calomel) (mV)	Passive current density, I_p ($\mu A/cm^2$)	Breakdown potential, E_b (mV)	Polarization current, and polarization Rc, at 37 °C		Repassivation time in 0.3 M NaCl resistance, (0.9%) (ms)	
						i ($\mu A/cm^2$)	Rc ($k\omega/cm^2$)	−500 mV	+500 mV
Stainless steel	ASTM F138	8.0	−400	0.56	200–770	0.006	1670	72000	35
Co-Cr-Mo alloys	ASTM F75	8.3	−390	1.36	420	0.004	2500	44	36
Ti Alloys cpTi	ASTM F67	4.5	−90 to −630	0.72–9.0	>2400	0.010	1000	43	44
Ti-6Al-4V	ASTM 136	4.43	−180 to −510	0.9–2.0	>1500	0.008	1250	37	41
Ti-5 Al-2.5 Fe	**	4.45	−530	0.68	>1500	*	*	110–130	120–160
Ti-45Ni	**	6.4–6.5	−430	0.44	890	*	*	*	*

* Data not available.
** No current ASTM designation.
Note: The corrosion potential represents the open circuit potential (OCP) between the metal and a calomel electrode. The more negative the OCP, the more chemically reactive and thus the less corrosion resistance. Generally low current density indicates greater corrosion resistance. The higher the breakdown potential, the better (i.e., the more elevated the breakdown potential, the more stable the protective layer).

<4.1 ng/mL titanium. Following total joint arthroplasty levels of circulating metal (Co, Cr, Ni, Al, and V) have been shown to increase (Table 5). Multiple studies have demonstrated chronic elevations in serum and urine cobalt and chromium following total primary joint replacement. Chronic elevations in serum Ti concentrations in subjects with well-functioning THR with Ti-containing components have also been reported without measurable differences in urine Ti concentrations, serum Al concentrations, or urine Al concentrations. Vanadium concentrations have not been found to be elevated in patients with TJA partially due to the technical difficulty associated with measuring the small concentrations present in serum (Table 5).

Metal ion levels within serum and urine of TJA patients can be affected by a variety of factors. The type of implant can affect metal serum concentrations. For example, patients with total knee replacement components containing Ti-based alloy and carbon fiber-reinforced polyethylene wear couples demonstrated tenfold elevations in serum Ti concentration at an average of 4 years after implantation. Substantial serum Ti elevations have also been reported in patients with failed metal-backed patellar components where unintended metal/metal articulation was possible. These individuals contained serum Ti levels up to a hundred times higher than normal. However even among these THA patients, there was no elevation in serum or urine aluminum, serum or urine vanadium levels, or urine titanium levels. Mechanically assisted crevice corrosion of modular total hip arthroplasty components have been associated with elevations in serum Co and urine Cr. It has been previously assumed that extensively porous coated cementless stems

Table 5 Upper Estimate Concentrations of Metal Reported in Human Body Fluids and in Human Tissue with and Without Total Joint Replacements

		Ti	Al	V	Co	Cr	Mo	Ni
Human body fluids ($\times 10^{-3}$ *mM*)								
Serum	Normal	0.06	0.08	>0.02	0.003	0.001	*	0.007
	TJA	0.09	0.09	0.03	0.007	0.006	*	<0.16
Whole blood	Normal	0.35	0.48	0.12	0.002	0.058	0.009	0.078
	TJA	1.4	8.1	0.45	0.33	2.1	0.104	0.50
Human tissue (μ*g/g*) *(roughly equivalent to:* $\times$*0.1–0.01 mM)*								
Skeletal muscle II	Normal	*	*	*	<12	<12	*	*
	TJA	*	*	*	160	570	*	*
Liver	Normal	100	890	14	120	<14	*	*
	TJA	560	680	22	15200	1130	*	*
Lung	Normal	710	9830	26	*	*	*	*
	TJA	980	8740	23	*	*	*	*
Spleen	Normal	70	800	<9	30	10	*	*
	TJA	1280	1070	12	1600	180	*	*
Kidney	Normal	*	*	*	30	<40	*	*
	TJA	*	*	*	60	<40	*	*
Lymphatic tissue	Normal	*	*	*	10	690	*	*
	TJA	*	*	*	390	690	*	*
Heart	Normal	*	*	*	30	30	*	*
	TJA	*	*	*	280	90	*	*

Normal: Subjects without any metallic prosthesis (not including dental).
TJA: Subjects with well-functioning total joint arthroplasty.
* Data not available [32,83,122,123].
Source: Refs. 33 and 121.

would give rise to higher serum and urine Cr concentrations due to the larger surface area available for passive dissolution. However some studies suggest that a predominant source of disseminated chromium degradation products is most likely the fretting corrosion of modular junctions.

To date, the form(s) of released metal in vivo remains relatively uncharacterized. What metal bioreactivity has been characterized in vitro using metal ions in culture medium with 10% serum (as is the case with most in vitro investigations) may differ from in vivo conditions of essentially 100% serum, where relatively inert compounds (e.g., metal–oxides) and complexes (e.g., metal–albumin) may more readily form and abrogate (or exaggerate) the toxic effects of metals. Metal ions released in vivo [25] and in vitro [26] are bound by specific serum proteins. Two molecular weight ranges of human serum proteins were determined to be associated with the binding of Cr from Co–Cr–Mo (ASTM F-75) implant alloy degradation (at approximately 68 and 180 kDa); only one range of serum protein(s) (at approximately 68 kD) is associated with the binding of Ti released from Ti-6Al-4V implant alloy. The role of serum or tissue proteins in the mediation of metal-induced effects remains largely unknown.

It has been repeatedly demonstrated that elevations in serum and urine metal content are associated with total joint replacement components. The toxicological importance of these findings are not known. Currently, there is limited information in the literature that describes the chemical form of the degradation products of metallic joint replacement prostheses. Ultimately, toxicological experiments using relevant chemical species identified by bioavailability studies will be used in animal models and cell cultures to define specific toxicities of the degradation products. However, at the present time this information is not available.

Homogenates of organs and tissues obtained postmortem from subjects with cobalt base alloy total joint replacement components have indicated that significant increases in cobalt and chromium concentrations occur in the heart, liver, kidney, spleen, and lymphatic tissue (Table 5). Similarly, patients with titanium base alloy implants demonstrated elevated titanium, aluminum, and vanadium levels around their metal implants (with up to 200 ppm of titanium, six orders of magnitude greater than that of controls; 880 ppb of aluminum; and 250 ppb of vanadium). Spleen aluminum levels and liver titanium concentrations can also be markedly elevated in patients with failed titanium alloy implants. It has been found that even in the absence of significant elevations in serum metal concentrations, deposition of metal can occur locally and in remote organ stores in association with a well-functioning device.

VII. PARTICLE RELEASE AND DISTRIBUTION

Polyethylene particles are generally recognized as the most prevalent particles in the periprosthetic milieu; however, metallic particulate species are also present in variable amounts and may have important sequelae. When present in sufficient amounts, particulates generated by wear, corrosion, or a combination of processes can induce the formation of an inflammatory, foreign body granulation tissue with the ability to invade the bone–implant interface. This can result in progressive, periprosthetic bone loss that threatens the fixation of both cemented and cementless devices, limiting the survivorship of total joint replacement prostheses.

Consequently, particulate wear debris of metal alloys used in prosthetic components has been the subject of intense study concerning their role in bone resorption and aseptic loosening. The clinical significance of corrosion at the modular head/neck junction lies, in part, in the effects that solid corrosion products increase the particulate burden within the joint and migrate along bone–implant interface membranes to sites remote from their origin. They can also migrate to the prosthetic bearing surface where they may result in three-body wear, thereby increasing the production of polyethylene debris. All of these factors can contribute to periprosthetic bone loss and aseptic loosening.

Numerous case reports document the presence of metallic, ceramic, or polymeric wear debris from hip and knee prostheses in regional and pelvic lymph nodes. Postmortem studies have demonstrated dissemination of wear particles to the liver, spleen, or abdominal lymph nodes in patients who have a total hip or knee replacement [27–29]. These studies also revealed both metallic and polyethylene wear particles in the para-aortic lymph nodes of approximately 90% of patients with a joint replacement prosthesis, whereas metallic wear particles alone were present in the para-aortic lymph nodes of approximately 70% of patients with a hip or knee implant. Of these approximately 40% of TJA patients were reported to have particles disseminated to the liver or spleen. Most disseminated metallic particles have been reported to be less than 1 μm in size, but the range of particle sizes is material dependent. Particles of commercially pure Ti and Ti–Al–V alloy may range from 0.1 μm to as large as 50 μm in the lymph nodes and as large as 10 μm in the liver and spleen. In contrast, particles of cobalt–chromium and stainless steel alloys rarely exceed 3 μm. The response to metallic (and polymeric) debris in lymph nodes includes immune activation of macrophages and associated production of inflammatory cytokines. Metallic and polyethylene wear particles in the liver or spleen are more prevalent in patients who have had a previously failed reconstruction when compared to patients with primary hip or knee arthroplasties [30]. While there have been numerous investigations concerning particulate debris in periprosthetic tissues, particularly with regard to the phenomenon of particle-induced, macrophage-mediated inflammation and osteolysis, relatively little is known about the dissemination of wear debris beyond the local tissues. Identification of orthopedic wear debris can be difficult, even in regional lymph nodes, due to the coexistence of particles from other sources.

The clinical significance of orthopedic wear debris accumulation at remote sites has been understood based largely on examination of lymph nodes biopsied at revision surgery or for cancer staging in patients who also happened to have a total joint replacement. Numerous case reports document the presence of metallic, ceramic, or polymeric wear debris from hip and knee prostheses in regional and pelvic lymph nodes (along with the findings of lymphadenopathy, gross pigmentation due to metallic debris, fibrosis, lymph node necrosis, and histiocytosis, including complete effacement of nodal architecture). The inflammatory response to metallic and polymeric debris in lymph nodes has been demonstrated to include immune activation of macrophages and associated production of cytokines. Accumulation of debris in remote organs and lymph nodes may explain, in part, past observations suggesting that circulating peripheral blood monocytes from patients with joint replacements are more reactive to particulate wear debris stimulation than monocytes from individuals without implants [31–36].

VIII. LOCAL EFFECTS OF IMPLANT CORROSION

A. Soluble Corrosion Debris

The tissues surrounding modern implants may include areas of bone ingrowth (osseointegration), fibrous encapsulation, and a variable presence of the foreign body responses. There are no generalizable types of metal release that are known to occur with all metallic implants. However, accelerated corrosion and a tissue response that can be directly related to identifiable corrosion products have been demonstrated in the tissues surrounding multipart devices.

In vitro investigations indicate that specific metals in ionic form can affect the functionality of a variety of peri-implant cells such as fibroblasts, osteoblasts, macrophages, and lymphocytes within the ranges of metal concentrations reported to exist in periprosthetic tissue. Generally, the most toxic metal ions have been found to be Ni, Fe, Cu, Mn, and V, while others such as Na, Cr, Mg, Mo, Al, Ta, and Co demonstrate relatively less cellular reactivity in vitro. Different metals act through different cellular mechanisms to induce distinct responses. There is mounting evidence that adverse local and remote tissue responses, which in the past have been entirely associated with metal particles, to some extent may be due to soluble forms of specific metal degradation products [37,38]. However, the effect of soluble metals on periprosthetic cells is a complex function of cell type, composition, and concentration of metal. In vitro investigation has determined the stimulatory effect of some metals (e.g., Al and V) on cells such as lymphocytes and fibroblasts, while the same metals (and concentrations) can suppress the viability and proliferation of other cell types (e.g., osteoblast-like cells). This differential impact of metal ions on some cell types and not others (particularly fibroblasts and osteoblasts) may potentially explain how fibrous membranes so readily form around implants initially placed in intimate contact with bone (i.e., osteoblasts). Adverse local and remote tissue responses purportedly associated with particulate debris may be due in part to specific soluble metals resulting from implant degradation.

High concentrations of metals negatively impact all types of cells at some level [39–50]. For certain cell types, such as human osteoblasts, these effects have been somewhat characterized. One of the main functions of osteoblasts (if not the main function) is to produce organic bone matrix, 90% of which is type I collagen. Type I collagen is comprised of three helical chains. Two of the three helical protein chains are $\alpha1[I]$. The third chain, $\alpha2[I]$, is similar in structure yet genetically distinct from $\alpha1[I]$. Metal particles and ions have been found to decrease gene expression of procollagen $\alpha1[I]$ before decreases could be observed in other more osteoblast-specific markers of bone deposition, such as gene expression of osteocalcin, osteonectin, and alkaline phosphatase [51–53]. Other metal-induced effects on osteoblasts have been noted, such as the production of cytokines which recruit, prime, and activate inflammatory cells. Interleukin-6 is secreted by osteblasts in response to Al, Fe, Mn, Na, Ni, and V chloride solutions (more toxic metals). The concentrations of metal ions associated with toxic osteoblast responses can be detected within some ranges of metal concentrations reported to exist in periprosthetic tissue (Table 4) [46]. Comparison of the effects of metal ions on osteoblasts to the effects of particles previously reported [54 demonstrates the potential of specific metal ions released from implants or particulate implant debris to play a clinical role in the pathogenesis of osteolysis. This contention is supported by past investigations where metal ions such as Al, V, and Ti have been shown to inhibit apatite formation in vitro by binding and blocking potential crystal growth sites. This poisoning of crystal growth sites by metal ions may thus act to interfere with normal in vivo osteoid mineralization and remodeling process of bone [55,56]. Whether through indirect osteoclast activation (i.e., IL-6 release via osteoblasts) or direct inhibitory effects on osteoblasts, it is apparent that metal ions released from implants have the potential to diminish bone formation, which previously has been largely attributed to particulate implant debris alone.

Important to the assessment of metal-induced osteolysis is the role of other peri-implant cells such as fibroblasts, osteoclasts, macrophages, and lymphocytes, which, after exposure to metal ions, may affect osteoblast function through paracrine mediators [40]. Although osteoclast activity has been reportedly impaired by exposure to metal ions at sublethal concentrations [43], these effects may be overridden by metal-induced autocrine and paracrine induction of IL-6, which can act to directly stimulate osteoclast activity. Thus, further study using mixed cell populations is required to more comprehensively assess released implant metal effects within the peri-implant milieu.

B. Particulate Debris Corrosion

Histological sections of the tissues surrounding stainless steel internal fixation devices generally show encapsulation by a fibrous membrane with little or no inflammation over most of the device. At screw–plate junctions, however, the membranes often contain macrophages, foreign body giant cells, and a variable number of lymphocytes in association with two types of corrosion products: iron containing granules and microplates of relatively larger particles of a chromium compound(s).

Chromium microplates are of variable morphology and are found within the tissues as closely packed, platelike particle aggregates ranging in size from 0.5 to 5.0 μm. They are often found free within acellular collagen or within frankly necrotic tissue. Several multinucleated foreign body giant cells are usually present within or bordering collections of these particles. In hematoxylin and eosin preparations, the majority of microplates are yellow or apple-green. Many microplates, however, stain darkly with hematoxylin and these microplates also react strongly to staining for iron. Electron microprobe energy dispersive x-ray analyses indicate that microplates are a chromium compound containing iron and a substantial amount of phosphorus.

Iron granules are often seen surrounding chromium microplates, but the granules are found alone as well. Iron granules are yellow-brown, mainly spherical, and 0.1 to 3 μm in diameter. They are predominantly intracellular, most often in macrophages, but may also be found in fibrocytes. X-ray diffraction indicates that the granules consist of a mixture of two or more of the iron oxides, αFe_2O_3 and σFe_2O_3, and the hydrated iron oxides $\alpha Fe_2O_3 \cdot H_2O$ and $\sigma Fe_2O_3 \cdot H_2O$.

Particles of the chromium–phosphate hydrate–rich material can be found at the bearing surface of the UHMWPE acetabular liners, suggesting they may participate in articular surface three-body wear resulting in an increased production of polyethylene debris. Particles of the chromium–phosphate hydrate–rich corrosion product found in tissues range in size from submicrometer to aggregates of particles up to 500 μm. These particles are similar in appearance to the chromium–containing microplates observed in association with corroded stainless steel implants when viewed through a light microscope. Larger particles are often found within areas of marked fibrosis or necrosis or associated with foreign body giant cells, although most of these particles are less than 5 μm in size and are found within macrophages.

The degradation products that can be observed in histologic sections of tissues adjacent to titanium base alloys are of a different nature than the precipitates associated with stainless steel and cobalt base alloys that have undergone accelerated corrosion. Titanium base alloys, although very corrosion resistant, typically discolor (darkly stain) adjacent tissue due to metallic debris. Examination has found that the ratios of titanium, aluminum, and vanadium concentrations in the periprosthetic tissues are similar to those of the bulk alloy. This suggests that the metallic debris represented wear particles from the head or stem rather than precipitated corrosion products. This wear debris presents enormous surface areas for electrochemical dissolution. This additional metal available for dissolution is likely a major factor contributing to observed systemic elevations in titanium (Table 5).

IX. SYSTEMIC EFFECTS OF CORROSION

A. General Considerations

The concern about the release and distribution of metallic degradation products is justified by the known potential toxicities of the elements used in modern orthopedic implant alloys—titanium, aluminum, vanadium, cobalt, chromium, and nickel. Broad reviews of the toxicology of these elements are available [57,58] and are summarized below. In general terms, metal toxicity may be by virtue of (1) metabolic alterations, (2) alterations in host–parasite interactions, (3) immunologic interactions of metal moieties by virtue of their ability to act as haptens (specific immunological activation) or antichemotactic agents (nonspecific immunological suppression) [59,60], and (4) by chemical carcinogenesis [61].

Cobalt, chromium, nickel, and vanadium are all essential trace metals in that they are required for certain enzymatic reactions [62]. In excessive amounts however, these elements are also toxic. Excessive cobalt may lead to polycythemia [63], hypothyroidism [63], cardiomyopathy, and carcinogenesis [64,65]. Chromium can lead to nephropathy, hypersensitivity, and carcinogenesis [66,67]. Nickel can lead to eczematous dermatitis, hypersensitivity, and carcinogenesis [68]. Vanadium can lead to cardiac and renal dysfunction, and has been associated with hypertension [69] and depressive psychosis [70]. The nonessential metallic elements also possess specific toxicities. Titanium, although generally regarded as inert, has been associated with pulmonary disease in patients with occupational exposure [71] and in animal models [72]. Aluminum toxicity is well documented in the setting of renal failure and can lead to anemia [73], osteomalacia [74], and neurological dysfunction, possibly including Alzheimer's disease [75]. However, considering the litany of documented toxicities of these elements, it is important to remember that the toxicities generally apply to soluble forms of these elements and may not apply to the chemical species that are the degradation products of prosthetic implants.

B. Immunogenicity of Metallic Implants

Some adverse responses to orthopedic biomaterial corrosion are subtle and continue to foster debate and investigation. One of these responses is ''metal allergy,'' or hypersensitivity to metallic biomaterial degradation. Dermal hypersensitivity to metal is common, affecting about 10–15% of the population [76–79]. Dermal contact and ingestion of metals have been reported to cause immune reactions, which most typically manifest as skin hives, eczema, redness, and itching [76,80,81]. As previously stated, all metals in contact with biological systems corrode [82,83 and the released soluble products, while not sensitizers on their own, can activate the immune system by forming complexes with native proteins [78,84,85]. These metal–protein complexes are considered to be candidate antigens (or, more loosely termed, allergens) for eliciting hypersensitivity responses. Metals known as sensitizers include beryllium [86], nickel [79–81,86], cobalt [86], and chromium [86], while occasional responses have been reported to tantalum [87], titanium [88,89], and vanadium [87]. Nickel is the most common metal sensitizer in humans followed by cobalt and chromium [76,79–81]. The prevalence of metal sensitivity among the general population is approximately 10–15%, with nickel sensitivity the highest (approximately 14%) [76]. Cross-reactivity between nickel and cobalt is most common [76,78]. The amounts of these metals found in medical grade alloys were shown in Table 2.

Hypersensitivity can be either an immediate (within minutes) humoral response (initiated by antibody or formation of antibody–antigen complexes of types I, II, and III reactions) or a delayed (hours to days) cell-mediated response [90,91]. It is the latter response with which implant related hypersensitivity reactions are generally associated, in particular type IV delayed type hypersensitivity (DTH).

Cell-mediated delayed type hypersensitivity is characterized by antigen activation of sensitized T_{DTH} lymphocytes releasing various cytokines, which result in the recruitment and activation of macrophages. T_{DTH} lymphocytes are subset populations of T helper (T_H) lymphocytes purported to be of the CD4+ $T_{H\text{-}1}$ subtype (and in rare instances CD8+, cytotoxic T cells, T_C). This $T_{H\text{-}1}$ subpopulation of T cells is characterized by their cytokine release profile, e.g., interferon-γ (IFN-γ), tumor necrosis factor α (TNF-α), interleukin-1 (IL-1), and interleukin-2 (IL-2). $T_{H\text{-}1}$ cells are generally associated with responses to intracellular pathogens and autoimmune diseases. Although $T_{H\text{-}1}$ cells mediate a DTH reaction, only 5% of the participating cells are antigen-specific T cells within a fully developed DTH response. The majority of DTH participating cells are macrophages.

Metals from implant corrosion have been shown in case studies to be temporally associated with specific responses such as severe dermatitis, urticaria, vasculitis [92–97], and/or nonspecific immune suppression [59,60,98,99].

Generally there are more case reports of hypersensitivity reactions associated with stainless steel and cobalt alloy implants than with titanium alloy components [77,78,92–95,100–104]. One such case report implicated cobalt hypersensitivity in the poor performance of cobalt alloy plates and screws used in the fracture fixation of a 45-year-old woman's left radius and ulna. In this case the patient presented with periprosthetic fibrosis, patchy muscular necrosis, and chronic inflammatory changes peripherally 7 years after implantation. After removal of all metal hardware, the swelling disappeared and eventually the patient became complaint-free. However, there remained a hypersensitivity to cobalt as evaluated by patch testing [97].

This and similar case reports prompted a number of larger patient cohort studies in the late 1970s and 1980s investigating the possible correlation between metal sensitivity and implant failure [77,105–116]. Data (from these different investigations) regarding the prevalence of metal sensitivity are compiled in Fig. 5. Unfortunately, these studies include heterogeneous patient populations and testing methodologies and consequently reach a variety of conclusions. However, all patient populations included in Fig. 5 were metal allergy tested to one or a combination of metals including nickel, cobalt, and/or chromium after receiving an implant. The prevalence of metal sensitivity among patients with well-functioning implants is approximately 25%, roughly twice as high as that of the general population [101,105,107,109,111,114,115,117,118]. Overall, the prevalence of metal sensitivity in patients with failed or failing implants is approximately six times that of the general population and approximately two to three times that of all patients with metal implants. This association does not prove cause and effect; that is, are these patients sensitive because the device has failed, or has the device failed because the patient had a preexisting metal sensitivity, or are alternative dominating mechanisms (e.g., genetic autoimmunity) responsible for both?

Specific types of implants with greater propensity to corrode and/or release metal in vivo may be more prone to induce metal sensitivity. Failures of total hip prostheses with metal-on-metal bearing surfaces have been associated with greater prevalence of metal sensitivity than similar designs with metal–on–ultrahigh molecular weight polyethylene bearing surfaces [105,118]

It is unclear whether hypersensitivity responses to metallic biomaterials affect implant performance in other than a few highly predisposed people [78,90,119]. It is clear that some patients experience excessive eczemic immune reactions directly associated with implanted metallic materials [77,92,94,95,97,104]. Metal sensitivity may exist as an extreme complication in only a few highly susceptible patients (i.e., less than 1% of joint replacement recipients), or it may be a more common subtle contributor to implant failure. Continuing improvements in immunologic testing methods will likely enhance future assessment of patients susceptible to hypersensitivity responses. The importance of this line of investigation is growing, as the use of metallic implants is increasing and as expectations of implant durability and performance increase [120].

C. Carcinogenesis

The carcinogenic potential of the metals used in TJA and other implants (e.g., nickel–titanium alloy arterial stents) remains an area of concern. Animal studies have documented the carcinogenic potential of orthopedic implant materials. Small increases in rat sarcomas were noted to

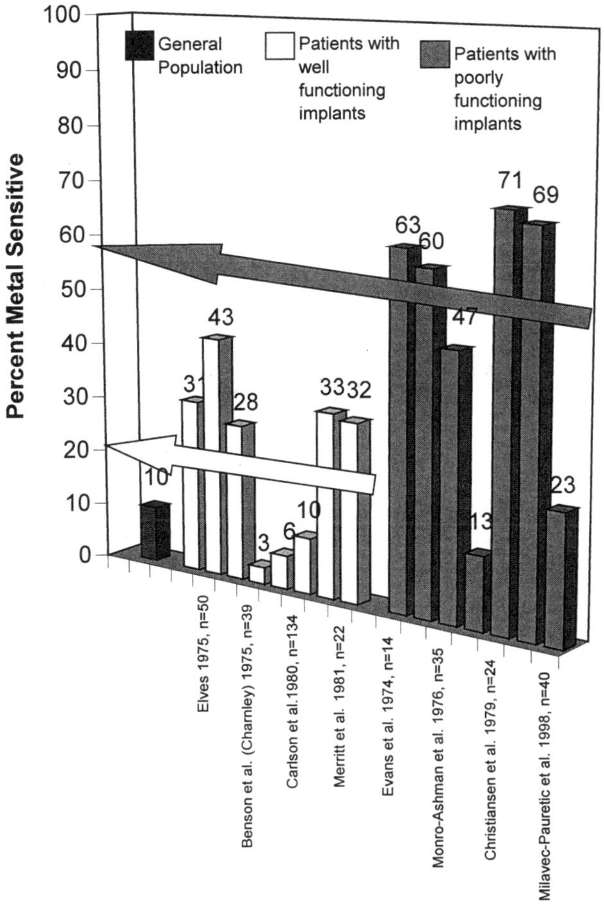

Figure 5 The bars indicate the averaged percentages of metal sensitivity (for nickel, cobalt, or chromium) among the general population and total arthroplasty patients with poor and well-functioning implants based on a number of published reports. Note that the average incidence of metal sensitivity is 10, 25, and 60% for the population at large, patients with well-functioning total joint prostheses, and patients with poorly functioning implants, respectively.

correlate with high serum cobalt, chromium, or nickel content from metal implants. Furthermore, lymphomas with bone involvement were also more common in rats with metallic implants. Implant site tumors in dogs and cats, primarily osteosarcoma and fibrosarcoma, have been associated with stainless steel internal fixation devices.

Initially, epidemiological studies implicated cancer incidence in the first and second decades following total hip replacement. However, larger more recent studies have found no significant increase in leukemia or lymphoma; however, these studies did not include as large a proportion of subjects with a metal-on-metal prosthesis. There are constitutive differences in the populations with and without implants that are independent of the implant itself, which confound the interpretation of epidemiological investigations.

The association of metal release from orthopedic implants with carcinogenesis remains conjectural since causality has not been definitely established in human subjects. The identification of such an association depends both on the availability of comparative epidemiology and on the ability to perform tests on the patient before and after device removal. Due to a number of factors such as patient age, the actual number of reported cases of tumors associated with orthopedic implants is likely underreported. However, with respect to the number of devices implanted on a yearly basis the incidence of cancer at the site of implantation is relatively rare. Continued surveillance and longer-term epidemiological studies are required to fully address these issues.

X. FUTURE DIRECTIONS AND CONCLUSIONS

Corrosion of orthopedic implants remains a significant clinical concern. Even though past implant alloys have been replaced with modern corrosion-resistant ''super alloys,'' deleterious corrosion processes have been observed in certain clinical settings. There is reason to believe that attention to (1) metallurgical processing variables, (2) tolerances of modular connections, (3) surface processing modalities, and (4) appropriate material selection, all can diminish corrosion and minimize the potential for adverse clinical outcome. The potential exists for future surface treatments (e.g., nitriding, ion implantation, etc.) to significantly reduce the magnitude of fretting corrosion of titanium alloy and other metal implant devices.

There remains a need to further investigate the mechanical-electrochemical interactions of metal surfaces. Characterization of the stresses and motion needed to fracture passivating oxide films as well as the effects of repeated oxide abrasion on the electrochemical behavior of the interface and ultimately the implant continue to be actively investigated. Evaluating the role of particulate corrosion products in adverse local tissue reactions also requires continuing investigation. Thus further clinical retrieval studies and in vitro cell culture experiments are needed to more fully characterize this relationship. Finally, the clinical significance of metal release and elevated metal content in body fluids and remote organs of patients with metallic implants needs to be elucidated. Considerable work will be required in discerning the chemical form(s) of released metal and the nature of its ligands to ultimately resolve questions of potential toxicity.

It is important to note that when evaluating the corrosion and biocompatibility of a particular metal component, the results do not necessarily apply to all implants made of the same material. The definition of ''biocompatibility'' remains the ability of a material to demonstrate host and material response appropriate to its intended application. Poor implant performance can be attributed to many factors, which include manufacturing errors, mechanical design errors, surgical errors, and inappropriate choice of material for a given application. Wise material selection cannot compensate for poor implant design or surgical error. It must be emphasized that currently there is no universal ''best'' metal for all implant applications. Ultimately, the most prudent choice of a corrosion-resistant metal for a particular application depends on careful evaluation of which specific mechanical properties of available materials (in addition to corrosion resistance) best satisfy the *in situ* demands and design characteristics of a particular implant component.

REFERENCES

1. Black J. Orthopaedic Biomaterials in Research and Practice. New York: Churchill Livingstone, 1988.

2. Jacobs JJ, Gilbert JL, Urban RM. Corrosion of metal orthopaedic implants. J Bone Joint Surg [Am] 1998; 80:268–282.
3. Lacombe P. Corrosion and oxidation of Ti and Ti alloys. In: . Williams JC , Belor AF, eds. Titanium and Titanium alloys. NY: Plenum Press, 1982:847–880.
4. Aladjem A. Review: anodic oxidation of titanium and its alloys. J Mater Sci 1973; 8:688–704.
5. Cabrera N, Mott NF. Physics 1948; 12:163–184.
6. Fraker A, Ruff AW, Yeager M. Studies of oxide film formation on titanium alloys in physiological solutions. Corrosion Sci 1973; 11:763–765.
7. Brown GM, Thundat T, Allison DA, Warmack RJ. Electrochemical and in-situ atomic force microscopy investigation of titanium in oxalic acid. J Vac Sci Tech 1992; 10:3001–3006.
8. Healy KE, Ducheyne P. The mechanisms of passive dissolution of titanium in model physiologic environment. J Biomed Mater Res 1992; 26:181–186.
9. Williams RL, Brown SA, Merritt K. Electrochemical studies on the influence of proteins on the corrosion of implant alloys. Biomaterials 1988; 9:181–186.
10. Wishey A, Gregson PJ, Peter LM. Effect of surface treatment on the dissolution of titanium based implant materials. Biomaterials 1991; 12:470–473.
11. MacDonald JR. Impedance Spectroscopy. NY: Wiley Interscience, 1987.
12. Gilbert JL, Smith SM, Lautenschlager EP. Scanning electrochemical microscopy of metallic biomaterials: reaction rate and ion release imaging modes. J Biomed Mater Res 1993; 27:1357–1366.
13. Bard AJ, Faulkner AJ. Electrochemical Methods. NY: John Wiley and Sons, 1980.
14. Lewis G, Daigle K. Electrochemical behavior of T-6Al-4V alloy in static biostimulating solutions. J Biomed Mater Res Appl Biomat 1993; 4:47–54.
15. Bundy K. Bone prosthesis and implant materials. In: editor. Cowin SC. Bone Mechanics. Boca Raton: CRC Press Inc., 1989:160–184.
16. Park JB. Biomaterials Science and Engineering. New York: Plenum Press, 1984.
17. Silver FH, Christiansen DL. Biomaterials Science and Biocompatibility. New York: Springer, 1999.
18. Breme J, Biehl V. Metallic biomaterials. In: . Black J , Hastings G, eds. Handbook of Biomaterial Properties. London: Chapman and Hall, 2001:135–214.
19. Black J. Biomaterials. New York: Marcel Dekker, 1992.
20. Black J. Prosthetic Materials. New York: VCH Publishers, 1996.
21. Gilbert JL, Buckley CA, Jacobs JJ, Bertin KC, Zernich MR. Intergranular corrosion-fatigue failure of cobalt-alloy femoral stems. A failure analysis of two implants. J Bone Joint Surg [Am] 1994; 76:110–115.
22. Collier JP, Suprenant VA, Jensen RE, Mayor MB, Suprenant HP. Corrosion between the components of modular femoral hip prosthesis. J. Bone Joint Surg. 1992; 74B:511–517.
23. Gilbert JL, Buckley CA, Jacobs JJ. In vivo corrosion of modular hip prosthesis components in mixed and similar metal combinations. The effect of crevice, stress, motion, and alloy coupling. J Biomed Mater Res 1993; 27:1533–1544.
24. Gilbert JL, Jacobs J. The mechanical and electrochemical processes associated with taper fretting crevice corrosion: a review. In:. ASTM STP 1301 Modularity of Orthopedic Implants. Philidelphia: ASTM, 1997:45–59.
25. Hallab NJ, Jacobs JJ, Skipor A, Black J, Mikecz K, Galante JO. Systemic metal–protein binding associated with total joint replacement arthroplasty. J Biomed Mater Res 2000; 49:353–361.
26. Hallab NJ, Mikecz K, Akrami J, Jacobs J. Differential metal release and protein binding associated with titanium and cobalt–chromium implant alloys. Trans 46th Orthop Res Soc. Orlando. Fl, 2000.
27. Urban RM, Jacobs J, Gilbert JL, Rice SB, Jasty M, Bragdon CR, Galante GO. Characterization of solid products of corrosion generated by modular-head femoral stems of different designs and materials. In: Marlowe DE , Parr JE editors, Mayor MB, eds. STP 1301 Modularity of Orthopedic Implants. Philadelphia: ASTM, 1997:33–44.
28. Urban RM, Jacobs JJ, Tomlinson MJ, Gavrilovic J, Black J, Peoc'h M. Dissemination of wear particles to the liver, spleen, and abdominal lymph nodes of patients with hip or knee replacement. J Bone Joint Surg [Am] 2000; 82:457–476.

29. Urban RM, Jacobs JJ, Gilbert JL, Galante JO. Migration of corrosion products from modular hip prostheses. Particle microanalysis and histopathological findings. J Bone Joint Surg [Am] 1994; 76:1345–1359.
30. Jasty M, Goetz DD, Bragdon CR, Lee KR, Hanson AE, Elder JR, Harris WH. Wear of polyethylene acetabular components in total hip arthroplasty. An analysis of one hundred and twenty-eight components retrieved at autopsy or revision operations [see comments]. J Bone Joint Surg Am 1997; 79: 349–358.
31. Jacobs JJ, Urban RM, Gilbert JL, Skipor AK, Black J, Jasty M, Galante JO. Local and distant products from modularity. Clin Orthop 1995; 319:94–105.
32. Jacobs JJ, Skipor AK, Urban RM, Black J, Manion LM, Starr A, Talbert LF, Galante JO. Systemic distribution of metal degradation products from titanium alloy total hip replacements: an autopsy study. Trans Orthop Res Soc 1994:838.
33. Stulberg BN, Merritt K, Bauer T. Metallic wear debris in metal-backed patellar failure. J Biomed Mater Res Appl Biomater 1994; 5:9–16.
34. Shanbhag AS, Jacobs JJ, Black J, Galante JO, Glant TT. Human monocyte response to particulate biomaterials generated in vivo and in vitro. J Orthop Res 1995; 13:792–801.
35. Urban RM, Jacobs J, Gilbert JL, Rice SB, Jasty M, Bragdon CR, Galante GO. Characterization of solid products of corrosion generated by modular-head femoral stems of different designs and materials. In: Marlowe DE , Parr JE editors, Mayor MB, eds. STP 1301 Modularity of Orthopedic Implants. Philadelphia: ASTM, 1997:33–44.
36. Urban RM, Jacobs JJ, Tomlinson MJ, Gavrilovic J, Black J, Peoc'h M. Dissemination of wear particles to the liver, spleen, and abdominal lymph nodes of patients with hip or knee replacement. J Bone Joint Surg [Am] 2000; 82:457–476.
37. Sunderman FW, Hopfer SM, Swift T, Rezuke WN, Ziebka L, Highman P, Edwards B, Folcik M, Gossling HR. Cobalt, chromium, and nickel concentrations in body fluids of patients with porous-coated knee or hip prostheses. J Orthop Res 1989; 7:307–315.
38. Hallab NJ, Jacobs JJ, Skipor A, Black J, Mikecz K, Galante JO. Systemic metal–protein binding associated with total joint replacement arthroplasty. J Biomed Mater Res 2000; 49:353–361.
39. Sun ZL, Wataha JC, Hanks CT. Effects of metal ions on osteoblast-like cell metabolism and differentiation. J Biomed Mater Res 1997; 34:29–37.
40. Wataha JC, Hanks CT, Sun Z. Effect of cell line on in vitro metal ion cytotoxicity. Dent Mater 1994; 10:156–161.
41. Wataha JC, Hanks CT, Craig RG. In vitro effects of metal ions on cellular metabolism and the correlation between these effects and the uptake of the ions. J Biomed Mater Res 1994; 28:427–433.
42. Wataha JC, Hanks CT, Craig RG. The effect of cell monolayer density on the cytotoxicity of metal ions which are released from dental alloys. Dent Mater 1993; 9:172–176.
43. Nichols KG, Puleo DA. Effect of metal ions on the formation and function of osteoclastic cells in vitro. J Biomed Mater Res 1997; 35:265–271.
44. Thompson GJ, Puleo DA. Ti-6Al-4V ion solution inhibition of osteogenic cell phenotype as a function of differentiation timecourse in vitro. Biomaterials 1996; 17:1949–1954.
45. Thompson GJ, Puleo DA. Effects of sublethal metal ion concentrations on osteogenic cells derived from bone marrow stromal cells. J Appl Biomater 1995; 6:249–258.
46. Puleo DA, Huh WW. Acute toxicity of metal ions in cultures of osteogenic cells derived from bone marrow stromal cells. J Appl Biomater 1995; 6:109–116.
47. Morais S, Dias N, Sousa JP, Fernandes MH, Carvalho GS. In vitro osteoblastic differentiation of human bone marrow cells in the presence of metal ions. J Biomed Mater Res 1999; 44:176–190.
48. Morais S, Sousa JP, Fernandes MH, Carvalho GS, de Bruijn JD, van Blitterswijk CA. Effects of AISI 316L corrosion products in in vitro bone formation. Biomaterials 1998; 19:999–1007.
49. Morais S, Sousa JP, Fernandes MH, Carvalho GS. In vitro biomineralization by osteoblast-like cells. I. Retardation of tissue mineralization by metal salts. Biomaterials 1998; 19:13–21.
50. Wagner M, Klein CL, van Kooten TG, Kirkpatrick CJ. Mechanisms of cell activation by heavy metal ions. J Biomed Mater Res 1998; 42:443–452.
51. Roebuck KA, Vermes C, Carpenter LR, Fritz EA, Narayanan R, Glant TT. Down-regulation of procollagen alpha1[I] messenger RNA by titanium particles correlates with nuclear factor kappaB

(NF-kappaB) activation and increased rel A and NF-kappaB1 binding to the collagen promoter. J Bone Miner Res 2001; 16:501–510.
52. Vermes C, Chandrasekaran R, Jacobs JJ, Galante JO, Roebuck KA, Glant TT. The effects of particulate wear debris, cytokines, and growth factors on the functions of MG-63 osteoblasts. J Bone Joint Surg Am 2001; 83:201–211.
53. Vermes C, Roebuck KA, Chandrasekaran R, Dobai JG, Jacobs JJ, Glant TT. Particulate wear debris activates protein tyrosine kinases and nuclear factor kappaB, which down-regulates type I collagen synthesis in human osteoblasts. J Bone Miner Res 2000; 15:1756–1765.
54. Vermes C, Chandrasekaran R, Carpenter C, Jacobs JJ, Galante JO, Roebuck KA, Glant TT. Altered osteoblast functions achieved by phagocytosis of particulate wear debris, proinflammatory cytokines and growth factors. J Bone Joint Surg [Am] 2000.
55. Blumenthal NC, Cosma V. Inhibition of apatite formation by titanium and vanadium ions. J Biomed Mater Res 1989; 23:13–22.
56. Blumenthal NC, Cosma V. The effect of aluminum and gallium ions on the mineralization process. Bull Hosp Joint Dis Orthop Inst 1989; 49:192–204.
57. Williams DF. Biological effects of titanium. In: , editor. Williams DF. Systemic Aspects of Biocompatibility. Boca Raton: CRC Press, 1981:169–177.
58. Gitelman H. Aluminum and Health, a Critical Review. New York: Dekker, 1989.
59. Gillespie WJ, Frampton CM, Henderson RJ, Ryan PM. The incidence of cancer following total hip replacement. J Bone Joint Surg [Br] 1988; 70:539–542.
60. Bravo I, Carvalho GS, Barbosa MA, de Sousa M. Differential effects of eight metal ions on lymphocyte differentiation antigens in vitro. J Biomed Mater Res 1990; 24:1059–1068.
61. Sinibaldi K, Rosen H, Liu SK, DeAngelis M. Tumors associated with metallic implants in animals. Clin Orthop 1976:257–266.
62. Luckey TD, Venugopal B. Metal Toxicity in Mammals. NY: Plenum, 1979.
63. Sederholm T, Kouvalainen K, Lamberg BA. Cobalt-induced hypothyroidism and polycythemia in lipoid nephrosis. Acta Med Scand 1968; 184:301–306.
64. Hartwig A. Carcinogenicity of metal compounds: possible role of DNA repair inhibition. Toxicol Lett 1998; 102–103:235–239.
65. Hayes RB. The carcinogenicity of metals in humans. Cancer Causes Control 1997; 8:371–385.
66. Costa M. Toxicity and carcinogenicity of Cr(VI) in animal models and humans. Crit Rev Toxicol 1997; 27:431–442.
67. Beyersmann D. Interactions in metal carcinogenicity. Toxicol Lett 1994; 72:333–338.
68. Sunderman FW. Carcinogenicity of metal alloys in orthopedic prosthesis: clinical and experimental studies. Fundam Appl Toxicol 1989; 13:205–216.
69. Jandhyala BS, Hom GJ. Minireview. Physiological and pharmacological properties of vanadium. Life Sci 1983; 33:1325–1340.
70. Naylor GJ, Smith AH, Bryce-Smith D, Ward NI. Tissue vanadium levels in manic-depressive psychosis. Psychol Med 1984; 14:767–772.
71. Garabrant DH, Fine LJ, Oliver C, Bernstein L, Peters JM. Abnormalities of pulmonary function and pleural disease among titanium metal production workers. Scand J Work Environ Health 1987; 13:47–51.
72. Lee KP, Henry NW, III, Trochimowicz HJ, Reinhardt CF. Pulmonary response to impaired lung clearance in rats following excessive TiO2 dust deposition. Environ Res 1986; 41:144–167.
73. Altmann P, Plowman D, Marsh F, Cunningham J. Aluminium chelation therapy in dialysis patients: evidence for inhibition of haemoglobin synthesis by low levels of aluminium. Lancet 1988; 1: 1012–1015.
74. Coburn JW, Norris KC, Nebeker HG. Osteomalacia and bone disease arising from aluminum. Semin Nephrol 1986; 6:68–89.
75. Good PF, Perl DP. Aluminium in Alzheimer's?. Nature 1993; 362:418.
76. Basketter DA, Briatico-Vangosa G, Kaestner W, Lally C, Bontinck WJ. Nickel, cobalt and chromium in consumer products: a role in allergic contact dermatitis?. Contact Dermatitis 1993; 28:15–25.
77. Cramers M, Lucht U. Metal sensitivity in patients treated for tibial fractures with plates of stainless steel. Acta Orthop Scand 1977; 48:245–249.

78. Merritt K, Rodrigo JJ. Immune response to synthetic materials. Sensitization of patients receiving orthopaedic implants. Clin Orthop Rel Res 1996; 326:71–79.
79. Gawkrodger DJ. Nickel sensitivity and the implantation of orthopaedic prostheses. Contact Dermatitis 1993; 28:257–259.
80. Kanerva L, Sipilainen-Malm T, Estlander T, Zitting A, Jolanki R, Tarvainen K. Nickel release from metals, and a case of allergic contact dermatitis from stainless steel. Contact Dermatitis 1994; 31: 299–303.
81. Haudrechy P, Foussereau J, Mantout B, Baroux B. Nickel release from nickel-plated metals and stainless steels. Contact Dermatitis 1994; 31:249–255.
82. Black J. Systemic effects of biomaterials. Biomaterials 1984; 5:12–17.
83. Jacobs JJ, Gilbert JL, Urban RM. Corrosion of metallic implants. In: Stauffer RN, ed. Advances in Orthopaedic Surgery, Vol. 2. St. Louis: Mosby, 1994:279–319.
84. Yang J, Black J. Competitive binding of chromium cobalt and nickel to serum proteins. Biomaterials 1994; 15:262–268.
85. Yang J, Merritt K. Production of monoclonal antibodies to study corrosion of Co–Cr biomaterials. J Biomed Mater Res 1996; 31:71–80.
86. Liden C, Wahlberg JE, Maibach HI. Skin. New York: Academic Press, 1995.
87. Angle C. Organ-Specific Therapeutic Intervention. New York: Academic Press, 1995.
88. Lalor PA, Revell PA, Gray AB, Wright S, Railton GT, Freeman MA. Sensitivity to titanium. A cause of implant failure. J Bone Joint Surg 1991; 73-B:25–28.
89. Parker AW, Drez-Jr D, Jacobs JJ. Titanium dermatitis after failure of a metal-backed patellas. Am J Knee Surg 1993; 6:129–131.
90. Hensten-Pettersen A. Allergy and hypersensitivity. In: Morrey BF, ed. Biological, Material, and Mechanical Considerations of Joint Replacements. New York: Raven Press, 1993:353–360.
91. Kuby J. Immunology. New York: W.H. Freeman and Company, 1991.
92. Merle C, Vigan M, Devred D, Girardin P, Adessi B, Laurent R. Generalized eczema from vitallium osteosynthesis material. Contact Dermatitis 1992; 27:257–258.
93. King J, Fransway A, Adkins RB. Chronic urticaria due to surgical clips. New Engl J Med 1993; 329:1583–1584.
94. Barranco VP, Solloman H. Eczematous dermatitis from nickel. JAMA 1972; 220:1244.
95. Thomas RH, Rademaker M, Goddard NJ, Munro DD. Severe eczema of the hands due to an orthopaedic plate made of Vitallium. Br Med J 1987; 294:106–107.
96. Abdallah HI, Balsara RK, O'Riordan AC. Pacemaker contact sensitivity: Clinical recognition and management. Ann Thoracic Surg 1994; 57:1017–1018.
97. Halpin DS. An unusual reaction in muscle in association with a vitallium plate: a report of possible metal hypersensitivity. J Bone Joint Surg 1975; 57-B:451–453.
98. Merritt K, Brown SA. Biological effects of corrosion products from metal. Philadelphia: American Society for Testing and Materials, 1985.
99. Poss R, Thornhill TS, Ewald FC, Thomas WH, Batte NJ, Sledge CB. Factors influencing the incidence and outcome of infection following total joint arthoplasty. Clin Orthop 1984:182.
100. Black J. Orthopaedic Biomaterials in Research and Practice. New York: Churchill Livingstone, 1988.
101. Elves MW, Wilson JN, Scales JT, Kemp HB. Incidence of metal sensitivity in patients with total joint replacements. Br Med J 1975; 4:376–378.
102. Gordon PM, White MI, Scotland TR. Generalized sensitivity from an implanted orthopaedic antibiotic minichain containing nickel. Contact Dermatitis 1994; 30:181–182.
103. Peters MS, Schroeter AL, Hale HM, Broadbent JC. Pacemaker contact sensitivity. Contact Dermatitis 1984; 11:214–218.
104. Rostoker G, Robin J, Binet O, Blamutier J, Paupe J, Lessana-Liebowitch M, Bedouelle J, Sonneck JM, Garrel JB, Millet P. Dermatitis due to orthopaedic implants. A review of the literature and report of three cases. J Bone Joint Surg 1987; 69-A:1408–1412.
105. Benson MK, Goodwin PG, Brostoff J. Metal sensitivity in patients with joint replacement arthroplasties. Br Med J 1975; 4:374–375.

106. Brown GC, Lockshin MD, Salvati EA, Bullough PG. Sensitivity to metal as a possible cause of sterile loosening after cobalt–chromium total hip-replacement arthroplasty. J Bone Joint Surg Am 1977; 59-A:164–168.
107. Deutman R, Mulder TH, Brian R, Nater JP. Metal sensitivity before and after total hip arthroplasty. J Bone Joint Surg [Am] 1977; 59-A:862–865.
108. Kubba R, Taylor JS, Marks KE. Cutaneous complications of orthopedic implants. A two-year prospective study. Arch Dermatol 1981; 117:554–560.
109. Mayor MB, Merritt K, Brown SA. Metal allergy and the surgical patient. Am J Surg 1980; 139: 477–479.
110. Merritt K, Brown S. Tissue reaction and metal sensitivity. Acta Orthop Scand 1980; 51:4103–4111.
111. Merritt K, Brown S. Metal sensitivity reactions to orthopedic implants. Int J Dermatol 1981; 20: 89–94.
112. Merritt K. Role of medical materials, both in implant and surface applications, in immune response and in resistance to infection. Biomaterials 1984; 5:53–57.
113. Pinkston JA, Finch SC. A method for the differentiation of T and B lymphocytes and monocytes migrating under agarose. Stain Technol 1979; 54:233–239.
114. Rooker GD, Wilkinson JD. Metal sensitivity in patients undergoing hip replacement. A prospective study. J Bone Joint Surg 1980; 62-B:502–505.
115. Carlsson AS, Macnusson B, Moller H. Metal sensitivity in patients with metal-to-plastic total hip arthroplasties. Acta Orthop Scand 1980; 51:57–62.
116. Fischer T, Rystedt I, Safwenberg J, Egle I. HLA-A, -B, -C and -DR antigens in individuals with sensitivity to cobalt. Acta Derm Venereol (Stockh) 1984; 64:121–124.
117. Munro-Ashman D, Miller AJ. Rejection of metal to metal prosthesis and skin sensitivity to cobalt. Contact Dermatiti 1976; 2:65.
118. Evans EM, Freeman MA, Miller AJ, Vernon-Roberts B. Metal sensitivity as a cause of bone necrosis and loosening of the prosthesis in total joint replacement. J Bone Joint Surg 1974; 56-B:626–642.
119. Boyan BD. Discussion of Toxicity and Allergy. New York: Raven Press, 1993.
120. Black J. Prosthetic Materials. New York: VCH Publishers, Inc., 1996.
121. Michel R, Nolte M, Reich M, Loer F. Systemic effects of implanted prostheses made of cobalt–chromium alloys. Arch Orthop Trauma Surg 1991; 110:61–74.
122. Jacobs JJ, Silverton C, Hallab NJ, Skipor AK, Patterson L, Black J, Galante JO. Metal release and excretion from cementless titanium alloy total knee replacements. Clin Orthop 1999; 358:173–180.
123. Jacobs JJ, Skipor AK, Patterson LM, Hallab NJ, Paprosky WG, Black J, Galante JO. Metal release in patients who have had a primary total hip arthroplasty. A prospective, controlled, longitudinal study. J Bone Joint Surg [Am] 1998; 80:1447–1458.

4

Technologies for the Surface Modification of Biomaterials

Aron B. Anderson, Anthony W. Dallmier, Stephen J. Chudzik, Lise W. Duran, Patrick E. Guire, Robert W. Hergenrother, Muhammad A. Lodhi, Amy E. Novak, Ronald F. Ofstead, and Klaus Wormuth
SurModics, Inc.
Eden Prairie, Minnesota, U.S.A.

I. INTRODUCTION

Technological advances in biomaterials have enhanced both medical and pharmaceutical applications. Many polymeric materials have been designed with the required mechanical properties necessary for manufacturing medical devices. However, these materials are often inadequate because of local and systemic reactions that occur at the harsh interface between biological systems and the biomaterial surface. Consequently, the biotechnology revolution of the past decade has spawned a vigorous interest in the emerging field of surface modification science and interfacial dynamics.

Surface modification is the process of changing the existing characteristics of a material surface to a more desirable characteristic. This could consist of changing a material from being hydrophobic to hydrophilic or to creating a more biocompatible product on a medical device. The concept of altering surface characteristics is not unique. Many methods have been examined for coating biomaterials with a variety of synthetic polymers or biologically active compounds. This can be done either by passivation to prevent biofouling or by activation to incorporate a specific functionality into the interfacial environment. Using this framework, our laboratory has developed many ways to modify the surfaces of various types of materials. This chapter describes the various methods and technologies used in surface modification at SurModics.

II. SURFACE MODIFICATION MATERIALS AND TECHNOLOGIES

A. Photochemical Coupling Chemistry for Biomaterials

Biomaterials, in general, are relatively inert to environmentally friendly thermochemical reactions since they are rich in hydrocarbon groups, namely, carbon–hydrogen bonds. High-energy, short-lived reactive species, such as free radicals and diradicals, are advantageous for bond formation with the surfaces of these materials. Desirable characteristics of these highly reactive species make the technology (1) capable of stable bond formation with hydrocarbon groups;

(2) rapid, versatile, and compatible with device manufacturing processes; (3) environmentally friendly since it is reactive in aqueous solution and ambient conditions; and (4) subject to reversible activation in useful reaction conditions (e.g., presence of water) for enhanced coupling yield.

Photochemistry and plasma chemistry are two useful processes for forming highly reactive chemical species that form covalent bonds with hydrocarbon groups. Plasma chemistry uses high radiofrequency electromagnetic energy to form reactive radicals from oxygen found in air and water, ammonia, low molecular weight hydrocarbons, and other gaseous species in a vacuum reaction chamber.

Photochemical covalent coupling of the coating molecules is a preferred approach to surface modification for two reasons. It can be used on preformed polymeric materials of varying geometries, and the high-energy intermediates involved in such processes are capable of reacting with a variety of polymer substrates. The diradical aromatic ketone groups used by SurModics are primarily from benzophenone, thioxanthone, and anthraquinone derivatives (1 in Fig. 1). Blue or ultraviolet illumination of these photogroups results in the initial formation of a singlet excited state (2 in Fig. 1). This short-lived species undergoes a rapid intersystem crossing to generate the longer-lived triplet state (3 in Fig. 1) [1]. This highly reactive intermediate is then capable of insertion into carbon–hydrogen bonds by abstraction of a hydrogen atom from the polymer surface, followed by collapse of the resulting radical pair (4 in Fig. 1) to form a new carbon–carbon bond (5 in Fig. 1). The high energy of the triplet state makes the photochemical coupling process relatively independent of the chemical composition of the surface, with the efficiency of the process being determined by the relative stability of the free radicals formed on the surface of the polymer. The photoactivated aryl ketones are not reactive with water molecules and can revert back to the ground state if no suitable substrate is available, thus substantially improving the efficiency of photogroup utilization [2].

Photochemical diradical generation from aromatic carbonyl groups provides numerous advantages for biomaterial surface modification, such as

1. The reactive species (the triplet carbonyl) can be reversibly generated by exposure

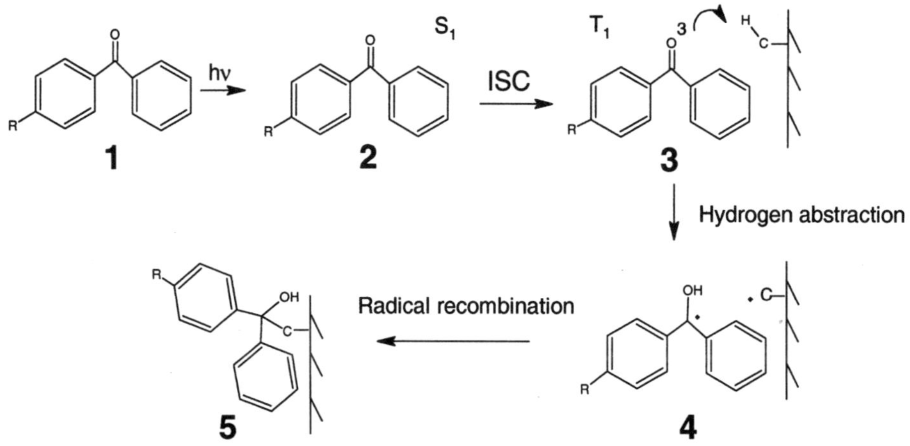

Figure 1 Aryl carbonyl photocoupling mechanism.

to visible or long-wavelength ultraviolet light from commonly available, relatively inexpensive light sources.

2. Good coupling yields may be procured by associating the photoactivatable coating derivatives with the target surface in water or volatile alcohol solvent before activating in the wet or dry state in ambient atmosphere.
3. Stable carbon–carbon bonds are formed between the activated carbonyl group and the hydrocarbon groups on the biomaterial surface, providing hydrolytic stability even in vivo.
4. The photoactivatable aromatic carbonyl group may be chemically incorporated into essentially all desired coating reagents, whether synthetic or biomolecular, independent of molecular weight, enabling the biomaterial device manufacturer to purchase the coating reagents and incorporate readily the coating step in its device manufacturing line.
5. The diradical reactive species enables covalent coupling, or crosslinking, within and between polymeric coating molecules on the surface, while being covalently coupled to the surface.

1. Classes of Photoreactive Coating Reagents

In order to incorporate photoreactive functionalities into surface-modifying molecules, they must possess both a thermochemically reactive group (X in Fig. 2) and the photoreactive component, which is the carbonyl group of the aryl ketones. Preferably, this attachment site is in the para position relative to the photoreactive functionality to minimize the possibility of intramolecular insertion reactions. For example, 4-benzoyl-benzoic acid (X $=$ CO_2H), anthraquinone, and thioxanthone derivatives are representative examples of the benzophenone, quinone, and xanthone classes of aryl ketones (Fig. 2). The carboxylic acid of the former reagent is suitable for a variety of coupling techniques, thereby coupling to the surface-modifying molecule or permitting introduction of a spacer containing another coupling group. In summary, numerous functional groups can be used for tethering the photoreactive group to the coating molecule, generating bonds such as esters, amides, ethers, carbamates, and ureas, the selection of which is made according to hydrolytic stability requirements in the desired application.

The photoreactive coating reagents can be divided into two major classes: heterobifunctional reagents and multifunctional reagents (Fig. 2). The heterobifunctional reagents are most often used to couple specific biomolecules to the surface, while the multifunctional reagents are used to change the surface properties, i.e., to achieve wettability, lubricity, or passivation. Surfaces can also be designed that both repel nonspecific biomolecule adsorption and covalently couple specific active biomolecules.

Heterobifunctional reagents consist of three important components: (1) the photoreactive group (P in Fig. 2); (2) a spacer group; and (3) a thermochemically reactive group (X in Fig. 2). The photogroup is typically an aryl ketone having a second group in the para position to permit attachment of the spacer group. The spacer group can be of varying length or composition, depending on the requirements of the application. In general, the spacer is used to hold the functional molecule away from the hydrophobic environment at the substrate surface. For example, a hydrophilic species such as polyethylene glycol (PEG) can be used to tether a biomolecule via flexible chains of bound soluble polymer. This method of attachment provides more protein mobility and hence greater opportunities for favorable interaction of the immobilized protein with its environment. While the spacer group can be terminated in a relatively nonreactive group, such as a hydroxy or methoxy group, to provide for surface passivation, the more common selection is a thermochemically reactive species that is not photochemically reactive. Selections

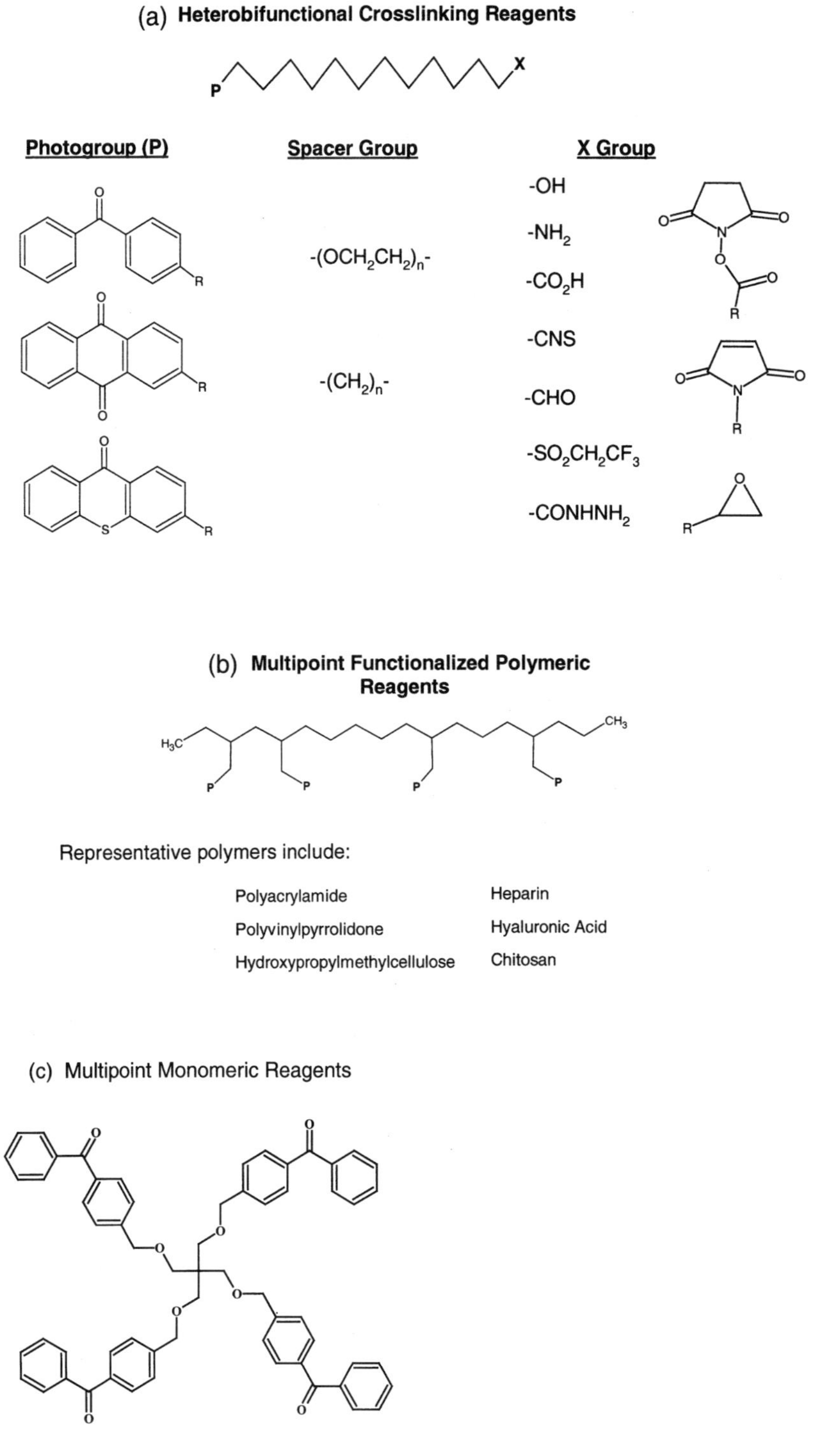

Figure 2 Classes of photoactive reagents: (a) heterobifunctional crosslinking reagents; (b) multipoint functionalized polymeric reagents; and (c) multipoint monomeric reagents.

for this group include species such as N-oxysuccinimide (NOS) esters, tresylate esters, isothiocyanates, aldehydes, and epoxides for reaction with amines; maleimides for reaction with sulfhydryls and amines; hydrazides for reaction with aldehydes and ketones; and amine groups for coupling to carboxylic acids. The three components of these heterobifunctional reagents can be substituted in unlimited combinations, providing a means for engineering reagents with precise characteristics and specifications.

Multipoint functionalized reagents are synthesized by reaction of these heterobifunctional reagents with polymerizable monomers, preformed polymeric, or multifunctional monomeric molecules. In this process, the heterobifunctional reagents are reacted with polymers possessing multiple reactive sites along their backbone. Thus, the resulting photoreagent possesses the physical and chemical properties of the polymer and with the photogroup attached to the backbone of the polymer via the spacer group of the heterobifunctional molecule. These polymeric molecules can be either synthetic or natural. For example, proteins frequently possess numerous lysine residues, which can be derivatized with the heterobifunctional photoreagents containing amine-reactive end groups such as NOS esters. The resulting protein can be analyzed to determine the photogroup content and the level of biological activity. Likewise, heparin and hyaluronic acid have been photoderivatized in several different ways to provide reagents capable of imparting anticoagulant and other activities to a polymeric surface.

In a similar manner, synthetic polymers may be prepared which possess multiple photoreactive groups along the backbone. In these syntheses, a base monomer is copolymerized with a controlled quantity of a second monomer possessing a reactive group activatable either thermally or photochemically, followed by dialysis to remove residual monomer and low molecular weight oligomers. The thermally reactive polymer so produced is then reacted with the selected heterobifunctional reagent to introduce the photogroups. This copolymerization approach to photopolymer preparation provides a multitude of opportunities to incorporate a variety of functionalities into the polymer, including cationic and anionic groups.

2. Photocoupling for Uniform Ultrathin Coatings

Another class of coatings has been developed specifically for ultrathin applications. These coatings are based on photoreactive diblock copolymers. Block copolymers are known for their ability to self-assemble into micelles in solution and into monolayers on surfaces. This is due to one block having a higher affinity for the surface and another having a higher affinity for the solvent, usually water. The structures they form are transient, however, changing with solvent composition or drying. By combining the ability of the block copolymers to spontaneously form a coating from a polar solvent on a hydrophobic device, with the ability of photochemistry to fix the structure in place, an ultrathin coating is obtained with minimum effort. This can be done by simply immersing a device in an aqueous solution of photoreactive diblock copolymer, allowing time for the monolayer to spontaneously form, then illuminating with activating light to bond it in place. Diblock copolymers are modified to contain the photoreactive group in only one of the blocks, typically the more hydrophobic block, as illustrated in Fig. 3. This leaves the other block free to interact with the aqueous solution, defining the surface properties. If this hydrophilic block is chosen to be poly(ethylene glycol), for instance, the surface will have the biocompatibility and wettability properties of poly(ethylene glycol). Each block may be tailored to fit a given application, allowing wide versatility and ease of processing for delivering ultrathin coatings for different devices.

3. Restrained Multifunctional Reagent for Surface Modification

Heterobifunctional reagents have also been used to produce restrained, multifunctional photoreagents. These reagents are useful for preparation of photoreactive surfaces for subsequent photo-

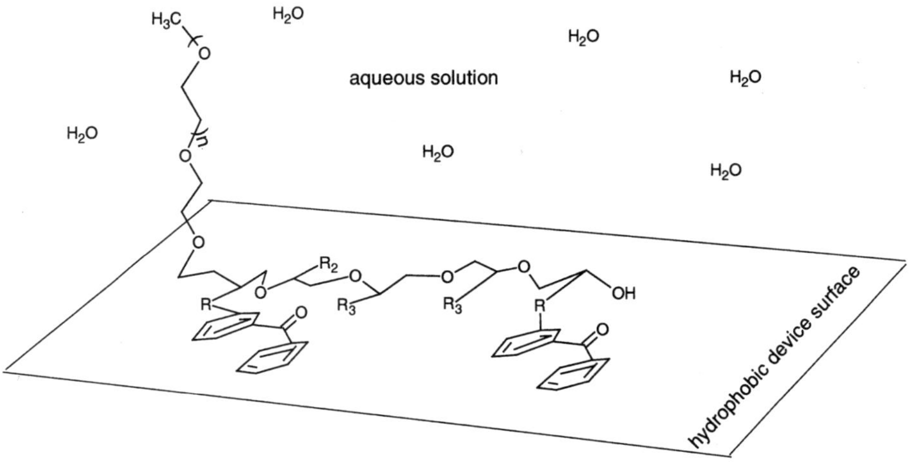

Figure 3 Photoreactive diblock polyether self-assembly on a hydrophobic surface immersed in water.

immobilization of added coating molecules, for example, self-assembling monolayer polymers. They are also used for simultaneous application with a coating molecule to a support surface, thereby providing both immobilization to the surface and crosslinking within the coating polymer layer [3]. A reactive reagent molecule containing multiple latent-reactive groups is ''restrained'' in that it is conformationally and/or chemically restricted from reacting with either itself or with other molecules of the same reagent. Upon activation, this feature causes the attachment of less than all of the reactive sites of the multifunctional reagent to a surface, thereby leaving the remaining sites free to react with molecules desired to be immobilized onto the surface or to be subjected to further crosslinking (Fig. 2c). This type of multifunctional photoreagent is useful for immobilizing molecules not readily converted to photoreactive derivatives and for stabilizing polymeric coatings against mechanical disruption.

B. Photochemical Graft Polymerization

Graft copolymers are generally defined as branched copolymers with a backbone of one or more monomers to which side chains of the same or different monomers are attached. They are generally prepared in order to impart dissimilar physical properties to an existing polymeric material. In the case of modification of biomaterials, changes in hydrophilicity, frictional properties, and blood or tissue compatibility illustrate property changes which may be sought by this approach.

Polymerization to form grafted chains may be initiated in a variety of ways. Chain transfer during the formation of the original backbone polymer may generate polymeric radical sites which lead to polymerization of new grafted chains. Ionizing radiation, such as gamma or electron beam exposure, may lead to polymeric radicals by numerous reaction pathways, and simultaneous or subsequent contact with suitable monomers leads to grafted chain formation. Redox techniques have been widely studied which usually require polymer backbone structures with readily oxidized functional groups. Alcohol groups on carbohydrate polymers can thus be converted to polymeric radical sites useful for the growth of grafted chains.

Graft polymerization can also be achieved by radiation involving ultraviolet light, often in the presence of a photochemical agent such as benzoin molecules. These photochemical agents can undergo processes such as fragmentation and hydrogen atom abstraction, resulting in polymeric radicals which may lead to grafted polymer chain formation. A highly versatile grafting technology has been developed which potentially allows graft copolymerization to be used for permanent surface modification of any polymeric surface bearing hydrogen atoms. This grafting technology may also be used with a wide range of finished articles for imparting changes in desired surface properties. This approach involves the use of a family of multifunctional compounds which contain two or more photosensitive groups (e.g., benzophenones) which are provided as anionic, cationic, or neutral photochemical reagents. The application of the multifunctional photoreagents to a polymer surface bearing abstractable hydrogen atoms is followed by illumination with ultraviolet light. This step brings about photoattachment of the multifunctional reagent to the polymer surface by the process of excitation, hydrogen atom abstraction, and collapse of the resulting radical pair to create a carbon–carbon covalent bond linking the reagent to the surface. Steric effects greatly reduce the probability that the additional photogroups on the reagent will become bonded to the polymer surface. Thus, the remaining photogroups are available for a second photochemical step which involves ultraviolet light illumination of the modified surface in the presence of the monomer(s) of choice, resulting in grafted polymer chain formation.

While a wide range of monomers may be used, this process typically employs monomers leading to hydrophilic grafted chains (e.g., acrylamide, AMPS) in order to impart properties such as wettability or lubricity to the surface of the article. The particular properties of the substrate polymer backbone or the intended use of the modified surface may influence the choice of anionic, cationic, or neutral multifunctional photoreagents. The versatility of the grafting process provides significant advantages relative to earlier methods. Suitable ultraviolet light sources are readily available, and brief ultraviolet light exposure is generally not detrimental to the stability or properties of polymeric materials. This is in contrast to the high-energy radiation methods, such as gamma radiation, where both equipment accessibility and material degradation may be significant issues. Chain transfer methods of graft polymer chain formation are generally not applicable to the modification of surfaces of polymeric articles, such as medical devices. Redox methods are only applicable to a relatively small number of polymeric backbone materials, most of which are not used for finished article fabrication. The use of the two-step photochemical grafting process also provides a high degree of assurance that the intended graft polymer chains are effectively covalently attached to the polymer surface, and will provide a more permanent modification of the surface of the biomaterial.

C. Tie Layers for Metal Surface Pretreatments

Many medical devices are prepared partially or wholly from metallic materials whose surface characteristics may not be appropriate or optimal for the intended use. Examples of such devices are guidewires, stents, pacemaker components, vena cava filters, and distal protection devices. These devices contain metallic components based on materials such as stainless steel, platinum, nitinol, titanium, or aluminum. Examples of needed or desired surface characteristics include wettability, lubricity, improved tissue or blood compatibility, or good adhesion of subsequent materials coatings on the metallic surfaces. It is often difficult or impossible to achieve such objectives by direct application of known coating materials to metallic surfaces due to the dissimilarity of most coating materials relative to the metallic material surfaces.

A great deal of technology has been developed for the purpose of providing uniform and durable surface coatings on metallic materials and is widely used in numerous industrial applications. Synthetic polymers of several types have been developed which incorporate carboxylic acid functionality in order to achieve interaction and bonding to metallic substrates. Notable examples include the ethylene/acrylic acid family of materials and the carboxyl-functional polyurethane (PU) materials. Alternatively, a very wide range of silane derivatives, typically trialkoxysilane compounds, have been developed by several manufacturers to undergo interaction and bonding to metallic surfaces, usually involving oxide or hydroxy functionality on the metal surface. Further developments have resulted in the hydrosiloxane materials, which are generally copolymer structures involving hydrosiloxane and dialkylsiloxane units. These materials are useful for treatment of clean, oxidizable, metal surfaces and allow attachment of silicone polymer segments to metallic surfaces.

In all of these approaches, the metallic surfaces have been modified in ways that provide abstractable hydrogen atoms attached to the organic components of these surface pretreatments and are therefore suitable surfaces for all of the photochemical surface treatments described earlier. SurModics has extensively examined the available technologies and materials for pretreatment of metallic substrates and has commercialized those systems found most suitable for the surface modification of medical devices, including carboxyl-functional urethanes, trialkoxysilanes, and hydrosiloxanes. In addition, SurModics provides the complete technology for subsequent modification of those treated surfaces to impart a wide range of surface properties.

D. Polymer Blends for Drug-Incorporation Coatings on Devices

Polymer blends have been widely examined in the fields of polymer science and engineering and have enjoyed considerable success in industrial commercial applications, particularly in the area of molded plastic materials. Blends are typically used in order to achieve combinations of properties that are unavailable in single materials or would require specific synthetic efforts to achieve. The majority of polymer blends are described as immiscible having distinct detectable separate phases made up of the component polymers. Examples of miscible polymer blends are also known, and blends have been developed in which a third component is included to stabilize the morphology of a polymer blend or serve as a "compatibilizer" material. Immiscible polymer blends often have one or more property deficiencies, such as a lack of optical clarity, reduced strength, or process-dependent variability of properties. Due to these property deficiencies, a great deal of attention needs to be given to selection and use of polymer blend components.

SurModics has extensively studied the use of polymer blends in the preparation of coatings for drug incorporation and to provide desired drug-release characteristics from the surfaces of coated medical devices. The polymer blend approach was selected because of the potential of being able to vary polymer component identity, ratio, and coating process parameters to give a wide range of mechanical properties, including drug diffusion and release characteristics in the resulting coatings. This approach has proven to be successful in controlling a wide variety of coating properties, while requiring the synthesis of only a small number of polymer component materials. For example, variation in identity and ratio of polymer components and coating process changes involving solvent choice, environmental conditions, and equipment parameters allow regulation of many characteristics of the resulting coatings. Coating modulus and durability may be enhanced by the selection of higher modulus blend components and by adjustment of the ratio of the polymer blend components. Partially phase-separated polymer blend morphologies may be influenced by careful control of coating process parameters, such as relative humidity, leading to control of both burst and controlled release aspects of the overall drug delivery properties of coatings. The polymer blend approach to providing very specific drug release characteristics has proven to be particularly useful in drug release from implanted medical devices with applicability to a wide range of drug materials and release profiles.

E. Processes for Surface Modification

The methods for reagent application can be classified as one-step or two-step, depending upon the number of processing steps required in the modification of the device. The one-step method involves prederivatization of the coating molecule of interest with the photoreactive moiety. The purified photoreagent is then brought into solution and applied to the device followed by illumination with suitable wavelength light to activate the photocoupling process. Alternatively, the two-step method can be employed in which the device itself can be prederivatized rather than the molecule of interest. For this method, the substrate is treated with the photoreactive heterobifunctional reagent (P^^^^^^X in Fig. 2a) in solution, illuminated, and finally the molecule of interest is coupled to the reactive surface using conventional coupling techniques well recognized in organic and protein chemistry.

The one-step approach is generally the preferred method for integration into device manufacturing since it minimizes the number of processing steps. In addition, the method is more efficient than the two-step approach, with minimal excess reagent being wasted. The primary limitation of this approach is the requirement that the coating reagent must not be chemically degraded by the illumination source. For example, some biologically active compounds can show significant inactivation under UV illumination.

While photochemical surface modification is a generic process across a broad range of coating molecules and polymer surfaces, each specific application typically requires some optimization. It is critical that the surface to be modified must be free of contaminants, such as plasticizers or low molecular weight oligomers, to insure coupling of the photoreagent to the base polymer. Such cleaning processes can consist of simple washing procedures, solvent extraction, or plasma cleaning of the surface.

The photoreagent is then applied in a solvent compatible with both polymer and photoreagent, typically an aqueous or alcohol-based system. The solution application can be accomplished in a number of ways. Dip coating works well for many devices, especially catheters and guidewires. Spraying, brushing, or ink jetting the coating solution onto the device can be used for parts with intricate geometries where dip coating may not reach portions of the device.

Illumination of the device can occur in either the wet or dry state and a variety of illumination systems, both UV and visible, can be used to accommodate differing geometries of the devices. For parts that do not allow a film to form easily, such as very hydrophobic substrates, the part can be illuminated in solution, which will result in the binding of a thin layer of the photoreagents to the substrate. The duration of the illumination is determined by the incident beam intensity (typically 1–3 min at $1 mW/cm^2$), the nature of the reagent being applied, and the type of polymer substrate. Finally, the device may be rinsed to remove any loosely adherent material, leaving only the tenaciously bound layer.

The two-step approach is used for those situations in which the user either wishes to prepare a more generic surface to which can be added any one of a number of molecules or has a molecule to be immobilized which is subject to UV degradation. This process can waste excess molecules that have not thermochemically reacted with the derivatized surface, and the user runs the risk of requiring an extra reaction to block exposed surface-reactive groups that were unable to couple the molecule, either through suboptimal reaction conditions or through steric hindrance.

Webbing and pooling of the coating solution can be common problems when using conventional spraying and dipping techniques, especially for complex geometries. This can lead to durability issues, occluded holes, uneven coating, and can limit the effectiveness of the medical device. Another two-step approach is to polymerize monomers directly onto the surface. The coating process consists of priming the surface with a photoreactive crosslinker followed by graft polymerization with the appropriate monomers. The outcome is a grasslike polymer coating that will not occlude holes and will evenly coat irregular surfaces.

III. HYDROPHILIC SURFACES

Three commonly required surface characteristics of biomaterials include wettability, lubricity, and passivation. Each of these attributes can be achieved by immobilizing hydrophilic photopolymers to the device surface. Although wettability, lubricity, and passivation are each imparted to the surface using similar hydrophilic reagents, these attributes are not necessarily related, nor are they mutually exclusive. In fact, each of these surface characteristics can be controlled by varying the polymer molecular weight, degree of crosslinking, and/or the thickness of the hydrogel matrix. For example, lubricity is best achieved by immobilizing high molecular weight, loosely crosslinked photopolymers; whereas lower molecular weight, more tightly crosslinked photopolymers are more effective for reducing protein fouling. When designing a particular medical device, it is often necessary to clearly define the surface qualities that would provide optimal benefit to the user. This section describes the importance of these characteristics, how each is measured, and the effect of a hydrophilic surface on device efficacy.

A. Wettability

Highly wettable surfaces are required for many biomaterials. Hydrophobic, non-wettable surfaces are undesirable due to adsorption and denaturation of proteins, adherence and damage of cells, and adherence of bacteria. Masking hydrophobic surfaces with hydrophilic polymers reduces these harmful effects. In addition, a hydrophilic material will hold an evenly distributed fluid film over the surface, enabling the device to deliver drugs, such as antibiotics, to the site of implantation. This is likely to result in a reduction of device-centered infections, and may reduce the need for systemic antibiotics. In addition to stabilizing a fluid film, a hydrophilic surface will also discourage adhesion of gas bubbles which is a common problem in purging tubing sets used in surgery, and in blood filters, oxygenators, and biosensors.

Sessile drop contact angle analysis is one commonly used method of determining surface wettability. When performed properly, the angle of contact between a small droplet (2.5 μl) of filtered deionized water and the substrate surface is a reliable measure of wettability. Lower contact angles result from the spreading of water across a surface, whereas higher contact angles are characteristic of water beading up on the hydrophobic surface. This method was used to compare unmodified and photopolymer-modified biomaterials (Figure 4). All of the biomaterials tested are frequently used in medical devices because of their desirable bulk properties. Without any surface modification, these materials demonstrate a wide range of relatively high contact angles; such hydrophobicity can compromise device performance and efficacy. Modification of the substrates with hydrophilic photopolymers results in significant reduction in contact angle (increase in wettability) for all materials tested, and all surface-modified materials approach a similar contact angle value. This is not surprising, since any surface modified with the same hydrogel should have essentially the same wetting characteristics, regardless of the substrate polymer.

B. Lubricity

Lubricity is a desirable attribute for medical devices that require movement against sensitive tissues, such as with a urinary tract catheter. Without this characteristic, blood vessel walls

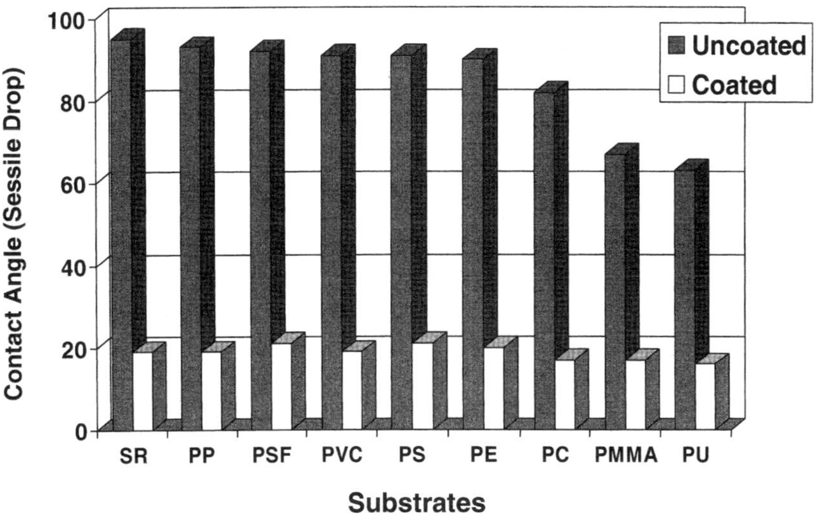

Figure 4 Improved wettability by surface modification with photoactivatable hydrophilic polymers. The materials were modified with a mixture of photopolymers containing polyacrylamide, polyvinylpyrrolidone and polyethylene glycol. The materials are silicone rubber (SR), polypropylene (PP), polysulfone (PSF), polyvinylchloride (PVC), polystyrene (PS), polyethylene (PE), polycarbonate (PC), polymethylmethacrylate (PMMA), and polyurethane (PU).

or mucous membranes can be damaged, resulting in pain and susceptibility to infections or thrombogenesis [4]. Hydrophilic, lubricious coatings reduce friction and significantly reduce tissue damage relative to hydrophobic materials. In addition to the damage caused by hydrophobic surfaces, the device itself may not function properly if it generates too much friction in use.

The wide variety of biomaterials used in medical devices has an equally wide distribution of surface frictional properties. These properties range from pliable, tacky, low-durometer silicones to the smooth hard surfaces of ultra high molecular weight polyethylene (UHMWPE) and Teflon®. In many cases, such as with intravascular catheters, the flexible materials are critical to device function since the material must yield to the shape of the vasculature as it is being fed into the body. Lubricious coatings can dramatically improve ease of use since they can slide more readily into place. The ideal biomaterial for these types of applications should be both flexible and slippery.

Lubricity is measured by determining the force needed to push or pull an object through a pathway that has some resistance. The coefficient of friction (ratio of pull or push forces to the normal, or load force) is one way of quantifying the lubricity values. Often the pull or push forces are stated by themselves along with a description of the test method. The lubricity of a coating is strongly dependent on the material against which the part is being tested. The biomaterial is typically hydrated prior to testing in order to provide the fluid film necessary to cushion the tissue–device interaction. There are several common ways to measure lubricity. A sled with a known weight applied to it can be pulled across the biomaterial to measure friction. Alternatively, the sample can be squeezed between two vertical pads and drawn through them. In other cases, a pathway can be created that simulates the appropriate anatomy (e.g., femoral artery, coronary arteries), and the force needed to move the device through the path can be measured. The testing can be cycled multiple times to determine if the frictional forces change over time.

Durability of a particular coating can be as important as the initial lubricity, especially for coatings applied to devices that are used for extended periods of time or are subject to extensive manipulation. Figure 5 shows the force needed to pull stainless steel wires through two submerged silicone pads that have been compressed with 500 g of force. The wire that was coated with Teflon, a lubricious yet very hydrophobic material, required about half the force of an uncoated wire to pull it through the fixture. In comparison, the wire that had a lubricious, hydrophilic, photoimmobilized polymer coating required less than a tenth of the force of an uncoated wire. The hydrophilic coating was durable for many cycles of the test. A wide range of biomaterials have been modified with photoimmobilized hydrophilic polymers and tested against various surfaces.

Lubricious surfaces on medical devices reduce the force required to manipulate medical devices during a surgical procedure. This in turn has the benefit of reducing the time required for any given procedure [5]. In addition, lubricious catheters can negotiate tortuosities in the vasculature that are not possible with unmodified devices, thus expanding the range of potential treatment sites within the body using procedures such as arthroscopic surgery, balloon catheter angioplasty, or site-directed drug delivery.

As an example of the benefits of the photoimmobilized hydrogel lubricious surface, a model in vitro system was devised using pinched fresh bovine aortas. This simulation of a narrowed artery was used to compare unmodified mock polyethylene (PE) catheters with those modified using a photoreactive mixture of polyacrylamide and polyvinylpyrrolidone. The force required to push catheters through the pinched aorta was carefully measured using a force gauge; the results (Fig. 6) show that the crossing force was reduced by 50% using the lubricious, hydrophilic coating.

C. Passivation

When a medical device is subjected to the harsh in vivo environment, its surface is exposed to the body's natural defense systems. Minerals, fatty acids, proteins, host cells, macromolecules,

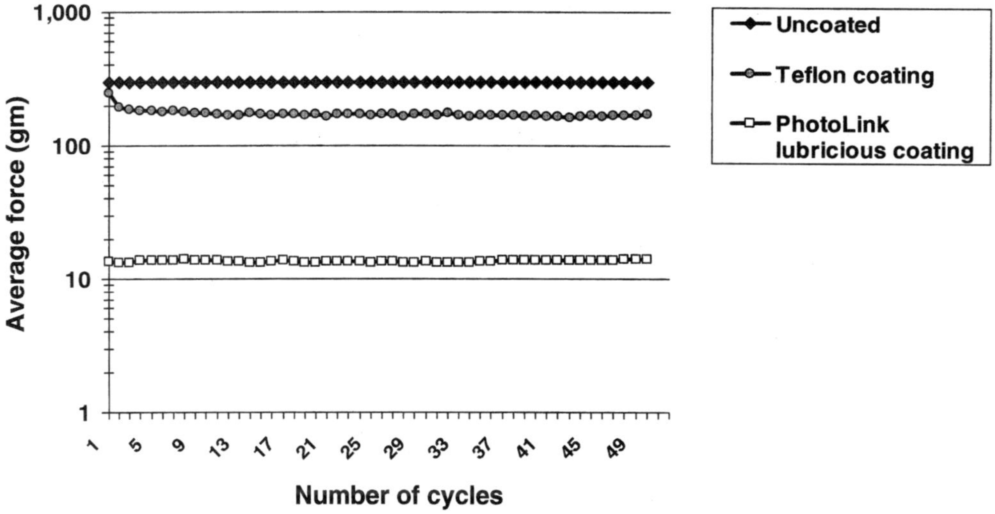

Figure 5 Reduction of friction by using a PhotoLink lubricious coating on a stainless steel wire compared to uncoated and Teflon coated wires. Forces were measured by pulling the wires through SR pads. Forces were recorded for 50 cycles of the test.

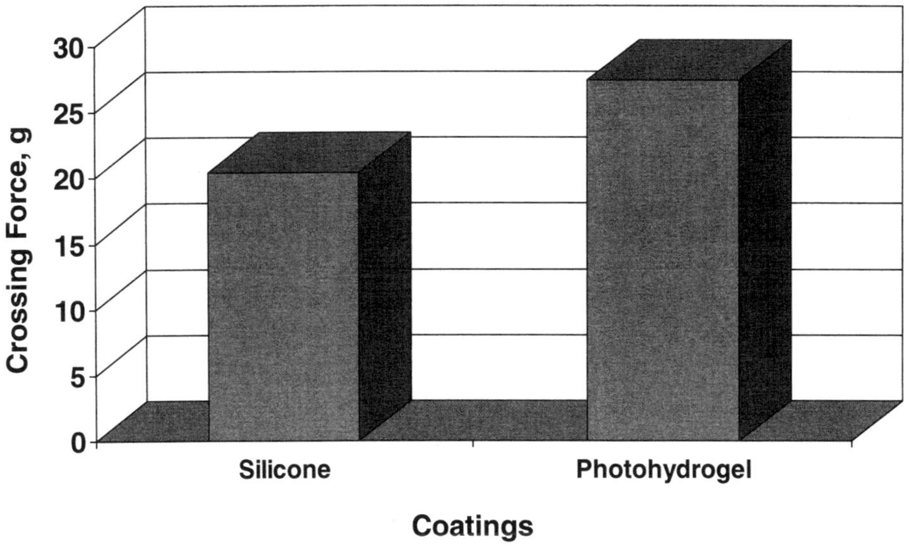

Figure 6 Reduction of crossing force using surface-modified catheters. The pinched aorta model for measuring lubricity was used to compare unmodified and modified catheters. Mock PE catheters were either coated with a common silicone lubricant or with photoreactive hydrogel polymer while submerged in heparinized bovine blood. Mean and standard deviation are shown ($n = 15$, $p<0.0001$).

and bacteria deposit on the surface. Adsorption of proteins or lipids can contribute to (1) initiation of blood coagulation, complement activation, and platelet adherence on intravascular devices; (2) fouling and plugging of microporous contact lenses or dialysis membranes; (3) opacification of the visual axis on IOLs, contact lenses, or keratoprostheses; and (4) bacterial colonization on almost any implantable product.

Surfaces modified with photoimmobilized hydrophilic polymers have been shown to effectively reduce protein adherence, bacterial colonization, and lipid deposition. By controlling the polymeric composition and the extent of crosslinking, hydrogel matrices can be designed to provide optimal passivation characteristics for enhancing the particular biomaterial and device performance. Further discussion of passivating coatings can be found in Section IV, ''Surface Modification to Improve Blood Compatibility of Medical Devices,'' and Section V, ''Antimicrobial Coatings.''

IV. SURFACE MODIFICATION TO IMPROVE BLOOD COMPATIBILITY OF MEDICAL DEVICES

Improved compatibility with blood is a desired feature for a variety of medical devices that must contact blood during clinical use. It is difficult, however, to achieve this improvement by altering the bulk materials used for fabricating the devices. Because of the diversity of blood-contacting medical devices, which range in complexity from syringe needles to artificial hearts and in duration of use from seconds to years, it is unlikely that sufficient ''blood-compatible'' bulk materials will be identified to serve the needs of the many devices currently used or in development. A more straightforward method is to apply a thin-film coating to the surface of the device to confer improved blood compatibility to the device. For many years, researchers have been developing coatings to improve the blood compatibility of medical devices. It is recognized that it is unlikely that a single surface modification will solve the broad spectrum of blood compatibility issues, and therefore it is essential to develop and test a variety of potential surface modification agents. Heparin-based coatings have been demonstrated to substantially improve the performance of a variety of blood-contacting medical devices. Recently, nonheparin-based, synthetic coatings have also been shown to be promising for medical device applications where biological molecules are not desired or appropriate.

The notion of using surface modifications to improve the blood compatibility of medical device materials is not new. Synthetic processes have been developed to immobilize (1) fibrin cascade inhibitors [6,7]; (2) passivating polymers, such as hydrogels [8–11] and polyethylene oxide [12,13]; (3) albumin binding agents [14–17]; (4) membrane mimics [18]; (5) platelet inhibitors [19]; and (6) sulfonated polymers [20–22] onto polymers and metals. However, these processes generally are limited to one or a few types of substrates, require harsh chemical reaction conditions, or are time or labor intensive. In contrast, surface modification using the photochemical coupling technology described above has several significant advantages over previous methods: (1) a broad spectrum of materials can serve as substrates for immobilization; (2) many types of molecules (including all of the classes listed above) can be immobilized using analogous protocols; and (3) the processing steps are simple, short, and devoid of severe chemical treatments.

A. Photoimmobilized Heparin Surface Modifications

Heparin is a pharmaceutical that has been used clinically for decades as an intravenous anticoagulant to treat inherent clotting disorders and to prevent blood clot formation during surgery and interventional procedures. Heparin molecules are polysaccharides with a unique chemical structure that gives them specific biological activity. When heparin is derivatized with photochemical coupling agents and immobilized onto the surface of a medical device material, it improves the performance of the material when in contact with blood in several ways: (1) providing local catalytic activity to inhibit several enzymes critical to the formation of fibrin (which holds thrombi together); (2) reducing the adsorption of blood proteins, many of which lead to undesirable reactions on the device surface; and (3) reducing the adhesion and activation of platelets, which are a primary component of thrombus.

Heparin-containing coatings have been assessed using a variety of in vitro and in vivo test systems. These tests include the verification of the heparin presence, uniformity, and stability on the surface as well as in vitro and in vivo function.

1. Presence and Uniformity

The presence and uniformity of photoheparin on material surfaces has been assessed by staining with toluidine blue, electron spectroscopy for chemical analysis (ESCA), static secondary ion mass spectrometry (SSIMS), and scanning electron microscopy with atomic x-ray analysis (SEM/EDAX). Toluidine Blue staining was used to determine macroscopic uniformity and extent of coverage of the heparin on the material surface. Coated samples were immersed in a 0.02% w/v solution of Toluidine Blue for 10–15 s, removed, and rinsed with DI water to remove unbound dye. Upon binding to the heparin, Toluidine Blue experiences a shift [23] in its absorbance spectrum such that the dye appears purple rather than blue. To confirm that the heparin is indeed exposed to the environment at the surface of a material, ESCA and SSIMS were used to identify the atomic composition and chemical fragmentation pattern of the outermost region of the interface. For example, Fig. 7 compares the SSIMS negative ion spectrum for unmodified Pellethane

PU and photoheparin-modified Pellethane and indicates for the latter the presence of fragments generated from the sulfonate pendant groups on the heparin carbohydrate backbone. These results, and comparable results with ESCA (data not shown), confirmed that the photoheparin was indeed at the surface of the substrates.

2. Stability

Photoimmobilized heparin-based coatings have also been tested for their physical characteristics to ensure their durability. Because the coatings are bound covalently to the device surface, they are expected to be durable to exposure to a variety of physical challenges. Figure 8 shows the results of exposure of heparin-coated surfaces to aqueous buffer solution at physiological pH and temperature for up to 33 days. These experiments demonstrate that the potency of the heparin coatings is not significantly altered by exposure to these conditions, suggesting that the coatings can maintain their durability and activity when subjected to in vivo conditions. SurModics

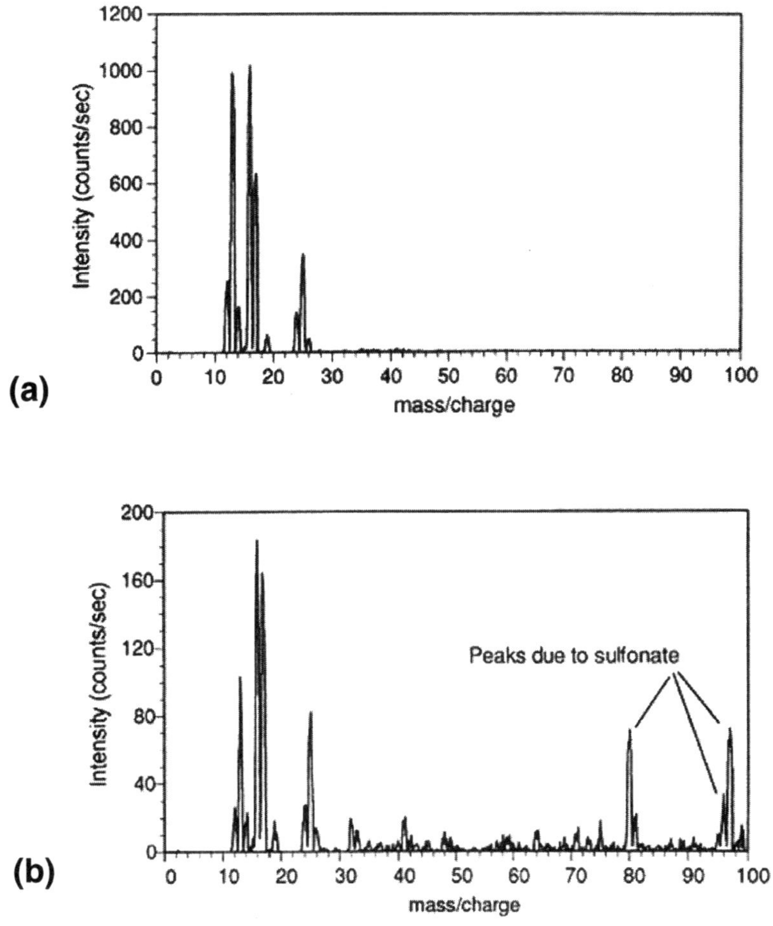

Figure 7 Static secondary ion mass spectrometry (SSIMS) of (a) unmodified and (b) photoheparin-modified PU. The peaks occurring at 80, 96, and 97 mass units in the negative ion spectrum correspond to SO_3, SO_4, and HSO_4, molecular fragments which arise from the sulfonate pendant groups on the heparin. No such peaks are observed in the spectrum for the unmodified PU.

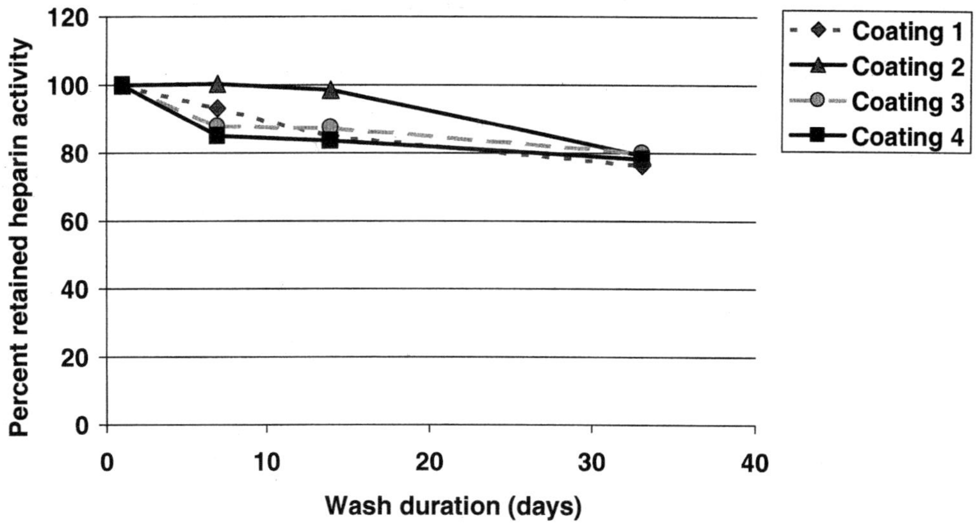

Figure 8 Durability testing of SurModics heparin-based coatings to saline washes. All four heparin-based coatings showed little decrease in heparin activity after 33 days in phosphate-buffered saline at 37°C with agitation.

heparin coatings have also been demonstrated to be stable under accelerated aging conditions designed to model long-term shelf storage. Figure 9 shows the results for heparin-coated materials held at 55°C under both 10 and 50% relative humidity environments, corresponding to 2-year shelf-life at ambient conditions. These results under accelerated aging conditions further highlight the physical durability of the coatings.

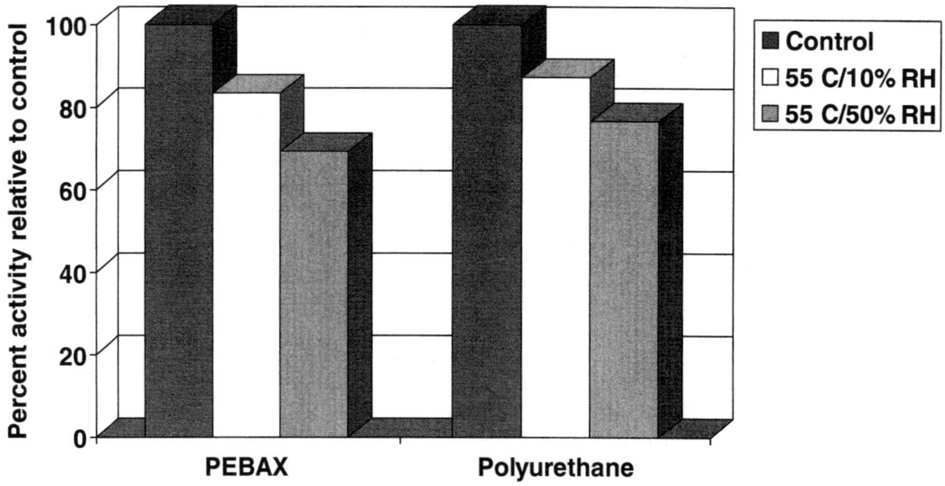

Figure 9 Durability testing of SurModics heparin-based coatings to accelerated aging conditions representing 2 years ambient storage. Heparin activity was tested after heparin-coated materials were held at 55°C under both 10 and 50% relative humidity. Accelerated aging conditions did not have a great effect on the heparin coatings.

Table 1 In vitro Activities of Photoheparin Immobilized on Polyurethane and Polyethylene

Polymer substrate	Thrombin inhibition activity (mU/cm^2)		Partial thromboplastin time (min)	
	Unmodified control	Photoheparin modified	Unmodified control	Photoheparin modified
Polyethylene	<1	21	1.3	>30
Polyurethane	4	22	1.3	>30

3. In Vitro Performance

Two in vitro assays have been used to assess the function of the immobilized photoheparin: a thrombin inhibition assay and platelet adhesion. In the thrombin inhibition assay, the photoheparin-treated material was incubated in a buffered protein solution containing thrombin, a chromogenic peptide substrate for thrombin (S2238, Kabi Diagnostica), and excess antithrombin III (ATIII). Activity was determined by comparing colorimetrically the solutions contacting the photoheparin-treated substrate with standard solutions containing known concentrations of soluble heparin. This assay gives the activity of the surface-bound heparin in terms of the activity of soluble heparin. Although the solutions are well agitated during the assay, mass transport limitations may exist which would hinder interaction between the immobilized heparin and the soluble thrombin and ATIII. Any transport limitations, however, would result in underestimation of the activity of the immobilized heparin. A comparison of the thrombin inhibition activity of one photoheparin reagent immobilized on PU and PE is shown in Table 1. It is clear from these results that the immobilized photoheparin retained its capacity to inactivate thrombin.

It has also been demonstrated that platelet adhesion and activation are greatly reduced when heparin-coated surfaces are exposed to platelet-rich plasma. From the photomicrographs in Figs. 10a and b, it can be seen that the presence of the heparin coating greatly reduces the number of bound platelets as well as the extent of activation of the platelets, as judged by their physical shape.

4. In Vivo Performance

Photoheparin reagents that yielded favorable results in the surface analysis and in vitro experiments were tested in an acute canine jugular vein implant model. In this system, Pellethane PU

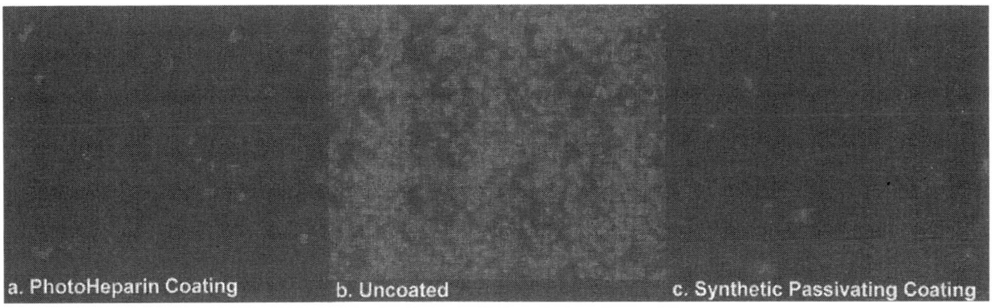

Figure 10 Photomicrographs of in vitro human platelet adhesion to (a) heparin-based coating, (b) uncoated PE, and (c) synthetic passivating coating. Note the differences in platelet number and morphology. The uncoated surface shows highly spread and activated platelets and SurModics coatings show reduced platelet adhesion with minimal platelet spreading.

rods, which serve as models for vascular catheters, were implanted for 1 h in the external jugular veins of dogs. In each experiment, one surface-modified rod and one control rod were implanted bilaterally. This strategy helps compensate for any dog-to-dog variations in the intensity of response to the implanted material. Attachment of ^{111}In-labeled platelets onto the rods was monitored in real time with spatial resolution using a digital gamma camera imaging system. At the conclusion of each experiment, the veins were isolated surgically and opened longitudinally to expose the rods for gross inspection. The rods were then removed and any adherent thrombus was stripped, lyophilized, and weighed. The gamma camera data was used to calculate a relative platelet attachment rate on the rods.

Shown in Fig. 11 is a representative gamma camera image from one of the jugular vein implant experiments. In this photograph of the gamma camera video monitor, the neck region of the dog is central, the head is at the top, and the shoulders and upper chest are at the bottom. The rectangles in the figure encompass the regions occupied by the implanted rods. The intensity of the pixels on the monitor corresponds to the intensity of the gamma radiation emanating from that location of the dog's anatomy. In this particular experiment, the photoheparin-treated rod was implanted in the left vein (indicated by the L; the dog is supine and the camera is positioned above the dog) and the unmodified control rod was implanted in the right vein (R). It is apparent from the image that the platelet attachment on the PU is reduced significantly by the immobilized heparin. Figure 12 shows the explanted rods from the same experiment as the gamma camera image above. It is clear that, in addition to the reduced platelet attachment on the heparin-modified surface, there was also substantial reduction in the formation of thrombus.

The results of jugular vein implants for two different photoheparin reagents are displayed quantitatively in Fig. 13, which compares the platelet attachment rates and thrombus weights for the heparin-modified and unmodified PU rods. ANOVA analysis of these data indicated that

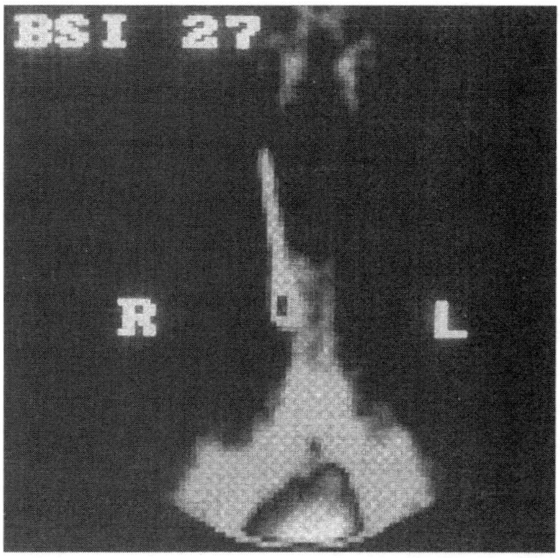

Figure 11 Gamma camera image of ^{111}In-labeled platelets binding to PU rods implanted in canine jugular veins. The photo encompasses the head, neck, and upper chest of the dog. The rectangles identify the locations of the implanted materials in the external jugular veins. The intensity of the video pixels is directly related to the intensity of gamma radiation collected by the camera. The control sample is on the right side (R) and the photoheparin modified sample is on the left side (L) of the dog's anatomy (dog is supine; view is from above).

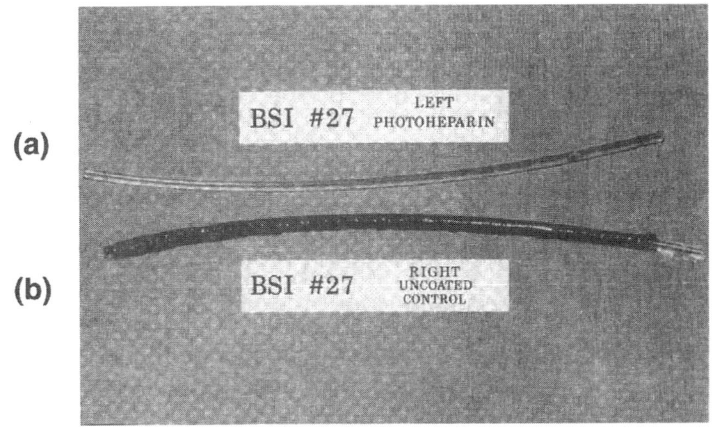

Figure 12 Samples explanted after jugular vein implant experiment show (a) photoheparin-modified material and (b) unmodified material.

for both photoheparin reagents the platelet attachment rates and the thrombus weights were reduced significantly compared to control ($p < 0.05$).

5. Conclusions

These results have shown that heparin can be immobilized using photochemical coupling techniques such that it (1) covers substrate surfaces completely and uniformly; (2) inhibits thrombin in a buffer solution; (3) retards the generation of fibrin in plasma in vitro; (4) withstands washing

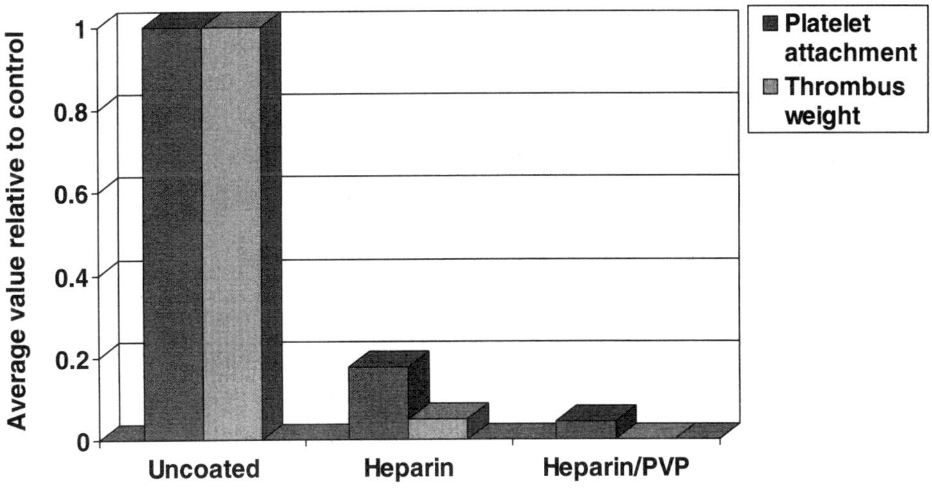

Figure 13 Comparison of the average platelet attachment rates and thrombus weights on photoheparin-treated and untreated PU. Samples were implanted for 1 h and the attachment of radiolabeled platelets was monitored quantitatively in real time using gamma camera imaging. The heparin-based coatings significantly decreased platelet attachment and thrombus formation.

in PBS for up to 33 days; and (5) reduces platelet attachment and thrombus formation in a dog jugular vein implant model.

B. Photoimmobilized Non-Heparin-Based Surface Modifications

As shown in the previous discussion, heparin-based coatings provide excellent antithrombogenic performance for medical device materials. There are, however, several potential regulatory hurdles associated with heparin-based coatings, especially in the European Union where heparin is considered a biological and medicinal product and a heparin coating may thereby change a device's classification and regulatory pathways. Therefore, there is a growing interest in non-heparin-based, nonbiological hemocompatible coatings.

After careful examination of the complex mechanism and pathways of blood clotting and the relevant scientific literature, novel photoimmobilized alternatives to heparin have been developed for preventing surface-induced thrombus formation. In these approaches, synthetic, nonbiological molecules are photoderivatized and immobilized onto a surface to create surfaces with improved blood compatibility. One approach employs the strategy of passivation with hydrophilic molecules to mask the underlying thrombogenic surface from the blood. The passivated surface reduces or prevents the adhesion of thrombogenic cells and proteins onto the underlying substrate or material, thereby preventing surface-induced blood clotting. Another approach involves coatings that actively recruit and bind native albumin from the patient's own blood onto the device surface. This albumin-binding coating acquires a thin, self-regenerating absorbed albumin layer on the surface. In turn, the albumin-covered surface minimizes and prevents the adhesion of unwanted thrombogenic cells and proteins.

To assess the performance of these non-heparin-based coatings, we have carried out a variety of in vitro and in vivo experiments, in some cases comparing directly with heparin-based coatings.

1. In Vitro Performance

Figure 10c shows the results of in vitro platelet adhesion experiments, analogous to those described above for heparin coatings. The hydrophilic, passivating coating showed results similar to those of heparin coatings in this experiment, providing substantially reduced platelet binding. Figure 14 shows the results of another in vitro test of blood compatibility. Hydrophilic passivating coatings, with and without heparin, were exposed to flowing blood in a recirculating loop model of the circulatory system [24]. In this experiment, the amount of platelet adhesion was quantified by using radiolabeled platelets. Both of these coatings showed greatly reduced platelet adhesion compared to uncoated surfaces.

The albumin-binding coatings have also been assessed using in vitro and in vivo test systems. Figure 15 shows the results of an experiment in which surfaces were exposed to human plasma, and the amount of albumin bound to the surface was determined using antibody binding techniques. As can be seen in this figure, the albumin-binding coating was capable of increasing the affinity of the surface to albumin four-fold compared to the uncoated surface. This result demonstrates the affinity that these coatings have to bind albumin.

2. In Vivo Performance

The photographs of explanted coated and uncoated polymer heart valves in Fig. 16 demonstrate the in vivo performance of photoimmobilized albumin-binding coatings. In this experiment, polymer heart valves were implanted in the mitral position in sheep for 5 months to compare an uncoated valve with a valve coated with an albumin-binding agent. The albumin-binding

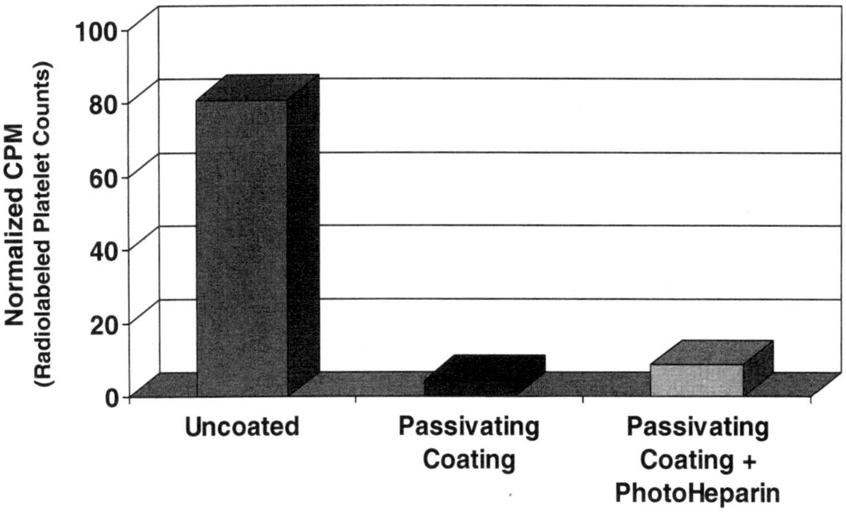

Figure 14 Platelet adhesion under flow condition in an in vitro circulating blood loop model. Uncoated PE and SurModics synthetic passivating coating, with and without heparin, were tested in the blood loop model for 1 h with an average flow velocity of 15 cm/s. The passivating coating clearly masked the underlying PE thrombogenic surface from the blood.

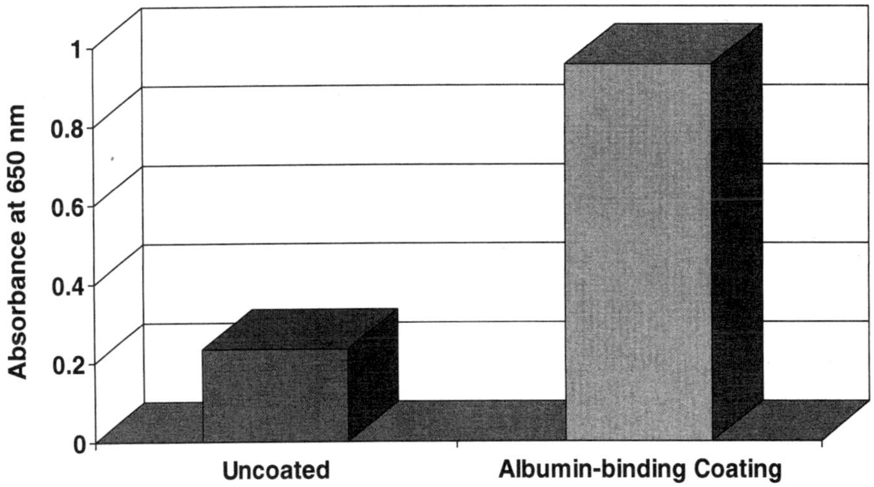

Figure 15 Albumin binding out of human plasma to uncoated and SurModics albumin-binding coating. Uncoated and coated SR rods were incubated in human plasma and washed. Albumin binding was detected using an ELISA technique. The albumin-binding coating increased albumin adsorption four-fold compared to that of uncoated SR.

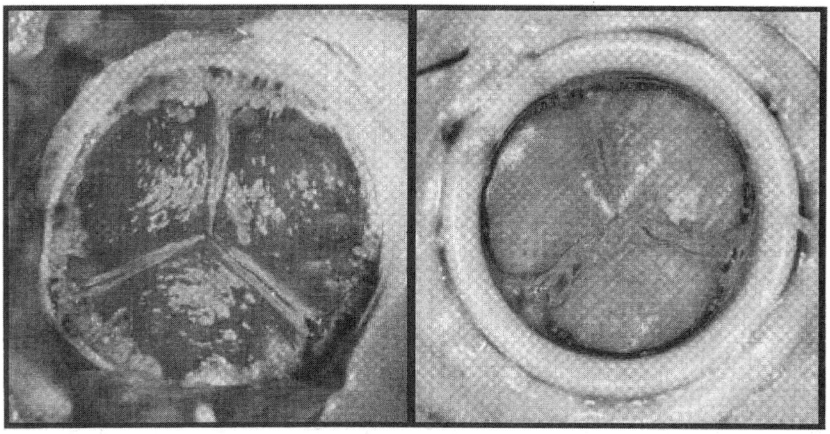

(a) (b)

Figure 16 Photographs of explanted (a) uncoated and (b) albumin-binding coated polymer heart valves. The heart valves were tested in 5-month mitral valve sheep implants. The albumin-binding coating clearly improved the hemocompatibility of the heart valve.

coating on the valves reduced the occurrence of thrombus accumulation on the flexible leaflets and other components of the valve.

3. Conclusions

The data generated from these experiments demonstrate that these non-heparin-based coatings can provide substantial improvements in the blood compatibility of medical devices. The results also validate the scientific principles behind these alternative approaches targeted at preventing thrombus formation and enhancing the function of blood-contacting medical device materials. Medical device manufacturers now have the option to choose from either heparin-based surface modification or a variety of nonheparin, non-biological surface modification approaches to create hemocompatible surfaces on medical devices.

C. Future Developments

The inherent flexibility of these photochemical coupling methodologies makes possible the immobilization of a broad variety of molecules with the potential to improve the compatibility of blood-contacting materials. Current efforts at SurModics in this area include immobilization of extracellular matrix proteins and peptides (see Section VII), hydrogels, plasminogen binding reagents, and cell membrane components.

V. ANTIMICROBIAL COATINGS

Medical devices are essential tools in modern patient care. Unfortunately, these useful implements can often lead to serious complications, such as device-related infection, if not managed properly. Nearly every medical device is prone to infection by a wide variety of pathogens. Implant devices account for about 45% of all nosocomial infections [25], with catheter-related urinary tract infections being the most common form [26]. Ninety percent of bloodstream infections are related to the use of an intravascular medical device [27]. Device-related infection occurs less frequently in other devices, such as penile implants, but if an infection is evident, the morbidity is dramatic [28,29]. In addition to those devices listed above, infection of fracture fixation devices, dental implants, joint prostheses, vascular grafts, cardiac pacemakers and leads, hip prostheses, artificial voice prostheses, intrauterine devices, mammary implants, mechanical heart valves, and heart assist devices are associated with millions of infections at rates between 1–50% [30,31].

The pathogenesis of device-related infection is not well understood. Several factors are thought to be involved. Host immune mechanisms are challenged soon after a device implantation or similar procedure reducing the normal infectious dose [32,33]. The medical device may itself act as a ''wick'' serving as a conduit for opportunistic pathogens. Even under the best aseptic conditions, studies found that 5000 to 50,000 skin particles are delivered daily from each physician's flora in intensive care units [34,35]. In the same study, 90% of clean wounds at the time of closure harbored pathogenic *Staphylococcus aureus*. A medical device's topographical features provide a niche for microorganisms to adhere and subsequently colonize forming biofilms. These biofilms, essentially a microbiological community housed within a exopolysaccharide matrix, are very tenacious and require as much as 1000 times higher antibiotic concentrations to kill the embedded microorganisms compared to free-floating cells [36,37].

Device-related infections cause significant morbidity and mortality. As with most maladies, prevention is the preferred approach. Simple interventions can reduce the device-related infection risk. For instance, training of health care workers (HCW) in the proper manipulation of medical devices reduces preinsertion contamination. Effective hand washing of HCW during patient care also reduces the risk of nosocomial infections. Full barrier protection during insertion and proper skin preparation greatly minimize the introduction of pathogens. Lastly, localizing an antimicrobial(s) at the likely infection site or at the device, or making the surface less conducive for microbial attachment, offers a more targeted approach compared to systemic dosing in order to prevent and/or control device-related infection.

Antimicrobial coatings are perhaps the last safeguard in this total system approach for prevention of device-related infections. SurModics' PhotoLink® photochemical coupling technology can be used in various ways to help reduce the risk of device-related infections. Some PhotoLink coatings provide antiadherent properties to devices; others can be engineered to serve as reservoirs for antimicrobial agents, which may be released at a measured rate from the device surface or immobilized.

A. Antiadherent Coatings

Biofilm formation is a complex process, which involves a variety of physical and chemical factors. In simplistic terms, biofilm genesis proceeds in this fashion: (1) organic molecules condition a device surface; (2) planktonic microbes migrate to the conditioned surface and attach; (3) genes are expressed to generate an exopolysaccharide matrix; (4) cell-to-cell communication occurs forming microcolonies; (5) the microcolonies propagate generating very high numbers of microorganisms housed within a growing protective shell of exopolysaccharide; and (6) sessile cells slough off and travel to other sites to regenerate this process and/or infect surrounding host tissue.

Prevention of biofilm formation is most likely to be successful if the multiple steps listed above are disrupted. Once the microcolonies start to propagate, it is very difficult for systemic antibiotics to stop the process. A means to prevent the onset of biofilm formation at the early stages is to passivate the surface with a hydrophilic coating. Keeping a medical device surface free from organics in vivo is highly unlikely; however, this surface can be coated to address some of the biofilm formation dynamics. Surface roughness, charge, material of construction, and hydrophobicity all contribute to biofilm initiation [38–40]. Photoactivated hydrophilic coatings ''smooth out'' the topography of many medical devices making them less susceptive to microbial attachment.

A study performed to test the antiadherent properties of the coatings was done as follows: High-density PE slides (5.0 × 1.0 × 0.16 cm) were coated with a photoactivated hydrophilic coating. Both coated and uncoated slides were then conditioned with human platelet-poor plasma diluted 1:4 in PBS. After a 2-h incubation period at room temperature, the slides were washed with tris NaCl Tween® 20 buffer. The plasma-conditioned slides were challenged by immersion in a 1×10^5 colony forming units (CFU)/mL suspension of *Staphylococcus epidermidis* and incubation for 24 h at 35°C. The contaminated slides were gently rinsed twice with sterile PBS to remove loosely adhered cells. The attached bacterial populations were stained with a fluorescent dye (BacLite, Molecular Probes, Eugene, OR), then imaged in a fluorescent microscope.

The photoactivated hydrophilic coating afforded protection against bacterial adherence in this test. Figure 17 shows coated and uncoated PE slides that were challenged with *S. epidermidis*. No bacteria were noted on the coated PE surface after 24 h, while the uncoated sample showed a sizable sessile population calculated to be approximately 3×10^6 CFU/cm^2. The observations noted in the micrographs depicted in Fig. 17 were consistent across numerous fields of view in the fluorescent microscope.

In a second study, polypropylene catheters were coated with photoactivated hydrophilic polymers. These catheters were inserted subcutaneously perpendicular to the dorsal midline into six anesthetized New Zealand white rabbits. Three coated catheters were placed equally spaced along the right dorsal side, while uncoated catheters were situated on the left dorsal side. Five

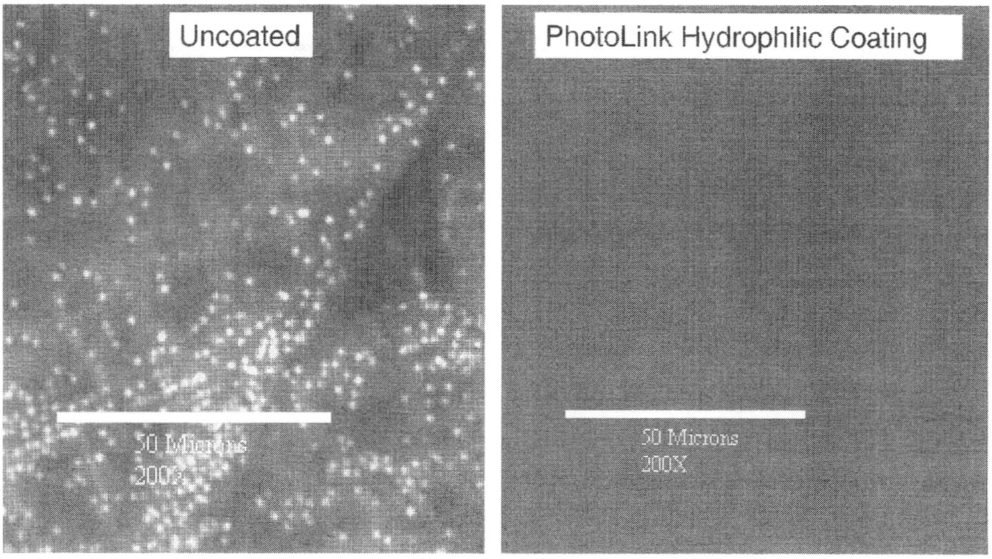

Figure 17 A PhotoLink hydrophilic coating greatly reduced adherence of *S. epidermidis* to PE following a 24-h exposure time. Surface analysis using imaging software determined that the sessile population on the uncoated sample was approximately 3×10^6 CFU/cm^2. No bacteria were observed on the hydrophilic-coated surface across numerous fields of view. Bars = 50 μm.

centimeters of each catheter were tunneled under the skin while, 1 cm remained exposed on the skin surface. The catheters were anchored to the rabbits' skin with adhesive tape and sutures. Cotton gauze sponges (5 $\times$ 10 cm) were secured over each exit site with silk sutures. The catheters were challenged by evenly inoculating the sponges with 10 mL of an overnight *S. aureus* suspension diluted to 3.5 $\times$ 10^8 CFU/mL in PBS. A bandage was placed over all the sponges and exit sites to prevent self-mutilation and disruption of the catheters. The rabbits were observed daily, and all rabbits recovered from the surgery without complications.

One week following the *S. aureus* challenge, all rabbits were euthanized with sodium pentobarbital, and the catheters and surrounding tissue were collected from each rabbit. One-centimeter segments of the catheter and surrounding tissue were clipped from the proximal and distal ends in relation to the insertion site. The segments were subjected to sonication and homogenization, then enumerated on tryptic soy agar.

The photoactivated hydrophilic coatings provided significant protection against bacterial adherence to the catheters (Fig. 18). The geometric means of the uncoated catheters were approximately two log units higher than the respective coated catheters. As expected, the proximal catheter segments exhibited a higher infection rate and a higher bacterial load compared to the distal segments. The photoactivatable hydrophilic coating markedly reduced bacterial adherence to the catheter surfaces, as well as minimizing ''wicking'' along the length of the catheter.

B. Active Antimicrobial Coatings

1. Matrices for Incorporating Antimicrobials

Antimicrobials can be incorporated into medical device coating matrices. This ''targeted'' approach is favored over systemically delivered antibiotics since the antimicrobial agent is concentrated at the site where infection is most likely to occur. Fewer side effects are realized for a targeted antimicrobial application compared to systemic dosing. Also, a broader range of antimicrobial agents may be used in medical device coatings compared to systemic application

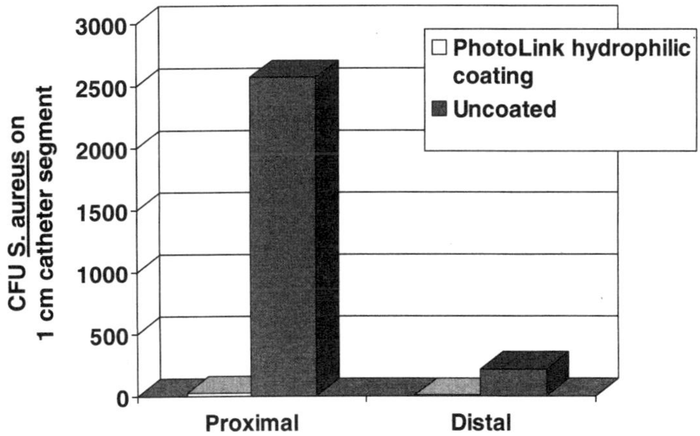

Figure 18 A PhotoLink hydrophilic coating afforded marked bacterial antiadherence protection to a polypropylene catheter. Coated and uncoated catheters were inserted perpendicular to the dorsal midline in rabbits. Ten milliliters of a 3.5 $\times$ 10^8 CFU/mL *S. aureus* suspension were applied to a gauze sponge covering the insertion site. Bacteria adhering to the catheter and surrounding tissue were enumerated following a 1-week exposure.

since far less active agent is dosed, and the agent is generally localized in the device microenvironment. Depending on the antimicrobial chosen, this may minimize the issue of microbial resistance associated with systemic antibiotics.

Antimicrobial agents, such as chlorhexidine, can easily be imbedded within a photoimmobilized hydrogel. A negatively charged hydrogel coating reservoir is particularly well suited for a positively charged antiseptic like chlorhexidine. The antimicrobial agent slowly percolates out of the hydrogel matrix into the microenvironment where it provides its antimicrobial benefit. Various methods exist for controlling the rate of release of active agents into the microenvironment. For instance, the application of a topcoat over an antimicrobial-containing coating extended the release of active agents over time (data not shown).

The data shown in Fig. 19 demonstrate that a photoactivatable hydrogel can serve as a reservoir for active antimicrobial agents. For this experiment, approximately 4×10^5 CFU/cm^2 of *S. aureus* was equally distributed onto the surface of a Mueller–Hinton (MH) agar plate and allowed to dry. *Staphylococcus aureus* is a pathogen found on the skin that is frequently associated with device-related infections. Coated and uncoated PU rods were placed onto separate *S. aureus* seeded plates and then placed into a 37°C incubator for 24 h. The zone of inhibition (ZOI), or clearing along the length of the PU rods, was measured on each plate with a caliper. If a ZOI was evident, the PU rod was transferred in the same orientation to a freshly seeded MH plate after each 24-h interval.

As seen in Fig. 19, the chlorhexidine-coated PU rod was efficacious against *S. aureus*. The uncoated PU rod did not afford any protection against the *S. aureus* challenge. Activity was maintained through 7 days of testing. The highest ZOI was seen at day 1, which is beneficial to the patient as the highest risk of infection typically occurs immediately following implantation. Days 2 through 7 showed a sustained release of chlorhexidine leaching consistently out of the coating to act against *S. aureus*.

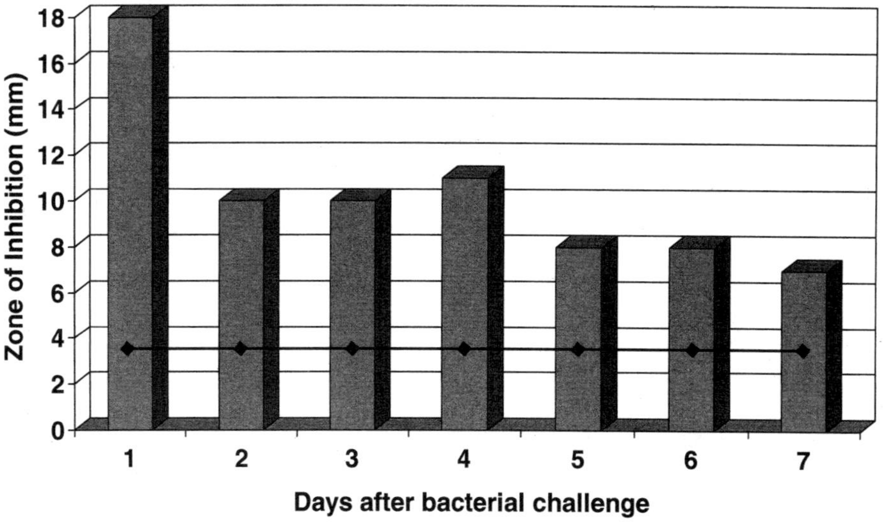

Figure 19 Activity of a PhotoLink antimicrobial coating containing chlorhexidine digluconate. This active antimicrobial coating on PU (3 mm diameter) exhibited sustained efficacy against methicillin-resistant *S. aureus* initially present on the PU surface at approximately 4×10^5 CFU/cm^2. The rod diameter depicted by the horizontal line is included in the bars.

As with most medical devices, infections are a frequent and a serious complication with orthopedic fixation devices. To address this problem, a photoimmobilized hydrogel drug reservoir containing chlorhexidine was evaluated using a 14-day rabbit tibial intramedullary model. In this study, stainless steel pins were incubated with 1×10^6 CFU of a clinical isolate of *S. aureus* from a chronic osteomyelitis infection. After the planktonic organisms were washed away, inoculated pins were inserted into the medular cavity of a rabbit leg through a hole drilled into the proximal tibia. Animals were sacrificed 2 weeks after the surgery. The explanted pins were sonicated to remove bacteria, which were enumerated by plate counting. In addition, the bone adjacent to each pin was collected and homogenized and the bacteria enumerated. Coated pins containing antiseptic reduced the number of bacteria by approximately 3 logs (Fig. 20) and prevented osteomyelitis in 80% of the animals (Fig. 21). Therefore, the photoimmobilized hydrogel coating containing chlorhexidine was shown to be efficacious in preventing osteomyelitis caused by *S. aureus*.

2. Immobilization of Antimicrobials

Another coating strategy is to immobilize antimicrobials directly to the medical device surface using a ''linker'' molecule. Immobilized antimicrobials on medical devices must have at least three characteristics: (1) a surface-active mode of action; (2) activity against a broad spectrum of pathogenic microorganisms; and (3) a chemical structure which allows linkage to the device surface while retaining its antimicrobial activity. There are some advantages and disadvantages associated with immobilizing antimicrobials directly to the device surface versus antimicrobials that are eluted from a coating matrix. A potential benefit of immobilization is longer lasting activity. A few drawbacks include (1) the microorganisms must intimately contact the immobilized antimicrobial for the surface-active agent to exert its effect; (2) if the device surface is fouled, biologically or otherwise, contact between the microbe and the immobilized antimicrobial may be masked; and (3) there may be a limited selection of surface-active antimicrobials which meet the criteria above.

However, a group of novel cationic antimicrobial peptides fulfill these requirements. These peptides have been isolated from a variety of sources and include peptide derivatives of human

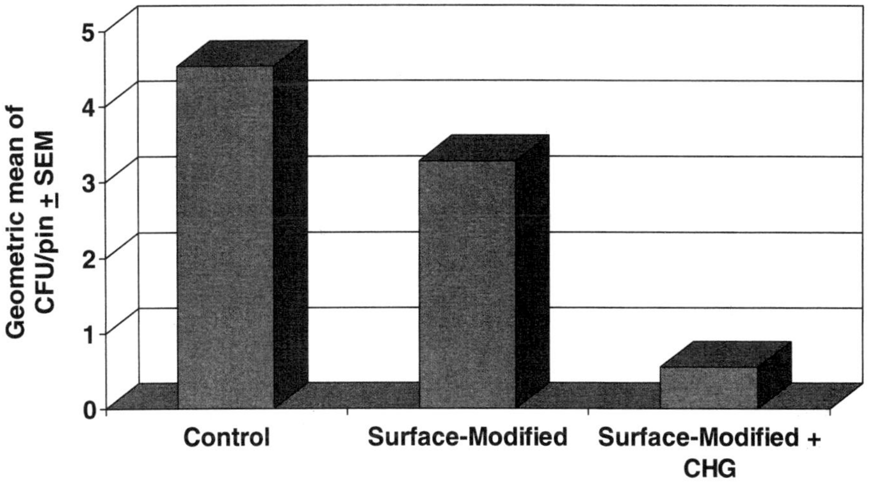

Figure 20 Efficacy of surface-modified pins in a rabbit tibial intramedullary model.

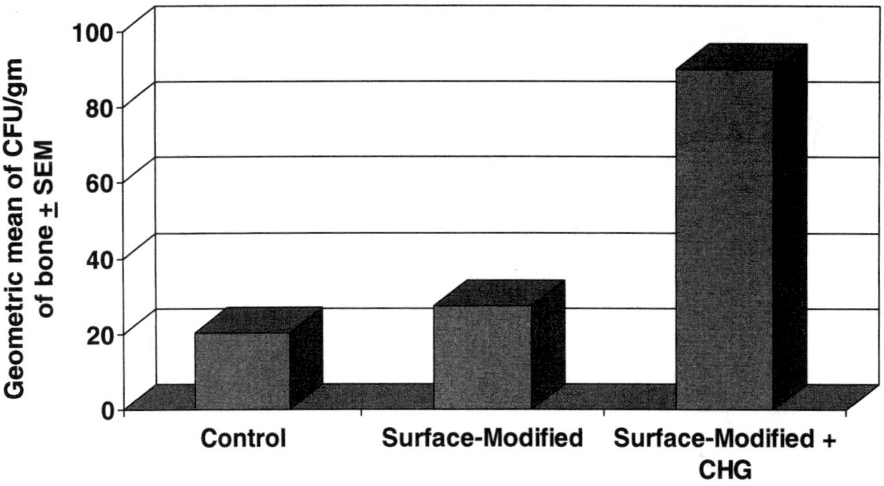

Figure 21 Osteomyelitis associated with surface-modified pins in a rabbit intramedullary model.

platelet factor IV [41]. These peptides, which kill a broad spectrum of bacteria, possess an amphiphilic α-helical structure and form holes in artificial membrane systems [41]. Hence, the mechanism of bactericidal activity of these peptides is probably by insertion of the helix into the bacterial membrane, causing osmotic lysis. Finally, these peptides (1) are surface-active [41–45]; (2) are microbicidal to variety of pathogens including *Staphylococcus epidermidis*, *Escherichia coli*, *Pseudomoas aeruginosa*, and *Pseudomonas mirabilis* [42,46,47], all of which play a role in device-centered infections [32,40,48,49]; (3) contain functional groups available for covalent coupling to support materials [43]; and (4) are available as synthetic peptide products with high specific activities [42,46,50].

To evaluate this coating strategy, W. L. Cosand, J. Blake, and R. P. Darveau (Bristol-Myers Squibb Pharmaceutical Research Institute, Seattle, WA) kindly provided a synthetic peptide analog of the bactericidally active 13-residue carboxy-terminal portion of human platelet factor IV (377VV). The 377VV peptide was photoderivatized, HPLC purified, and evaluated by microdilution assay to determine the minimal inhibitory concentration (MIC) against a variety of microorganisms. The results shown in Table 2 indicate that a photoreactive derivative of 377VV could be synthesized with little or no loss in microbicidal activity against a variety of gram-negative and gram-positive bacteria.

Table 2 Minimal Inhibitory Concentration of Peptides on Various Organisms

Microorganism	377VV (μg/mL)	Photo-377VV (μg/mL)
E. coli	3.125	12.5
S. epidermidis	1.56	12.5
K. pneumoniae	6.25–12.5	12.5
Proteus vulgaris	3.125	3.125
S. aureus	25	25–50
P. aeruginosa	6.25	12.5
Serratia marcescens	100	>100

In addition, an in vitro assay was used to assess the antimicrobial activity of the peptides immobilized on silicone rubber (SR) disks. Briefly, the SR disks were placed individually in the wells of a microtiter plate. Subsequently, approximately 100 CFU of a *S. epidermidis* suspension was added to each disk by centrifugation. Bacteria-coated disks were incubated overnight at 37°C on tryptic soy agar, and the number of colonies growing on the surface of each disk was enumerated. The results shown in Fig. 22 indicate that 377VV adsorbed onto the surface was ineffective in reducing the growth of *S. epidermidis*, whereas photoimmobilized 377VV completely abrogated bacterial growth.

C. Conclusions

Device-related infections are a significant problem in the health care industry. Controlling the sterility of an implant area and device is extremely difficult to maintain before and after implantation. Proper insertion site preparation, maintaining a strict sterile field, appropriate hygiene of HCW, and postwound site cleanliness are crucial elements in this continuous battle against a vast variety of pathogens. Systemic antibiotic dosing has many disadvantages associated with it, including but not limited to the buildup of resistant organisms. Coating medical devices with antiadherent and/or antimicrobial agents offers a value-added benefit to medical devices serving as perhaps the last line of defense in minimizing the risk of nosocomial infections. The photoactivatable processes described in this chapter offer an environmentally friendly, flexible, and economical means to coat many medical devices on the market.

VI. COATINGS FOR LOCAL DRUG DELIVERY

Most medical devices have limitations in terms of their efficacy or longevity because of the inherent response of the body or pathogens to the implant. For example, tissue ingrowth, infec-

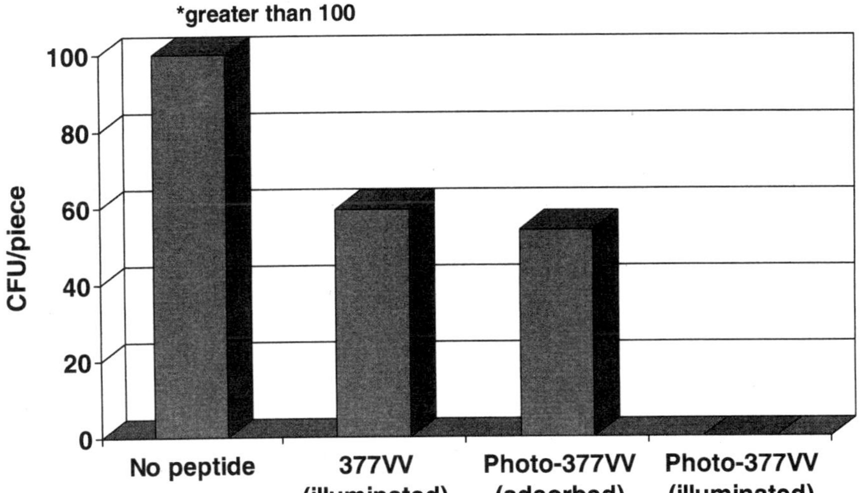

Figure 22 Antimicrobial activity of an immobilized peptide against *S. epidermidis*. Bacterial growth was evaluated on SR disks that were untreated; coated with underivatized peptide and illuminated with UV light; coated with photoderivatized peptide, but not illuminated; or coated with photoderivatized peptide and illuminated with UV light.

tion, inflammation, immune system targeting, and thrombogenesis can all inhibit the function of devices [30,51,52]. In many situations, pharmaceuticals or antimicrobial agents are given to a patient systemically to improve or enable the performance of the implanted medical device. Although generally helpful, in many cases this approach is insufficient or entirely ineffective in treating the undesirable events associated with the implant [37,53–55]. A different approach to this problem is to provide a local dose of the pharmaceutical or antimicrobial agent delivered directly from a coating on the surface of the implanted device [56]. This approach has several advantages over systemic delivery of the drug: (1) it circumvents the toxicity associated with many drugs which must be given at high doses to achieve sufficient local concentrations at the implant site; (2) the drug is targeted at the intended site of action, ensuring that only the tissue needing the drug receives it; and (3) much less drug is needed, providing a cost savings. In this section, the development and application of drug delivery coatings for medical devices will be discussed. The specific example of coatings for vascular stents to inhibit restenosis will be presented. Antimicrobial drug delivery coatings have been described in a separate section.

A. Applications of Drug Delivery Coatings

There are a variety of devices that could benefit from the inclusion of a local drug delivery coating. For example, pacing leads frequently result in excessive tissue ingrowth and fibrous capsule formation at the point of electrical contact with the heart muscle tissue, which interfere with sensing and the delivery of the pacing signal. If a coating could be applied to the surface of the pacing lead to deliver a drug to reduce the tissue ingrowth, the functionality and useful lifetime of the lead could be increased. As a second example, coronary stents, which are implanted during interventional cardiology procedures, result in the phenomenon of smooth muscle proliferation, which is stimulated by the vessel wall damage caused by the inflation of the balloon during the procedure and the continuing presence of the stent thereafter. Such cellular proliferation results in the accumulation of tissue in the lumen of the stent causing decreased blood flow through the vessel, which often necessitates repeat interventional procedures. Other devices, such as indwelling catheters, orthopedic implants, and heart valves could also benefit from local drug delivery approaches.

B. Development of Drug Delivery Coatings

Local drug delivery coatings can be fabricated in several ways using biodegradable or biostable materials. Biodegradable coatings have the advantage of being present in the body only transiently and thus do not pose a long-term biocompatibility risk. Such coatings are capable of being loaded with a high concentration of drug and can have release rates tailored to a variety of durations. The degradation products of a biodegradable coating, however, must be tolerated by the host [57]. The chemical breakdown products must not provoke any significant inflammatory response, nor should the coating create unwanted particulates during the degradation process. A biostable coating must possess good long-term biocompatibility that meets or exceeds the biocompatibility of the implant itself. If a coating must be exposed to the bloodstream during implantation, the coating should be no more thrombogenic than the bare implant device.

In addition to biocompatibility issues, which are of concern for any implant material, there are several challenges that are unique in designing and implementing a drug delivery coating for a medical device. For example, there are limits to the thickness of the coating that can be applied to the device. This will constrain the quantity of drug that can be incorporated onto the surface of the device. Some devices have very small surface areas, such as coronary stents and ophthalmic devices, which also limit the amount of drug which can be incorporated onto the surface of the device. Another challenge in creating drug delivery coatings is that the chemical properties (e.g., size, solubility, charge) of the drug may limit the quantity of drug that can be formulated in the coating. Because of the limitations on the amount of drug that can be incorporated in a drug delivery coating, it is important that drugs of sufficient potency be available for specific applications.

Drug delivery coatings must adhere well to the device surface with the desired goal of no delamination during use of the device. These coatings must be capable of surviving the physical challenges of manufacturing, sterilization, and implantation. For example, stent coatings must survive balloon mounting procedures; sterilization by gamma, ethylene oxide, or e-beam methods; travel through the lumen of a guide catheter; and the balloon expansion process in the target vessel. In addition, drug delivery coatings must not interfere with the primary function of the device.

Drug delivery coatings must provide a sufficient dose of drug during a desired time frame. As discussed previously, there are limitations to the amount of drug that can be incorporated in a coating. This sets an upper limit to the amount and duration of delivery. Therefore, for effective performance, the delivery of the drug should be timed and controlled by the components of the coating to match with the timing of the biological process the drug is intended to modulate. Because of the variety of drug candidates available, this emerges as a new challenge with each drug which must be delivered. It is probably not possible to design a coating system that works with all drugs, but coating systems should be flexible enough to work with many compounds. Drugs of interest include salts, small water-soluble compounds, small organic-soluble compounds, polysaccharides, peptides, proteins, oligonucleotides, DNA, and RNA.

C. Example of SurModics' Drug Delivery Coatings: Drug Delivery Stents

Vascular stents are small, hollow, metal scaffolds that are placed in blood vessels to maintain lumen patency after interventional procedures. There are two major complications associated with stent use: thrombosis and restenosis [58,59]. Thrombosis occurs when blood coagulation initiates on the surfaces of the stent, which leads subsequently to severe narrowing or complete occlusion of the blood vessel. Restenosis is a progressive narrowing of the blood vessel, which occurs over weeks to months, caused by the injury to the vessel during stent expansion and the continued presence of the metal stent. Operator technique and anticoagulant drugs given to patients have substantially reduced, although not eliminated, the occurrence of stent thrombosis, with rates currently in the neighborhood of 1%. The phenomenon of restenosis, however, continues to be a substantial problem, resulting in repeat intervention in about 20–30% of cases [60].

Several approaches to treating restenosis in stents have been attempted, including administration of drugs systemically, catheter-based local delivery of drugs, and local radiation treatment (brachytherapy). None of these have been substantially effective at reducing restenosis rates in patients receiving stents, although brachytherapy has shown success in reducing the reoccurrence of restenosis in patients who received angioplasty to treat a stent that failed previously due to restenosis. A different approach to preventing restenosis is to deliver a drug from the surface of the stent into the vessel tissue. The current understanding of the mechanism of restenosis suggests that the primary contributor to renarrowing is the proliferation and migration of smooth muscle cells from the injured artery into the lumen of the vessel [61,62]. Therefore, the types of drugs that have been studied to treat restenosis include agents that inhibit the proliferation and migration events, as well as agents that inhibit inflammation, resulting from the arterial injury caused by balloon expansion, which is suspected to contribute to the cell infiltration.

Preparation of drug delivery coatings on stents has several challenges, including some of those mentioned above. Stents have very low surface area, typically in the range of 1–2 cm^2, which only allows small quantities (hundreds of micrograms) of drug to be incorporated into a coating. Consequently, potent drugs must be used. The release of the drug from the coating must be coordinated with the intended biological targets, which may be upregulated anywhere from minutes to weeks after the stent implantation procedure. The coating must survive several physical challenges. The coating must also not interfere with the primary function of the stent, which is to physically support the walls of the blood vessel.

There are several ways to prepare drug delivery coatings and to control the release of a drug from the stent surface. The method SurModics has employed is to form coatings using combinations of nonbiodegradable polymers and drug, with the objective of achieving a homogeneous mixture on the surface of the stent. The release of the drug is controlled by the loading of the drug and the composition of the polymer components, both of which influence the rate at which the drug diffuses out of the coating. Additional control of the release can be achieved by applying barrier coatings onto the surface of the drug-containing polymer layer.

As an example of the work done at SurModics with drug delivery coatings, we will describe examples involving several model drug compounds. Due to proprietary constraints, these drugs will not be named. Polymer coatings containing drug were prepared by depositing films on stainless steel disks or laser-cut, stainless steel, balloon-expandable stents. Coating solutions were prepared using a blend of two polymers, polyethylenevinylacetate and polybutylmethacrylate (PEVA and PBMA), and drug. The concentrations of components (polymers and drug) were varied to obtain different loading levels of drug, different ratios of drug to polymer, and different ratios of the polymer components.

The coatings were investigated using several types of surface analytical techniques. Optical microscopy and scanning electron microscopy were used to assess uniformity and surface texture of coatings. Raman IR spectroscopy was used to obtain spatial information on the chemical composition of the films, including location and relative concentration of polymer and drug components.

The surfaces of most of the drug-containing coatings were smooth when viewed under the visible microscope (Fig. 23) and SEM. In some cases, coatings showed indications of greater roughness or cloudiness. Texture in the surface of the coatings, however, could be minimized by coating process changes. The results of the Raman microscopy were used to delineate the thicknesses of the coatings and the distribution of the components within the coating. It was possible to locate unique peaks in the spectra, which identified the drugs and each of the polymer components in the surface coatings. In some cases, it was possible to visualize the distribution of drug within the coating, indicating the coating homogeneity. In Fig. 24, the distribution of polymers and one particular drug in a stent coating are shown.

Mechanical integrity of the coatings was tested by crimping coated stents onto balloon catheters, hydrating the balloon-mounted stents, and then expanding the stents by inflating the balloons. The durability of the coatings was assessed by optical microscopy. Coatings were developed which withstood the crimping and balloon expansion with no cracking or delamination. Coatings with lower concentrations of drug showed the best tolerance to balloon expansion. Optimal ratios of polymer to polymer and of polymer to drug in the coating were defined for each drug.

Elution of drug from the coatings into solution was assessed by incubating the coated disks or stents in capped vials containing solvent stirred at 37°C. At various intervals, the stent or disk was removed from the vial and placed in fresh solvent, and the incubation was continued. The drug concentration in each sample was measured using UV/visible spectrophotometric or HPLC methods. Samples were taken at intervals until a predetermined time point or until drug no longer eluted from the coating. Solvents used included PBS and mixtures of PBS containing organic solvent.

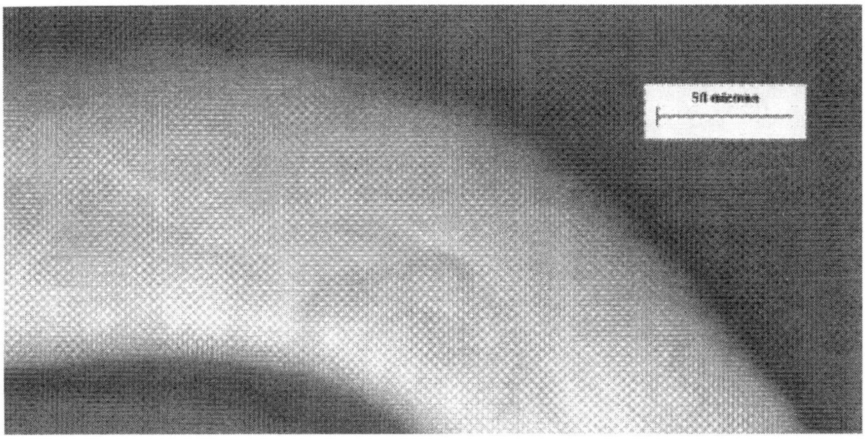

Figure 23 Visible microscope image of drug delivery coating. The coating covers the stent completely and uniformly. The surface of the coating is generally smooth and even.

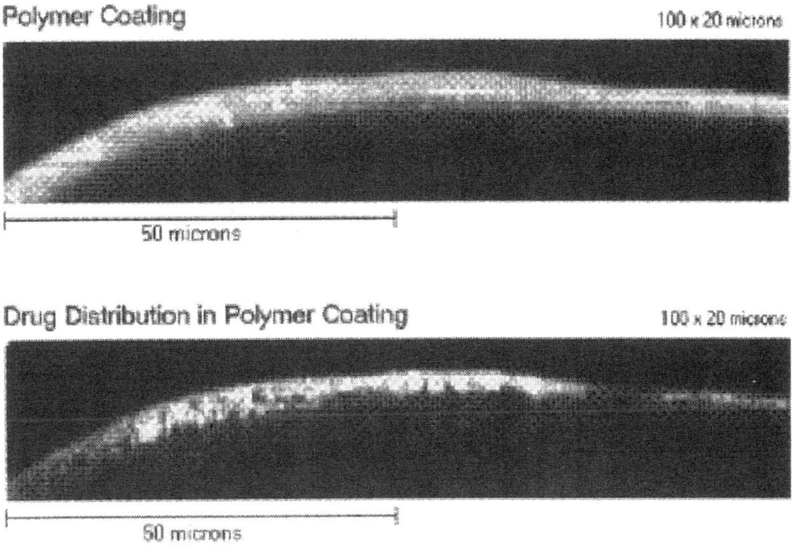

Figure 24 Raman infrared microscope images of a drug delivery coating on a stent. The coating is viewed in cross-section. In the top image, a Raman peak specific to the polymer components is used to indicate the location of the polymers in the coating. It can be seen that the polymer components are uniformly distributed. In the bottom image, a Raman peak specific to the drug is used to map the location of the drug in the coating. It can be seen that the drug is well distributed in the polymer coating, but with some areas of higher and lower concentration.

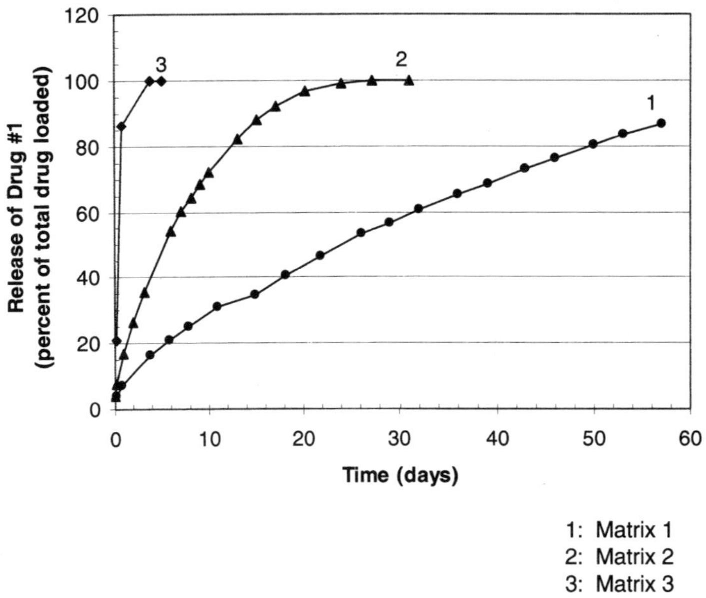

Figure 25 Example of in vitro drug release from several coating variations. Three different polymer coatings containing drug 1 were prepared on stainless steel disks. The same drug loading was used for each coating type, but the composition of the polymer matrix holding the drug was modified to control release. Elution of the drug was controlled to yield fast or slow release.

Experiments with the coatings on disks or stents showed that all three of the drugs could be successfully loaded into the polymer coatings. For stents, it was possible to achieve levels of drug as high as about 500 μg/cm^2, with a more typical loading of 100–200 μg. Elution of the drug was controlled by varying the ratios of the polymers constituting the coating and by varying the ratio of drug to polymer in the coating. As an example, release of drug 1 from the surface of coated disks could be controlled to elute quickly, in a few days, or slowly, in about 2 months (Fig. 25). The control over release was achieved using different coating compositions, with the same total loading of drug. Figure 26 shows the results for drug 2 eluting from similar coating formulations. As can be seen from the data, a wide range of release profiles can be achieved with this coating system. Figure 27 shows the release of drug 3 from stents coated with similar polymer formulations. Although data are not shown here, it is also possible to control the release kinetics by varying the drug loading in the coating.

In addition to these in vitro release studies conducted by SurModics, this polymer coating system containing the drug Sirolimus has been tested extensively in vivo. The release of Sirolimus from coated stents has been demonstrated in both animal [63] and human clinical [64,65] studies to substantially reduce the restenosis rate of coronary stents. In addition to demonstrating the effectiveness of local drug delivery at treating restenosis, these results indicate that the coating system is well tolerated in the arterial tissue, confirming the biocompatibility of the polymer materials.

D. Summary

Coatings can be formed on the surfaces of medical devices to contain and release drug. The delivery of the drug can be designed to enhance the in vivo performance of devices. In a particular

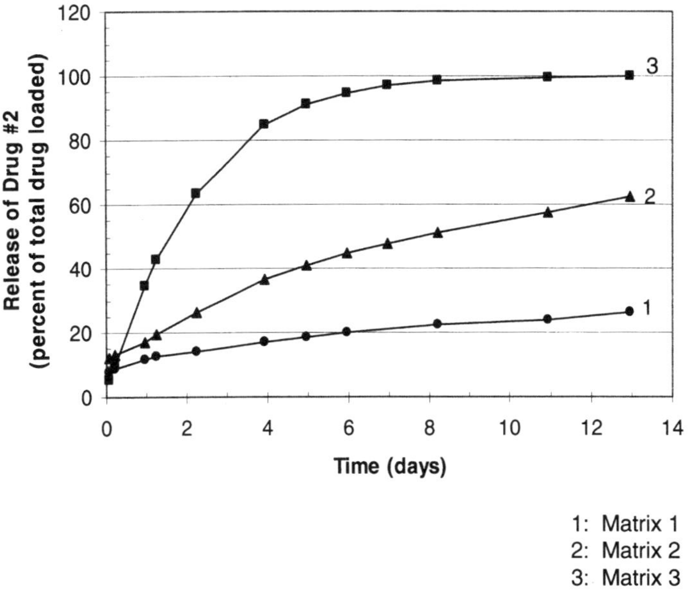

Figure 26 Example of in vitro drug release with a different drug candidate. Three coatings containing drug 2 were prepared using methods similar to those in Fig. 25. Again a broad range of release rates was achieved by varying the composition of the polymer matrix, while holding the drug loading constant.

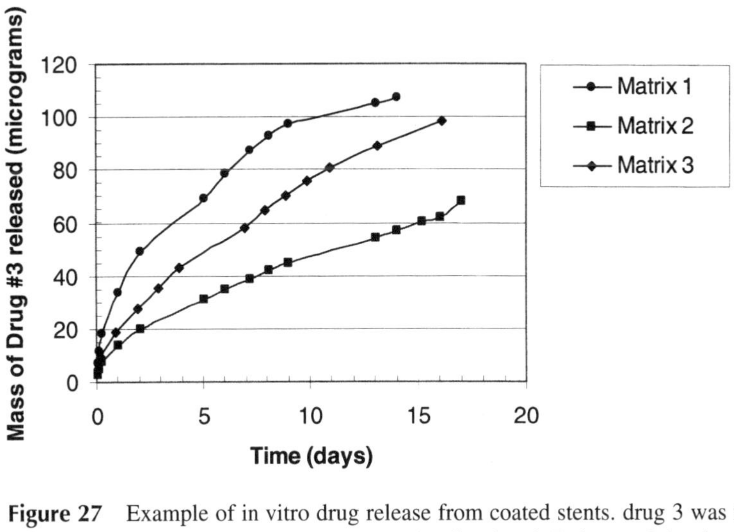

Figure 27 Example of in vitro drug release from coated stents. drug 3 was formulated in three different polymer matrix coatings at the same drug loading level. Release of the drug from the coating into PBS at 37°C was monitored over time. As seen with the coatings on disks above, the release rate of the drug could be controlled by adjusting the coating formulation.

example, coronary stents, coatings have been prepared that can bind up to 500 μg/cm^2 of drug and can release the drug for durations ranging from days to several months. The drug delivery coatings were homogeneously distributed on the stent and were found to be durable to crimping and balloon expansion of the stents, as demonstrated by surface analytical techniques. This coating method is feasible with a variety of drug types. In vivo testing of coated stents containing drug has demonstrated efficacy in animal and clinical studies. Also, although the studies described in this section were focused on coatings for stents to release antirestenosis compounds, the same type of coating approach could be applied to a variety of medical devices.

VII. HIGH-THROUGHPUT BIOCHEMICAL ASSAYS

With the completion of the human genome, several new ''omes'', such as proteome, transcriptome, metabolome, ligandome, and physiome, have been introduced to address various functional components of an organism in its entirety. The traditional approach of studying a single gene, protein, or biomolecule has been complemented and, in many cases, supplanted with the entire set. As an example, capturing a single protein in an organism is important for identification and characterization, but it is the interaction of this protein with numerous other proteins and cellular factors that determines its function and effect in the overall activity of that organism. Studies that challenge the whole genome or proteome are becoming more common and relevant. Therapeutic and pharmaceutical industries are also adapting such platforms for expediting target identification and validation schemes.

In order to analyze the whole genome or proteome, thousands of molecules are processed simultaneously in an orderly fashion. For higher throughput, these processes have to conform to the same standards as was developed for the computer industry decades ago, i.e., miniaturization, automation, and parallelism. Arrays or microarrays provide such an integrated platform for the analysis of biomolecules [66,67]. Microarrays are generally small glass microscope slides conditioned for immobilization of nucleic acids, proteins, carbohydrates, tissues, or living cells and running subsequent assays for identification or detection of targets of interest. These arrays are used to run nucleic-based expression analyses (DNA arrays) [68–70], run immunoassays, and protein–protein and protein–DNA interactions (protein arrays) [71–74]; analyze small molecular weight ligands to capture drug targets (chemical arrays) [75,76]; study gene effects, functions and phenotypes in living cells (cell arrays) [77,78]; monitor molecular recognition and anti-infection responses (carbohydrate arrays) [79,80]; and to obtain a molecular profile of tissues (tissue arrays), [81–83].

DNA arrays are still the most commonly used arrays. These arrays have been used for monitoring gene cataloging, gene discovery, genotyping, mutation detection, exon mapping, and resequencing analyses [84–89]. There are two types of DNA microarrays: cDNA and short oligonucleotides (oligos). Short oligo–based microarrays are further divided into two groups. In one group, microarrays are developed by in situ synthesis of the oligos on the solid support by photolithography (Affymetrix, Rosetta); whereas in the second type arrays contain oligos that are presynthesized, purified, and immobilized on the glass surface (Amersham Biosciences, Agilent). Arrays with presynthesized oligos provide better quality control and flexibility of designing any array of choice without redesigning the whole chip.

Proteins are the ultimate functional units in any organism. The RNA level, which is studied to understand the expression of different genes, is not truely representative of the cellular activity in a certain tissue at a certain time and condition. There are close to a million proteins, versus about 30,000–40,000 genes, in humans [90,91]. Several proteins are translated from a certain gene; therefore, the ultimate goal of a functional analysis is to get a hold of proteins, directly or indirectly. Protein arrays, though still in infancy, are opening doors to address challenges faced by protein research. Proteins are quite sensitive to the reaction conditions; slight variation in the buffer and environmental conditions can render them inactive.

A. Matrices for Nucleic Acid and Protein Immobilization

Immobilization of nucleic acids onto surfaces is quite challenging. A desirable surface supports the attachment of biomolecules as well as being suitable for running biochemical assays. In a majority of the cases, glass or plastic surfaces as well as the oligos have to be chemically modified for immobilization of the oligos. PhotoLink technology is used to modify glass surfaces for producing three-dimensional porous surfaces (see Figs. 1 and 2; Section II). Silanized glass slides are coated with polymers containing both photoactivatable groups and active esters. Upon illumination, the polymer forms an insoluble, porous, lightly crosslinked hydrogel matrix with reactive groups that are covalently bound to the surface. Amine-terminated oligos are spotted on the polymer layer where the ester reacts to form the structure shown in Fig. 28a. To enhance the binding of the oligos and hydrolyze the reactive esters, spotted slides are incubated at higher humidity conditions followed by soaking in ethanolamine to purge residual esters.

Unlike DNA, proteins are quite sensitive to immobilization on planar surfaces and lose activity if proper conditions are not maintained. In their simplest form, proteins are adsorbed on surfaces that have high affinity for them, e.g., polystyrene. These surfaces are simple to use but can cause protein inactivation and steric occlusion [92]. In addition, coating on such surfaces results in a patchy and heterogeneous immobilization. Another method relies on genetically adding a tag on one end of the protein to covalently bind proteins to the activated surface. However, this approach is expensive and not practical for high throughput. For our protein immobilization research, we use an aldehyde surface (Fig. 28b) followed by a reduction step. Aldehydes bind with primary amines of proteins and provide a gentler surface that maintains a high level of protein activity. High immobilization and signal-to-noise ratio are produced, which makes it suitable for protein–protein and DNA–protein interaction as well as for immunoassays [73].

B. Characteristics of Surfaces Coated with PhotoLink Technology

1. Three-Dimensional Configuration

Immobilization of a biomolecule on a planar surface results in lower sensitivity and consistency in high-throughput biochemical assays. Membranous and polymer-coated slides provide the

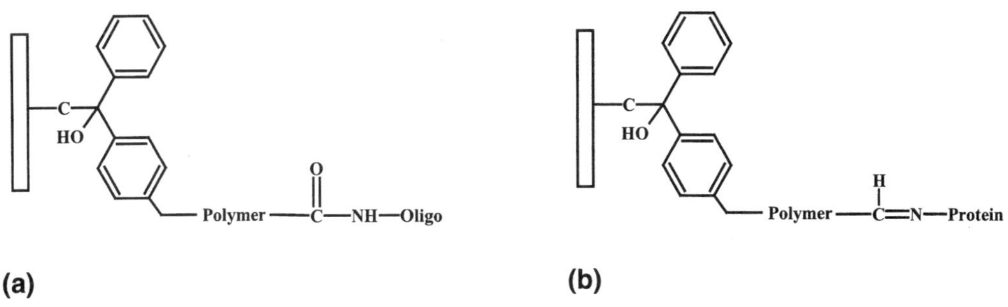

Figure 28 Photocoupling of polymers to the glass surface for (a) oligo and (b) protein immobilization.

porosity needed for higher loading of the biomolecule as well as for running the subsequent assays. Having a three-dimensional supporting matrix is advantageous since it generates a uniform surface capable of homogeneous immobilization. Matrices generated at SurModics conform to these needs (Fig. 29).

With the increase in coating thickness, higher hybridization signals were obtained with lower oligo usage in DNA microarrays; 1.5- to 2-fold more signal was detected in thick-coated slides than that in thin-coated slides printed at the same oligo concentration (Fig. 30). Thus, thicker coated surfaces can reduce the cost of the microarray fabrication considerably. Moreover, a thicker hydrogel matrix is well suited for preserving protein activity by maintaining suitable buffer conditions. This matrix could also be modified to support live cells for functional analysis and identification of drug targets in arrays of living cells [78]. Protein immobilization experiments for enzyme-linked immunosorbent assay (ELISA) on slides coated with PhotoLink reagents have shown improved sensitivity and minimal nonspecific binding of the antibodies or antigens.

2. High Attachment Efficiency

Binding capacity is indicative of the reactivity of the surface conferred by active esters. Several factors affect this capacity, especially print buffers, thickness of the matrix, and environmental factors. A linear relationship was observed when a fluorescently tagged oligo was printed at various concentrations (Fig. 31). However, this binding is not reflective of the hybridizability of the oligo; steric hindrance limits the amount of the target that is hybridized. Maximum hybridization signal is seen when the oligo is printed at a concentration between 5–10 μM,

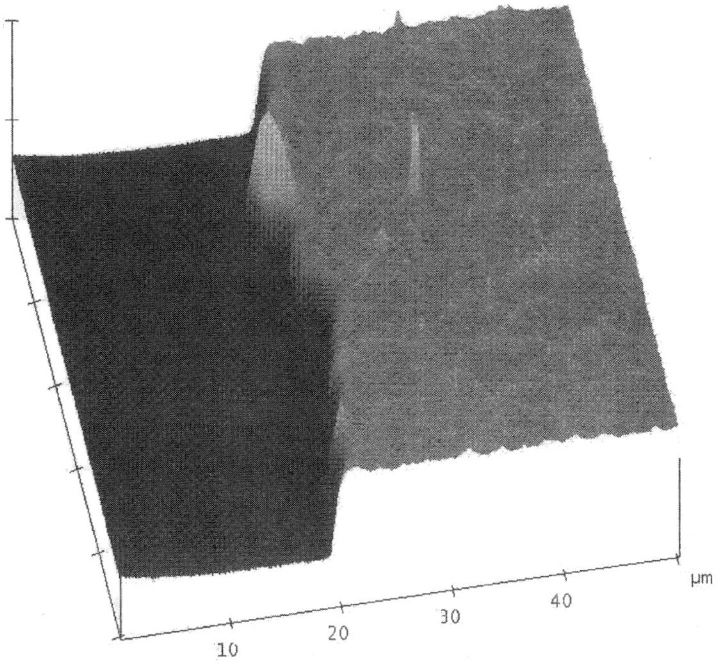

Figure 29 Three-dimensional image of a section of a coated slide taken by atomic force microscopy. Part of the surface has been scratched to measure thickness by tapping mode and is shown in black.

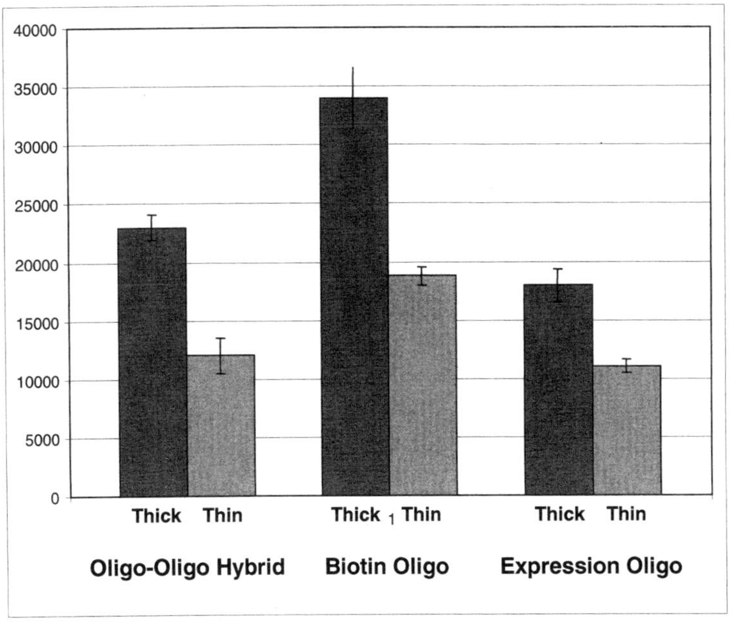

Figure 30 Effect of thick versus thin coated slides on print density and hybridization signal. Oligos were printed at a concentration of 20 μM in 50 mM phosphate buffer using a contact printer. Oligo in the oligo–oligo system was a 30-base-long probe picked from a human gene. To determine differences in thin- and thick-coated slides, the oligo-printed slides were hybridized to a 30-base-long, fluorescently tagged target in a standard oligo hybridization procedure. Biotin and expression were also 30-base-long oligos from human genes designed to determine attachment density and hybridization pattern, respectively. The target for the biotin oligo was a 30-base-long oligo terminated with a biotin, while for the expression, the target was biotinylated cRNA amplified through in vitro transcription methods from human liver RNA. Hybridization was carried out in Grace bioLab Chambers for 16 h at 37°C with constant shaking at 300 rpm. Finally, the hybridized slides were coupled with Streptavidin Alexa Fluor-647 and imaged on a GenePix Axon scanner.

irrespective of the target concentrations used (data not shown). Binding capacity of a PhotoLink slide matrix was determined using ^{35}S-labeled 30-base-long oligos. The oligo was printed at a concentration of 20 μM in standard printing buffer and allowed to react over night at 75% relative humidity. Slides were run through a mock hybridization. At the end, oligos were detached from the slides and the activity measured in a scintillation counter. These and other results obtained at SurModics indicate that 5–6 pmol/cm^2 were immobilized (Fig. 32).

Oligo binding is dependent on the concentration applied (Fig. 29). This information, combined with NMR data on the reactive esters and a ^{3}H-lysine study to determine the amount of polymer deposited on the slide, indicates that immobilization of 5–6 pmol/cm^2 results from reaction with a small fraction of NOS groups on the surface. Higher oligo immobilization results when higher concentrations are deposited. Nevertheless, this enhanced immobilization of oligos does not affect the hybridization pattern of the system when larger molecules, like RNA, are used a targets. Maximum hybridization is detected at an oligo print concentration between 5–10 μM (Fig. 33). A drop in hybridization signal beyond 10 μM oligo print concentration is suggestive of the steric hindrance.

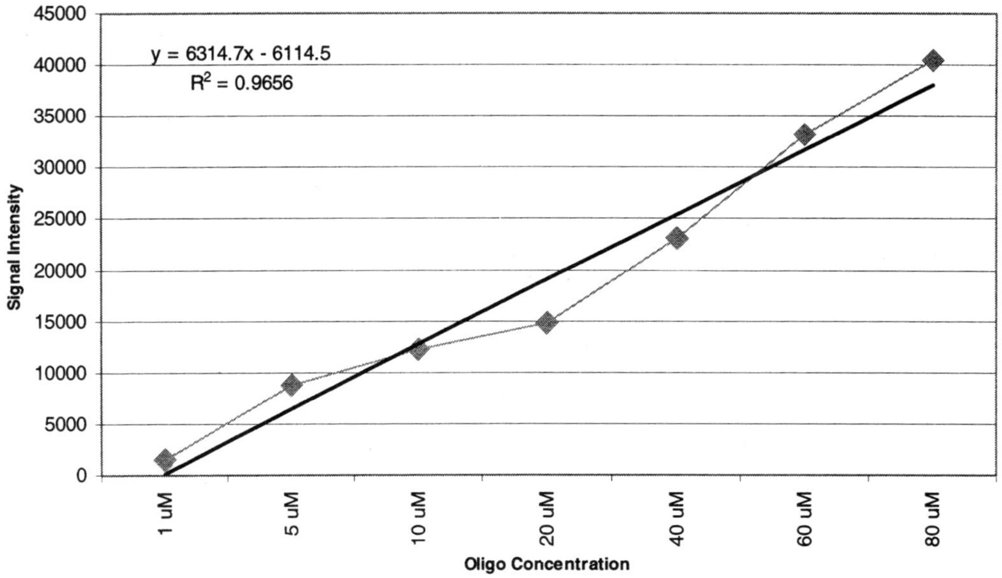

Figure 31 Signal intensity of an oligo printed at different concentrations. A 30-base-long oligo was labeled with Cy3 on the 3′ and C6-amino linker on the 5′ end. Signal was detected using an Axon scanner in the appropriate channel.

3. Linear Dynamic Range and Sensitivity

A comparison of amine- and hydroxyl-terminated oligos revealed that several-fold less attachment efficiency is observed in the latter. If active esters are not hydrolyzed, such nonspecific binding could result in enhanced background when hybridization assays are run on them for an extended period of time. To minimize this binding, printed slides are incubated at higher relative humidity (75%) and soaked in ethanolamine, which helps improve the signal-to-noise ratio. This has been confirmed in several studies done in house as well as by others [93]. Lindroos, et al. [93] compared several commercially available coated slides with the ones coated in their labora-

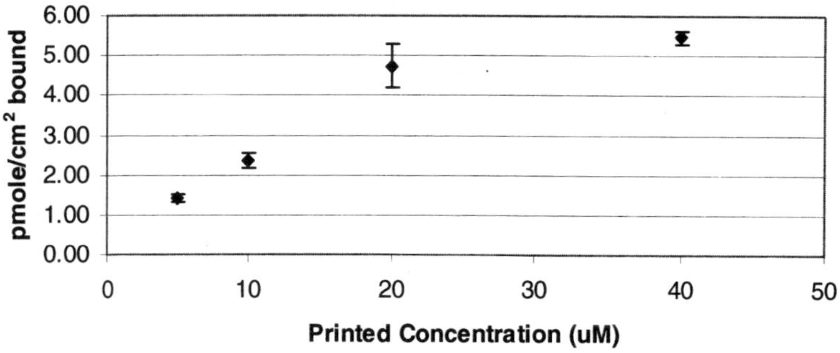

Figure 32 Signal intensity of an oligo printed at different concentrations. A 30-base-long oligo was labeled with Cy3 on the 3′- and C6-amino linker on 5′ end. Signal was detected using an Axon scanner in the appropriate channel.

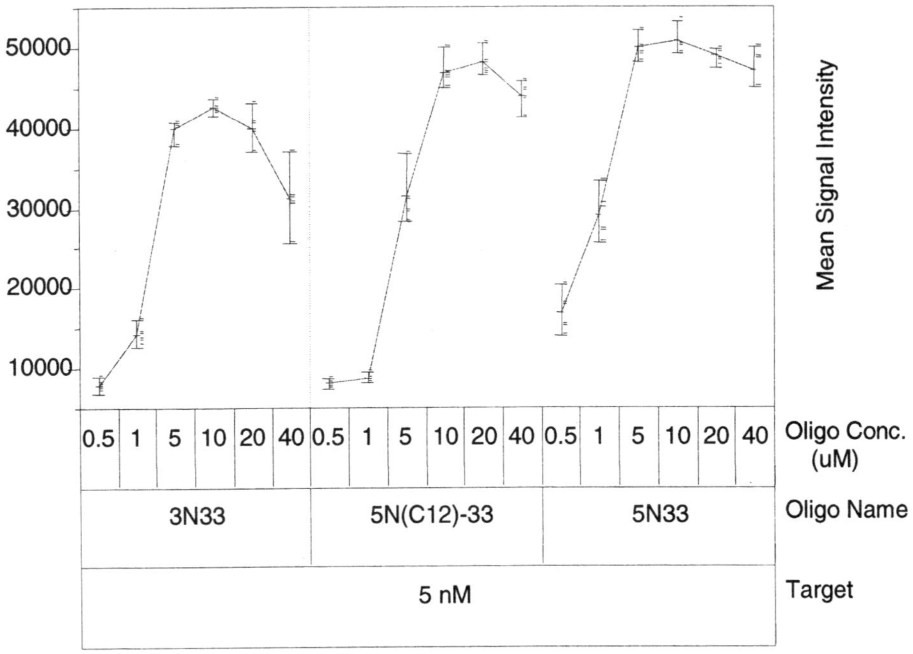

Figure 33 Effect of print concentrations of capture oligo on the hybridization signal. Hybridization target was done with a synthetic fluorescently tagged target at a concentration of 5 nM. All the oligo sequences are the same and 30 bases long; 3N33 was modified with C6 amino linker at the 3′ end, while 5N33 was modified with the same linker at the 5′ end, and 5N(C12)-33 was modified with C12 amino linker on the 5′ end. Hybridization was carried out at 45°C for 2 h with shaking at 350 rpm in 5X SSC, 0.1% SDS.

tory using different chemistries. Parameters used for the evaluation of slides consisted of slide background, attachment efficiency, and performance in detecting single nucleotide polymorphism. In this study, overall signal intensities from 3D-Link™ Activated Slides (amine functional SurModics slides, sold as CodeLink™ Bioarray Slides by Amersham Biosciences) were 6–40 times higher than the second best performer among the selected group of slides.

Ramakrishnan et al. [94] assessed the performance of the CodeLink slides for performing gene expression analysis in terms of specificity, reproducibility, and linear dynamic range. Based on results from transcript spiking and hybridization, sensitivity of the system was determined to be 1:900,000. According to some estimates, a mass ratio of 1:300,000 [95] or 1:100,000 [96] corresponds to the gene expression of one copy per cell. Specificity of the CodeLink system was also quite high with a 90% sequence homology, which corresponds to a 3-base mismatch in a 30-base-long fragment. Similarly in this study, coefficient of variation of 10% was observed across slides and target preparations. Hybridization signal was linear over 2.5–3.0 orders of magnitude. Results from this study underscore the importance of a high quality matrix along with attachment chemistry and target preparation methods.

4. Overall High Performance

In a recent study conducted at SurModics, three activated surfaces were compared with 14 other commercially available activated slides, all generated through PhotoLink technology (Fig. 34). Poly-L-lysine slides coated in house were used as a control. All slides were printed according

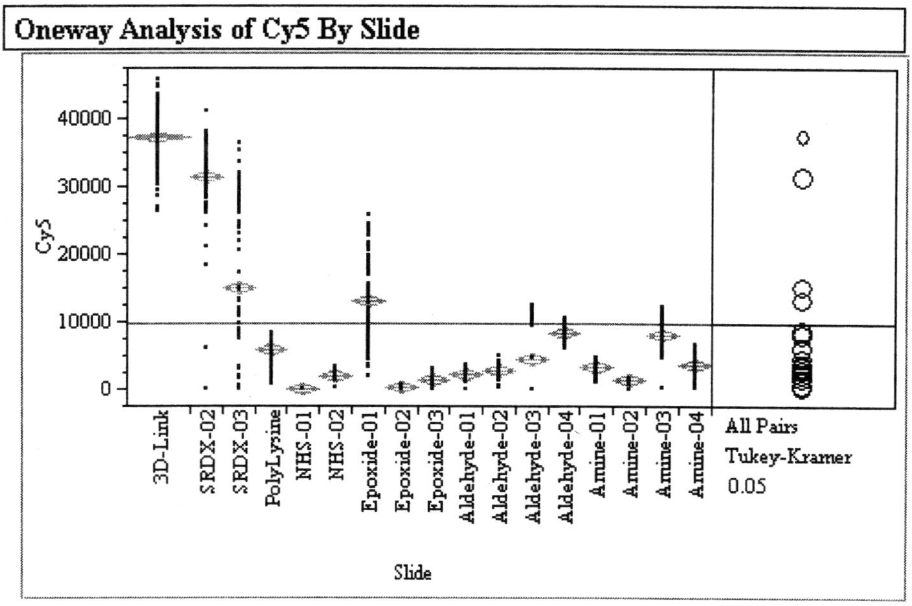

Figure 34 Comparison of commercially available activated slides with the ones produced at SurModics using PhotoLink technology. 3D-Link is the standard amine functional slide; SRDX-02 and SRDX-03 are two slides produced with different chemistries. Other slides are arbitrarily numbered and grouped as NHS, Epoxide, Aldehyde, and Amine, based on their reactivity. Poly-L-lysine is used as a control. Slides were printed with several expression and control oligos and hybridized with biotinylated cRNA from human liver for 16 h at 37°C with constant shaking at 300 rpm in Grace BioLabs chambers in 4X SSC, 0.1 % SDS. Streptavidin Alexa Fluor-647 was coupled to the hybridized targets and images were scanned on an Axon Scanner.

to the manufacturers' recommendations for buffers and subsequent processing. The slides were printed with several positive and hybridization controls as well as with oligos from known human genes. Hybridization was done with biotinylated cRNA generated from human liver total RNA at 37°C for 16 h at 300 rpm. The results showed a significantly high attachment efficiency in slides produced at SurModics (data not shown). The hybridization signal in 3D-Link slides was also significantly higher than any other products included in the study, even though the spot size was very similar in all the products. Similar results have been reported in personal communications.

Protein immobilization experiments with aldehyde and other slides resulted in high efficiency as well. Several sandwich ELISA assays were designed to understand the sensitivity, reproducibility, and overall performance of activated slides with IL-1α, IL-2, IL-4, TNF-α, and IFN-γ proteins. Higher sensitivity for antibody–antigen coupling and low background was detected when compared with other similar systems.

VIII. BIOMIMETIC SURFACE MODIFICATIONS FOR CELL GROWTH AND TISSUE INTEGRATION

One approach to improving the performance of tissue culture products and implant devices consists of modifying their surfaces with either extracellular matrix (ECM) proteins or ECM peptides derived from these proteins. Surfaces modified with appropriate proteins or peptides are less likely to be recognized as foreign than the original device surface and will promote the attachment and overgrowth of specific desirable cell types. Previous attempts to enhance the performance of biomedical products with adsorbed ECM proteins or peptides have produced only marginal improvements. However, as is described below, when such proteins or peptides were covalently immobilized at monolayer or greater levels (via photochemistry), the resultant surfaces greatly improved cell attachment and growth in vitro and tissue integration in vivo. Specific ECM proteins that were photoimmobilized onto surfaces and shown to improve the in vitro and/or it in vivo performance of pecific devices include fibronectin (FN), laminin (LM), type I collagen (COL I), and type IV collagen (COL IV). Also, 16 peptides derived from these proteins have been evaluated with in vitro assays.

A. Methods for Reagent Synthesis and Photocoupling to Surfaces

Photoreactive ECM proteins or peptides were added to the substrates and photoactivated to produce covalent coupling as described previously for other photoactivatable reagents. Adsorbed controls were generated by adding nonphotoreactive ECM proteins or peptides to the same materials under similar conditions (concentration, incubation time, etc.) and not activating with light. To remove loosely adherent reagents, each surface-modified material was then washed overnight with continuous agitation in four sequential solution changes of PBS which contained 1% Tween 20. The samples were then sterilized by soaking 30–60 min in 70% ethanol, and residual Tween 20 and ethanol were removed by four sequential washes in PBS (15–20 min each).

The proteins and peptides were radiolabeled with tritium and used to quantitate immobilized levels on each substrate. Tritium was added to each reagent by reductive methylation [97], which consisted of reacting formaldehyde with a small portion of the primary amines on each protein or peptide and then reducing the resultant Schiff base with ^{3}H-sodium borohydride (5–20 Ci/mmol). Each peptide was synthesized with at least two lysine moieties to provide the necessary primary amines, and the resultant radiolabeled proteins and peptides had specific activities of $2–8 \times 10^4$ dpm/μg. This tritium-labeling procedure was chosen over iodination as the routine labeling method because (1) tritium produces sufficient specific activity for the experiments described below; (2) tritium has a longer decay half-life, which allows labeled reagents to be stored and used for longer times; (3) tritium is less hazardous to use since it is both less energetic and has a 12-fold shorter biological half-life; and (4) the tritiation procedure is less likely to degrade the proteins or peptides.

Evidence supporting the latter concern is presented in Fig. 35, which describes a preliminary experiment in which an ECM peptide (F-9) was radiolabeled via both tritiation and iodination and then evaluated by reverse-phase HPLC. When the eluant was monitored for absorbance at 215 nm (to detect peptide bonds), unlabeled F-9 and the products of each labeling method produced a single major peak at 16–17 min. The tritium counts also eluted at the same time (when adjusted for the lag time between the absorbance and tritium detectors) and as a single peak of similar width to the absorbance at 215 nm, which is consistent with the tritiation method labeling the peptide with no significant peptide degradation. In contrast, the ^{125}I counts eluted later and as a broad double peak, which is consistent with peptide degradation or preferential labeling of minor components. Results from the experiments in Fig. 35 support the preference for tritiation instead of iodination as a radiolabeling method. The major disadvantage of using tritium to quantitate the immobilized levels of protein or peptide lies in the need to remove the labeled reagent from the biomaterial surface and to disperse it uniformly in scintillation cocktail to achieve accurate counting. Depending on the biomaterial substrate being evaluated, this is

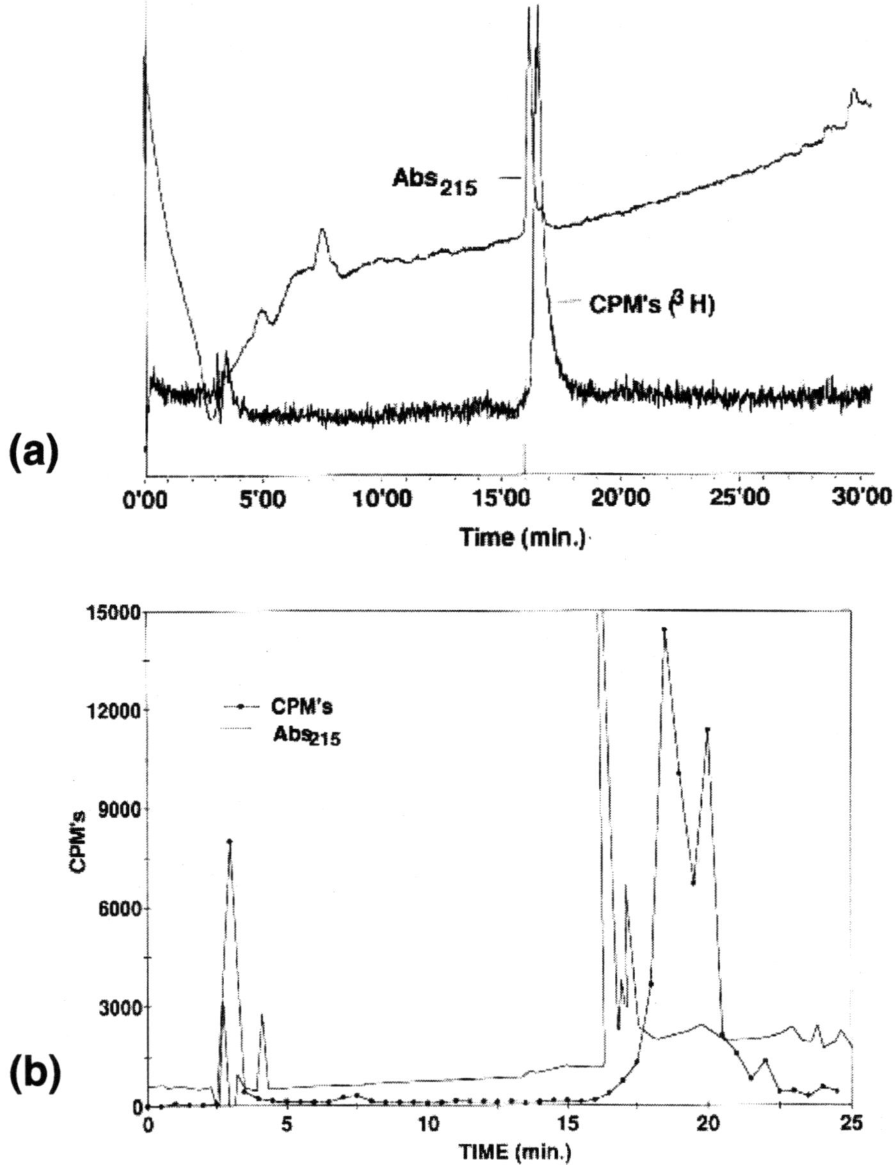

Figure 35 HPLC elution of an ECM peptide after being radiolabeled by either (a) tritiation or (b) iodination. The tritiation followed the reductive methylation procedure described by Means and Feeney [97]. The iodination used IODO-GEN™ and procedures provided by the supplier (Pierce, Rockford, IL). F-9 is a peptide derived from laminin [98], and the amino acid sequence used here was RYVVLPRPVCKK. The HPLC procedure used a Vydac C_{18} reverse-phase column, eluted with 0.05% trifluoroacetic acid in a linear water/acetonitrile gradient that started with 95% water/5% acetonitrile and ended with 40% water/60% acetonitrile at 28 min. Absorbance at 215 nm was monitored with a Perkin-Elmer model LC 85-B variable wavelength detector. Tritium was monitored with a flow-through radioactivity detector (-RAM MOdel 1B) from IN/US Systems, Inc. (Tampa, FL). ^{125}I was monitored by collecting aliquots at 0.5-min intervals and counting in a liquid scintillation analyzer (Packard Tri-Carb® model 1900 CA).

achieved by either dissolving the underlying substrate (if it is readily dissolved in common organic solvents) or else hydrolyzing the protein or peptide (e.g., with 6 N HCl at 75°C for 18 h). For each of the materials discussed below, the proteins or peptides were photoimmobilized at monolayer or greater levels ($\geq$100 ng/cm^2 for the peptides and $\geq$500 ng/cm^2 for the proteins).

B. ECM Proteins and Peptides Immobilized on Polystyrene Cell Culture Plates

For this series of experiments, proteins and peptides were photoimmobilized onto previously unmodified 24-well polystyrene (PS) plates and evaluated with each of four cell lines: CPAE endothelial cells, NRK-49F fibroblasts, PA-1 or MDBK epithelial-like cells, and MG-63 osteoblasts. As controls, cell attachment and proliferation were also evaluated on unmodified PS, standard tissue culture polystyrene plates (TC-PS), and PS modified with photoimmobilized COL IV. The biological activity of each immobilized protein or peptide was evaluated for cell attachment (at 2 h) and cell proliferation (through $\geq$6 days). At the end of each experiment, relative cell numbers on each surface were quantitated with a tetrazolium metabolic dye (MTT) [99].

With each cell line, photoimmobilized COL IV promoted cell attachment, spreading, and proliferation that was equal or superior to that observed with the other photoimmobilized ECM proteins or peptides. Also, no peptides were identified that promoted uniquely superior attachment, spreading, or proliferation of any one cell type. Instead, eight of the most active peptides promoted similar attachment, spreading, and proliferation of each cell type. When compared to unmodified PS, PS coated with these peptides promoted 5- to 20-fold greater attachment and proliferation of each cell type, and when compared to TC-PS the same peptides coated onto PS promoted two- to four-fold greater attachment. However, when proliferation on TC-PS was compared to that on peptide-coated PS, high passage cells grew equally well on both surfaces. In contrast, low passage human umbilical vein endothelial cells (HUVECs) showed threefold greater cell numbers at 8 days when cultured on COL IV or a peptide derived from COL IV as compared to TC-PS (Fig. 36). A similar trend was observed with low passage bovine cornea epithelial cells, where 30–60% greater cell numbers were observed at 8 days on COL IV and several ECM peptides as compared to TC-PS. These results indicate that low passage cells respond more favorably to protein or peptide-modified surfaces than do established cell lines.

With a single exception, no peptide showed greater than threefold selectivity for one cell type over another. However, when RGD peptide was immobilized on PS via a short (7-atom) spacer, it promoted attachment and proliferation of nonepithelial cells (CPAE and NRK-49F), but not of epithelial cells (PA-1 and MDBK). However, with longer spacers, RGD promoted attachment and proliferation of both epithelial and nonepithelial cell lines. Typical cell attachment results are shown in Fig. 37. Cell proliferation assays showed similar results. Such selection against epithelial cells should be useful for percutaneous implant devices where inhibition of epithelial cell marsupialization (downgrowth along device surfaces) is desirable.

C. ECM Proteins Immobilized on Implant Devices

Extracellular matrix proteins were photoimmobilized at monolayer or greater levels onto implant devices and improved cell attachment and proliferation in vitro and tissue integration and device performance in vivo.

For one application, bovine skin collagen (95% COL I plus 5% COL IV) was photoimmobilized onto the surface of a PMMA intracorneal lens and implanted into rabbits for up to 15 months [102]. The unmodified lenses showed considerable stromal inflammation and corneal

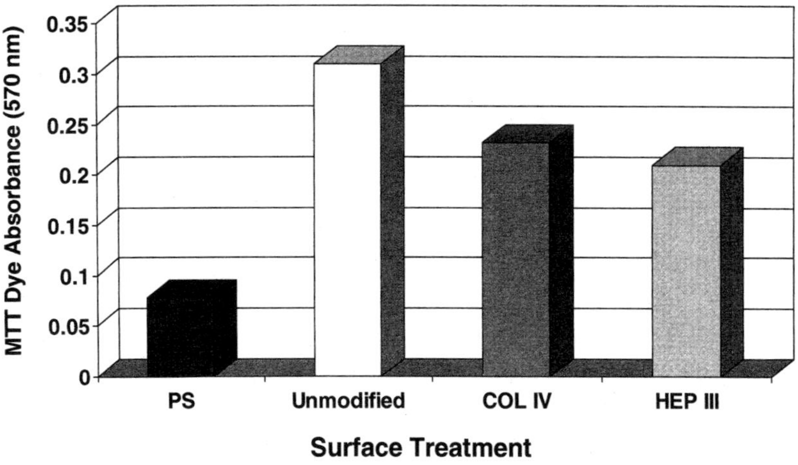

Figure 36 Growth of human umbilical vein endothelial cells on polystyrene 24-well plates. Tissue culture PS = standard cell culture PS. The remaining surfaces were unmodified PS or PS-coated with photoderivatized protein (COL IV) or a photoderivatized COL IV peptide (HEP III). The amino acid sequence of the HEP III was CKGEFYFDLRLKGDK [100]. The plates were seeded with 1500 cells per well, cultured for 8 days, and relative cell numbers were quantitated with a tetrazolium metabolic dye (MTT). Each bar is the average of four determinations.

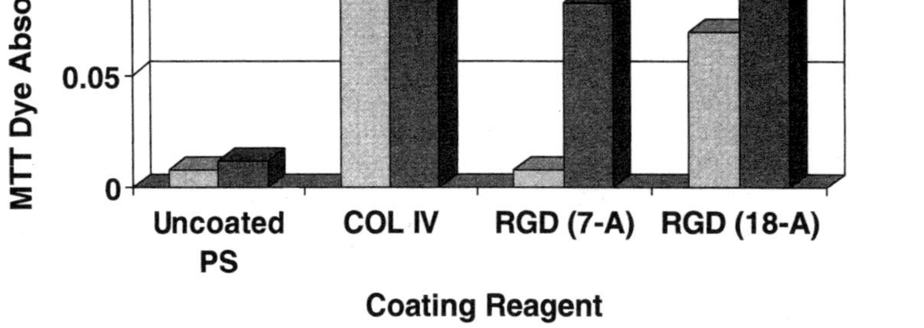

Figure 37 Cell attachment to RGD peptide. RGD (7-A) and (18-A) = RGD photoimmobilized via 7- and 18-atom spacers, respectively. RGD is a peptide derived from fibronectin [101], and the amino acid sequence used here was GRGDSPKKC. The plates were seeded with 50,000 cells per well (of 48-well plates), incubated for 2 h, and relative cell numbers were quantitated with MTT. Each bar is the average of four determinations.

necrosis by 2 months. In contrast, the collagen-modified lenses showed decreased inflammation, strong bonding of stromal tissue to the device, and retention of normal corneal morphology and clarity at all time points from 1 through 15 months. These lenses have been implanted into two patients who could not be treated by other methods and who had each been blind for more than 5 years. Their respective visions at 7 and 12 months were 20/50 to 20/100.

For a second application, ECM proteins were immobilized onto two types of vascular graft materials: PU and expanded polytetrafluoroethylene (ePTFE). Figure 38 shows that in vitro endothelial cell growth was 10- to 20-fold greater on photoimmobilized FN than on adsorbed FN or unmodified ePTFE. Similar in vitro results were observed with photoimmobilized COL IV and a combination of FN plus COL IV. Then unmodified and protein modified grafts were implanted intrafemorally into dogs for 30 days, after which the patent grafts were retrieved and evaluated by SEM for the percentage of the luminal surface of each graft that was covered with endothelial cells [103]. As is shown in Table 3, similar trends were observed with both types of graft materials. Patency rates (i.e., the fraction of grafts retaining unblocked blood flow) averaged 33% for unmodified controls (one of four PU grafts plus three of eight ePTFE grafts) versus 83% for grafts modified with a combination of FN + COL IV (three of four PU grafts plus seven of eight ePTFE grafts). Luminal coverage by endothelial cells averaged 31% for unmodified controls and 86% for protein modified grafts. It is important to note that when FN and COL IV were immobilized individually, the patency and/or endothelialization was less than that observed when both proteins were immobilized, therefore indicating that each protein is providing a critical activity.

In a third application, ECM proteins were photocoupled to 100-mL model SR breast prostheses that were filled with saline and implanted subcutaneously in pigs for 4 months. The prostheses were explanted and evaluated histologically for thickness of fibrous capsule adjacent to the surface (data not shown). As compared to unmodified controls, implants with photoimmobilized COL IV produced a 48% thinner fibrous capsule. For each implant variation, the average capsule thickness, standard error of the mean, and number of replicates were as follows: unmodi-

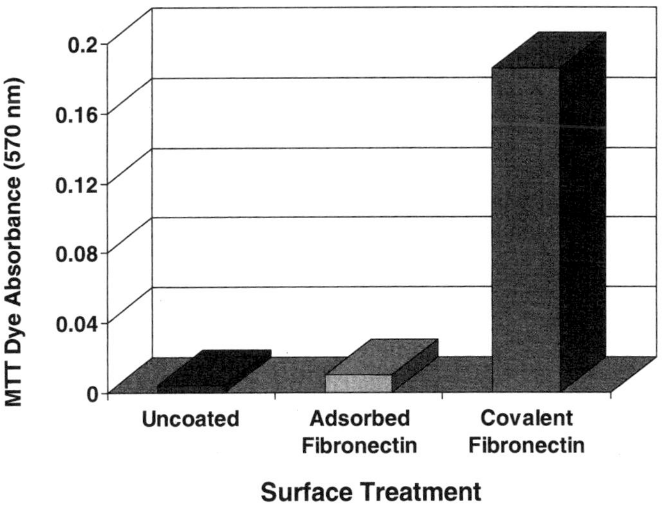

Figure 38 Growth of calf pulmonary artery endothelial (CPAE) cells on ePTFE. Coated and uncoated ePTFE disks were placed in 24-well plates, seeded with 1500 CPAE cells per disk, and cultured for 7 days. Relative cell numbers were quantitated with MTT. Each bar is the average of four determinations.

Table 3 Patency and Luminal Coverage of Endothelial Cells on Unsodded 4-mm Grafts Implanted 30 Days in Dogs

	Polyurethane			ePTFE		
Surface modification	No. implanted [a]	No. patent grafts	Average L.C.[b]	No. implanted[a]	No. patent grafts	Average L.C.[b]
Unmodified	4	1	25	8	3	33 ± 15
Type IV collagen	4	1	100	4	2	54 ± 6
Fibronectin	4	1	100	–	–	–
Fibronectin plus type IV collegen	4	3	82[c] ± 18	8	7	88[c] ± 6

[a] No. implanted = number of grafts implanted.
[b] L.C. = luminal coverage; the percentage of the luminal surface of patent grafts that was covered with endothelial cells (as evaluated by scanning electron microscopy). Averages ± standard error of the means are shown.
[c] As compared to uncoated controls, grafts modified with fibronectin plus type IV collagen produced statistically significant improvements in both patency ($p = 0.01$) and endothelial cell coverage ($p<0.001$). Two-tailed t-tests; data from PU and ePTFE grafts were pooled for each comparison.

fied control, 1.06 ±0.13 mm (n = 7); COL IV surface modification, 0.55 ± 0.07 mm (n = 7); and FN + COL IV surface modification, 1.30 ± 0.27 mm (n = 8). Interestingly, the FN + COL IV surface modification that performed best for vascular grafts produced the thickest fibrous capsule when coupled onto breast implants.

Finally, a preclinical study evaluating the effects of a COL I coating on the in vivo performance of endovascular stent grafts was performed (S. Williams et al., unpublished results). Endovascular stenting is an emerging technology that is being evaluated for the treatment of aneurysms and arterial occlusive disease. The use of stent grafts to accomplish aneurysmal exclusion provides many potential benefits and improvements to the current treatments. Unfortunately, blood seepage (i.e., endoleaks) at the site of the graft attachment, through the implant material, and through the backflow of lumbar branches can all contribute to the possible rupture of the aneurysm sac.

Type I Collagen was photocoupled to Dacron-covered endoluminal stent grafts and the grafts, along with uncoated controls, were implanted in the femoral arteries of dogs. The arteries were catheterized and the endovascular grafts were inserted and advanced to a position below the renal arteries. Pre- and postangiograms were performed to observe collateral flow through any of the branches adjacent to the implant. At 12 weeks, the dogs were anesthetized and the grafts exposed. An explant angiogram was performed, and the prostheses were removed and prepared for histological examination.

Angiographic analysis following implantation revealed the absence of collateral blood flow around the collagen-coated grafts when compared to uncoated controls. All grafts remained patent, indicating that the collagen coating did not contribute to the occlusion of the graft lumens.

Histological examination showed a well-formed neointimal lining with an endothelial monolayer spanning the entire length of all the grafts. Cells in the subendothelial layer reacted with antibodies specific for alpha-smooth muscle cell actin. The results of this study demonstrate that a PhotoLink collagen coating may help prevent endoleak formation in endovascular implants and that such coatings do not interfere with the normal healing response.

D. Summary

These in vitro and in vivo results demonstrate that photochemistry is a gentle process that immobilized fragile biomolecules onto several biomedical device polymers with retention of critical activities. The resultant modified surfaces greatly improved the culture of low passage cells in vitro and implant device performance in vivo. As one would expect, the studies discussed previously demonstrate that different ECM proteins or peptides are required for different device applications.

IX. THREE-DIMENSIONAL MATRICES FOR TISSUE ENGINEERING APPLICATIONS

The use of synthetic polymers to improve the biocompatibility of implantable medical devices has become a well-established method in the industry. For interventional applications requiring the use of a device, the development of effective surface modification technology has accelerated the growth and expanded the use and the number of applications that can be addressed through the use of these devices.

There remain, however, a number of therapeutic applications in which the use of a device is either not possible or less beneficial than other approaches. Several of these applications fall within the scope of the discipline that is currently known as tissue engineering. Tissue engineering is commonly defined as the combination of synthetic materials with tissue or cellular materials which is implanted to treat an injury or disease state. An expanded definition of tissue engineering includes the implantation of a wholly synthetic construct, which may contain bioactive compounds, for the purposes of inducing a tissue response to repair an injury or treat a disease. Typically, these constructs are preformed porous matrices into which cells of interest are seeded and the matrix plus cells implanted. While these constructs are potentially useful for the treatment of many conditions, it is often the case that this method is not ideal. The preformed matrices do not necessarily fit into the implantation site; they do not conform intimately to the tissue at the implantation site; and their implantation is not amenable to minimally invasive techniques.

A potentially superior methodology involves the in situ formation of the matrix construct directly at the tissue site of interest. One way to accomplish such a construct formation is to begin with matrix precursor reagents which can be applied to the tissue site as a liquid and subsequently solidified, rapidly and on command, to effect matrix formation directly at the site of interest. The use of PhotoLink technology allows the preparation of matrix precursor reagents which can be applied to a tissue defect, wound site, aneurysm, or any anatomical or ex vivo site in liquid form, and subsequently solidified into a matrix possessing the necessary morphological properties, by a brief exposure to visible light. These matrix precursor reagents can be prepared from synthetic biocompatible polymers or from naturally occurring biopolymers.

A. Reagents

Reagents useful for matrix-forming tissue-contacting applications can be prepared from a variety of starting materials. The use of PhotoLink technology to effect matrix formation permits a large degree of flexibility in the design of the reagent system.

Polymeric matrices can be prepared using a variety of techniques and starting materials. Polymers can be prepared containing crosslinkable groups that will form matrices when exposed to the appropriate reagents. Polymer networks can be prepared by exposing the correct mixture of monomers to the appropriate polymerization initiating conditions. When forming polymeric matrices in situ, in the presence of living tissue, many of these techniques and materials suffer from severe drawbacks. Biocompatibility and cytotoxicity concerns significantly limit the number of matrix-forming approaches appropriate for these applications.

One approach that avoids these problems incorporates the use of preformed polymeric reagents, or macromers, which are capable of additional polymerization reactions. These macromers are generally polymers to which additional polymerizable groups have been incorporated. Polymerizable groups are those which can undergo initiation and partake in free radical polymerization reactions. These groups generally possess unsaturated bonds and are exemplified by groups such as vinyls and acrylates. When exposed to the proper initiating conditions, macromers of this type rapidly form matrices. Since all free radicals are associated with a polymer, these matrices can be formed in the presence of living tissue or cells with little or no toxicity. Macromers can be prepared by synthesizing copolymers containing reactive monomers to which polymerizable groups can be subsequently added. Alternatively, macromers can be prepared from naturally occurring biopolymers by the incorporation of polymerizable groups into these compounds. In this way biopolymers such as hyaluronic acid and collagen, which possess inherent desirable bioactivity, can be derivatized and formed into matrices.

The other necessary component of this matrix-forming technology is the initiating system. This technology distinguishes itself by the use of polymeric initiating systems. These are systems in which the chemical groups responsible for the formation of available free radicals are polymer bound. By incorporating the initiating groups into polymeric reagents several advantages can be realized. Polymeric initiators are more efficient than their low molecular weight analogs, and the use of polymer-bound initiators insures that all free radicals are associated with polymers. These features both speed matrix formation and reduce toxicity, making them ideal for matrix-forming applications where living tissue or cells are present.

Initiator groups useful in the presently described systems include those that can be used to initiate polymerization of the macromer system by free radical generation to a desired extent within a desired time frame. For many applications it is desirable that these initiator groups be photosensitive (i.e., groups which capture light energy and produce free radicals), but it is sometimes desirable that these initiator groups be thermosensitive and produce free radicals in response to the input of thermal energy.

B. Applications

1. Cellular Encapsulation

The American Diabetes Association estimates that diabetes mellitus is the seventh leading cause of death in the United States. Despite many years of research, the effectiveness of current therapeutic approaches has remained unsatisfactory due to the difficulties of duplicating the glycemic control provided by a healthy pancreas.

Pancreatic transplantation has been studied as a method of achieving normoglycemia in diabetic patients without the need for exogenous insulin. There are many problems associated with whole or segmental pancreatic transplantation, however, including the limited availability of donor organs and the requirement for generalized immunosupression.

Since each pancreatic islet cell is an autonomous unit, the concept of cellular transplantation has been explored as an alternative to segmental pancreatic or organ grafts. The recent successes of investigations following the Edmonton Protocol [104] have magnified interest in the feasibility of islet transplantation as a treatment for diabetes. The requirement for lifelong immunosuppressive drug therapy is responsible for dampening the enthusiasm for this approach.

It is possible, however, that alternative immunomodulation of the graft or the host can be utilized to prevent the rejection of islet allografts or xenografts. Encapsulation of islets in a semipermeable membrane is one type of immunomodulation that can prevent rejection. Many materials have been evaluated for the preparation of immunoprotective membranes around pancreatic islets, but the stringent permeability, morphology, and biocompatibility requirements of these membranes make their development very difficult. Some of the issues that hamper this development include (1) lack of permeability control, (2) lack of biocompatibility, (3) lack of the appropriate morphological characteristics, and (4) large increases in graft volume, which leads to limitations to graft location.

One way of forming immunprotective membranes around pancreatic islets that overcomes the obstacles that result from these issues is to directly form matrix membranes on the surfaces of the islets via interfacial polymerization of biocompatible macromers. The first step to effect such an encapsulation is to incubate isolated islets in a solution of polymeric initiator. For this application, a polymeric initiator is especially appealing. It not only provides the benefits enumerated above, it also allows the incorporation of affinity groups into the reagent. These groups have affinity for the islet surface and thus provide a means of concentrating the initiator at the islet surface.

A polymer that contains both pendent eosin groups and pendent positively charged groups is ideal as the photoinitiator. Eosin is an efficient photoinitiator with an absorption maximum at 517 nm and positively charged groups that possess affinity for binding to the islet surface. A solution of the photoinitiator is applied to a preparation of isolated pancreatic islets. After a brief incubation, the unbound initiator is washed away. The islets with bound initiator are suspended in a macromer solution and illuminated with an argon ion laser. The laser has a line of strong intensity at 520 nm. The polymer-bound eosin groups on the islet surface produce free radicals that initiate a free radical chain reaction causing the polymerization of the macromer. By carefully controlling the macromer concentration, molecular weight, and polymerizable group content, as well as the illumination time and intensity, it is possible to form thin, semipermeable matrices around each islet.

Macromers that are useful for cellular encapsulation can be prepared from many different polymeric materials. Polysaccharides, such as hyaluronic acid can be used, but in most instances synthetic, hydrophilic polymers have been evaluated. The use of synthetic polymers permits absolute control over the molecular weight and the polymerizable group content of the macromers. A variety of synthetic polymers has been evaluated for their utility as cell-encapsulating macromers, but poly(ethylene glycol) (PEG) polyacrylates have received the most attention. The biocompatibility of PEG has been thoroughly evaluated, and it can be synthesized in virtually any molecular weight.

A number of studies have evaluated islets encapsulated in interfacially photopolymerized PEG diacrylate matrices [60,105,106]. The results from these studies show both in vitro and in vivo function of PEG-encapsulated islets and the ability of PEG matrices to prevent immune rejection in allograft and xenograft models.

2. Tissue Repair

Since macromers can be prepared from bioactive polymers, and solutions of these macromers can be applied to the sites of tissue defects and subsequently solidified into durable, bioresorbable matrices by the application of visible light, their use in tissue repair applications is logical. There are many tissue repair applications amenable to therapeutic intervention involving in situ matrix formation. The following are examples.

Chronic Cutaneous Wound Repair. Chronic cutaneous wounds are skin wounds that either will not heal or are very slow to heal as a result of an underlying disease state or other physiologic insufficiency. The three major types of chronic cutaneous wounds are decubitus ulcers, diabetic ulcers, and venous stasis ulcers. As the median age of the American population increases, the incidence of all three types of chronic skin ulcers increase as well.

There are several approaches currently being investigated for the treatment of chronic cutaneous wounds. These approaches range from traditional wound dressings to gene therapy approaches to total artificial skin. An approach that utilizes in situ formation of a bioactive matrix directly in the wound bed offers many potential advantages. The dressing material is applied to the wound bed as a liquid that conforms completely and intimately to the wound surface. The solution is subsequently solidified by a brief illumination. The resulting ''bandage'' is composed of bioactive materials that are in intimate contact with the wound. The bioactivity of the matrix material is maximized, and bacteria are prevented from infiltrating into the wound. In addition, bioactive compounds can be added to the dressing formulation and slowly released into the wound.

Wound healing formulations have been prepared from collagen, hyaluronic acid, and synthetic macromer solutions and evaluated in both partial and full thickness cutaneous wounds using a pig model. Results from these evaluations show no increased inflammatory or other host response of any formulation over controls covered with standard occlusive dressings. Some formulations showed enhanced reepithelization over controls receiving standard occlusive dressings.

Cartilage repair. Cartilage is a unique tissue, strong and durable, but once damaged is unable to heal itself. This is particularly true for articular, or hyaline, cartilage, the material that lines the articulating joints. Defects caused by overuse or trauma, common in athletes and other physically active people, will not heal. Until recently, the only treatment for this condition was joint replacement. Currently, there are some therapies being evaluated that show great promise.

Filling the defect with autologous chondrocytes (cartilage cells), matrices containing gene therapy agents, growth factors, bone morphogenetic proteins (BMPs), or combinations of these materials are some of the approaches being evaluated. One of the issues unique to hyaline cartilage repair is the tendency of the repaired tissue to form fibrocartilage instead of hyaline cartilage. Fibrocartilage is not as durable as hyaline cartilage and eventually deteriorates, forming a new defect. One way to direct the formation of hyaline cartilage instead of fibrocartilage is to deliver the bioactive substance to the defect in a hyaluronic acid matrix. A hyaluronic acid–rich environment favors the formation of hyaline cartilage.

Cells, growth factors, and other bioactive substances can be added to a solution of hyaluronic acid macromer, applied to the defect as a viscous liquid and solidified by a brief illumination thereby fixing the matrix in place. The matrix is resorbed as new hyaline cartilage is formed.

3. Prevention of Adhesions

The formation of postsurgical adhesions is a common cause of surgical complications. Adhesions formed after abdominal surgeries are the leading cause of bowel obstruction in the United States. Likewise, the leading cause of infertility in women is the formation of adhesions following pelvic surgery. In addition, the formation of adhesions is a cause of complication in many types of surgeries including spinal and cardiovascular.

For many years, the use of barriers and adjuvants has been evaluated. These materials essentially prevent tissues from interacting until the tissue healing response is complete, and then, ideally, they go away. In general, the film barriers are efficacious but they are difficult to use. The liquid or gel adjuvants are easy to use but suffer from efficacy issues. In both instances, the use of modified hyaluronic acid as the adhesion-preventing material has received the most notable attention.

Using a solution of hyaluronic acid macromer, it is possible to apply the solution to the affected area as you would a liquid or gel adjuvant, and subsequently solidify the material via a brief illumination to form a matrix with residence times similar to the film products. It is thereby possible to have the best of both worlds, ease of use and efficacy.

ACKNOWLEDGMENTS

The authors wish to thank Nancy Hupfer and Connie Hecker for their assistance in the preparation of this chapter. The authors thank Dr. Merle Olson and his research team at the University of Calgary, who collaborated with us on the antimicrobial evaluations in rabbits. We would also like to acknowledge the technical staff at SurModics, who were instrumental in helping us devise and conduct the work contained within this chapter.

REFERENCES

1. March J. Advanced Organic Chemistry. 3rd Edition. New York: Wiley, 1985:216,1078.
2. Galardy RE, Craig LC, Jamieson JD, Printz MP. J. Biol. Chem 1974; 249:3510.
3. Swan DG, Hastings CA. U.S. Patent 5,414,075, 1995.
4. Nagaoka S, Akashi R. J. Bioactive Compatible Polymers 1990; 5:212.
5. Kallmes DF, McGraw JK, Evans AJ, Mathis JM, Hergenrother RW, Jensen ME, Cloft HJ, Lopes M, Dion JE. Am. J. Neuroradiol 1997; 18:1243–1251.
6. Park KD, Okano T, Nojiri C, Kim SW. J. Biomed. Mater. Res 1988; 22:977.
7. Rollason G, Sefton MV. J. Biomater. Sci. Polym Ed 1989; 1:31.
8. Hoffman AS, Horbett TA, Ratner BD. Ann. NY Acad. Sci 1977; 283:372.
9. Marchant RE, Johnson SD, Schneider BH, Agger MP, Anderson JM. J. Biomed. Mater. Res 1990; 24:1521.
10. Ip WF, Sefton MV. J. Biomed. Mater. Res 1991; 25:875.
11. Hsuie GH, Yang JM, Wu RL. J. Biomed. Mater. Res 1988; 22:405.
12. Mori Y, Nagaoka S, Takiuchi H, Noguchi N, Tanzawa H, Noishiki Y. Trans. Am. Soc. Artif. Intern. Organs 1982; 28:459.
13. Nojiri C, Okano T, Jacobs HA, Park KD, Mohammad SF, Olsen DB, Kim SW. J. Biomed. Mater. Res 1990; 24:1151.
14. Munro MS, Quattrone AJ, Ellsworth SR, Kulkarni P, Eberhardt RC. Trans. Am. Soc. Artif. Intern. Organs 1981; 27:499.
15. Eberhardt RC, Munro MS, Williams GB, Kulkarni PV, Shannon WA, Brink BE, Fry WJ. Artif. Organs 1987; 11:375.
16. Grasel TG, Pierce JA, Cooper SL. J. Biomed. Mater. Res 1987; 21:815.
17. Keogh JR, Velander FF, Eaton JW. J. Biomed. Mater. Res 1992; 26:441.
18. Hayward JA, Durrani AA, Shelton CJ, Lee DC, Chapman D. Biomaterials 1986; 7:126.
19. Jacobs HJ, Okano T, Kim SW. J. Biomed. Mater. Res 1989; 23:611.
20. Ito Y, Iguchi Y, Kashiwagi T, Imanishi Y. J. Biomed. Mater. Res 1991; 25:1347.
21. Okkema AZ, Visser SA, Cooper SL. J. Biomed. Mater. Res 1991; 25:1371.
22. Santerre JP, ten Hove P, VanderKamp NH, Brash JL. J. Biomed. Mater. Res 1992; 26:39.
23. Smith PK, Mallia AK, Hermanson GT. Anal. Biochem 1980; 109:466.
24. Sukavaneshvar S. In vitro assessment of device thrombosis and thromboembolism from roadmap to successful development and regulatory approval of medical devices with hemocompatible coatings workshop handbook, Scottsdale, AZ, 2001.
25. Stamm WE. Infections related to medical devices. Ann. Int. Med 1978; 89:764–769.
26. Stamm WE. Prevention of infections. Infections related to medical devices. Ann. Intern. Med 1978; 89:764–769.

27. Donelli G, Francolini I. Efficacy of antiadhesive, antibiotic and antiseptic coatings in preventing catheter-related infections: review. J. Chemotherapy 2001; 13:595–606.
28. Carson CC. Management of prosthesis infections in urologic surgery. Urol. Clin. North. Am 1999; 26:829–839.
29. Mulcahy JJ. Long-term experience with salvage of infected penile implants. J. Urol 2000; 163: 481–482.
30. Darouiche RO. Device-associated infections: a macroproblem that starts with microadherence. Clin. Infectious Diseases 2001; 33:1567–1572.
31. Donlan RM. Biofilm formation: a clinically relevant microbiological process. Clin. Infectious Diseases 2001; 33:1387–1392.
32. Christensen GD, Simpson WA, Bisno AL, E.H. AL, Beachey AL. Infect. Immun 1983; 40:407–410.
33. Zimmerli W, Waldvogel FA, Vaudaux P, Nydegger UE. J. Infect. Dis 1982; 146:487–496..
34. Oakes JA, Wood AJJ. Infections in surgery. N. Engl. J. Med 1986; 315:1129–1138.
35. Nelson CL. Environmental bacteriology in the unidirectional (horizontal) operating room. Arch. Surg 1979; 114:778–782.
36. Anwar H, Strap JL, Chen K, Costerton JW. Dynamic interactions of biofilms of mucoid Pseudomonas aeruginosa with tobramycin and piperacillin. Antimicrobial Agents Chemother 1992; 36: 1208–1214.
37. Costerton JW, Lewandowski Z, Caldwell DE, Korber DR, Lappin-Scott HM. Microbial biofilms. Ann. Rev. Microbiol 1995; 49:711–745.
38. McEldowney S, Fletcher M. J. Gen. Microbiol 1986; 132:513–523.
39. Pasmore M, Todd P, Smith S, Baker D, Silverstein J, Coons D, Bowman CN. Effects of ultrafiltration membrane surface properties on Pseudomonas aeruginosa biofilm initiation for the purpose of reducing biofouling. J. Mem. Sci 2001; 194:15–32.
40. Gristina AG. Science 1987; 237:1588–1595.
41. Darveau RP, Blake J, Seachord CL, Cosand WL, Cunningham MD, Cassiano-Clough L, Maloney G. J. Clin. Invest 1992; 90:447.
42. Zasloff M. Proc. Natl. Acad. Sci. USA 1987; 84:5449.
43. Giovannini MG, Poulter L, Gibson BW, Williams DH. Biochem. J 1987; 243:113–120.
44. Williams RW, Starman R, Taylor KMP, Gable K, Beeler T, Zasloff M, Covell D. Biochemistry 1990; 29:4490–4496.
45. Cannon M. Nature 1987; 328:478.
46. Gwadz RW, Kaslow D, Lee J, Maloy WL, Zasloff M, Miller LH. Infection and Immunity 1989; 57:2628–2633.
47. Soravia E, Martini G, Zasloff M. FEBS Lett 1988; 228:337–340.
48. Dankert J, Hogt AH, Feigen J. CRC Crit. Rev. Biocompat 1986; 2:219–301.
49. Peters G, Gray ED, Johnson GM. In: Bisno AL, Waldvogel FA, eds. Infections Associated with Indwelling Medical Devices. Washington, DC: American Society for Microbiology, 1989:61–74.
50. Zasloff M, Martin B, Chen HC. Proc. Natl. Acad. Sci. USA 1988; 85:910–913.
51. Ratner BD. The blood compatibility catastrophe. J. Biomed. Mater. Res 1993; 27:283–287.
52. Mehran R, Dangas G, Abizaid AS, Mintz GS, Lansky AJ, Satler LF, Pichard AD, Kent KM, Stone GW, Leon MB. Angiographic patterns of in-stent restenosis: classification and implications for long-term outcome. Circulation 1999; 100:1872–1878.
53. Riessen R, Isner JM. Prospects for site-specific delivery of pharmacologic and molecular therapies. J. Am. Coll. Cardiol 1994; 23:1234.
54. Lincoff AM, Topol EJ, Ellis SG. Local drug delivery for the prevention of restenosis. Circulation 1994; 90:2070.
55. Hanson SR, Ratner BD. Testing of blood–materials interactions. In: Ratner BD, Hoffman AS, Schoen FJ, Lemons JE, eds, Biomaterials Science, Academic Press. 1996:228–238.
56. Camenzind E, Kutryk MJB, Serruys PW. Use of locally delivered conventional drug therapies. Semin. Intervent. Cardiol 1996; 1:67.
57. van der Giessen WJ, Lincoff AM, Schwartz RS, van Beusekom HMM, Serruys PW, Holmes DR, Ellis SG, Topol EJ. Marked inflammatory sequelae to implantation of biodegradable and nonbiodegradable polymers in porcine coronary arteries. Circulation 1996; 94:1690–1697.

58. Hårdhammar PA, van Beusekom HMM, Emanuelsson HU, Hofma SH, Albertson PA, Verdouw PD, Boersma E, Serruys PW, van der Giessen WJ. Reduction in thrombotic events with heparin-coated Palmaz–Schatz stents in normal porcine coronary arteries. Circulation 1996; 93:423.
59. Serruys PW. A comparison of balloon expandable stent implantation with balloon angioplasty in patients with coronary artery disease. N. Engl. J. Med 1994; 331:489.
60. Hill RS, Cruise GM, Hager SR, Lamberti FV, Yu X, Garufis CL, Yu Y, Mundwiler KE, Cole JF, Hubbell JA, Hegre OD, Scharp DS. Ann NY Acad. Sci 1997; 831:332–343.
61. Casscells W. Migration of smooth muscle and endothelial cells. Critical events in restenosis. Circulation 1992; 86:723.
62. Bauters C, Meurice T, Hamon M, McFadden E, Lablanche J-M, Bertrand ME. Mechanisms and prevention of restenosis: from experimental models to clinical practice. Cardiovasc. Res 1996; 31:835.
63. Suzuki T, Kopia G, Hayashi S, Bailey LR, Llanos G, Wilensky R, Klugherz BD, Papandreou G, Narayan P, Leon MB, Yeung AC, Tio F, Tsao PS, Falotico R, Carter AJ. Stent-based delivery of Sirolimus reduces neointimal formation in a porcine coronary model. Circulation 2001; 104: 1188–1193.
64. Sousa JE, Costa MA, Abizaid AC, Rensing BJ, Abizaid AS, Tanajura LF, Kozuma K, Van Langenhove G, Sousa AGMR, Falotico R, Jaeger J, Popma JJ, Serruys PW. Sustained suppression of neointimal proliferation by Sirolimus-eluting stents: one-year angiographic and intravascular ultrasound follow-up. Circulation 2001; 104:2007–2011.
65. Morice M-C, Serruys PW, Sousa JE, Fajadet J, Ban Hayashi E, Perin M, Colombo A, Schuler G, Barragan P, Guagliumi G, Molnàr F, Falotico R. A randomized comparison of a Sirolimus-eluting stent with a standard stent for coronary revascularization. N. Engl. J. Med 2002; 346:1773–1780.
66. Fodor SP, Read JL, Pirrung MC, Stryer L, Lu AT, Solas DG. Science 1991; 251:767–773.
67. Southern EM, Maskos U, Elder JK. Genomics 1992; 13:1008–1017.
68. Brown PO, Botstein D. Nature Genet 1999; 21:33–37.
69. Lipshutz RJ, Fodor SPA, Gingeras TR, Lockhart DJ. Nature Genet 1999; 21:20–24.
70. Eisen MB, Brown PO. Methods Enzymol 1999; 303:179–205.
71. Haab BB, Dunham MJ, Brown PO. Genome Biol, 2001.
72. Walter G, Büssow K, Cahill D, Lueking A, Lehrach H. Curr Opin Microbiol 2000; 3(3):2998–302.
73. MacBeath G, Schreiber S. Science 2000; 289:1761–1763.
74. Heng Z, Bilgin M, Bangham R, Hall D, Casamayor A. Science 2001; 293:2101–2105.
75. Lam KS, Renil M. Curr Opin Chem Biol 2002; 6(3):353–358.
76. Seeskin E. Drug Discovery Today, Spring 2002:43–47.
77. Xu CW. Genome Research 2002; 12(3):482–486.
78. Ziauddin J, Sabatini DM. Nature 2001; 411:107–110.
79. Wang D, Liu S, Trummer BJ, Deng C, Wang A. Nature Biotech 2002; 20:275–281.
80. Houseman BT, Mrksich M. Chem. Biol 2002; 9(4):443–454.
81. Andersen CL, Monni O, Wagner U, Kononen J, Barlund M. Am. J Pathol 2002; 161(1):73–79.
82. Skacel M, Skilton B, Pettay JD, Tubbs RR. Appl Immunohistochem Mol Morphol 2002; 10(2): 1–6.
83. Terris B, Bralet MP. Ann Pathol 2002; 22(1):69–72.
84. Shoemaker DD, Schadt EE, Armour CD, He YD, Garrett-Engele R. Nature 2001; 409:922–927.
85. Hacia JG, Brody LC, Chee MC, Fodor SP, Collins FS. Nature Genet 1996; 14:441–447.
86. Cutler DJ, Zwick ME, Carraquillo MM, Yohan CT, Tobin KP. Genome Res 2001; 11:1913–1925.
87. Chee M, Yang R, Hubbell E, Berno A, Huang XC. Science 1996; 277:610–614.
88. Schena M, Shalon D, Heller R, Chai A, Brown PO, Davis RW. Proc Natl Acad Sci, USA 1996; 93:10614–10619.
89. DeRisi JL, Iyer V, Brown PO. Science 1997; 278:680–686.
90. Ewing B, Green P. Nature Genet 2000; 25:232–234.
91. International Genome Sequencing Consortium. Nature 2001; 409:860–921.
92. Butler JE, Ni L, Brown WR, Joshi KS, Chang J, Rosenberg B, Voss EW, Jr. Mol Immunol 1993; 30(13):1165–1175.

93. Lindroos K, Liljedahl U, Raito M, Syvanen A-C. Nucl Acids Res 2001; 29(13):e69.
94. Ramakrishnan R, Dorris D, Lublinsky A, Nguyen A, Domanus M. Nucl Acids Res 2002; 30(7): e30.
95. Lockhart DJ, Winzler EA. Nature 2000; 405:827–836.
96. Hughes TR, Mao M, Jones AR. Nature Biotech 2001; 19:342–347.
97. Means GE, Feeney RE. Biochemistry 1968; 7:2192–2201.
98. Charonis AS, Skubitz APN, Koliakos GG, Reger LA, Dege J, Vogel AM, Wohlhueter R, Furcht LT. J. Cell Biol 1988; 107:1253–1260.
99. Green LM, Reade JL, Ware CF. J. Immunol. Methods 1984; 70:257–268.
100. Koliakos GG, Kouzi-Koliakos K, Furcht LT, Reger LA, Tsilibary EC. J. Biol. Chem 1989; 264: 2313–2323.
101. Pierschbacher MD, Ruoslahti E. Nature 1984; 309:30–33.
102. Kirkham SM, Dangel ME. Ophth. Surg[(&&&)] 1991; 22:455–461.
103. Clapper DL, Hagen KM, Hupfer NM, Anderson JM, Guire PE. Trans. Soc. Biomater 1993; 16:42.
104. Shapiro PE. N. Engl. J. Med 2000; 343:230–238.
105. Cruise GM, Hegre OD, Lamberti FV, Hager SR, Hill R, Scharp DS, Hubbell JA. Cell Transplantation 1999; 8:293–306.
106. Cruise GM, Hegre OD, Scharp DS, Hubbell JA. Biotech. Bioeng 1998; 57:655–665.

5

Rational Design of Absorbable Polymers for Orthopedic Repair

James B. Beil
University of Illinois, Urbana-Champaign
Urbana, Illinois, U.S.A.

Jorge Heller
AP Pharma, Inc.
Redwood City, California, U.S.A.

Kirk P. Andriano
MacroMed, Inc.
Sandy, Utah, U.S.A.

I. INTRODUCTION

The first absorbable sutures were developed in the 1960s [1,2]. These materials were made from synthetic polymers that fulfilled their function within the body and then were absorbed. With this advance, removal of sutures from a patient once the wound healed became unnecessary.

Investigators have developed absorbable polymers similar to those used as sutures for the fixation of bone defects or fractures. In many cases, for a fractured bone to heal correctly, a patient will usually require implantation of hardware such as pins, rods, screws, or plates or injection of cements [2–5]. Many of the commercially used materials for these applications are nonabsorbable polymers, metals, and ceramics, which often have serious side effects such as inflammation [6,7] or irritation [7]. These side effects sometimes necessitate removal of the implant. Nonabsorbable implants also prevent the bone form growing into the space where the implant is. This can lead to weakened bones and often to refracture [1,3–5]. Implants fabricated from absorbable polymers have many advantages over their nonabsorbable counterparts [1]. They eliminate the need for second surgery to remove the implant after it has served its function. They allow bones to grow into the polymer matrix so that as the polymer absorbs, the bone maintains better integrity and can begin load bearing earlier. Also, absorbable implants may contain bioactive molecules that enhance bone growth and accelerate fracture healing [8,9]. The distinct advantages of absorbable polymer implants have led to considerable interest and research in this area.

The usefulness of an absorbable polymer for bone repair applications depends on its ability to fulfill the following criteria [1]: (1) the absorption rate of the polymer must match the rate at which the bone grows into the polymer; (2) the polymer should facilitate bone growth or at least not inhibit it; (3) the polymer and its degradation products must be biocompatible to prevent adverse reaction; and (4) the mechanical properties of the polymer must fit the application and remain until the final stages of degradation.

II. POLYMER DEGRADATION

A key consideration in the design and evaluation of an absorbable polymer is its degradation mechanism (Fig. 1) [1,10]. Bulk degradation is the mechanism invoked for most traditional absorbable polyesters. It is characterized by water entering the polymer bulk where hydrolysis causes degradation. The degradation products are usually carboxylic acids, which decrease the pH of the bulk polymer. This leads to autocatalysis and further drop in pH. As the polymer is hollowed out it will finally break apart releasing the acid groups that had been confined to the interior. This is called acid bursting and can result in inflammation and a decrease in mechanical properties in earlier stages of degradation. On the other hand, surface erosion (Fig. 1) results in the thinning of the device over time while retaining integrity of the bulk polymer [1,10]. Surface-erosion usually occurs in cases where the polymer is significantly hydrophobic, a property that keeps water from entering the bulk. Surface-eroding polymers are also less likely to result in inflammation due to acid burst because the degradation products are washed away from the surface at a constant rate.

A. Polyesters

Polylactide (PLA) is an absorbable polyester used clinically in medical devices such as sutures and some orthopedic implants [1]. Copolymerization of lactide with other hydrophilic monomers is a way to tune the properties of the resulting polyester [1,8]. PLA is formed by ring opening polymerization with a metal catalyst such as aluminum isopropoxide (Fig. 2) [11]. Absorbable polyesters such as **1** (see figure) have some distinct advantages over nonabsorbable materials. For example, polymer **1** is used commercially as orthopedic pins and screws. PLA also has sufficient mechanical strength for in vivo uses. In addition, polymer **1** and poly(lactide-*co*-glycolide) support sufficient cell growth to be used commercially. Polyesters such as **1** also have

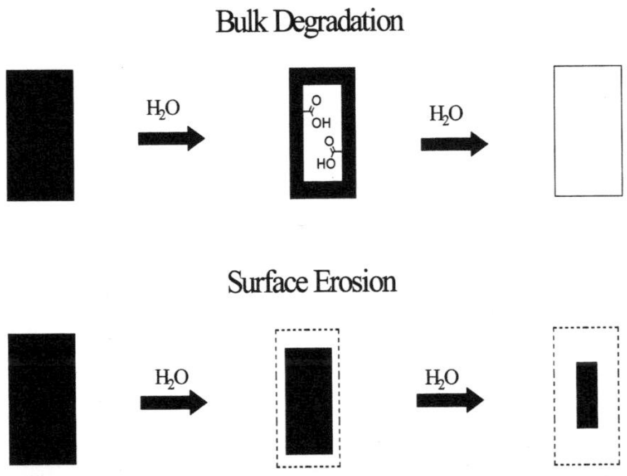

Figure 1 Schematic of polymer degradation mechanisms.

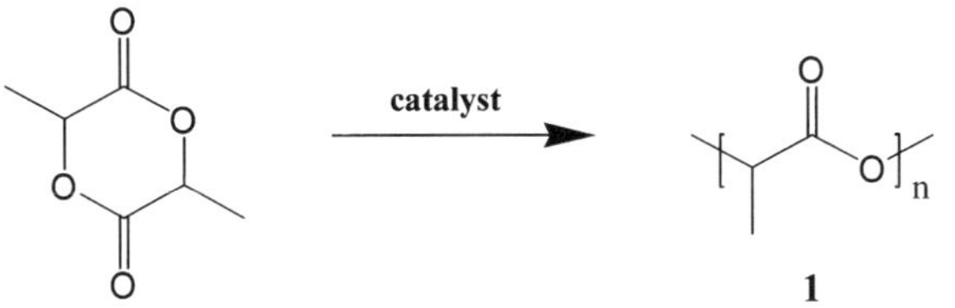

1

Figure 2 Synthesis of poly(lactide).

a major disadvantage [7]. Most of these hydrophilic polyesters degrade by a bulk degradation mechanism [15,16].

A. Tyrosine-Derived Polycarbonates

Recognizing the problems associated with the bulk erosion polyesters, current efforts are focused on polymers that degrade by surface erosion [1,10]. For example, Kohn and coworkers have synthesized tyrosine-derived polymers with carbonate linkages in the polymer backbone (Fig. 3) [5,17,18]. Polymer **2** has three groups per repeating unit that can potentially be hydrolyzed: amide, ester, and carbonate. Studies have shown [17] that the carbonate group hydrolyzes at a faster rate than the ester group, and the amide bond is not labile in vitro. The products of hydrolysis of the carbonate groups are two alcohols and carbon dioxide. Thus significant degradation is achieved without a corresponding decrease in pH. This alleviates the problem of acid bursting [5] seen in polyesters. The pendant ester group on the chain accomplishes two objectives:

1. In the later stages of degradation, the ester bonds are hydrolyzed into acids. Hydrophilic acid groups provide the necessary solubility for the degradation products to be eliminated from the implant area.
2. The pendant ester moiety also contains a hydrophobic R group that leads to a surface erosion mechanism.

With degradation times as long as 900 days, no significant water uptake was measured indicating the polymer's hydrophobicity and its strong resistance to water entering the bulk polymer. Polymer **2** has been reported to have sufficient mechanical properties for small bone fixation applications [5,18] and has been shown to be biocompatible in vivo [5]. In one study, polymer **2** was implanted into canine specimens where it promoted significant bone ingrowth [5].

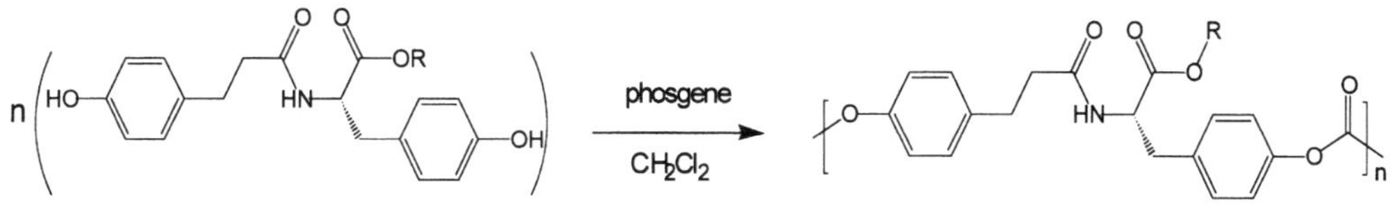

2

R = ethyl, isopropyl, butyl, hexyl, or octyl

Figure 3 Synthesis of tyrosine-derived polycarbonates.

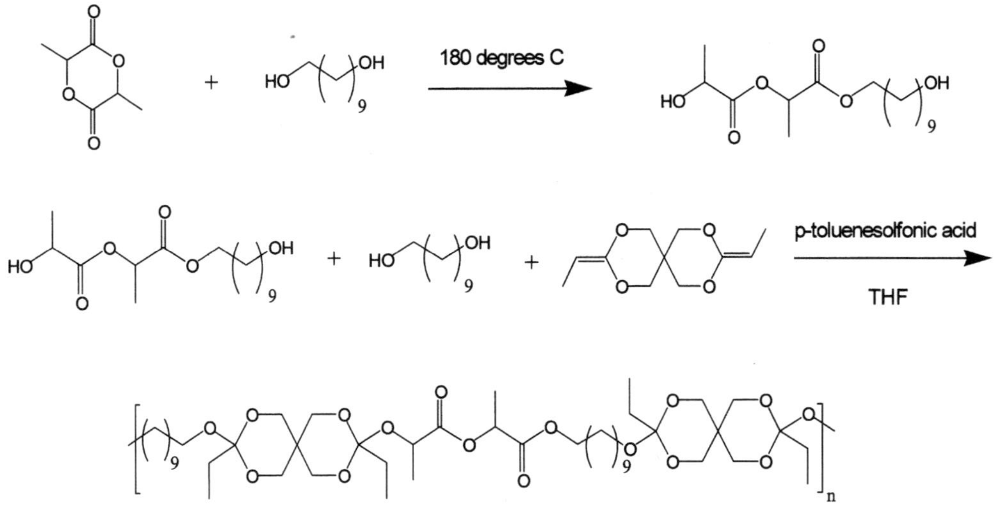

3

Figure 4 Synthesis of self-catalytic poly(ortho esters).

B. Poly(Ortho Esters)

Poly(ortho esters) (POEs) also have been identified as absorbable polymers for orthopedic applications. Heller and coworkers have synthesized a specific family of POEs (Fig. 4) [19]. Polymer **3** degrades by surface erosion. With the addition of lactide and/or glycolide segments to polymer **3** tunable degradation times ranging from 15 to hundreds of days can be achieved [19]. Most previous POEs that have been synthesized contained acidic additives that were used to aid in degradation. This was necessary because POEs degrade at slightly acidic pH, whereas the body is slightly basic. Such additives could lead to toxicity concerns. This family of polymers, however, is self-catalyzing, and it can degrade easily without the presence of acidic additives. This is due to the lactic acid and/or glycolic acid segments in the chain that produce carboxylic acids upon hydrolysis catalyzing the hydrolysis of the ortho ester groups. Preliminary results of electron paramagnetic resonance studies using pH-sensitive nitroxide radicals indicate that the internal pH of this polymer is about 6.4 and the pH of the eroding front is about 5.4 [20]. This will be extremely useful for delivery of acid-sensitive osteoinductive agents for repair and/or regeneration applications that cannot be delivered with polyesters. Polymer **3** has also been shown to increase bone ingrowth in comparison with poly(DL-lactide-*co*-glycolide), thus suggesting improved biocompatibility [15].

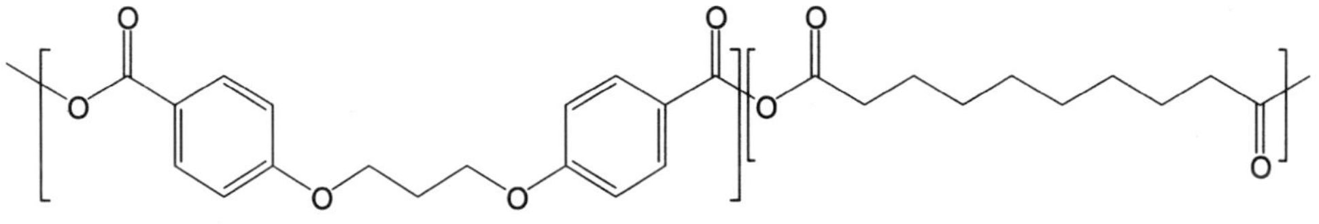

4

Figure 5 Synthesis of polyanhydrides.

C. Polyanhydrides

Langer and coworkers have synthesized polyanhydrides for drug delivery applications (Fig. 5) [21,22]. Since polymer **4** has been approved by the FDA, it is the basis of design for some of the polyanhydrides in the following sections. Polymer **4** is used to deliver carmustine, an anticancer drug, to sites in the brain where a tumor has been removed. The polymer is processed into wafers and then placed in the brain cavity during surgery. The polymer and its degradation products are nontoxic and have a controlled surface erosion degradation mechanism that allows them to deliver drugs at a known rate. Polyanhydride **4** has many characteristics that would make it suitable for orthopedic repair; however, it does not have the necessary mechanical properties. This has led Langer and coworkers to synthesize modified polyanhydrides with enhanced mechanical properties and surface erosion degradation.

D. Poly(Anhydride-*co*-Imides)

Polyimides are polymers that are well known for their high thermal and mechanical properties. Langer and coworkers have combined the strength of polyimides and degradation characteristics of polyanhydrides to obtain suitable materials for orthopedic repair. Poly(anhydride-*co*-imides) derived from amino acids such as **5** (Fig. 6) [23] have been synthesized for this purpose. By varying the R group in the polymer, controlled degradation times can be achieved [24], ranging from 1 to 63 days. It has been reported that the polymer degrades via a surface erosion mechanism [16,25]. This particular poly(anhydride-*co*-imide) has been shown to be biocompatible in vitro [16] for bone repair and drug delivery applications. Increased compression strength of this polymer compared to polyanhydrides not containing imide linkage is sufficient to warrant further research of this material for orthopedic repair.

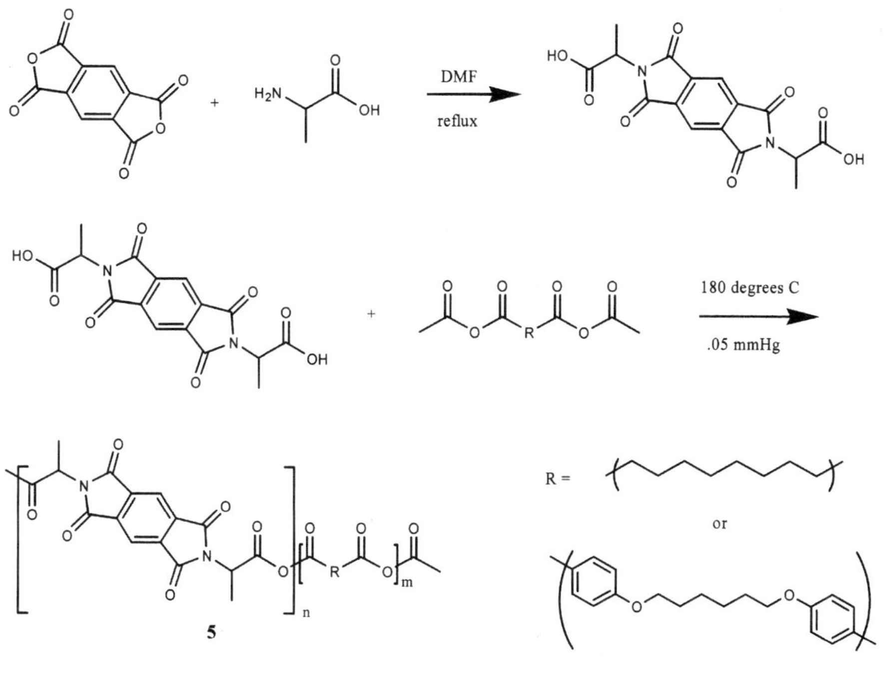

Figure 6 Synthesis of poly(anhydride-*co*-imides).

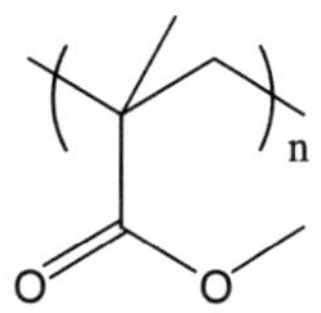

Figure 7 Molecular structure of polymethylmathacrylate.

III. BONE CEMENT

Bone cement is presently made mostly of the nondegradable polymer polymethylmethacrylate (PMMA) (Fig. 7) [26]. Surgeons internally cement fractures and total joint prostheses with PMMA by an in situ polymerization reaction. PMMA has several advantages. Injection of the monomer mixture and photopolymerization allows for a less invasive surgical technique. The mechanical properties of PMMA are sufficient to bear the stress of in vivo loads. There are also some clear disadvantages [26]. For example, it is difficult to control the temperature rise associated with exothermic polymerization. Temperatures at the bone–cement interface often reach up to 90°C causing cell necrosis [26]. In addition, PMMA cement often causes reduced blood flow irritation [6].

A. Poly(Propylene-Furmarate) Networks

To overcome the disadvantages of PMMA, researchers are developing polymers that can be polymerized in situ during surgery. Mikos and coworkers have synthesized degradable poly(propylene-fumarate) networks for bone cement applications (Fig. 8) [27]. In general, network polymers exhibit greater mechanical strength than their linear counterparts. Network **8** is formed by initiation with benzoyl peroxide. The polymer is mechanically weaker than PMMA [27,28]. However, by creating networks with higher crosslinking densities, higher strengths may be achieved. Also, the degradation products of this polymer **8** are nontoxic [27]. In addition, the polymer has been shown to deliver bone growth–enhancing bioactive drugs to the fracture site upon degradation [29].

B. Polyanhydride Networks

Anseth and coworkers have reported polyanhydride networks that are formed by photoinitiation and are hydrolytically degraded (Fig. 9) [26]. Polymer **11** is based on the scheme in Fig. 4 [26]. The polymer degrades by a surface erosion mechanism [26]. Photoinitiation can also take place through tissue for an even less invasive surgical technique. By using a mixture of monomers, **9** and **10**, controlled degradation times can be achieved ranging from 3 to 500 days [26]. Also, by using a CQ/TEA initiating system, Anseth has achieved network thickness of approximately 3 cm. Histological studies have also shown that the monomers, **9** and **10**, and polymer network **11** are biocompatible [26]. By shuttering the light source used to polymerize, temperature at the bone–polymer interface has been controlled to 42°C. Finally, these networks have been shown to form on acceptable time scales, from 500 to 1500 s, depending on the intensity of the light used to initiate polymerization.

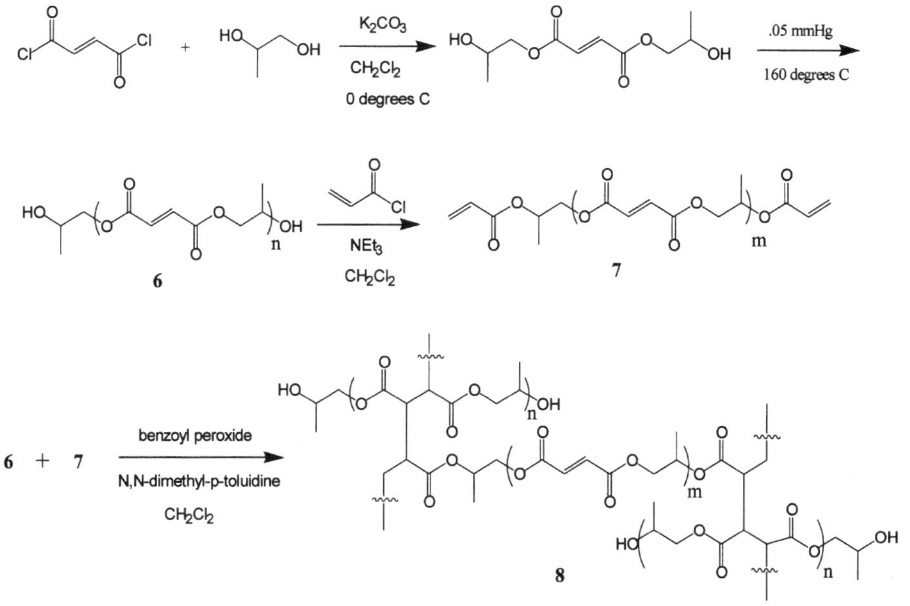

Figure 8 Synthesis of poly(propylene-fumarate) networks.

IV. CONCLUSIONS

Linear and crosslinked networks, absorbable polymers, and copolymers have been studied for possible use in absorbable orthopedic hardware and cements. These materials have many advantages over current nondegrading technologies, including their ability to degrade at a controlled rate by a surface erosion mechanism. This allows researchers the opportunity to synthesize

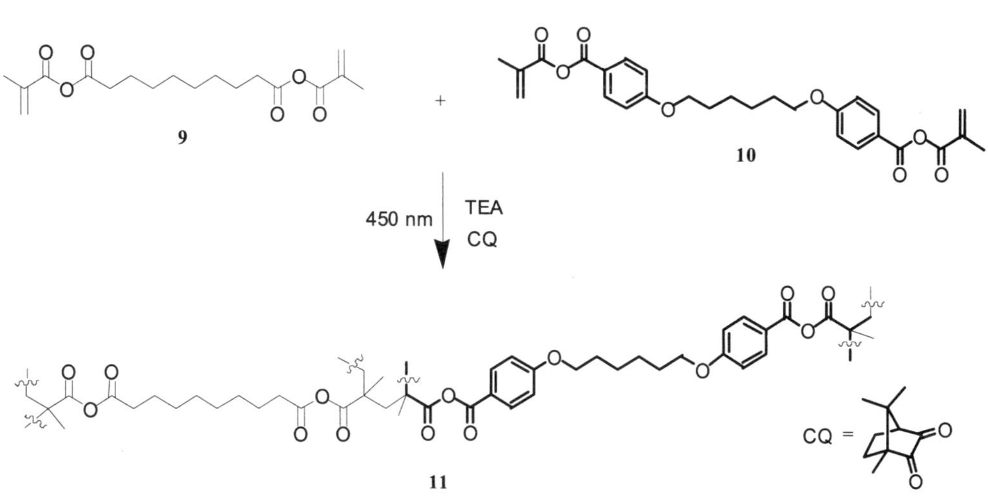

Figure 9 Synthesis of polyanhydride networks.

polymers with a degradation rate similar to that of bone growth. These polymers also appear to be biocompatible, and some even support bone growth. Finally, most of these polymers are shown to possess appropriate mechanical properties for orthopedic applications and, due to their surface erosion mechanism, maintain these properties through the initial stages of degradation until the bone is ready to accept additional mechanical load.

REFERENCES

1. Middleton JC, Tipton AJ. Synthetic biodegradable polymers as orthopedic devices. Biomaterials 2000; 21(23):2335–2346.
2. Gilding DK, Reed AM. Biodegradable polymers for use in surgery—polyglycolide/poly(lactic acid) homo- and copolymers. Polymer 1979; 20:1459–1464.
3. Bergsma JE, de Bruijn WC, Rozema FR, Bos RR, Boering G. Late degradation tissue response to poly(L-lactide) bone plates and screws. Biomaterials 1995; 16(1):25–31.
4. Muggli DS, Burkoth AK, Anseth KS. Crosslinked polyanhydrides for use in orthopedic applications: degradation behavior and mechanics. J. Biomed. Mater. Res 1999; 16(1):271–278.
5. Tangpasuthadol V, Pendharkar SM, Peterson RC, Kohn J. Hydrolytic degradation of tyrosine-derived polycarbonates, a class of new biomaterials. Part II: 3-year study of polymeric devices. Biomaterials 2000; 16(1):2379–2387.
6. Tams J, Joziasse CA, Bos RR, Rozema FR, Grijpma DW, Pennings AJ. High-impact poly(L/D-lactide) for fracture fixation: in vitro degradation and animal pilot study. Biomaterials 1995; 16(1):1409–1415.
7. Muggli DS, Burkoth AK, Keyser SA, Lee HR, Anseth KS. Reaction behavior of biodegradable, photo-cross-linkable polyanhydrides. Macromolecules 1998; 31:4120–4125.
8. Burg KJ, Porter S, Kellam JF. Biomaterial developments for bone tissue engineering. Biomaterials 2000; 16(1):2347–2359.
9. Lu L, Stamatas GN, Mikos AG. Controlled release of transforming growth factor beta 1 from biodegradable polymer microparticles. J. Biomed. Mater. Res 2000; 16(1):440–451.
10. Gopferich A. Mechanisms of polymer degradation and erosion. Biomaterials 1996; 16(1):103–114.
11. Kricheldorf HR, Mang T, Jonte MJ. Polylactones. 1. Copolymerization of glycolide and L-caprolactone. Macromolecules 1984; 17:2173–2181.
12. Dubois PH, Jacobs C, Jerome R, Teyssie PH. Macromolecular engineering of polymerization of l-lactide and 1,5-dioxepon-2-one forming triblock copolymer. Macromolecules 1991; 24:2266–2270.
13. Stridsberg K, Albertsson AC. Preparation and characterization of poly(L-lactic acid) foams. J. Polym. Sci.: Part A: Polym. Chem 2000; 38:1774–1784.
14. Mikos AG, Thorsen AJ, Czerwonka LA, Bao LA, Bao Y, Langer R. Preparation and characterization of poly(L-lactic acid) foams. Polymer 1994; 35:1068–1077.
15. Andriano KP, Tabata Y, Ikada Y, Heller J. In vitro and in vivo comparison of bulk and surface hydrolysis in absorbable polymer scaffolds for tissue engineering. J. Biomed. Mater. Res 1999; 48: 602–612.
16. Attawia MA, Herbert KM, Uhrich KE, Langer R. Laucrencin CT. Proliferation, morphology, and protein expression by osteoblasts cultured on poly(anhydride-*co*-imides). J. Biomed. Mater. Res 1999; 16(1):322–327.
17. Tangpasuthadol V, Pendharkar SM, Kohn J. Hydrolytic degradation of tyrosine-derived polycarbonates, a class of new biomaterials. Part I: study of model compounds. Biomaterials 2000; 21(23): 2371–2378.
18. Pulapura S, Kohn J. Tyrosine-derived polycarbonates: backbone-modified ''pseudo''-poly(amino acids) designed for biomedical applications. Biopolymers 1992; 16(1):411–417.
19. Ng S, Taylor MS, Heller J. Synthesis and erosion studies of self-catalyzed poly(ortho esters). Macromolecules 1997; 30:770–772.
20. Heller J, Barr J, Ng S, Shen H, Schwach-Abdellaoui K, Gurny R. Development of poly(ortho esters). Drug Delivery Systems & Sciences 2001; 16(1):79–84.

21. Domb AJ, Langer R. Polyanhydrides. 1. Preparation of high molecular weight polyanhydrides. J. Polym. Sci.: Part A: Polym 1987; 25:3337–3386.
22. Brem H, Piantadosi S, Burger PC, Walker M, Selker R, Vick NA, Black K, Sisti M, Brem S, Mohr G, Muller P, Morawetz R, Schold CS. Placebo-controlled trial of safety and efficacy of intraoperative controlled delivery of biodegradable polymers of chemotherapy for recurrent gliomas. The Lancet 1995; 345:1008–1012.
23. Staubli A, Ron E, Langer R. Hydrolytically degradable amino acid containing polymer. J. Am. Chem. Soc 1990; 112:4419–4424.
24. Uhrich KE, Thomas TT, Laurencin CT, Langer R. In vitro degradation characteristics of poly(anhydride-imides) containing trimellitylimidoglycine. J. Appl. Polym. Sci 1997; 63(11):1401–1411.
25. Seidel JO, Uhrich KE, Laurencin CT, Langer R. Erosion of poly(anhydride-*co*-imides): a prelininary mechanistic study. J. Appl. Polym. Sci 1996; 62:1277–1283.
26. Burkoth AK, Anseth KS. A review of photocrosslinked polyanhydrides: in situ forming degradable networks. Biomaterials 2000; 16(1):2395–2404.
27. He S, Timmer MD, Yaszemski MJ, Yasko A, Mikos AG. Synthesis of biodegradable poly(propylene fumarate) networks with poly(propylene fumarate)—diacrylate macromers as crosslinking agents and charaterization of three degradation products. Polymer 2001; 42:1251–1260.
28. He S, Yaszemski MJ, Yasko A, Engel PS, Mikos AG. Injectable biodegradable polymer composites based on poly(propylene-fumarate) crosslinked with poly(ethylene glycol)-dimethacrylate. Biomaterials 2000; 16(1):2389–2394.
29. Peter SJ, Lu L, Kim DJ, Stamatas GN, Miller MJ, Yaszemski MJ, Mikos AG. Effects of transforming growth factor beta 1 released from biodegradable polymer microparticles on marrow stromal osteoblasts cultured on poly(propylene fumarate) substrates. J. Biomed. Mater. Res 2000; 16(1):452–462.

6

Synthesis and Evaluation of a Poly(Propylene Glycol-*co*-Fumaric Acid) Bone Graft Extender

Stephen A. Doherty, David D. Hile, Donald L. Wise, and Debra J. Trantolo
Cambridge Scientific, Inc.
Cambridge, Massachusetts, U.S.A.

Kai-Uwe Lewandrowski
Massachusetts General Hospital
Boston, Massachusetts, U.S.A.

I. INTRODUCTION

Bone grafting procedures are used for a variety of purposes ranging from spinal fusion to reconstructive surgery. Spinal fusion is the most common bone grafting procedure conducted in the United States, comprising some 50% of the estimated half-million bone grafting procedures performed annually [1]. It is estimated that worldwide over two million bone grafting procedures are performed annually [2]. As the demand for these procedures continues to increase, it will place a strain on the limited supply of graft material.

Successful bone regeneration requires several critical elements including osteoinduction and osteoconduction. Osteoinduction involves the use of growth factors and proteins to cause osteoprogenitor cells to differentiate and stimulate new bone growth. The process of osteoconduction involves the scaffold that provides the necessary support on which new bone can grow to fill the defect area. As new bone continues to grow to fill the defect area, the process of resorption and remodeling occurs.

The development of synthetic materials that can extend the limited supply of natural bone grafts while providing these critical elements has been the focus of much research. One synthetic material that shows promise in this regard is a porous formulation based on the biopolymer poly(propylene glycol-*co*-fumaric acid) (PPF). This polymer can be crosslinked in the presence of effervescent agents and osteoconductive filler to create a scaffold that mimics the physical and chemical properties of human cancellous bone. A combination of this formulation and an osteoinductive material, such as morselized cancellous autograft, venous blood, or bone marrow, comprises a material with improved handling characteristics and the potential for improved clinical results relative to other composite materials. This paper reviews a series of in vitro and in vivo studies on the use of a PPF-based material as a bone graft extender for use in clinically relevant repair scenarios.

II. AUTOGRAFT BONE

Iliac crest autograft is the gold standard for bone graft material. Autograft bone provides the critical elements required for bone growth. It is histocompatible, i.e., as it is a native material there is no risk of disease transfer and it is not rejected by the body [3]. The material is also able to maintain the strength and structural integrity of the defect area.

Despite these advantages, the material suffers from several disadvantages. The primary disadvantage is the need for a secondary surgical site. Autografts are predominantly removed from the iliac crest of the patient. The procedure of harvesting material from the iliac crest has been shown to cause complications in 4–49% of patients [4]. These complications can range from increased pain—observed in 25% of patients [5]—to infection, sensory loss, and neurological and vascular injury [6]. Further complications include increased anesthesia time and operative blood loss as well as prolonged recovery time [3]. In addition to the procedural complications, the supply of bone available for harvest is limited. These complications, along with the risk of donor site morbidity and the limited supply of material available for harvest, have pushed the search for alternative materials.

III. ALLOGRAFT BONE

Allograft bone is the substitute most often used for autograft bone. It is typically available fresh, frozen, or freeze-dried. This material undergoes extensive processing and donor screening in an effort to reduce the risk of disease transmission. This processing decreases but does not eliminate the risk of disease transmission. In addition, the processing decreases the mechanical and biological properties of the bone, while adding to the cost of the material [2]. In summary, although a viable alternative to autograft bone, allograft bone suffers from concerns of possible disease transmission, recipient rejection, increased cost, and limited availability.

IV. SUBSTITUE BONE GRAFT MATERIALS

The problems associated with autograft and allograft materials described above, not the least of which is the increasing demand for a limited supply, have acted to fuel the development of substitute materials. Substitute materials have primarily been developed as a replacement for graft materials and currently are used in approximately 10% of the bone graft procedures performed worldwide [7]. Calcium salt ceramics and glasses, such as calcium phosphates and hydroxyapatites, are widely used synthetics. Demineralized bone and collagen-based materials have also been used as substitute bone graft materials.

Calcium phosphate materials such as tricalcium phosphate (TCP) and hydroxyapatite (HA) are among the most common synthetics. These materials have been successfully applied in orthopedic and dental applications for decades [2]. Calcium salt materials have compositions similar to that of native bone and provide osteoconductive surfaces on which new bone can form. Although somewhat similar in composition, the materials elicit different biological responses [2]. The TCP material is readily resorbed, whereas the HA is a more or less permanent material, taking several years to be removed and replaced by native bone. TCP typically is more porous than HA, helping it to be resorbed more rapidly. Most TCP and HA materials are able to provide a highly osteoconductive structure; however, the materials lack or have limited osteoinductive properties [8]. In addition, the materials are often brittle, making them unable to support physiological loads without additional internal or external support. Commercially available products include Interpore Internationals ProOsteon™ (Irvine, CA), which is a coral-derived HA material. The material was the first synthetic approved by the Food and Drug Administration (FDA) [1].

As with most materials of this type, it is highly osteoconductive, with limited osteoinductive properties and limited mechanical strength [8]. Norian [9] has developed an injectable form of calcium phosphate, the Norian SRS® (skeletal repair system). The calcium and phosphate powder are mixed with solution to form a paste that can be injected or packed into a defect site. External or internal reinforcement is usually required to provide adequate support.

Demineralized bone matrix (DMB) is a highly processed allograft material. Treatment of bone with mild acid removes the mineral component of bone, while leaving growth factors and proteins [8]. These growth factors and proteins are mixed with a substrate, such as glycerol, to form a workable material. Materials of this type typically have limited osteoconductive properties, but good osteoinductive properties. The level of these properties is dictated by the extent of processing of the material [8]. In addition, insufficient mechanical strength has limited the applications of the material. Osteotech Inc. (Eatontown, NJ) produces a DMB matrix, Grafton®, in both a putty and sheet form that has been shown successful in oral [10] and lumbar spine arthrodesis [11].

V. IDEAL STRUCTURE OF A BONE REPLACEMENT MATERIAL

The open and interconnected porosity of a bone replacement material should allow body fluids to circulate throughout its entire extent. The range of pore sizes should encourage tissue ingrowth. Pores that are very small will allow fluid transport to take place but will not be large enough to accommodate cells or allow them to grow and divide. On the other hand, pores that are relatively large will accommodate ingrowth of blood vessels, but may be too large for effective cell recruitment and migration. If regions of the implanted bone replacement were not fully accessible to nearby blood vessels, such as would be the case if the porosity of the material were partially blocked off or if the pore size distribution lacked larger pores, then biologic repair can be impaired. An ideal pore size distribution enables and enhances all factors leading to bone formation: cell seeding, vascular ingrowth, extravascular fluid transport, and cellular resorption of the scaffold.

VI. PPF-BASED BONE GRAFT EXTENDERS

A bioresorbable bone graft extender appears to be a viable alternative to autografts and allografts. In light of currently approved synthetic products having significant drawbacks (e.g., lack of resorbability, inclusion of animal- or marine-derived components, and poor handling characteristics [12]), the challenge becomes to create a bone graft extender material that behaves—both biologically and biomechanically—more like bone. In addition, these materials could offer a reasonable solution to the clinical dilemma of deficient autologous bone stocks.

Studies have demonstrated the development of a resorbable bone repair material that does not contain biological material (either collagen or protein). This material is made from the unsaturated polyester, poly(propylene glycol-*co*-fumaric acid), and is referred to in the literature as poly(propylene fumarate) (PPF)[13–15]. The material can be mixed with cancellous autograft and crosslinked in the presence of a hydroxyapatite filler and a sodium bicarbonate/citric acid (CA) effervescent agent. The autograft–extender formulation can then be used to fill the void created by removal of a cyst or infected bone or from trauma. However, little is known about the appropriate autograft/extender ratio for the poly(propylene glycol-*co*-fumaric acid) bone graft extender at which an osteoinductive effect is seen. In addition, debate still remains as to how much cancellous autograft bone can be diluted with the PPF-based bone graft extender. Ideally, the design of the most appropriate PPF bone graft extender formulation should consider the effects of porosity, autograft content, and the contributions of the recipient tissue bed on the process of osteoinduction. Furthermore, extender degradation should occur at a rate supportive of bone cell ingrowth and deposition of new bone at defect site.

In previous in vitro and in vivo studies, development of porous bone repair scaffolds have primarily relied on the hypothesis that a more rapid ingrowth of bone cells will occur in these types of materials [12,16,17]. Addition of autologous bone graft to the formulation should not alter these material properties in vivo after implantation and in situ curing. However, these material properties have not been investigated. It is generally assumed that a material with such properties would initially provide structural support to the defect site. Thereafter, as the implant degrades, the net result of newly formed bone plus residual implant, the *repair-composite*, must continue to provide support to the defect reconstruction while yielding to the establishment of native bone.

A number of in vitro studies have examined the relationship between porosity and mechanical strength and determined the amount of new bone formation in the rat tibial defect model following implantation of the biodegradable PPF bone graft extender material at varying autograft/extender mixing ratios [18]. The ultimate objective of these studies was to determine whether new bone formation observed at a low autograft/extender mixing ratio is equivalent to the amount of new bone formation observed at a high autograft/extender.

A. Formulation of a PPF-Based Bone Graft Extender and Its Evaluation

The PPF polymer [(MW~5000 by gel permeation chromotography (GPC)] was synthesized from equimolar fumaric acid and propylene glycol in the presence of p-toluene sulfonic acid [19]. Utilizing the unsaturated carbon double bond of the fumaric acid component, the polymer can be crosslinked into a solid three-dimensional polymer construct. A typical polymer construct formulation utilized in evaluation of the PPF material as a bone graft extender is shown in Table 1.

The final form and utility of the polymer construct can be dictated by altering the relative amounts of the various formulation components. The polymer can be prepared as a puttylike consistency that is packed into a bony void [13,18,21] or as viscous quasisolid that can be injected through a needle into a defect site[18,21]. Sodium bicarbonate (SB) and citric acid (CA) can be included in the formulation as effervescent agents. The reaction of CA and SB produces carbon dioxide, which is responsible for foam expansion and development of porosity throughout the polymer construct. Porosity is developed with relative pore sizes of 100–1000 μm. Scanning electron microscopy (SEM) revealed that the PPF foam was characterized by a few large interconnecting pores measuring approximately 0.5–3.0 μm in diameter and a large number of small pores ranging from 50–400 μm in diameter (see Fig. 1). In addition, the PPF foam was noted to have a wide pore size distribution (median pore size 70 μm) with at least

Table 1 Sample Composition of PPF Foam Formulation

Chemical	Amount (%w/w)
Poly(propylene fumarate) (PPF)	50.9
Hydroxyapatite (HA)	14.7
Vinyl pyrrolidone (VP)	13.7
Sodium bicarbonate (SB)	1.1
Benzoyl peroxide (BP)	2.3
Citric acid (CA)	0.9

Source: Ref. 20.

30% of pores with an average diameter greater than 200 μm (as confirmed by mercury intrusion porosimetry).

The concentration of effervescent agents affects the porosity of the polymer construct as well as the overall expansion of the material. The expansion of the material is used to provide intimate contact between the construct and surrounding native bone. This close contact, along with the porosity of the material, acts to encourage bone ingrowth into the polymer. The development of porosity using effervescent agents is more advantageous than the development of porosity using soluble salts [14,22]. Effervescent agents allow for porosity to be developed during placement of the graft, unlike soluble salts, which require time for the salts to dissolve and porosity to develop. Bony ingrowth can therefore begin immediately following implantation. Increasing the effervescent agent concentration from 1 to 5% causes the void fraction to increase 1.5 times. Figure 2 shows the increase in expansion that accompanies an increase in effervescent agent concentration.

Yaszemski et al. [14] and Mikos et al. [22] reported using a mixture of poly(propylene fumarate) crosslinked by N-vinyl-pyrrolidone in the presence of composite material consisting of a porogen material (either gelatin microspheres or salt particles such as sodium chloride) and beta-tricalcium phosphate. In comparison, Domb et al. [15] utilized calcium carbonate and tricalcium phosphate as a particulate filler, and Gerhart et al. [23] employed a composite matrix consisting of gelatin, water, and sodium salicylate.

A different technology to generate porosity of the PPF-based bone graft extender material was applied by the authors. The addition of citric acid and sodium bicarbonate to the formulation led to formation of carbon dioxide, which is ultimately responsible for foam formation and

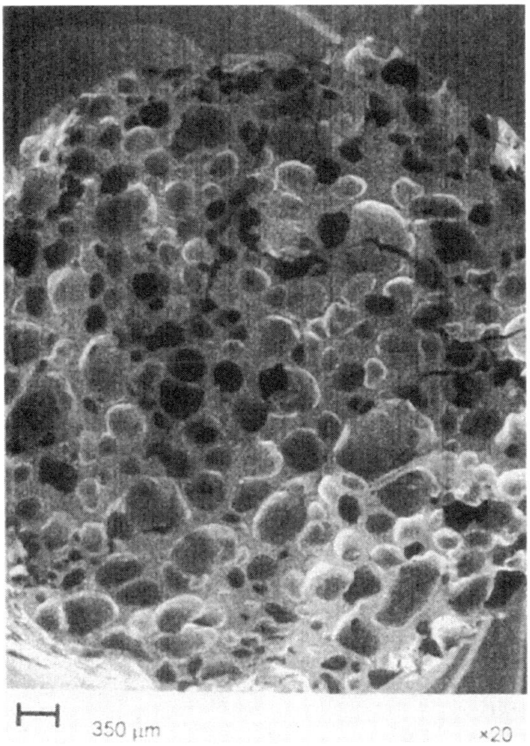

Figure 1 SEM of PPF foam.

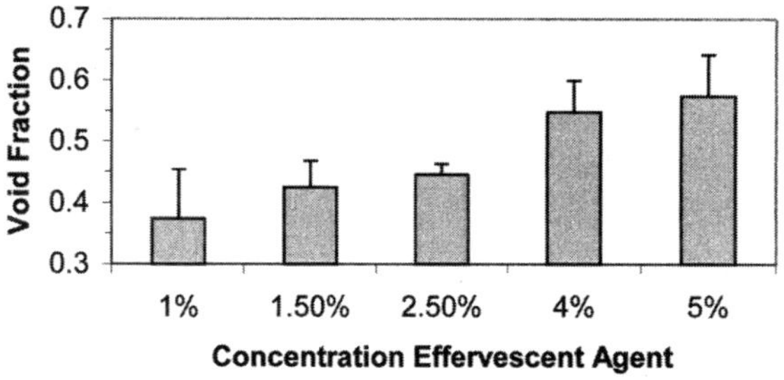

Figure 2 Void fraction of PPF-based bone graft extender as a function of effervescent agent concentration.

expansion with respective pore sizes of 100–1000 μm. The PPF-foam was noted to have a wide pore size distribution (median pore size 70 μm) with at least 30% of pores having an average diameter greater than 200 μm as confirmed by mercury intrusion porosimetry. These material properties, combined with practical handling characteristics and working times on the order of 15 min, made PPF-based foaming scaffold an ideal bone graft extender carrier material that easily mixed with cancellous autograft bone.

B. In Vitro Evaluation of a PPF-Based Bone Graft Substitute

An initial in vitro experiment examined the mechanical strength and handling of the graft material mixed with ground freeze-dried human bone (Lifenet Virginia Beach, VA). The allograft bone was mixed with the PPF material, crosslinking agent, and effervescent agents to form a XL-PPF pellet, following curing at 37 °C for 48 h.

Cured XL-PPF extender pellets were removed from the mold and subjected to compressive stress tests before and after in vitro degradation. Peak compressive strength was measured on an Instron (Canton, MA) Model 8511 materials tester at a strain rate of 0.1 mm/min. Initial mechanical properties were conducted on samples conditioned in saline at 37 °C for 60 min, and temporal properties were assessed following in vitro degradation in phosphate buffered saline (PBS) at 1, 3, and 6 weeks. Temporal mechanical properties were measured for XL-PPF formulations with either 0 or 25% human allograft bone.

Peak compressive forces for both XL-PPF formulations were measured through 6 weeks of in vitro degradation (see Fig. 3). Initial compressive strengths for XL-PPF with 0% allograft bone (3.5 ± 1.5 MPa) and XL-PPF with 25% (w/w) allograft bone (3.5 ± 4.7 MPa) were comparable to values for human cancellous bone (5.0 MPa) measured under similar conditions [24]. In addition, both formulations retained approximately 50% of their initial mechanical strength following 3 weeks of in vitro degradation.

Although both formulations had similar compressive strengths through 3 weeks of in vitro degradation, the XL-PPF–based extender mixed with 25% bone had a significantly higher strength (1.9 vs. 0.5 MPa) after 6 weeks as determined by a one-way analysis of variance ($p = 0.003$). Thus, mixing the XL-PPF–based graft extender with 25% bone graft did not compromise the mechanical integrity of the implant.

C. In Vivo Analysis of Autograft Extension

An in vivo study conducted by the authors assessed new bone growth within an XL-PPF–based implant at varying autograft/extender ratios. The osteoconductive effect of the PPF-based bone

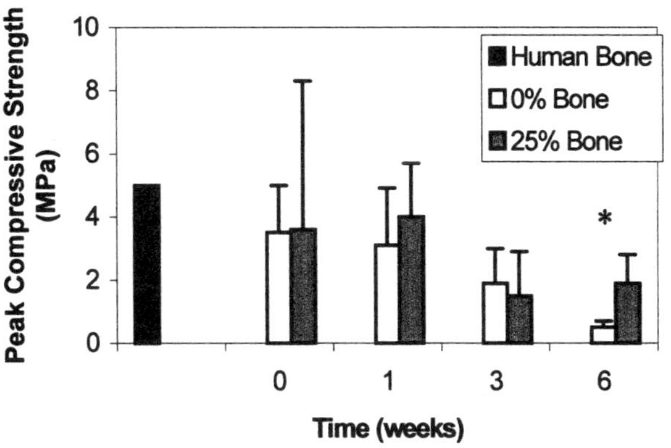

Figure 3 In vitro temporal mechanical strengths of XL-PPF extenders mixed with either 0 or 25% allograft human freeze-dried bone.

graft extender was evaluated using the rat tibial metaphysis implantation model according to Lewandrowski et al. [7]. Extender implants mixed with autograft bone at ratios of 3:1 and 1:3 (autograft/extender) were implanted within noncritical defects created in rat tibiae. New bone growth within the autograft/extender implants were compared to positive (100% autograft) and negative (100% extender) controls. Healing of the defect was assessed qualitatively by histology at 6 weeks postimplantation. New bone growth and osteoconduction within the implant was quantified by histomorphometry.

Histomorphometric evaluation of new bone formation was done by acquiring images of serial longitudinal sections of the specimen using a CCD video camera system (TM-745, PULNiX, Sunnyvale, CA) that was mounted on a Zeiss microscope. Images were digitized and analyzed. The approximate absolute volume of the newly formed bone was presented as an average (mean ± standard deviation) of these volume measures for each bone specimen. This parameter was given as a percentage rate and is presented as the average of all sections of eight grafted animals per graft type. This was defined as the *New Bone Volume Index*. Differences in the amount of new bone formed in response to implantation of the various types of grafts were analyzed for statistical significance by employing an ANOVA test. A p-level of less than 0.05 was considered statistically significant.

In the extender formulation, which was analyzed at a high autograft/extender (75% autograft/25% extender) mixing ratio, the implanted area was clearly identifiable. The bone graft extender material was dispersed throughout newly formed bone. New bone formation was primarily noted to occur within pores and seemingly proceeded from the endosteal surfaces and the adjacent tibial metaphysis into the implant.

The depth of bone ingrowth could not be reliably determined as the autologous bone graft had been mixed with the extender graft material prior to implantation. No significant inflammatory changes were noted. New bone formation within and around the implant took place without interposition of fibrous tissue. In addition to bone ingrowth into the implant, there was evidence of enchondral bone formation between bone graft extender particles. Furthermore, osteoclastic and osteoblastic activity at the implant–bone interface suggested that the newly formed bone was undergoing active remodeling. Occasionally, macrophages were present that contained intracellular inclusions of what appeared consistent with polymer.

In the extender formulation, which was analyzed at a low autograft/extender (25% autograft/75% extender) mixing ratio, histologic findings were similar to the bone samples that were retrieved from the group of animals that received extender formulation mixed with autograft at a high autograft/extender (75% autograft/25% extender) mixing ratio. PPF-based bone graft extender material was found in the metaphyseal implantation site with intimate contact to the surrounding newly formed woven bone. Although the absolute amounts of the PPF-based bone graft extender were higher and those of autologous bone graft lower in this group than in the other experimental group, the amount of newly formed woven bone did not seem to be different between the two experimental groups.

Quantitative histomorphometric volume measures of new bone formation in response to the implantation of any of the four types of grafting materials corroborated histologic findings. That is, the amount of new bone formation was significantly less in the positive and negative control groups by quantitative volume measures as expressed by the new bone volume index (see Table 2; $p < 0.02$). However, there was no statistically significant difference in the new bone volume index between the two experimental groups suggesting that mixing the PPF-based bone graft extender with a small amount of autologous bone graft resulting in mixed formulations with an autologous bone graft content as low as 25% is just as effective as using higher amounts of cancellous autograft.

VII. DISCUSSION

The reviewed in vivo study focused on the quantitative assessment of osteoinduction with the PPF-based bone graft extender at varying mixing ratios with cancellous autograft bone. This was achieved by analyzing the extender formulation at a high autograft/extender (75% autograft/ 25% extender) and at a low autograft/extender (25% autograft/75% extender) mixing ratio and by comparing these formulations to negative (extender alone) and positive (autograft alone) controls. To investigate these material properties the osetointegration and biocompatibility of the various formulations was evaluated in a rat tibial defect.

Results of in vitro and in vivo studies demonstrated maintenance of the structural integrity of the bone graft extender material used alone or in combination with cancellous autograft. There was no evidence of implant failure or disintegration at 4 weeks postoperatively. In addition, it was clearly evident that these control grafts, which did not contain any autograft, were extremely osteoconductive.

The PPF bone graft extender formulations, which were mixed with cancellous autologous bone graft and presumably had enhanced osteoinductive properties in vivo did not disintegrate. New bone ingrowth was shown to reside within and around particles of the PPF-based bone graft extender material. In the mixed groups, the amount of new bone, which formed at the implantation sites was significantly higher. When compared to the mixed groups, the amount of new bone formation was significantly less in the positive and negative control groups The

Table 2 New Bone Volume Index for Each Graft Type Based on Eight Rats per Group and 4 Weeks Postoperative Follow-up

	New bone volume index (%)
Negative control (PPF alone)	24 ± 3
Positive control (cancellous autograft alone)	15 ± 6
75% autograft/25% PPF extender (high)	48 ± 7
25% autograft/75% PPF extender (low)	46 ± 5

fact that there was no statistically significant difference in the new bone volume index between the two experimental groups containing PPF-based extender and cancellous autograft suggested that mixing the PPF-based bone graft extender with a small amount of autologous bone graft with an autologous bone graft content as low as 25% is just as effective as using higher amounts of cancellous autograft. These results clearly demonstrated that the porous PPF- based scaffold, in fact, can function as a true bone graft extender. These findings have immediate applicability to the development of bone graft extender formulations for clinical use.

VIII. CONCLUSIONS

Bioresorbable bone graft extenders could eliminate disadvantages associated with the use of autografts, allografts, and other synthetic materials currently used in clinical bone graft procedures. The major clinical application for this resorbable bone graft extender includes its use as an adjunct to filling of defects that arise from surgical removal of cysts and tumors, trauma, osteolytic defects, or surgical debridement of infections. Because autologous bone stocks are not sufficient to deal with extensive bony defects in clinical applications, a biodegradable bone graft extender requiring only minimal amount of autograft bone to produce an equivalent osteoinductive response as seen with autografts is increasing in demand.

Initial studies, employing both ex situ and in situ cured bone graft extender materials, indicated that new bone formation at the implantation site appears to be closely coupled to the addition of autograft. Therefore, optimization of autograft/PPF ratios by controlled in vitro and in vivo studies seems critical for the understanding of healing of larger bony voids after implantation in a clinical setting. Such optimization results from the consideration of the ratio of cancellous autologous bone to the polymeric extender, as well as the porosity of the extender. It is theorized that degradation of polymeric formulations that are structurally stable yet capable of initially developing in vivo porosities for bony ingrowth could be synchronized with the sequence of histologic events of the bone healing process.

The investigations demonstrated that mixing a PPF-based bioresorbable graft extender with graft material increased the working volume of autograft or allograft bone material. Significant increases in new bone growth were observed in cases when as little as 25% autograft was mixed with the PPF-based extender material. These findings could have a significant impact on addressing the increasing problem of limited graft material supply. Moreover, the substitute material has the potential ability to reduce or eliminate the need for a secondary surgical site and the associated complications.

ACKNOWLEDGMENTS

The authors wish to thank Dr. Joseph Alroy, DVM, Associate Professor in Pathology, Tufts University Schools of Medicine and Veterinary Medicine for his assistance in the histologic analysis of this study. This work was supported in part by NIH/NIAMS Grant No. 1 R43 AR46970-01A1 (to Debra J. Trantolo).

REFERENCES

1. Bucholz RW. Nonallograft osteoconductive bone graft substitutes. Clin. Orthop. Related Res 2002; 395:44–52.
2. Vaccaro AR. The Role of the Osteoconductive Scaffold in Synthetic Bone Graft. Orthopedics 2002; 25:s571–s578.

3. Gazdag AR, Lane JM, Glaser D. Alternatives to autologous bone graft. Efficacy and indications. J. Am. Acad. Orthop. Surg 1995:3:1–8.
4. Younger EM, Chapman MW. Morbidity at bone graft donor sites. J. Orthop. Trauma 1989; 3:192–95.
5. Summers BN, Eisenstein SM. Donor site pain from the ilium: a complication of lumbar spine fusion. J. Bone Joint Surg. Br 1989; 71:677–80.
6. Sandhu HS, Grewal HS, Parvataneni H. Bone grafting for spinal fusion. Orthop. Clin. North. Am 1999; 30:685–98.
7. Lewandrowski KU, Gresser JD, Wise DL, Trantolo DJ. Bioresorbable bone graft substitutes of different osteoconductivities: a histological evaluation of osteointergration of poly(propylene glycol-co-fumaric acid)–based cement implants in rats. Biomaterials 2000; 21:757–764.
8. Khan SN, Tomin E, Lane JM. Clinical applications of bone graft substitutes. Tiss Eng. Orthop. Surg 2000; 31:389–398.
9. Whitman MSK. Calcium phosphate bone cement-the Norian skeletal repair system in orthopedic surgery. AORN J 1997; 65:958–962.
10. Callan DP, Salkeld SL, Scarborough N. Histologic analysis of implant sites after grafting with demineralized bone matrix putty and sheets. Implant Dent 2000; 9(1):36–44.
11. Martin GJ, Boden SD, Titus L, Scarborough NL. New formulations of demineralized bone matrix as a more effective graft alternative in experimental posterolateral lumbar spine arthrodesis. Spine 1999; 24(7):637–745.
12. Cook DS, Salkeld SL, Rueger DC. Comparison of osteoinductive and osteoconductive biomaterials. In Healing Large Segmental Bone Defects: American Academy of Orthopaedic Surgeons, 1995: Paper No. 005.
13. Lewandrowski KU, Cattaneo MV, Gresser JD, Wise DL, Trantolo DJ. Effect of a poly(propylene fumarate) foaming cement on healing of critical size bone defects. Tissue Engineering 1999; 5(4): 305–316.
14. Yaszemski MJ, Payne RG, Hayes WC, Langer R, Mikos AG. in vitro degradation of a poly(propylene fumarate)–based composite material. Biomaterials 1996; 17(22):2127–2130.
15. Domb AJ, Manor N, Elmalak O. Biodegradable bone cement compositions based on acrylate and epoxide terminated poly(propylene fumarate) oligomers and calcium salt compositions. Biomaterials 1996; 17(4):411–417.
16. Thomson RC, Yaszemski MJ, Powers JM, Mikos AG. Hydroxyapatite fiber reinforced poly(alpha-hydroxy ester) foams for bone regeneration. Biomaterials 1998; 19(21):1935–1943.
17. Ishaug-Riley SL, Crane-Kruger GM, Yaszemski MJ, Mikos AG. Three–dimensional culture of rat calvarial osteoblasts in porous biodegradable polymers. Biomaterials 1998; 19(15):1405–1412.
18. Lewandrowski KU, Hile DD, Thompson BMJ, Sonis S, Wise DL, Trantolo DJ. Evaluation of a porous biodegradable biopolymer scaffold for mandibular reconstruction. Int. J. Oral Maxillofacial Implants, In press.
19. Gresser JD, Hsu SH, Nagaoka H, Lyons CM, Nieratko DP, Wise DL, Barabino GA, Trantolo DJ. Analysis of a vinyl pyrrolidone/poly(propylene fumarate) resorbable bone cement. J. Biomed. Mater. Res 1995; 29:1241–1247.
20. Lewandrowski KU, Hile DD, Thompson BMJ, Wise DL, Tomford WW, Trantolo DJ. Quantitative measures of osteoinductivity of a porous poly(propylene fumarate) bone graft extender. Tissue Engineering: In Press.
21. Lewandrowski KU, Hile DD, Kowaleski MP, Doherty SA, Wise DL, Trantolo DJ. An injectable porous poly(propylene glycol-co-fumaric acid) cement as an adjunct for treatment of metacarpal and phalangeal injuries. Submitted to Applied Biomaterials 2002.
22. Mikos AG, Sarakinos G, Leite SM, Vacanti JP, Langer R. Laminated three-dimensional biodegradable foams for use in tissue engineering. Biomaterials 1993; 14(5):323–330.
23. Gerhart TN, Miller RL, Kleshinski SJ, Hayes WC. in vitro characterization and biomechanical optimization of a biodegradable particulate composite bone cement. J. Biomed. Mater. Res 1988; 22(11): 1071–1082.
24. Carter DR, Hayes WC. Bone compressive strength: the influence of density and strain rate. Science 1976; 194:1174–1176.

7

Self-Reinforced Bioabsorbable Devices for Osteofixation of Craniofacial Bones

Nureddin Ashammakhi
Oulu University Hospital, and Oulu University, Oulu, Finland, and Tampere University of Technology, Tampere, Finland

Timo Waris
Tampere University Hospital
Tampere, Finland

Willy Serlo
Oulu University Hospital
Oulu, Finland

Pertti Törmälä
Tampere University of Technology
Tampere, Finland

I. INTRODUCTION

Classically, internal fixation of bones has been performed using metallic devices. However, metals are much more rigid than bone itself. Rigid fixation inhibits callus formation [1], leading to bone atrophy due to stress protection [2,3]. Metallic implants may also harbor infection which is difficult to eradicate without implant removal. Metals that are commonly used include stainless steel and titanium, and vitalium to a lesser extent. Stainless steel, however, has low resistance to corrosion. Titanium has also been shown to ''degrade,'' and it has been found in the soft tissues surrounding titanium plates and also in the regional lymph nodes.

In craniomaxillofacial (CMF) surgery the problem of translocation of metallic implants to the inner side of the cranial table and interference with the growth of the skull in children are the main serious concerns. Metallic implants may also interfere with radioimaging. They may also be associated with cold sensitivity in the facial skeleton, which is an especially important consideration in cold climates. It is recommended that metallic implants be removed, and this entails a second surgical procedure with consequent financial and psychological consequences as well as an increased risk of complications, especially in pediatric surgery. Continuous research and development in the field of biomaterials has led to the introduction of reliable bioabsorbable internal fixation devices. At the advancing front of successful implants are the new self-reinforced (SR) devices, which will replace metallic devices, at least in certain osteofixation indications. We review here the developments that have led to the current state-of-the-art use of bioabsorbable devices in bone surgery.

II. PROBLEMS WITH METALS

A. Pseudomigration

Pseudo-migration of metals [4–10] can be a problem, especially in the growing skulls of children, because cranial growth takes place by the laying of new bone on the outer surface and resorption of old bone on the inner side of the skull. Consequently, metals that are applied on the outer surface of the skull can occasionally sink into the bone, be found on the inner side of the skull [4], or even against the dura [7]. Although no neurological complications have been reported yet (follow-up of 2–6 years), there is a potential risk of damage to the brain should implant dislodgement occur in conditions such as trauma [4]. In minipigs [6], total invagination of titanium plates with initial intracranial translocation occurs after 12 to 16 weeks postimplantation. Hence, it is recommended that all metals used in infant crania should be removed as early as possible (within 3 months). Currently, such devices are not removed unless they give rise to problems. Pseudomigration more often complicates the use of longer plates and application in the temporal region [5].

B. Growth Disturbance

In pediatric CMF surgery, metals may lead to growth disturbance of the skull bones. This can occur with the placement of rigid plates, especially across sutures (in rabbits) [11–14], but it may also occur even when sutures are not crossed (in monkeys) [15]. The more complex the fixation, the greater the magnitude of growth changes seen, and hence it is advised that the least amount of fixation needed to provide three-dimensional stability should be used [16].

C. Stress Protection

Because of the rigidity of metals (E = 100–200 GPa), which is higher than that of bone itself (e.g., cortical bone, E = 10–20 GPa) [17], they are associated with stress concentration [3–18,19], which leads to bone atrophy or osteopenia, weakened bone, and increased risk of bone and implant fractures [2,3,20–23]. Bone mineral density was found to be lower when metallic screws were used to fix distal femoral osteotomies in rabbits, compared with bioabsorbable screws [24].

In experimental studies, stress shielding has been shown to occur in grafted mandibular bone [25,26], and bone resorption has been reported clinically in association with the use of rigid plate fixation. However, the clinical significance of this problem remains unclear.

D. Radio-opacity

The radio-opacity of metals can lead to obliteration of the view of tissues lying behind them, interfering with adequate radiological evaluation [27–31] such as CT scanning [32,33] and MRI [34]. This is a disadvantage in the field of CMF surgery, where neuroimaging is needed, e.g., in cancer patients. In addition, radio-opacity of metals may interfere with radiotherapy used in the treatment of cancer patients [35].

E. Infection

Implants may be colonized by bacteria that can form a biofilm, as may occur with *Staphylococcus epidermidis* [36]. A biofilm is known to interfere with the reach of antibodies and phagocytes to the residing bacteria, and with antibiotic sensitivity [37–39,42], leading to a status of persistent infection [40]. Various solutions are being explored to enhance the resistance of biomaterials to bacterial adhesion [41].

F. Corrosion

Stainless steel may suffer from corrosion in body fluids. With titanium, corrosion is limited by the formation of an oxide film, but titanium particles have been reported to be found at distant locations [42]. Corrosion products can accumulate in tissues encapsulating the implant [3,43] or in the draining lymph nodes [44]. The effect of metal ions on osteoblasts has been investigated and it was found that they may alter osteoblast behavior even at subtoxic concentrations [45]. These effects may not be very apparent clinically, but it is a matter of concern and there is a need to develop better and ''smarter'' materials.

G. Costs

As mentioned earlier, in pediatric CMF surgery, the problems of implant pseudomigration and interference with growth of the skull warrant the removal of metal implants. Removal operations are associated with potential complications and extra costs, for obvious reasons, as the second operation is usually technically more difficult, consumes more time, and requires more facilities.

III. DEVELOPMENT OF BIOABSORBABLE OSTEOFIXATION DEVICES

A. General

Research has focused on a class of aliphatic polymers called poly-α-hydroxy acids. This class has been under intensive research in the development of osteosynthesis devices since the 1960s [46–57]. Devices made of polyglycolide (PGA) or polylactide (PLA) are the strongest obtained in this class.

Bioabsorbable polyesters are broken down by hydrolysis, with progressive attack on their ester bonds, leading ultimately to release of monomers. In vivo, their degradation may also be accelerated due to the effect of enzymes [58]. The resulting monomers are metabolized and eliminated without the risk of toxic organ accumulation.

An important polymer that is used for manufacturing surgical devices is the polymer of glycolic (α-hydroxy acetic) acid. The polymer is manufactured by ring opening of the dimer (glycolide) [54,59] hence the name polyglycolide. Both names, polyglycolic acid and polyglycolide, are used interchangeably. Glycolide, the cyclic dimer condensation product, is formed by dehydrating glycolic acid. PGA of high molecular weight (20, 000 to 145, 000) [60] is a hard, tough, crystalline polymer melting at about 224–228°C, with a glass transition temperature (T_g) of 36°C. PGA can be spun into fibers which increase the strength properties of the polymer because of higher preferred molecular orientation of the polymer in the fiber [46,61].

The other important polymer used in manufacturing surgical devices is the polymer of lactic (α-hydroxycarboxylic) acid. The polymer is called polylactic acid or polylactide (because of synthesis from the dimer of lactic acid, the lactide). PLA has two enantiomeric isomers, the L and the D isomers. PLLA is a crystalline polymer with a prolonged degradation time. Addition of the D isomer into the PLA polymer chain makes it more amorphous, with faster degradation. Various degradation rates of the copolymer can be achieved by changing the percentages of L and D components. PLA with plastic properties was introduced as early as 1955 [62]. Later, surgical applications of materials made of either poly-L/D-lactide (PDLA) or poly-L-lactide (PLLA) were tested in the 1960s and 1970s [56,63]. PLLA was initially more commonly used in clinical applications, but it is becoming more and more replaced by rapidly degrading PLLA and PLGA copolymers. PLLA has a melting point of about 174–184°C and a glass transition temperature of about 57°C [49], and it can be spun to form fibers when its average molecular weight is 20,000 to 145,000 [60]. Getter et al., in 1972, used PLLA plates and screws [64] for the treatment of mandibular fractures in beagles (n = 6). After 32–40 weeks, the fracture sites were indistinguishable from the adjacent bone. Roed-Peterson, in 1974 [65], used PGA sutures (Dexon^R) to fix mandibular fractures in two young patients. However, at that time, with the available materials, it was not possible to rely completely on bioabsorbable implants for bone fixation. As far as bone healing was concerned, the results were successful and encouraging. Later, in 1989, Illi et al. [66] used PLLA screws and polydioxanone (PDS) bands in children for cranioplasties and cranial reconstructions (n = 8). After a follow-up time of 2–20 months, aesthetic and radiological results were as good as those obtained with conventional materials.

To improve the strength retention properties of polymeric devices, one way is to increase the bulk of polymer in the device. Another way is to use a polymer with a higher molecular weight. High molecular weight, so called ''as-polymerized,'' PLA plates and screws have been used for the fixation of fractures of the zygomatic arch in humans. Although healing was uneventful, the patients developed swelling at the site of implantation 3 years postoperatively [67]. This occurred in all 10 of the treated patients. Eight of them had to undergo exploratory surgery. The retrieved tissue revealed PLA particles encapsulated in thick fibrous tissue.

B. Reinforced Bioabsorbable Materials

The focus of research at that stage was aimed at the development of strong devices with better strength retention properties [47–61,68]. One approach involved the use of elements such as fibers, fibrils or oriented polymer chains to reinforce the matrix of the device. These reinforcing elements were made of a material different from the material of the matrix of the device, such as carbon fibers or other polymers. PGA-reinforced PLLA plates were used in sheep by Christel et al. [69]. Five plates were used on sheep tibiae (no osteotomies), and they exhibited cracks at the screw holes. One of two plates implanted on osteotomized tibiae broke after 1 month and the other was still under evaluation at the time of the report. To produce stronger materials, more and more reinforcing elements are needed, and the use of chemical agents called adhesion promoters is also needed. The degree of reinforcement, however, may be limited because of a limit to the amount of adhesion promoters that can be used, as those normally used are toxic [70].

C. Self-Reinforced Bioabsorbable Materials

Another approach to develop stronger bioabsorbable devices was developed by Törmälä and Rokkanen [71], and it comprises the use of reinforcing elements in the device that are of the same material as that of the matrix of the device itself. The technique was therefore called self-reinforcement (Fig. 1A). When the self-reinforcing method is used, adhesion promoters are not needed, their side effects can be avoided, and a higher degree of reinforcement can be achieved, resulting in the production of stronger devices. So far, various self-reinforced osteofixation devices have been produced, such as rods, pins, screws, plates, and tacks (Fig. 1B), and they have proved to be successful.

Over the last 10 years the clinical application of SR devices in CMF surgery has evolved. The first applications involved SR-PLLA, with SR-PLDLA and SR-PLGA devices being used nowadays. Experience gained in other fields such as orthopedics and trauma surgery, as well as experience gained by other groups, have added to our own experience in the development

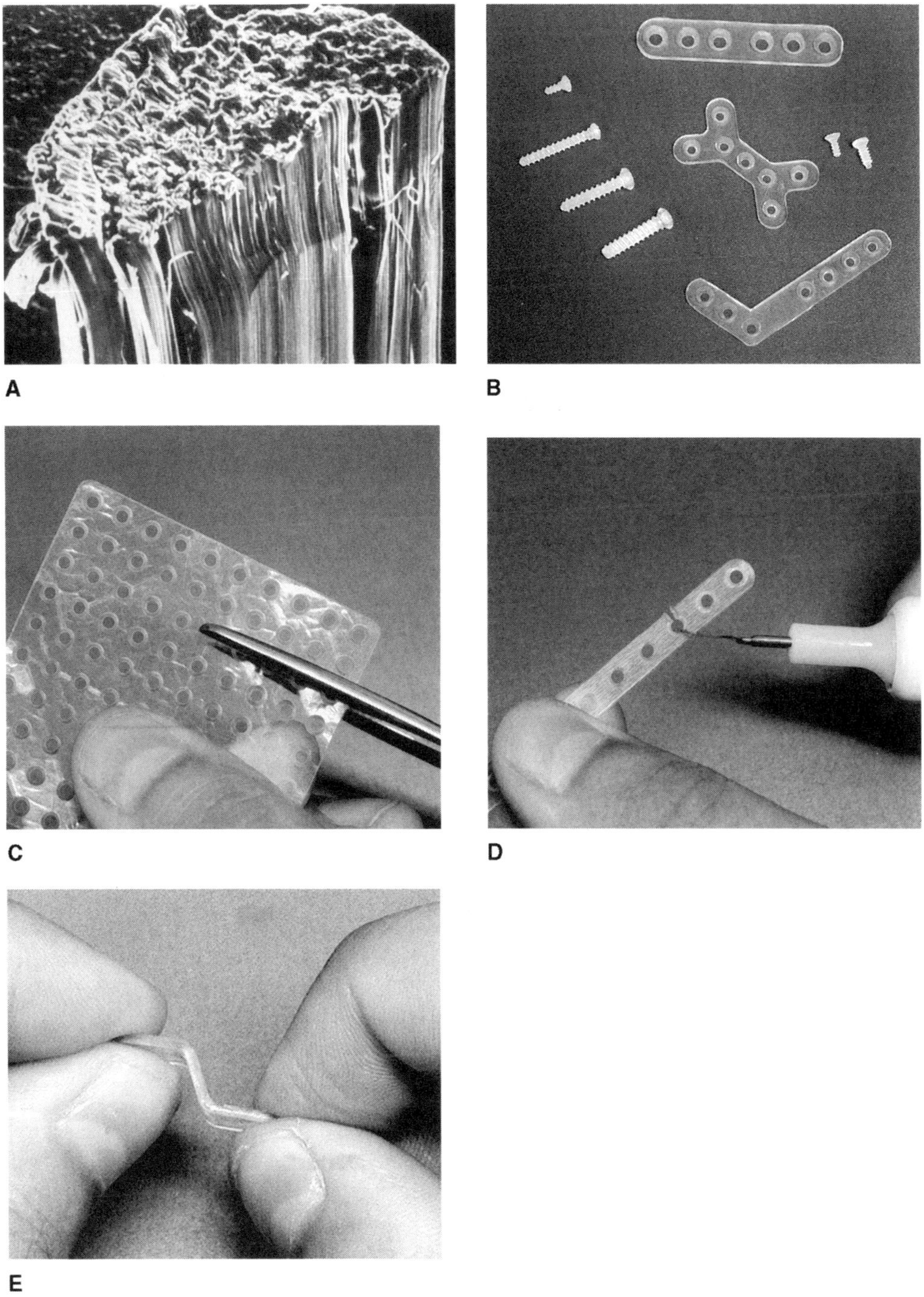

Figure 1 Cross-section of self-reinforced (die-drawn) rod showing the structure SR composite that comprises oriented fibrils and binding matrix (A). Various types of devices such as plates and screws can be produced from SR material (B). SRplates can be cut using scissors (C) or heat loop (D). SR plates can be bent twice fore and back (E) in room temperature without affecting significantly their strength properties. Bending them more than twice may lead to weakening of their strength.

of appropriate materials for various applications. We recently reported our long-term results [72], which were encouraging. We then undertook a multicenter study in an EU project that involved centers from Innsbruck, Paris, Oulu, and London. In all, we have operated on more than 180 CMF patients (mostly for correction of congenital deformities) using SR devices [73]. The results have been encouraging; they are being analyzed further and will be reported in detail in a later publication.

D. Advantages of Bioabsorbable Osteofixation Devices

1. Elimination from the Body

Bioabsorbable materials undergo degradation by hydrolysis and to a lesser extent by enzymes [58]. As they are degraded into smaller products, they are engulfed by phagocytes and eliminated through natural body metabolism in the form of water and CO_2 without toxic tissue accumulation [74]. Self-reinforced devices have a ductile mode of deformation during their degradation in vivo.

2. Plastic Handling

Bioabsorbable devices have plastic handling properties. One can cut or trim them using scissors (Fig. 1C), saw, or heat loop (Fig. 1D). Screw holes can be added to the plates whenever necessary. Should a screw break during the insertion procedure, there is no need to remove it — it can be drilled through and another screw can be applied. Accordingly, plate repositioning is not needed. Self-reinforced devices have high initial strength (appropriate modulus). They can also be made into small yet strong devices to minimize the risk of complications such as palpability, as seen with other types of implants. They can also be reshaped in the operating theater at room temperature without significantly affecting their strength properties (Fig. 1E). Use of time-consuming heating systems (heat gun, heating bags [75], thermal packs [76], or electric heating devices [77]) is consequently not needed. After heating PLGA plates in excess of their T_g, it has been reported that their flexural modulus falls by 20% [78]. The plates need 2 min of cooling down to regain 50% of their stiffness, but after 1 week (in vitro), the mechanical properties of heated and nonheated plates are identical. Because of ''memory," bent plates may return to their prebent shape, which may lead to insecure fixation. Such a memory effect is minimal with amorphous SR-PLDLA plates (2–3%). Moreover, memory can be balanced by overmolding the plates to the desired shape at the time of operation, without problems.

3. Plastic Behavior

Osteofixation with bioabsorbable devices is not as rigid as that provided by metals. Some degree of micromotion can occur. While the polymer loses its strength gradually, the load is transferred to the bone. The strength of PLGA copolymer, for instance, is maximal when healing of bone is just beginning, whereas when bone healing is well advanced, the strength of the device is less [79]. Such an effect may lead to better quality of bone. Bioabsorbable devices should not lead to growth disturbances when used in children, except if they are too slow to degrade or are placed across sutures [11]. Compensatory bone lengthening (normalizing bone response) may, however, take place, leading to maintained overall skeletal morphology [80]. Another determining factor is the size of the implant. Hence, small devices should be used whenever possible to avoid growth disturbances. More rapidly degrading materials such as PLGA are preferred because they soften with time and allow for device elongation without interfering with cranial bone growth [81].

4. Reduced Time of Operation

Recently, we introduced bioabsorbable tacks that can be applied using a special instrument, a tack-shooter (Fig. 2A). Use of tacks can lead to reduction of operative time because the tacks do not require tapping or tightening (Fig. 2) [82–84]. We have found that the time of operation can be reduced by 10–15% when using bioabsorbable tacks instead of screws [85], after having passed the early phase of the learning curve, where the time was somewhat prolonged.

5. Radiolucency

Bioabsorbable devices are, by nature, radiolucent and they do not preclude subsequent use of MRI or cause artefacts in CT scanning. We believe that this is an advantage, although the need to visualize implants is also recognized [86]. It is also possible to manufacture cannulated bioabsorbable screws to allow introduction of K-wires and radiological monitoring when indicated. This may be needed in orthopedic and trauma surgery.

6. Postoperative Radiotherapy

The fact that bioabsorbable devices are radiolucent allows for greater x-ray penetration for postoperative radiotherapy [87]. Studies have indicated that PLA material can be regarded as a tissue equivalent and can be safely used to fix osteotomies when postoperative irradiation is anticipated [87]. Hofmann et al. have studied the effect of radiation on PGA in vitro [88]. Following irradiation, PGA specimens were incubated in a chamber simulating the humid milieu of a living organism. Weekly bending strength testing of specimens for 3 weeks showed no significant difference compared with control specimens.

7. Cost Effectiveness

Costs can be reduced by using one large bioabsorbable panel plate that can be cut into several small plates [89]. They can be cut with scissors or a hot loop and tailored as required. However, precise cost-effectiveness needs to be evaluated in future studies. It is worth noting that with the advent of bioabsorbable tacks and their application device (tack-shooter) (Fig. 2) costs can be further reduced due to reductions made in operative time and reduction in the risk of complications. Multifunctional osteofixation devices that contain antibiotics [90,91] may also contribute to long-term cost-effectiveness by reducing the rate of infections. However, this issue has to be addressed in a future clinical study.

E. Disadvantages

1. Size Limitations

It is now possible to produce relatively small yet strong miniplates employing the self-reinforcing technique. However, it would be an advantage to have even smaller microplates and microscrews, as in many cases plate thickness is often a problem where soft tissue cover is thin, with the risk of palpability, extrusion, or sinus formation. Biomaterials technologists may circumvent this problem, either by finding new methods to produce microimplants using the same materials or by using new materials in the future.

2. Fluid Accumulation and Sinus Formation

Fluid accumulation has been reported, even with nonabsorbable devices [92]. However, the fluid accumulation (and consequent swelling) that we refer to here is that related to the use of

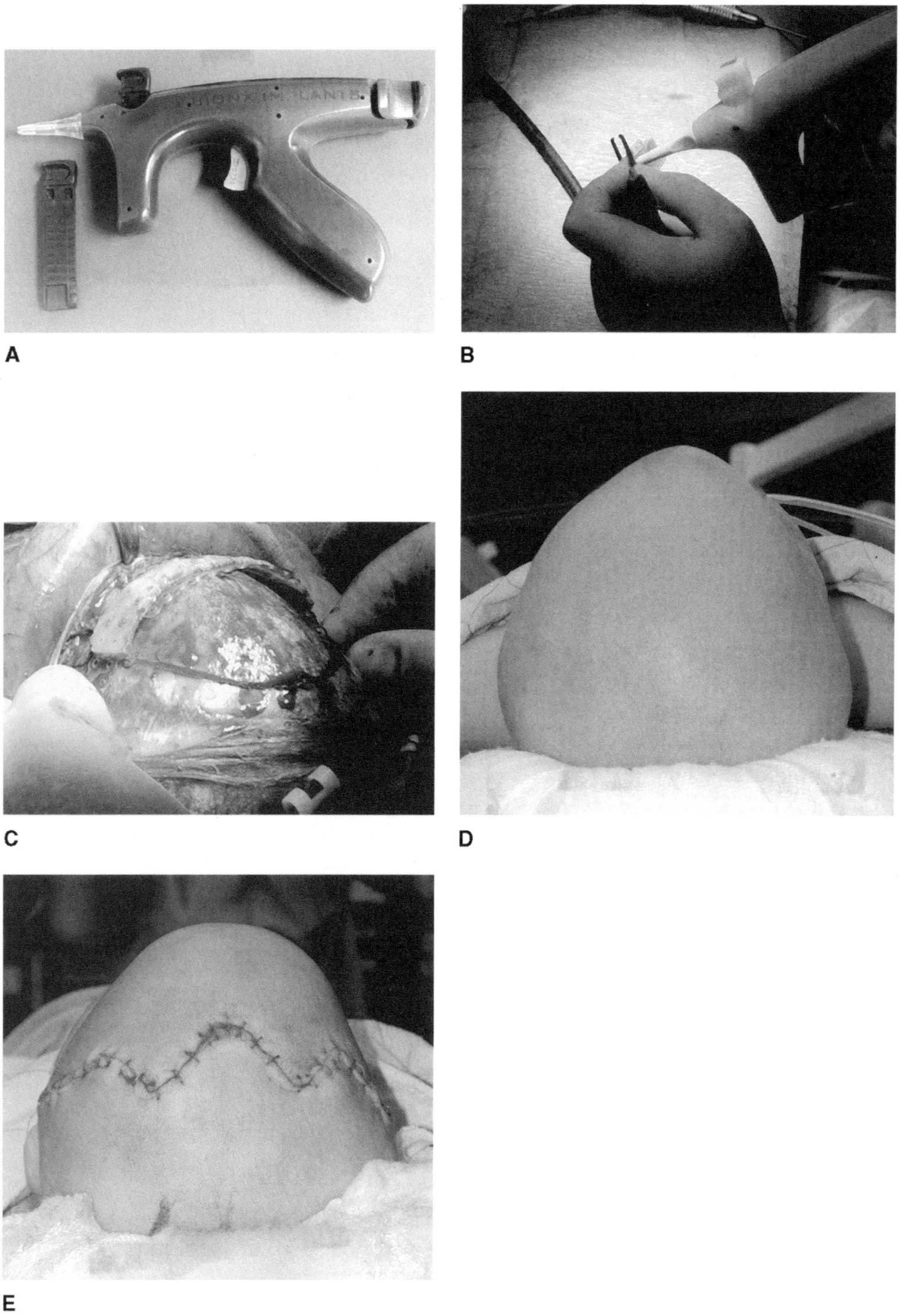

Figure 2 Tack-shooter (A) is used to apply tacks (B) that are provided ready preloaded in a magazine containing up to ten tacks. To maintain corrected shape of the frontal bone a plate fixed with tacks was applied endocranially (C). Preoperative trigonocephaly deformity in infant (D) and immediate postoperative view (E) are shown.

bioabsorbable implants. It is a sterile fluid that contains polymer degradation products and it occurs because the rate of degradation exceeds the rate of absorption (elimination). The fluid may track to the exterior in the form of a sinus (often wrongly called a fistula). It can be treated by aspiration or excision and it does not interfere with the healing of the bone. It was particularly observed to occur when pure homopolymeric devices such as relatively rapidly degrading PGA implants [93] and as-polymerized slowly degrading PLLA [67] were used. High molecular weight is also a factor that can be implicated in retarded resorption [94]. Immunological studies of accumulated fluid after the implantation of PGA revealed nonspecific lymphocytic activation secondary to inflammatory mononuclear cell migration and adhesion [95]. Currently, copolymeric rather than homopolymeric devices are used and the risk should be remote. However, it should still be kept in mind, and the smallest possible amount of polymer should be used, to ensure tissue absorption.

In CMF surgery, the clinical application of SR devices has been relatively gradual and no such complications have yet been reported in short- and long-term follow-up studies [48,52,72,82, 96–104]. Other (nonreinforced) bioabsorbable devices have been used as large implants [105] to compensate for their brittleness and low strength [49]. Nonreinforced, 2-mm-thick as-polymerized PLLA plates that were used for fixation of zygomatic fractures in ten patients [67] caused remarkable persistent swelling [105]. At 3 years postimplantation, analysis revealed PLA crystals encapsulated in thick fibrous tissue. Copolymers rather than homopolymers are preferred for clinical application and they are thought to be associated with diminished risks of inflammatory reactions. Copolymeric devices made of amorphous PLDLA or PLGA have not caused clinically significant foreign body reactions.

IV. FUTURE PROSPECTS

A. New Indications

Our surgical thinking has long been dominated by metals. With the introduction of bioabsorbables, osteofixation procedures need to evolve in a parallel fashion to accommodate new ideas and innovations. One should always remember that bioabsorbables will not completely replace metals in every indication or every situation. Bioabsorbables, on the other hand, make new applications possible because of their capacity to carry drugs, cells, and biomolecules. They can be used inside the bone [106] or in body cavities because they are resorbed and do not require removal. In cranioplasty, for instance, carried out to correct trigonocephaly, absorbable plates have been endocranially placed behind the frontal bone to avoid extracalvarial application over the thin frontal soft tissue cover, with successful outcome (Fig. 2) [97,107].

Surgical speciality training now has to include bioabsorbables as they have particular properties and requirements for successful use. This is very important to reduce the rate of failed screw and tack application and the rates of complications. Attention has to be paid to the type of material used, its mode of deformation, and its degradation and absorption profiles. Enough soft tissue cover of the implant, and adequate vascularity of the area should be ensured to avoid postoperative problems and to enable absorption of the polymer once degraded.

B. Multifunctional Materials

Our group has recently succeeded in the development and introduction of a new generation of multifunctional bioabsorbable devices. These devices are composed of bioabsorbable polymers, bioactive glass (for osteoconduction), and drugs (in one example, antibiotics) [90–91,108]. Once implanted in the bone, these devices should be replaced with bone rather than fibrous tissue and will be more resistant to bacterial attachment, proliferation, and biofilm formation on their surfaces. We have observed reduced bacterial attachment and biofilm formation on these devices in our preliminary in vitro studies (Fig. 3), and we are currently studying them in experimental animals in regard to tissue reactions. These multifunctional devices (Fig. 4) [90,91,108–110] can combat infection and should have a particular use in cases of trauma and in contaminated operation fields.

Other bioactive devices are also under intensive research and development [111]. Such bioactive devices can be made to contain biomolecules such as growth factors [112], chemotherapeutics, hormones, or vaccines. The inclusion of growth factors may very well help to treat difficult-to-heal bone defects, especially when such biomaterials contain osteoconductive ceramics such as bioactive glass. As regards growth factors, we still need to learn more about the threshold values [27] and long-term clinical outcome to be able to provide more profound information.

ACKNOWLEDGMENTS

Research funds from The Technology Development Center in Finland (TEKES, 90220), The Special Government Subsidy for Medical Research and Education (EVO TYH 1331), and The

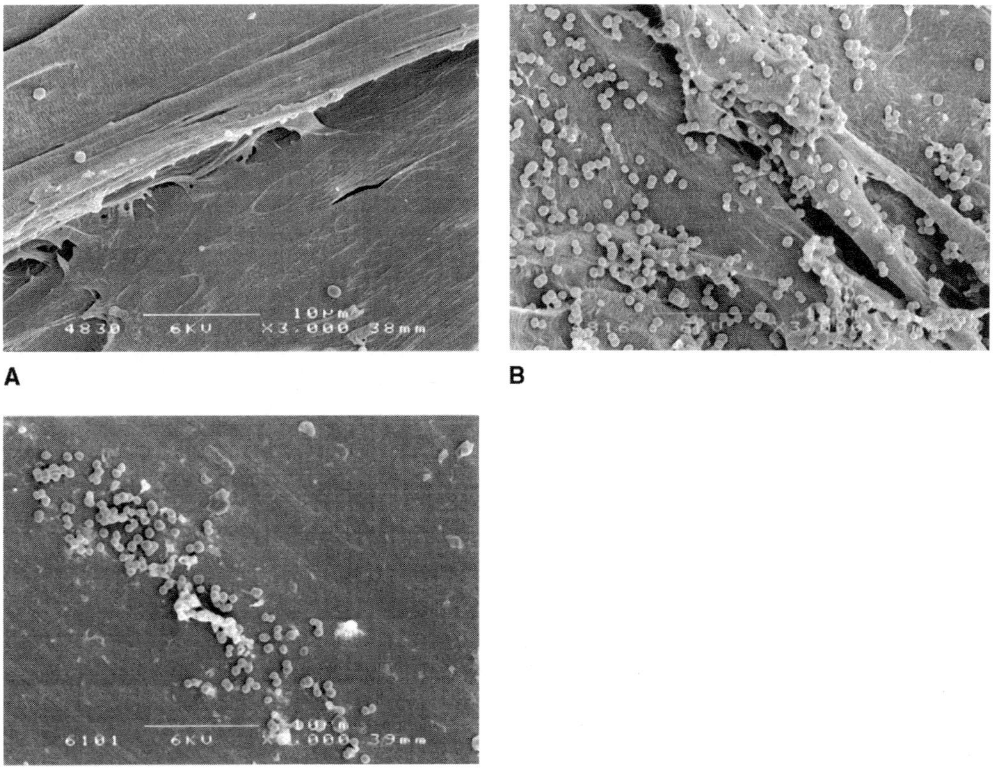

A B C

Figure 3 SEM pictures showing bacterial attachment and biofilm formation on various multifunctional bioabsorbable materials, 3 days after culture with *S. epidermidis.* Note that bacterial attachment is least with ciprofloxacin-releasing PLGA material (A) as compared to control PLGA material (not containing ciprofloxacin) (B), and to titanium (C).

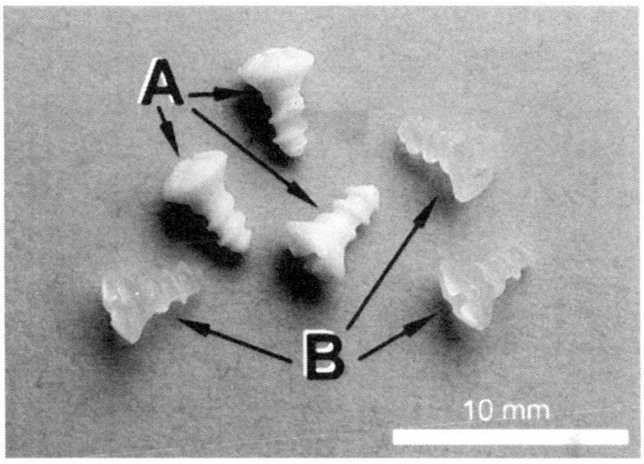

Figure 4 New multifunctional ciprofloxacin-releasing SR-PLGA 80/20 screws (A) as compared to corresponding screws that do not contain ciprofloxacin (B).

European Commission (BMH4-98-3892 and QLRT-2000-00487) are greatly appreciated. The authors thank Nick Bolton, PhD, for checking the language of the manuscript, Anja Jussila for administrative assistance, Sanna-Mari Niemelä, Jouko Ilomäki, and Minna Veiranto for the photographs, and Liisa Kärki and Seija Leskelä for the photographic artwork.

REFERENCES

1. Tayton K, Johnson-Nurse C, McKibbin B, Bradley J, Hastings G. The use of semi-rigid carbon-fibre–reinforced plastic plates for fixation of human fractures. Results of preliminary trials. J. Bone Joint Surg 1982; 64B(1):105–11.
2. Tonino AJ, Davidson CL, Klopper PJ, Linclau LA. Protection from stress in bone and its effects. J. Bone Joint Surg 1976; 58B(1):107–113.
3. Paavolainen P, Karaharju E, Slätis P, Ahonen J, Holmstrom G. Effect of rigid plate fixation on structure and mineral content of cortical bone. Clin. Orthop 1978; 136:287–293.
4. Fearon JA, Munro IR, Bruce DA. Observations on the use of rigid fixation for craniofacial deformities in infants and young children. Plast. Reconstr. Surg. 1995; 95:634–637.
5. Goldberg DS, Bartlett SP, Yu JC, Hunter JV, Whitaker LA. Critical review of microfixation in pediatric craniofacial surgery. J. Craniofac. Surg. 1995; 6:301–307.
6. Hönig JF, Merten HA, Luhr HG. Passive and active intracranial translocation of osteosynthesis plates in adolescent minipigs. J. Craniofac. Surg. 1995; 6:292–298.
7. Papay FA, Hardy S, Morales L, Walker M, Enlow G. "False" migration of rigid fixation appliances in pediatric craniofacial surgery. J. Craniofac. Surg. 1995; 6:309–313.
8. Stelnicki EJ, Hoffman G. Intracranial migration of microplates versus wires in neonatal pigs after frontal advancement. J. Craniofac. Surg. 1998; 9:60–64.
9. Yaremchuk MJ, Posnick JC. Symposium. Resolving controversies related to plate and screw fixation in the growing craniofacial skeleton. J. Craniofac. Surg. 1995; 6:525–538.
10. Yu JC, Bartlett SP, Goldberg DS, Gannon F, Hunter J, Habecker P, Whitaker LA. An experimental study of the effects of craniofacial growth on the long-term positional stability of microfixation. J. Craniofac. Surg. 1996; 7:64–68.
11. Antikainen T, Pernu H, Törmälä P, Waris T, Serlo G. Development of the neurocranium after transsutural fixing by new, resorbable poly-L-lactide miniplates. A comparison to fixing with the common titanium miniplates. Acta Neurochir 1994; 128:26–31.

12. Resnick JI, Kinney BM, Kawamoto HK. The effect of rigid internal fixation on cranial growth. Ann. Plast. Surg. 1990; 25:372–374.
13. Wong L, Dufresne CR, Richtsmeier JT, Richtsmeier JT, Manson PN. The effect of rigid fixation on growth of the neurocranium. Plast. Reconstr. Surg. 1991; 88:395–403.
14. Wong L, Richtsmeier JT, Manson PN. Craniofacial growth following rigid fixation: suture excision, miniplating, and microplating. J. Craniofac. Surg. 1993; 4:234–244.
15. Yaremchuk MJ, Fiala TG, Barker F, Ragland G. The effects of rigid fixation on craniofacial growth in rhesus monkeys. Plast. Reconstr. Surg. 1994; 93:1–10.
16. Yaremchuk MY. Experimental studies addressing to rigid fixation in craniofacial surgery. Clin. Plast. Surg. 1994; 21:517–524.
17. Vert M, Christel P, Chabot F, Leray G. Bioresorbable plastic materials for bone surgery. In: Hastings GW, Ducheyne P, eds. Macromolecular Biomaterials. Boca Raton. Florida: CRC Press, 1984: 119–142.
18. Pilliar RM, Cameron HU, Binnington AG, Szivek J, Macnab G. Bone ingrowth and stress shielding with a porous surface coated fracture fixation plate. J. Biomed. Mater. Res 1979; 13(5):799–810.
19. Woo SL, Akeson WH, Coutts RD, Rutherford L, Doty D, Jemmott GF, Amiel G. A comparison of cortical bone atrophy secondary to fixation with plates with large differences in bending stiffness. J. Bone Joint Surg 1976; 58A(2):190–195.
20. Uhthoff HK, Dubuc FL. Bone structure changes in the dog under rigid internal fixation. Clin. Orthop 1971; 81:165–170.
21. Uhthoff HK, Jaworski ZF. Bone loss in response to long-term immobilisation. J. Bone Joint Surg. Br 1978; 60B(3):420–429.
22. Iizuka T, Lindqvist C, Hallikainen D, Mikkonen P, Paukku G. Severe bone resorption and osteoarthrosis after miniplate fixation of high condylar fractures. A clinical and radiologic study of thirteen patients. Oral Surg. Oral Med. Oral Pathol 1991; 72:400–407.
23. Hench LL. Biomaterials. Science 1980; 208:826–831.
24. Viljanen J, Kinnunen J, Bondestam S, Majola A, Rokkanen P, Törmälä G. Bone changes after experimental osteotomies fixed with absorbable selfreinforced poly-L-lactide screws or metallic screws studied by plain radiographs, quantitative computed tomography and magnetic resonance imaging. Biomaterials 1995; 16:1353–1358.
25. Kennady MC, Tucker MR, Lester GE, Buckley MJ. Histomorphometric evaluation of stress shielding in mandibular continuity defects treated with rigid fixation plates and bone grafts. Int. J. Oral Maxillofac. Surg. 1989; 18:170–174.
26. Kennady MC, Tucker MR, Lester GE, Buckley MJ. Stress shielding effect of rigid internal fixation plates on mandibular bone grafts. A photon absorption densitometry and quantitative computerized tomographic evaluation. Int. J. Oral Maxillofac. Surg. 1989; 18:307–310.
27. Habal MB. Regeneration osteogenesis: the challenges that face bone graft surgery in the new millinium. Perspectives in Plastic Surgery 1998; 11(2):1–18.
28. Ebraheim NA, Coombs R, Rusin JJ, Jackson WT. Reduction of postoperative CT artifacts of pelvic fractures by use of titanium implants. Orthopedics 1990; 13:1357–1358.
29. Ebraheim NA, Savolainen ER, Stitgen SH, Jackson WT. Magnetic resonance imaging after pedicular screw fixation of the spine. Clin. Orthop 1992; 279:133–137.
30. Sullivan PK, Smith JF, Rozzelle AA. Cranio-orbital reconstruction: safety and image quality of metallic implants on CT and MRI scanning. Plast. Reconstr. Surg. 1994; 94:589–596.
31. Weese JL, Rosenthal MS, Gould G. Avoidance of artefacts on computerized tomograms by selection of appropriate surgical clips. Am. J. Surg. 1984; 147:684–687.
32. Anastakis DJ, Antonyshyn OM, Cooper PW, Yaffe MJ, Bush K, Mawdsley GE. Computed tomography artifacts associated with craniofacial fixation devices: an experimental study. Ann. Plast. Surg. 1996; 37:349–355.
33. Fiala TG, Novelline RA, Yaremchuk MJ. Comparison of CT imaging artifacts from craniomaxillofacial internal fixation devices. Plast. Reconstr. Surg. 1993; 92:1227–1231.
34. Fiala TG, Paige KT, Davis TL, Campbell TA, Rosen BR, Yaremchuk MJ. Comparison of artifact from craniomaxillofacial internal fixation devices: magnetic resonance imaging. Plast. Reconstr. Surg. 1994; 93:725–730.

35. Scher N, Poe D, Kuchmir F, Reft C, Weichselbaum R, Panje WR. Radiotherapy of the resected mandible following stainless steel plate fixation. Laryngoscope 1987; 98:561–563.
36. An YH, Friedman RJ, Draughn RA. Bacterial adhesion to biomaterial surfaces. In: Wise Donald L, Trantolo Derba J, Altobelli David E, Yaszemski Michael J, Gresser Joseph D, eds. Human Bioamterials Applications. NJ: Humana Press, 1996:19–57.
37. Gray ED, Peters G, Verstegen M, Regelmann WE. Effect of extracellular slime substance from *Staphylococcus epidermidis* on the human cellular immune response. Lancet 1984; 1(8373): 365–367.
38. Gristina AG, Hobgood CD, Webb LX, Myrvik QN. Adhesive colonization of biomaterials and antibiotic resistance. Biomaterials 1987; 8(6):423–426.
39. Gristina AG, Jennings RA, Naylor PT, Myrvik QN, Webb LX. Comparative in vitro antibiotic resistance of surface-colonizing coagulase-negative staphylococci. Antimicrob. Agents Chemother 1989; 33(6):813–816.
40. Francois P, Vaudaux P, Foster TJ, Lew DP. Host–bacteria interactions in foreign body infections. Infect. Control Hosp. Epidemiol 1996; 17(8):514–520.
41. Koerner RJ, Butterworth LA, Mayer IV, Dasbach R, Busscher HJ. Bacterial adhesion to titanium-oxy-nitride (TiNOX) coatings with different resistivities: a novel approach for the development of biomaterials. Biomaterials 2002; 23(14):2835–2840.
42. Schliephake H, Reiss G, Urban R, Neukam FW, Guckel G. Metal release from titanium fixtures during placement in the mandible: an experimental study. Int. J. Oral Maxillofac. Implants 1993; 8(5):502–511.
43. Jacobs JJ, Gilber JL, Urban RM. Corrosion of metal orthopaedic implants. J. Bone Joint Surg 1998; 80A:268–282.
44. Yuen JC, Klitzman B, Serafin D. Biomaterials in plastic surgery. In: Greco RS. Implantation Biology: The Host Response and Biomedical Devices. Boca Raton, Florida. USA: CRC Press, Inc., 1994: 191–228.
45. Sun ZL, Wataha JC, Hanks CT. Effects of metal ions on osteoblast-like cell metabolism and differentiation. J. Biomed. Mater. Res 1997; 34:29–37.
46. Vert M, Chabot F, Leray J, Christel G. Stereoregular bioabsorbable polyesters for orthopaedic surgery. Makromol. Chem 1981; (Suppl.)5:30–41.
47. Vainionpää G. Biodegradation of polyglycolic acid in bone tissue, an experimental study on rabbits. Arch. Orthop. Trauma Surg 1986; 104:333–338.
48. Waris T, Pohjonen T, Törmälä G. Self-reinforced absorbable polylactide (SR-PLLA) plates in craniofacial surgery. A preliminary report on 14 patients. Eur. J. Plast. Surg. 1994; 17:236–238.
49. Törmälä P, Pohjonen T, Rokkanen P. Bioabsorbable polymers: materials technology and surgical applications. Proc. Instn. Mech. Engrs 1998; 212(Pt H):101–111.
50. Suuronen R, Manninen MJ, Pohjonen T, Laitinen O, Lindqvist G. Mandibular osteotomy fixed with biodegradable plates and screws: an animal study. Br. J. Oral Maxillofac. Surg. 1997; 35(5):341–348.
51. Suuronen R, Laine P, Sarkiala E, Pohjonen T, Lindqvist G. Sagittal split osteotomy fixed with biodegradable, self-reinforced poly-L-lactide screws. A pilot study in sheep. Int. J. Oral Maxillofac. Surg. 1992; 21(5):303–308.
52. Suuronen R, Kallela I, Lindqvist G. Bioabsorbable plates and screws: current state of the art in facial fracture repair. Craniomaxillofac. Trauma 2000; 6(1):19–27.
53. Suuronen G. Comparison of absorbable self-reinforced poly-L-lactide screws and metallic screws in the fixation of mandibular condyle osteotomies: an experimental study in sheep. J. Oral Maxillofac. Surg. 1991; 49(9):989–995.
54. Schmitt EE, Poilistina RA. Polyglycolic acid prosthetic devices, 1969.
55. Rokkanen PU. Bioabsorbable fixation devices in orthopaedics and traumatology. Ann. Chir. Gynaecol 1998; 87:13–20.
56. Kulkarni RK, Pani KC, Neuman C, Leonard G. Polylactic acid for surgical implants. Arch. Surg. 1966; 93:839–843.
57. Cutright DE, Hunsuck EE. The repair of fractures of the orbital floor using a biodegradable material, polylactic acid. J Oral Surg 1972; 33(1):28–34.

58. Williams DF. Some observation on the role of cellular enzymes in the in vivo degradation of polymers. In: Syrett BC, Acharya A, eds. ASTM Special Technical Publications, Corrosion and Degradation of Implant Materials. Philadelphia: American Society for Testing and Materials, 1979: 61–75.
59. Higgins NA. Condensation polymers of hydroxyacetic acid. U.S. Patent 2,676,945, 1954.
60. Frazza EJ, Schmitt EE. A new absorbable suture. J. Biomed. Mater. Res., Symp 1971; 1:43–58.
61. Törmälä G. Biodegradable self-reinforced composite materials: manufacturing structure and mechanical properties. Clin. Mater 1992; 10:29–34.
62. Schneider AK. Polymers of high melting lactide. US Patent No. 2,703,316, 1955.
63. Kulkarni RK, Moore EG, Hegyeli AF, Leonard G. Biodegradable poly(lactic acid) polymers. J. Biomed. Mater. Res 1971; 5:169–181.
64. Getter L, Cutright DE, Bhaskar SN, Augsburg JK. A biodegradable intraosseous appliance in the treatment of mandibular fractures. J. Oral Surg 1972; 30:344–348.
65. Roed-Paterson G. Absorbable synthetic suture material for internal fixation of fractures of the mandible. Int. J. Oral Biomed. Mater. Res 1974; 3:133–136.
66. Illi OE, Sailer H, Stauffer UG. Résultats préliminaires de l'ostéosynthèse biodégradable en chirurgie cranio-faciale chez l'enfant. Chir. Pédiatr 1989; 30:284–287.
67. Bos RRM, Boering G, Rozema FR, Leenslag JW. Resorbable poly(Llactide) plates and screws for fixation of zygomatic fractures. J. Oral Maxillofac. Surg. 1987; 45:751–753.
68. Törmälä P, Vasenius J, Vainionpää S, Laiho J, Pohjonen T, Rokkanen G. Ultra-high-strength absorbable self-reinforced polyglycolide (SR-PGA) composite rods for internal fixation of bone fractures: in vitro and in vivo study. J. Biomed. Mater. Res 1991; 25:1–22.
69. Christel P, Chabot F, Leray JL, Morin C, Vert G. Biodegradable composites for internal fixation. In: Winter GD, Gibbon DF, Plenk H, eds. Biomaterials. New York: John Wiley and Sons Ltd., 1980:271–280.
70. Törmälä P, Pohjonen G. Ultra-high strength biodegradable polymeric composites for surgical applications. In: Rokkanen P, Törmälä P, eds. Self-Reinforced Bioabsorbable Polymeric Composites in Surgery. Tampere. Finland: Tampereen Pikapaino Oy, 1995:1–23.
71. Törmälä P, Rokkanen P, Laiho J, Tamminmäki M, Vainionpää G. Material for osteosynthesis devices. U.S. Patent 4,743,257, 1988.
72. Serlo W, Kaarela OI, Peltoniemi HH, Merikanto J, Ashammakhi NA, Lassila K, Pohjonen T, Törmälä P, Waris TH. Self-reinforced polylactide osteosynthesis devices in craniofacial surgery. a long-term follow-up study. Scand J Plast. Reconstr. Surg. Hand Surg 2001; 35(3):285–292.
73. Ashammakhi N, Waris G. Use of bioabsorbale osteofixation devices in treatment of craniofacial syndromes (www.biocranio.com). 2002.
74. Miller RA, Brady JM, Cutright DE. Degradation rates of oral resorbable implants (polylactates and polyglyconates): rate modification with changes in PLA/PGA copolymer ratios. J. Biomed. Mater. Res 1977; 11:711–719.
75. Weissberger EC, Eppley BL. Resorbable fixation in the head and neck surgery. Laryngoscope 1997; 107:716–719.
76. Wiley RC, Sarver DR. Method for heating biocompatible implants in a thermal packaging line. US Patent 5,263,991, 1993.
77. Tschakaloff G. Surgical system. US Patent 5,290,281, 1994.
78. Pietrzak WS, Sarver DR, Bianchini SD, D'Alessio G. Effect of simulated intraoperative heating and shaping on mechanical properties of a bioabsorbable fracture plate material. J. Biomed. Mater. Res. (Appl. Biomater.) 1997; 38:17–24.
79. Habal MB, Pietrzak WS. Key points in the fixation of craniofacial skeleton with absorbable biomaterial. J. Craniofac. Surg. 1999; 10:491–499.
80. Eppley BL, Sadove AM. Effects of resorbable fixation on craniofacial skeletal growth: a pilot experimental study. J. Craniofac. Surg. 1992; 3:190–195.
81. Eppley BL, Sadove AM. Effects of resorbable fixation on craniofacial skeletal growth: modifications in plate size. J. Craniofac. Surg. 1994; 5:110–114.

82. Serlo W, Ashammakhi N, Waris T, Törmälä P. A new technique for cranial bone osteofixation: use of absorbable tacks and plates to fix parietal bone split grafts for post-traumatic frontal bone defect reconstruction. J. Craniofac. Surg. 2002; 13(2):331–336.
83. Leinonen S, Tiainen J, Kellomäki M, Törmälä P, Waris T, Ninkovic M, Ashammakhi N. Holding power of bioabsorbable Self-reinforced Poly-L/D-lactide 70/30 tacks, and miniscrews in human cadaver bone. J. Craniofac. Surg. 2003; 14(2):171–175.
84. Tiainen J, Leinonen S, Ilomäki J, Suokas E, Törmälä P, Waris T, Ashammakhi N. Comparison of the pull-out forces of bioabsorbable polylactide/glycolide screws (Biosorb and Lactosorb) and tacks: a study on the stability of fixation in human cadaver parietal bones. J. Craniofac. Surg. 2002; 13(4): 538–543.
85. Spanio S, Ashammakhi N, Ilomäki J, Välimaa T, Rainer C, Waris T, Törmälä P, Ninkovic M. Use of new bioabsorbable tacks and a tack-shooter in cranial bone osteofixation saves operative time. J. Craniofac. Surg. 2002; 13(5):693–696.
86. Hofmann GO. Biodegradable implants in traumatology: a review on the state-of-the-art. Arch. Orthop. Trauma Surg. 1995; 114:123–132.
87. Rozema FR, Levendag PC, Bos RR, Boering G, Pennings AJ. Influence of resorbable poly(L-lactide) bone plates and screws on the dose distributions of radiotherapy beams. Int. J. Oral Maxillofac. Surg. 1990; 19:374–376.
88. Hofmann GO, Wagner FD, Keller A, Feist G. Postoperative irradiation treatment and bioresorbable implants in orthopaedic surgery: an experimental in vitro study. Biomaterials 1996; 17(11): 1149–1153.
89. Tharanon W, Sinn DP, Hobar PC, Sklar FH, Salomon G. Surgical outcomes using bioabsorbable plating systems in pediatric craniofacial surgery. J. Craniofac. Surg. 1998; 9:441–444.
90. Leinonen S, Suokas E, Veiranto M, Törmälä P, Waris T, Ashammakhi G. Holding power of bioabsorbable ciprofloxacin-containing self-reinforced poly-L/DL-lactide 70/30 bioactive glass 13 miniscrews in human cadaver bone. J. Craniofac. Surg. 2002; 13(2):212–218.
91. Tiainen J, Veiranto M, Suokas E, Törmälä P, Waris T, Ninkovic M, Ashammakhi G. Bioabsorbable ciprofloxacin-containing and plain self-reinforced polylactide-polyglycolide 80/20 screws: pull-out strength properties in human cadaver parietal bones. J. Craniofac. Surg. 2002; 13(3):427–433.
92. Habal MB, Leake DL, Maniscalco JE, Kim G. Repair of major cranio-orbital defects with an elastomer-coated mesh and autogenous bone paste. Plast. Reconstr. Surg. 1978; 61(3):394–404.
93. Böstman OM. Intense granulomatous inflammatory lesions associated with absorbable internal fixation devices made of polyglycolide in ankle fractures. Clin. Orthop 1992; 278:193–199.
94. Rozema FR, Bos RRM, Pennings AJ, Jansen HW. Poly(L-lactide) implants in repair of defects of the orbital floor: an animal study. J. Oral Maxillofac. Surg. 1990; 48:1305–1309.
95. Santavirta S, Konttinen YT, Saito T, Gronblad M, Partio E, Kemppinen P, Rokkanen G. Immune response to polyglycolic acid implants. J. Bone Joint Surg 1990; 72-B(4):597–600.
96. Suuronen R, Laine P, Pohjonen T, Lindqvist G. Sagittal ramus osteotomies fixed with biodegradable screws: a preliminary report. J. Oral Maxillofac. Surg. 1994; 52:715–720.
97. Serlo W, Ashammakhi N, Törmälä P, Waris G. Case report: a new technique for correction of trigonocephaly in an infant: application of an absorbable endocranial plate. Child's N. System 2000; 16:595–597.
98. Haers PE, Suuronen R, Lindqvist C, Sailer G. Biodegradable polylactic acid plates and screws in orthognathic surgery. J. Craniomaxillofac. Surg. 1998; 25:87–91.
99. Haers PE, Sailer HF. Biodegradable polylactide acid plates and screws in orthognathic bimaxillary surgery. Short term skeletal stability and material related failures. J. Craniomaxillofac. Surg. 1998; 26:363–372.
100. Fuente del Campo AF, Waris T, Vargas A, Törmälä P, Pohjonen G. Placas de autofijacion interna absorbibles en la inmovilizacion de osteotomias maxilares. Revista Colombiana de Cirurgia Plastica y Reconstructiva 1995; 4:74–78.
101. Fuente del Campo AF, Waris T, Vargas A, Törmälä P, Pohjonen G. Estabilidad a largo plazo con placas absorbibles para fijacion interna. Anales Medicos 1995; 40:54–58.

102. Fuente del Campo A, Pohjonen T, Törmälä P, Waris G. Fixation of horizontal maxillary osteotomies with biodegradable self-reinforced absorbable polylactide plates. Preliminary results. Eur. J. Plast. Surg. 1996; 19:7–9.
103. Ashammakhi N, Peltoniemi H, Waris E, Suuronen R, Serlo W, Kellomaki M, Törmälä P, Waris G. Developments in craniomaxillofacial surgery: use of self-reinforced polyglycolide and polylactide osteofixation devices. Review. Plast. Reconstr. Surg. 2001; 108(1):167–180.
104. Peltoniemi H, Ashammakhi N, Waris T, Kontio R, Salo A, Lindqvist C, Grätz K, Suuronen G. current issues on biodegradable osteofixation devices in craniomaxillofacial surgery. Oral Surg. Oral Med. Oral Pathol. 2002; 94(1):5–14.
105. Bergsma JE, De Bruijn WC, Rozema FR, Bos RR, Boering G. Late degradation tissue response to poly(L-lactide) bone plates and screws. Biomaterials 1995; 16:25–31.
106. Peltoniemi H, Tulamo R-M, Pihlajamäki H, Törmälä P, Pohjonen T, Toivonen T, Waris G. Intraosseous plating: a new method for biodegradable osteofixation in craniofacial surgery. J. Craniofac. Surg. 1998; 9(2):171–176.
107. Serlo W, Ashammakhi N, Länsman S, Törmälä P, Waris G. A new technique for correction of trigonocephaly using bioabsorbable osteofixation tacks, tack-shooter and plates. J. Craniofac. Surg. 2002: Accepted.
108. Overbeck JP, Winckler ST, Meffert R, Törmälä P, Spiegel HU, Brug G. Penetration of ciprofloxacin into bone: a new bioabsorbable implant. J. Inves. Surg. 1995; 8:155–162.
109. Nie L, Nicolau DP, Tessier PR, Kourea HP, Browner BD, Nightingale CH. Use of bioabsorbable polymer for the delivery of Ofloxacin during experimental osteomyelitis treatment. J. Orthop. Res. 1998; 16:76–79.
110. Wei G, Kotoura Y, Oka M, Yamamuro T, Wada R, Hyon SH, Ikada G. A bioabsorbable delivery system for antibiotic treatment of osteomyelitis. The use of lactic acid oligomer as a carrier. J. Bone Joint Surg. 1991; 73B:246–452.
111. Habal MB. Polymers and copolymers science in the operating room. J. Craniofac. Surg. 2001; 12: 407–408.
112. Tielinen G. The effect of bioabsorbable polylactide pins with polymer paste containing transforming growth factor -β1, on the healing of osteotomies and bone defects. An experimental study. PhD thesis, Helsinki University. 2000.

8

Osseous Grafting Materials for Periodontal Defects

David D. Hile, Stephen A. Doherty, Donald L. Wise, and Debra J. Trantolo
Cambridge Scientific, Inc.
Cambridge, Massachusetts, U.S.A.

Stephen T. Sonis
Harvard School of Dental Medicine
Boston, Massachusetts, U.S.A.

Kai-Uwe Lewandrowski
Massachusetts General Hospital
Boston, Massachusetts, U.S.A.

I. INTRODUCTION

Long-term success in the treatment of periodontal defects depends upon both the growth of new bone and periodontal regeneration at the defect site. A wide range of materials has been studied for the treatment of osseous periodontal defects. Materials used in these treatments offer a number of desirable properties such as biocompatibility, osteoconductivity, osteoinductivity, availability, mechanical strength, resorbability, and ease of application. Although several of these properties may be found in certain materials, it is difficult to meet all of these criteria with a single treatment method.

Treatment materials for the repair of periodontal defects may be divided into categories, such as bone graft, ceramics, or synthetic polymers. Bone graft may be obtained from the patient (autograft), a donor (allograft), or from another species (xenograft). Bone graft treatments have the longest clinical history and have inherent biological advantages, such as osteoinductivity, which synthetic materials do not provide. However, availability and safety concerns of bone graft sources limit their clinical use. Synthetic materials include ceramics and polymers that are readily available, but these materials are mostly osteoconductive and demonstrate clinical outcome limitations [1]. Biocompatible synthetic compounds, such as ceramics prepared from calcium phosphates, do not always result in reliable periodontal regeneration. Polymeric bone graft substitutes based on homo- and copolymers of lactide and glycolide, poly(methyl methacrylate), and a new material based on poly(propylene glycol-*co*-fumaric acid) (PPF) are often used in conjunction with bone graft to promote osteoinduction.

II. BONE GRAFT

A. Autograft

Bone graft procured from the patient offers biological advantages over other materials, as it is a natural source of osteogenic factors and progenitor cells. Relatively large amounts of autograft may be obtained from the iliac crest. In addition to new bone growth, iliac graft has been linked to new periodontal regeneration [2]. However, collection of the graft necessitates a second surgical procedure and creates complications such as donor-site morbidity and potential fracture risk [3]. Furthermore, clinical use of iliac bone for treatment of periodontal defects is limited due to resorption at the defect site [4]. Clinically relevant autograft is collected from intraoral cancellous bone and marrow. Intraoral bone grafts typically promote reliable new bone regeneration within the periodontal defects [5–7]. However, regeneration of the periodontal attachment apparatus using intraoral grafts is not always predictable, suggesting that new bone formation does not always result in periodontal regeneration [8]. The limited availability of autograft may be circumvented by using a relatively small volume of graft to augment a synthetic bone graft substitute.

B. Allograft

New bone formation in periodontal defects has also been stimulated with the use of donor bone. Allograft provides an alternative to autograft as a source for osteoinductive bone [9], but the allograft must be processed to reduce the risks of disease transmission. Furthermore, allograft supply is not always adequate or may not be available when needed [10]. Clinical studies using allograft bone to fill periodontal defects resulted in comparable new bone formation to defects treated with autograft [11,12]. The ability to stimulate growth of new bone and regeneration of periodontal tissues has been evaluated. Periodontal defects treated with demineralized, freeze-dried bone allograft (DMB) enabled new bone growth and periodontal attachment [13]. Regeneration of alveolar bone using DMB may be limited or unpredictable depending upon the size of the defect site. Preclinical studies of large allogeneic cortical strips supported cementum regeneration, but did not stimulate reliable bone formation in supra-alveolar periodontal defects [14,15].

C. Xenograft

Large quantities of graft material have been produced from bovine-derived anorganic bone. Xenograft promotes new bone growth at the defect site, but periodontal regeneration is limited. A canine preclinical study demonstrated new bone formation, but the graft did not maintain periodontal tissues [16]. Tissue responses in human trials have not been significantly improved using xenogeneic materials. Periodontal tissue outcomes using xenograft in conjunction with guided tissue regeneration (GTR) were not enhanced after 6 months in comparison to patients treated with GTR alone [17]. The use of autograft and allograft materials is preferred for both bone and periodontal regeneration versus xenogeneic sources.

III. CERAMICS

A. Hydroxyapatite

Calcium phosphate compounds with chemical formulas similar to inorganic bone have been studied as defect fill materials. Hydroxyapatite (HA) may be of natural origin (coral derived) or synthesized. These materials have demonstrated ability to form new alveolar bone in patients with periodontal defects. Coral-derived and synthetic porous hydroxyapatite materials were compared to debridement alone for bone repair. Percent bone fill and clinical attachment assessed after 12 months demonstrated advantages using either hydroxyapatite material versus debridement alone [18]. However, recent clinical studies indicate that hydroxyapatite materials do not promote adequate periodontal regeneration. In a direct comparison between hydroxyapatite and demineralized bone, patients treated with hydroxyapatite demonstrated increased probing depth and decreased clinical attachment gain [19]. In addition, hydroxyapatite implants have been associated with significant loss of alveolar bone and granulation tissue at the defect site [20].

B. Bioactive Glass

Bioactive ceramics have been produced as a synthetic graft replacement material intended to promote new bone formation. Specifically, bioactive glass materials stimulate new bone formation at the implant–bone interface [21]. Preclinical studies demonstrated that bioactive glass particles achieved reduction in probing depths and gain in clinical attachment versus open debridement treatments [22]. Although periodontal regeneration associated with bioactive glasses was improved when compared to hydroxyapatite, this material still has limitations. A study of bony defects treated with commercially available bioactive glass particles did not produce significant regeneration of cementum, periodontal ligament, or bone [23]. However, the biocompatibility and bone regenerative properties of bioactive glasses provide a useful material for treatment of osseous periodontal defects.

IV. POLYMERIC BONE GRAFT SUBSTITUTES

A. Hard Tissue Replacement Polymer

Polymeric bone graft substitutes such as hard tissue replacement (HTR) polymer have been used to fill periodontal defects. HTR polymers are prepared from a core of poly(methyl methacrylate) and poly(hydroxy ethyl methacrylate) and a coating of calcium hydroxide. Defects treated with this nonresorbable material have demonstrated clinical outcomes comparable to GTR techniques [24]. A long-term clinical study evaluated probing depth and clinical attachment in maxillary and mandibular furcations treated with HTR [25]. Clinical measures were again comparable to results obtained from GTR treatment. These results indicate that HTR polymers may be used as an alternative to bone graft for the treatment of similar defects.

B. Polylactide and Poly(Lactide-*co*-Glycolide)

Bioabsorbable polymers have advantages when compared to other nondegradable synthetic materials because a second surgical procedure is not required to remove the device. Membranes consisting of homo- and copolymers of lactide and glycolide have been investigated for guided tissue engineering applications [26]. The membranes stabilize the defect site, while permitting regeneration of tissue. The ability to replace expanded polytetrafluoroethylene (ePTFE) barrier membranes with polylactide (PLA) was tested in a class II furcation defect [27]. New alveolar bone formation and periodontal regeneration were demonstrated in this case study. Clinical results indicate that resorbable PLA barriers may be an attractive alternative to ePTFE for the treatment of small osseous defects.

C. PPF-Based Bone Graft Substitutes

A new bone repair material based on the degradable polymer poly(propylene glycol-*co*-fumaric acid) has also been investigated for the treatment of osseous periodontal defects. The bone repair material was prepared by crosslinking unsaturated PPF in the presence of an osteoconductive filler, hydroxyapatite, and effervescent agents [28]. The cured composite provides a porous structure that provides dimensional stability to defects and an osteoconductive scaffold for new bone formation. Mixing osteoinductive cancellous autograft enhances bony ingrowth within osseous defects with new bone formation evident within 7 days postoperatively [29]. The PPF-based bone graft subsitute has been evaluated for the treatment of periodontal defects based on morphological, mechanical, and histomorphometric outcomes.

The primary goal of a bone graft material is to stabilize the periodontal defect, while promoting recovery of native bone. Engineering porosity into the degradable construct enhances cellular ingrowth and osteointegration within bone graft substitutes [28,30–32]. Average pore sizes on the order of 150 to 500 μm are desired for optimal bone cell ingrowth [33,34]. Bony ingrowth enables integration of the synthetic material within host tissue and promotes replacement of the degrading material with native bone. Porosity and mechanical properties of PPF composites as a function of material formulation were optimized for treatment of maxillofacial and mandibular defects.

Formulation of the PPF-based bone graft substitutes enables easy application to osseous defects. Mixing of PPF and crosslinkers creates a putty that may be applied to defect sites and cured in situ. Viscosity of the resultant mixture permits injection through needles as small as 16 gauge. The set time of the material ranges from 1 to 10 min, during which time the material expands to fill void spaces and may be molded or placed by the surgeon. The crosslinking reaction is not highly exothermic, with temperatures between 37 and 40°C measured within the curing graft substitute [35]. Highly exothermic reactions, such as those observed in PMMA bone cements, have been associated with local tissue necrosis. The PPF-based bone graft substitute approaches peak mechanical properties 18–24 h after application.

Developing porosity of the cured graft substitute has been used to promote osteointegration and incorporation of the synthetic material. Engineered porosity of PPF-based bone graft substitutes has been created prior to implantation via a porogen leaching process. Controlled pore sizes are prepared by leaching porogen materials (either gelatin microspheres or salt particles like sodium chloride) from crosslinked PPF scaffolds [31,36]. Porous PPF scaffolds have also been produced using calcium carbonate and tricalcium phosphate as porogen fillers [37]. These techniques are useful in engineering porosity in the scaffold as a function of particulate size and solubility.

Generation of pores during the curing process was used as a method to induce immediate porosity and expansion of the graft substitute in situ. The addition of effervescent agents to the formulation controls expansion of the test material during curing. Expansion enables the material to intimately fill the defect site and creates pores that encourage tissue ingrowth. Porosity of the cured PPF composites was analyzed by scanning electron microscopy (SEM) (Fig. 1). Tuning the concentration of effervescent agents—sodium bicarbonate (SB) and citric acid (CA)—effectively controls the porosity of the cured test material yielding average pore sizes ranging from 50 to 500 μm. Porosity of the bone graft substitute promotes osteointegration and new bone formation within the graft substitute [38].

Porosity of the PPF-based bone graft substitute affects the mechanical integrity of the material. Compressive mechanical properties were determined for highly porous, photocross-linked PPF scaffolds [39]. Porosities greater than 80%, as produced by salt leaching, yielded a compressive strength of 0.11 ± 0.02 MPa and an elastic modulus of 2.3 ± 0.5 MPa. The highly porous scaffolds had mechanical properties less than the average value for cancellous bone. Flexural properties of nonporous PPF and beta-tricalcium phosphate composite cylinders were evaluated using four-point bending [40]. Bending strengths (4–17 MPa) and moduli (0.6–1.7 GPa) were consistent with trabecular bone. Design of these bone graft substitutes can balance porosity and mechanical considerations for clinical applications.

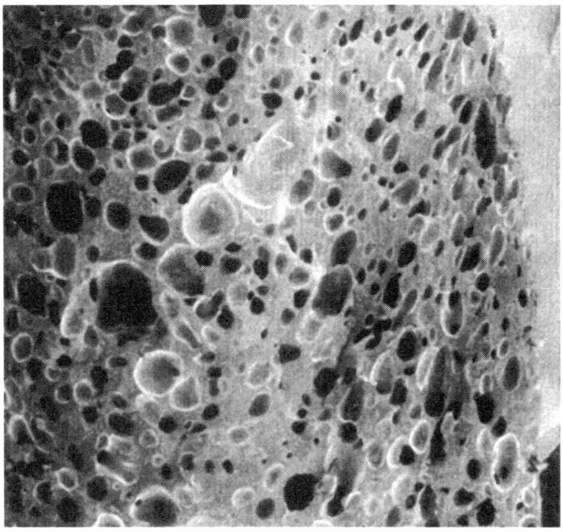

Figure 1 Scanning electron micrographs of XL-PPF bone graft substitute with an average pore size of 200 μm. Nucleation of gas molecules within the curing PPF-based material results in controlled expansion and creates porosity in situ.

Initial mechanical properties of porous PPF-based bone graft substitutes prepared with effervescent agents are comparable to cancellous bone. Peak compressive forces were measured through 12 weeks of in vitro degradation (Fig. 2). Initial compressive strengths of the bone repair material (7.2 ± 3.4 MPa) are comparable to values for human cancellous bone tested under similar conditions [41]. The average compressive strength of the PPF-based bone graft substitute decreases approximately 30% during the first 3 weeks of degradation. Compressive strengths averaged 5.5 MPa through 12 weeks of degradation. The porous bone graft substitute is mechanically comparable to cancellous bone and may be used in conjunction with secondary fixation devices in load-bearing applications. The pore size of the tested materials was on the order desired for osteointegration (200–300 μm), thereby addressing morphological and mechanical graft substitute design objectives.

V. MANDIBULAR RECONSTRUCTION: FEASIBILITY STUDY

The PPF-based bone graft substitute enabled repair of osseous defects as demonstrated in preclinical studies. In vivo analysis focused on the qualitative and semiquantitative assessment of new bone formation within mandibular defects treated with PPF-based bone repair material. A radiographic, histologic, and histomorphometric analysis was performed by comparing the various mixing ratios of the PPF-based repair material with autograft, demineralized bone, autograft alone, and PPF alone. Clinical applications may require mixing of the bone graft repair material with autograft to achieve an optimal relationship between shape and function of mandibular reconstruction.

Investigation of grafts and bone substitutes in the mandible has been difficult in the past because most mandibular defects heal spontaneously [42,43]. The animal model of Kaban and Glowacki [44] was used, not only because it was convenient, but also because it was relevant

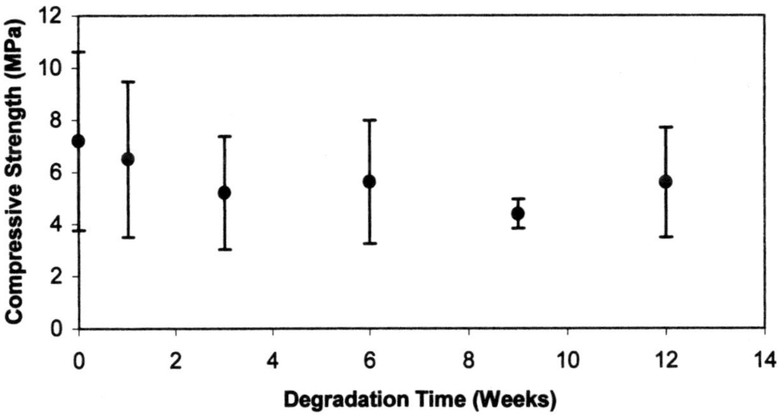

Figure 2 Peak compressive strengths of crosslinked PPF bone graft substitutes following degradation in vitro. Initial mechanical properties of the synthetic test material were comparable to cancellous bone.

(in no instance did the control defect heal). The hole was large (4 mm in diameter is a significant defect for the ramus of a rat mandible) and of full thickness, with the periosteum totally disrupted and not replaced. In this model, therefore, healing must be attributed to some effect of the implant material. The study allowed comparative histologic and histomorphometric assessments of the degradation and bone cell ingrowth.

A. Results of Histological and Histomorphometric Evaluation

Histologic analysis revealed superior healing of the mandibular defects with PPF-based bone repair material when compared to the control defects, which were left empty. New bone formation, reported as the *new bone volume index* was more pronounced when the graft substitute material was mixed with autograft (Table 1). However, there was no statistically significant difference in the new bone volume index between the two experimental PPF groups mixed with autograft. These results suggest that mixing the PPF-based bone graft substitute with a small amount of autologous bone graft is just as effective as using higher amounts of cancellous autograft for this limited rodent model.

Table 1 New Bone Volume Index for Each Implant Type Based on 8 Rats per Group and 4 Weeks Postoperative Follow-Up

Test material	New bone volume index
PPF	51 ± 8
DMB	87 ± 14
Autograft	95 ± 17
PPF/autograft (75/25)	61 ± 10
PPF/autograft (25/75)	67 ± 12
Empty control	33 ± 5

Histologic analysis of the bone–implant interface demonstrated biocompatibility of the PPF-based bone graft substitute. Inflammatory cells were only noted in the 1-week groups and were not related to type of material used for mandibular reconstruction. Inflammatory infiltrates were absent in all other groups evaluated at later postoperative follow up times. Results of this study demonstrated both biocompatibility and osteoconductive properties of the porous PPF-based scaffold in a mandibular defect model. These findings have applicability to the further development of bone repair material for reconstructive applications of the mandible.

B. Biomechanical Analysis of Defect Sites

Macroindentation testing was conducted to evaluate the stiffness of the test material–bone complex at 7 weeks postoperative. The stiffest materials were the positive controls (DMB and autograft) as shown in Table 2. Both PPF and PPF/autograft with a mixing ratio of 25/75 were stiffer than untreated defects. The stiffness of PPF/autograft at a low autograft mixing ratio (25/75) was measured and found to be slightly less than untreated defects. Biomechanical results were similar to the histomorphometric findings indicating that the bone–material composite mechanical properties were related to new bone formation.

VI. CONCLUSIONS

Several materials have been tested for the treatment of osseous periodontal defects. Materials have been evaluated based upon their ability to not only support bony recovery, but also periodontal regeneration. The ideal properties of a repair material for periodontal defects are biocompatibility, osteoconductivity, osteoinductivity, availability, resorbability, ability to stabilize the defect site under mechanical stresses, and ease of application. Although several of these properties may be found in certain materials, it is difficult to find a material where all of these criteria are satisfied. A new material based upon the resorbable polymer poly(propylene glycol-*co*-fumaric acid) has been investigated for the treatment of periodontal defects. The PPF-based bone graft substitute supported new bone formation in a mandibular defect model. The material is osteoconductive, but may be mixed with a relatively small percentage of autograft to enhance its bone regenerative properties. Additional studies will be necessary to evaluate the PPF-based graft substitute's ability to support periodontal regeneration. However, the results indicate that the PPF-based material may provide a new treatment method for oral, maxillofacial, and periodontal defects.

Table 2 Stiffness of Test Material/Bone Complex at 7 Weeks Postoperative

Test material	Stiffness (N/mm)
PPF	99 ± 6
DMB	176 ± 40
Autograft	126 ± 39
PPF/autograft (75/25)	54 ± 34
PPF/autograft (25/75)	102 ± 24
Empty control	80 ± 20

ACKNOWLEDGMENTS

The authors would like to recognize the support provided by NIH/NIAMS Grant No. 1 R43 AR46130 (to David D. Hile) and NIH/NIAMS Grant No. 1 R43 DE14272 (to Debra J. Trantolo).

REFERENCES

1. Nasr HF, Aichelmann-Reidy ME, Yukna RA. Bone and bone substitutes. Periodontology 2000 1999; 19:74–86.
2. Dragoo MR, Sullivan HC. A clinical and histological evaluation of autogeneous iliac bone grafts in humans. Part I. Wound healing 2–8 months. J Periodontology 1973; 44:599–612.
3. Hu RW, Bohlmann HH. Fracture at the iliac bone graft harvest site after fusion of the spine. Clin Orthop 1994; 309:208–213.
4. Schallhom RG. Postoperative problems associated with iliac transplants. J Periodontology 1972; 43: 3–9.
5. Hiatt WH, Schallhom RG. Intraoral transplants of cancellous bone and marrow in periodontal lesions. J Periodontology 1973; 44:194–208.
6. Froum SJ, Ortiz M, Witkin RT, Thaler R, Scoop IW, Stahl SS. Osseous autografts. III. Comparison of osseous coagulum–bone blend implants with open curettage. J Periodontology 1976; 47:287–294.
7. Renvert SGarrett, Schallhom R, Egelberg J. Healing after treatment of periodontal intraosseous defects. III. Effect of osseous grafting and citric acid conditioning. J Clin Periodontology 1985; 12: 441–455.
8. American Academy of Periodontology: Scientific, Clinical, and Educational Affairs Department, Approved May 1993 Periodontal Regeneration: Research, Science, and Therapy Committee, Chicago 1993.
9. Urist MR. Bone: formation by autoinduction. Science 1965; 150:893–899.
10. Henman P, Finlayson D. Ordering allograft by weight: suggestions for the efficient use of frozen bone-graft for impaction grafting. J Arthroplasty 2000; 15:368–371.
11. Sanders JJ, Sepe WW, Bowers GM, Koch RW, Williams JE, Lekas JS, Mellonig JT, Pelleu GB, Jr, Gambill V. Clinical evaluation of freeze-dried bone allografts with and without autogeneous bone grafts. J Periodontology 1983; 54:1–8.
12. Schrad SC, Tussing GJ. Human allografts of iliac bone and marrow in periodontal osseous defects. J Periodontology 1986; 57:205–210.
13. Bowers GM, Chadroff B, Carnevale R, Mellonig J, Corio R, Emerson J, Stevens M. Histologic evaluation of new attachment apparatus formation in human. Part I. J Periodontology 1989; 60: 664–674.
14. Kim CK, Cho KS, Choi SH, Prewett A, Wikesjo UME. Periodontal repair in dogs: effect of allogeneic freeze-dried demineralized bone matrix implants on alveolar bone and cementum regeneration. J Periodontology 1998; 69:26–33.
15. Caplanis N, Lee MB, Zimmerman GJ, Selvig KA, Wikesjo UME. Effect of allogeneic freeze-dried demineralized bone matrix on regeneration of alveolar bone and periodontal attachment in dogs. J Clin Periodontology 1998; 25:801–806.
16. Sonis ST, Williams RC, Jeffcoat MK, Black R, Shklar G. Healing of spontaneous periodontal defects in dogs treated with xenogeneic demineralized bone. J Periodontology 1985; 56:470–479.
17. Simonpietri JJ, Novaes AB, Batista EL, Filho EJ. Guided tissue regeneration associated with bovine-derived anorganic bone in mandibular class II furcation defects: 6-month results at re-entry. J Periodontology 2000; 71:904–911.
18. Mora F, Ouhayoun JP. Clinical evaluation of natural coral and porous hydroxyapatite implants in periodontal bone lesions: results of a 1-year follow-up. J. Clin Periodontology 1995; 22:877–884.
19. Brown GD, Mealey BL, Nummikoski PV, Bifano SL, Waldrop TC. Hydroxyapatite cement implant for regeneration of periodontal osseous defects in humans. J Periodontology 1998; 69:146–157.

20. Rupprecht RD, Horning GM, Towle HJ. A clinical evaluation of hydroxyapatite cement in the treatment of class III furcation defects. J Periodontology 2001; 72:1443–1450.
21. Ducheyne P, Cuckler JM. Bioactive ceramic prosthetic coatings. Clin Orthop 1992; 276:102–114.
22. Froum SJ, Weinberg MA, Tarnow D. Comparison of bioactive glass synthetic bone graft particles and open debridement in the treatment of human periodontal defects: a clinical study. J Periodontology 1998; 69:698–709.
23. Nevins ML, Camelo M, Nevins M, King CJ, Oringer RJ, Schenk RK, Fiorellini JP. Human histologic evaluation of bioactive ceramic in the treatment of periodontal osseous defects. Int J Periodontics Restorative Dent 2000; 20:458–467.
24. Calongne KB, Aichelmann ME-, Yukna RA, Mayer ET. Clinical comparison of microporous biocompatible composite of PMMA, PHEMA, and calcium hydroxide grafts and expanded polytetrafluoroethylene barrier membranes in human mandibular molar class II furcations: a case series. J Periodontology 2001; 72:1451–1459.
25. Yukna RA, Yukna CN. Six-year clinical evaluation of HTR synthetic bone grafts in human grade II molar furcations. J Periodontal Res 1997; 32:627–633.
26. Buckley MJ, Agarwal S, Gassner R. Tissue engineering and dentistry. Tissue Eng 1999; 26:657–662.
27. Stoller NH, Johnson LR, Garrett S. Periodontal regeneration of a class II furcation defect utilizing a bioabsorbable barrier in a human. A case study with histology. J Periodontology 2001; 72:238–242.
28. Cattaneo MV, Gresser JD, Wise DL, White RL, Bonassar L, Trantolo DJ. Effect of a poly(propylene fumarate) foaming cement on healing of critical size tissue defects. Tissue Eng 1999a; 5:305–316.
29. Hile DD, Thompson BMJ, Wise DL, Trantolo DJ. Quantitative measures of osteoinductivity of a porous poly(propylene fumarate) bone graft extender. Tissue Eng 2002: In press.
30. Wake MC, Patrick CW, Mikos AG. Pore morphology effects on the fibrovascular tissue growth in porous polymer substrates. Cell Transplant 1994; 3(4):339–343.
31. Ishaug-Riley SL, Crane-Kruger GM, Yaszemski MJ, Mikos AG. Three-dimensional culture of rat calvarial osteoblasts in porous biodegradable polymers. Biomaterials 1998; 19:1405–1412.
32. Lewandrowski KU, Bondre S, Gresser JD, Silva AE, Wise DL, Trantolo DJ. Augmentation of osteoinduction with a biodegradable poly(propylene glycol-*co*-fumaric acid) bone graft extender. Bio-Medical Materials and Engineering 1999b; 9:325–334.
33. Gazdag AR, Lane JM, Glaser D, Forster RA. Alternatives to autogenous bone graft: efficacy and indications. J Am Acad Orthop Surg 1995; 3:1–8.
34. Boyan BD, Lohmann CH, Romero J, Schwartz Z. Bone and cartilage tissue engineering. Tissue Eng 1999; 26:629–645.
35. Lewandrowski KU, Gresser JD, Wise DL, White RL, Trantolo DJ. Osteoconductivity of an injectable and bioresorbable poly(propylene glycol-*co*-fumaric acid) bone cement. Biomaterials 2000; 21: 293–298.
36. Crane GM, Miller MJ, Yasko AW, Yaszemski MJ, Mikos AG. Bone formation by three-dimensional stromal osteoblast culture in biodegradable polymer scaffolds. J Biomed Mater Res 1997; 36(1): 17–28.
37. Domb AJ, Manor N, Elmalak O. Biodegradable bone cement compositions based on acrylate and epoxide terminated poly(propylene fumarate) oligomers and calcium salt compositions. Biomaterials 1996; 17(4):411–417.
38. Lewandrowski KU, Bondre SP, Hile DD, Thompson BMJ, Wise DL, Tomford WW, Trantolo DJ. Porous poly(propylene fumarate) foam coating of orthotopic cortical bone grafts for improved osteoconduction. Tissue Eng 2002b: In press.
39. Fisher JP, Holland TA, Dean D, Engel PS, Mikos AG. Synthesis and properties of photocross-linked poly(propylene glycol-*co*-fumaric acid) scaffolds. J Biomater Sci Polym Ed 2001; 12:673–687.
40. Porter BD, Oldham JB, He SL, Zobitz ME, Payne RG, An KN, Currier BL, Mikos AG, Yaszemski MJ. Mechanical properties of a biodegradable bone regeneration scaffold. Trans ASME 2000; 122: 286–288.
41. Hile DD, Kirker-Head C, Doherty SA, Lewandrowski KU, Kowaleski MP, McCool J, Wise DL, Trantolo DJ. Mechanical evaluation of a porous bone graft substitute based on poly(propylene glycol-*co*-fumaric acid). J Biomed Mater Res 2002; In press.

42. Hjørting-Hansen E, Andreasen O. Incomplete bone healing of experimental cavities in dog mandibles. Br J Oral Surg 1971; 9:33–40.
43. Frame JW. A convenient animal model for testing bone substitute materials. J Oral Surg 1980; 38: 176–80.
44. Glowacki J. Induced osteogenesis in the repair of experimental mandibular defects in rats. J Dent Res 1981; 60:1356–1364.

9
Guided Diaphysis Regeneration

C. E. Olson
UroMetrics, Inc.,
St. Paul, Minnesota, U.S.A.

S. D. Wagner and T. D. McGee
Iowa State University
Ames, Iowa, U.S.A.

I. INTRODUCTION

Bone grafting and joint replacement are important areas in orthopedic research because of the necessity for both methods in treating animal and human orthopedic problems [1]. A recent review of bioactive ceramics reported that 1,230,000 fractures were treated with osteosynthesis materials in the United States in 1988. The basic philosophy in that chapter was to address the effect of synthetic graft materials on bone tissue formation and bone cell function, not on the problems of bone replacement on guided bone regeneration where function is maintained to provide clinical success. This philosophy in the materials community has encouraged the use of resorbable materials and the study of particulate materials in cancellous bone, and resulted in the identification of bulk or particulate calcium phosphates ceramics and glasses as bioconductive, not bioinductive [2–4]. Only recently have some materials scientists realized that osteoinductive performance is possible [5–7], although a few have reported such behavior earlier [8,9]. Most materials evaluations, however, continue the basic philosophy to the present [10,11].

Current research includes new materials and new designs in tremendous variety, usually aimed at either replacing tissue to achieve repair, at controlling tissue response, or at avoiding existing mechanisms of failure [1]. Existing research spans possible approaches from the theoretical mechanics to the cellular level to actual clinical practice. In that research there are always underlying hypotheses about what is important in repairing the skeleton. These are usually not tied directly to clinical trials but to the understanding of the area in which the hypothesis is generated. For example, the cell biologists might ask what are the fundamental processes that control cellular response and how can they be manipulated?. The engineer in continuum mechanics might ask what is the stress distribution in the tissue around the implant and how can the geometry or the materials properties be manipulated to ensure the tissue will respond favorably to the stress. And the materials scientists might ask what new material will enhance tissue regeneration to achieve bonding. Unfortunately, without benefit of clinical trials many hypotheses can be produced and evaluated that do not actually apply to clinical success. This chapter is an attempt to show how clinical evaluation can give insight into the confrontation of theory with practice, with emphasis on biologically active materials. It is based on over 30 years experience with biologically active ceramic materials for load-bearing applications in dental and orthopedic repair. The ultimate goal is to produce an artificial material and design such that tissue will be regenerated in such a way that it will remain normally healthy for the life of the patient. While the number of hypotheses is enormous, certain hypotheses are common to many of the research areas. Some will be discussed below.

II. FOUR BONE REPAIR HYPOTHESES

A. H1: Bone Will Bond to Porous Implants and Infiltrate Them

This is currently popular with repliform structure of other species; rapid prototype forming of implant structures, usually based on hydroxyapatite (HA), fluoroapatite (FA), tricalcium phosphate (TCP), or mixtures thereof; and pores produced by introducing gases or by fugitive phases in producing the implant [12].

This hypothesis is reinforced by in vitro experiments with cellular growth, often with morphologic components, where osteoblasts often are shown to deposit and grow on materials, including polymers, metals, and ceramics. Unfortunately, when conditions are right for osteoblast proliferation, unless there are toxic reactions from the substrate, such cell growth usually occurs [13,14]. Based on results of such an experiment the scientist may even conclude that Wolff's law has been repealed [15,16].

Early support of the hypothesis was given by Hulburt et al. through in vivo experiments in which they reported that Haversian systems would grow into large pores, and that 200-μm or larger pores should be present for this to occur [17].

There are three problems with this hypothesis [18]:

1. Bone does not grow into interconnected pores because:

   Interconnected pores may have connections less than 100 μm. The labyrinth nature of such pores do not lend themselves to complete filling.

   Bone grows in response to physical stress. Microstrain is needed to maintain bone, and remodeling is continuous in response to stress. The stress shielding of the matrix of a body, as strong and stiff as bone in its porous condition, prevents bony ingrowth throughout because the interior bone is not stressed. If the porous body is not as strong and stiff as bone, then macrostrain may occur, and bone cannot form under macrostrain conditions.

   Unfilled pores are ideal for the proliferation of bacteria. They are not accessible to the usual defense mechanisms. They constitute a biohazard to the possibility of deep infections.
2. Each part of the body has a blood supply system of a particular structure. If the pores do not allow the same structure, then healthy bone is not to be expected. That structure is oriented to stress, not random.
3. Pores weaken structure. Porous metals, such as Ta foam, cannot support load. Porous ceramics become very weak because the pores act as critical flaws. Porous polymers have low strength and modulus. None are suitable as load-bearing structures because of the size and volume of the pores that are required.

B. H2: Resorbable Materials Are Needed to Serve as a Scaffold to Guide Bone to Repair Itself

This hypothesis is reinforced by in vitro experiments in which soluble, biocompatible polymers serve as a scaffold for osteoblast generation and by in vivo experiments in which trabecular bone infiltrates a scaffold area more rapidly than for the control. It is also reinforced by the use of calcium phosphate structures or granules in which the same phenomenon occurs. The rates are enhanced when osteogenic factors are also included.

The idea is not new. Graves et al., in 1971, proposed calcium aluminates could act to enhance bone regeneration as they simultaneously dissolved and supported bone [19]. They conducted in vivo experiments that appeared positive in the short term.

Unfortunately it is almost impossible for a load-bearing resorbable material to function in a general way in vivo. The rate of resorbtion depends upon the tissue, and response varies greatly. And to support load the tissue must have strength at the resorbtion interface, and that interface must be strong enough to support the load. The problem of maintaining a strong bond at the same time the scaffold is being resorbed seems insurmountable. That is the reason that bioglasses have never been successful in load-bearing situations. The surface silica gel transition zone is too weak to maintain a functional bond and is subject to catastrophic failure.

Bone never repairs itself under macrostrain conditions. So bone grafts must be stabilized with plates, screws, wires, nails, or casts to allow the bone to repair. The problem, then, is to transfer the load from the scaffold structure to the bone under conditions that resorb the scaffold, an impossible task.

C. H3: The Modulus of Elasticity of an Implant, or an Appliance, Must Be the Same as That of Bone

This hypothesis is supported by observations that metal femoral components, that are much stiffer than bone, often have weak bone support in the proximal medial location [20]. The stiff femoral stem is often supported by trabecular bone at the distal apex; so that the bone in the proximal medial location may not receive enough stress to be strong and healthy. (Stress shielding has occurred and is responsible for local resorbtion and subsequent loosening of the stem.) If there were a bond between the stem and the bone, shearing stresses would occur, and this would prevent the weakness. Stepped geometry, attached metal spheres or mesh, roughened surfaces, and calcium phosphate coatings have been used to try to correct this. However, if the implant had the same elastic properties as bone, both it and the bone would be stressed the same and stress shielding would not occur.

This hypothesis is especially appealing to those representing bone structures and implants with continuum mechanics theories. There are alternative hypotheses. For example, appliances cemented with PMMA cement have inhibited blood supply, and this could be responsible for the weak bone in that location. Mammalian skeletons have low modulus cartilage, trabecular bone, and cortical bone in series at the joints of all long bones. Yet the difference in elastic modulus is accommodated by tissue response and structure. Because tissue with a healthy blood supply responds to stress level and orientation there is the possibility that a stiffer implant can be accommodated. A lower stiffness may lead to macrostrain and failure. Orthopedic surgeons are familiar with trabecular bone surrounding bone screws in the medullary cavity of long bones where the trabeculae are oriented to lines of stress produced by the loads carried by the bone screws.

D. H4: Prevention of a Foreign Body Response Can Be Achieved Using Calcium Phosphates as the Tissue Contact Material

This hypothesis is the basis for much research on calcium phosphates in crystalline forms and amorphous forms such as bioglasses [1]. Most, but not all, authors agree that the current metals, polymers, and ceramics used in joint replacement are walled off by the foreign body defense mechanism of the body to produce a thin fibrous capsule. This prevents a direct bond between the implant and the tissue. When the capsule is thin and threads, holes, or other anchoring shapes are provided, the mechanical fixation has been effective in stabilizing the implants. They are often described as exhibiting osteointegration to indicate that functional anchoring has been achieved.

A fibrous capsule is not as rigid as bone, so the capsule thickness is important. The amount of relative motion depends on the capsule thickness and geometry. If the motion is excessive, bone resorbtion can occur. This results in a progressively thicker capsule and has been identified as a mechanism of failure.

Demineralized tissue sections can be used to examine the tissue–implant interface if the implant can be demineralized enough to allow microtoming. It is generally agreed that the calcium phosphates of Ca/P ratio between 1.0 and 2.0 do not have a fibrous capsule. Composites and bioglasses can be difficult to evaluate. Many authors have reported excellent tissue compatibility when the local tissue chemistry is dominated by calcium phosphates. Tissue compatibility must be combined with tissue attachment for this approach to be successful unless the calcium phosphates can be used to cause tissue regeneration under load-bearing conditions.

Unfortunately the bulk calcium phosphates do not have enduring strength, even when prestressed with metals [21]. They must be combined with other materials in such a way that the calcium phosphates control the local tissue chemistry; but a second phase, or some other mechanism, is used to provide enduring strength. We have discovered a way to provide enduring strength and a way of inducing regeneration of tissue.

III. TWO GUIDED-REGENERATION EXPERIMENTS

Two experimental guided-regeneration techniques will be described. The first is a conventional bone graft of special geometry made from the ISU Osteoceramic composite. The geometry includes grooves for guiding the regeneration of the cortex, where the tissue supports the implant. The second method uses an intermedullary tubular guide to induce diaphysis regeneration across a supercritical ostectomy. First the osteoceramic will be described, followed by the two guided-regeneration experiments.

A. The Osteoceramic

The Osteoceramic combines 50 vol% of tricalcium phosphate with 50 vol% of a strong, inert ceramic, magnesium aluminate spinel [22]. The calcium phosphate gives the implant biological activity and the spinel provides strength. The calcium phosphate component induces bone regeneration, causing the bone to repair itself. The Ca/P ratio for these experiments was similar to the 1.62 ratio of natural bone. The flaw size in the strong, spinel skeleton of the Osteoceramic is controlled by the size of the calcium phosphate grains as first produced. After dissolution of surface calcium phosphate, the flaw size is unaltered so the Osteoceramic has enduring strength. This is necessary for permanent bone grafts.

The Osteoceramic has the properties shown in Table 1 when fired to maturity at 1450°C. It is a classically brittle material.

The structure of the Osteoceramic after firing to 1450°C is like that of a sponge. The skeleton of the sponge is interconnected spinel. The holes are filled with interconnected calcium phosphate (Figs. 1–3). The local chemistry at the tissue implant is dominated by the calcium phosphate because the spinel is insoluble. The high solubility of the calcium phosphate in

Table 1 Properties of the Composite

Composition:	$\alpha Ca_3(PO_4)_2$ and $MgAl_2O_4$
Compressive strength	199 MPa
Tensile strength	70 MPa
Young's modulus	114 GPa
Reversible thermal expansion	$10.7 \times 10^{-6}/°C$
Bulk density	3.09 gm/cc
True density	3.37 gm/cc
Ca/P ratio	1.62

the implant provides a reservoir as a source of calcium and phosphorous for hydroxyapatite formation.

The composite is manufactured from finely powdered tricalcium phosphate and magnesium-aluminate spinel. The powders are mixed with dispersants and organic binders and formed into suitable shapes by extrusion, casting, or pressing.

B. Guided Bone Replacement

This experiment has been reported in more detail previously [23,24]. This presentation provides additional understanding of the results that were obtained after the death of the dog from natural causes. The implant was made to comply with the considerations discussed above as an Osteoceramic tubular cylinder with end notches for tissue attachment, longitudinal grooves to induce critical bone regeneration across the implant, and holes connecting the grooves to the medullary cavity for blood supply to bone induced to grow in the grooves (Fig. 4). No open pores were present for tissue attachment. Instead, the dove-tailed end notches were large enough to be filled with bone and with normal endosteal and periosteal blood supply to form and maintain bone

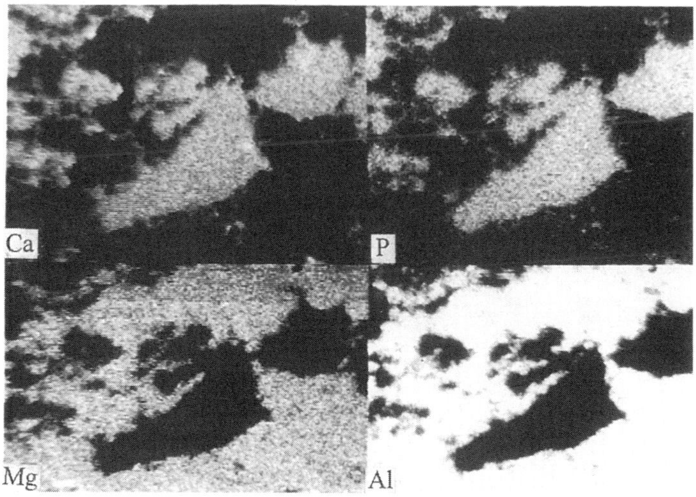

Figure 1 Energy dispersive x-ray map of the polished Osteoceramic showing the Ca and P phase separated from the Mg and Al phase.

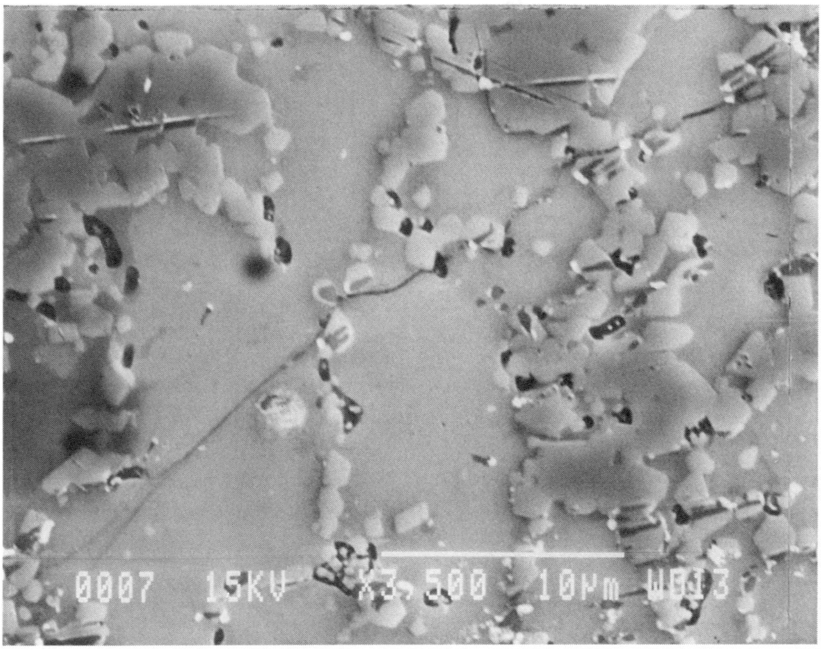

Figure 2 Surface of the Osteoceramic polished to produce relief of the hard, fine-grained $MgAl_2O_4$ relative to the soft larger Ca_3PO_4 grains.

Figure 3 Structure of the spinel skeleton when the Ca_3PO_4 is removed with EDTA.

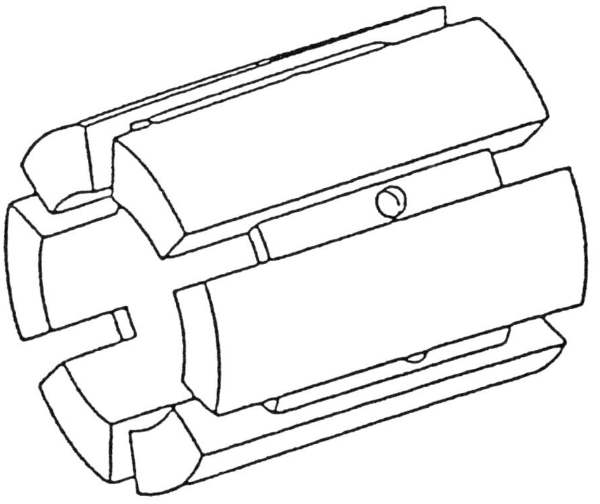

Figure 4 Geometry of a guided bone replacement implant showing the external longitudinal grooves for guiding the cortex and the radial holes for blood vessels from the medullary supply.

in the notches. After dissolution of calcium phosphate at the surface the pores in the spinel are the same size as the dissolved calcium phosphate. They are smaller than 100 μm so Haversian systems cannot grow into them. However, calcium phosphates could be deposited at the surface by osteoblast activity.

The longitudinal grooves are provided to guide cortical bone to regenerate across the implant. If this is successful, and a blood supply is encouraged by the regeneration grooves, then cortical bone strong enough to surround and support the implant should be possible. These experiments were conducted to test the hypotheses that tissue attachment could be achieved and that the grooves would guide the regeneration process.

1. Experimental Evaluation

Seven implants with variation in wall thickness, size of the end notches, size of the longitudinal grooves, and placement of the radial holes were implanted. The number of notches was set arbitrarily at six, enough so that random orientation would provide fixation for bending forces (Table 2).

Surgical Procedure. One implant was placed in one femur of each mongrel dog. (The surgical procedure was conducted in accordance with the federal Animal Welfare Act. The procedures, and the justification for the research, were approved by the Iowa State University Committee on Animal Care.)

After sedation, preparation and exposure of the femur, the surgeon contoured a compression bone plate to the femur, removed a section with a reciprocating saw and fastened the implant in place using bone screws to hold the compression plate axially against the implant (Fig. 5).

Circlage wire was used to attach the implant to the bone plate. One method was to pass a single wire through the axial holes at the center of the implant, bending it around the bone plate and twisting the wire tight enough to stabilize the implant without breaking it. A second way was to circle the implant and the plate with the wire and twist it to stabilize the wire. The third method was to use two circling wires, placed at about one-third of the length from each end. Sixteen-gauge and eighteen-gauge wires were used. The stiffer wire was more difficult to

Table 2 Dimensions of the Bone Bridges

Dog number	Implant length (mm)	Inside diameter (mm)	Outside diameter (mm)	Outside tenon width (mm)	Inside tenon width (mm)	Dovetail slot width (mm)	Transverse hole width (mm)		Dovetail slot depth (mm)
							(d_1)	(d_2)	
1	27.10	6.80	16.70	5.7	1.2	2.6	1.7	3.4	3.6
2	27.05	10.15	15.70	2.5	1.9	2.4		1.8	3.8
3	26.95	6.65	16.60	6.4	2.1	2.6		2.2	5.0
4	26.25	8.85	16.70	6.5	3.2	2.1		1.9	3.5
5	27.35	5.45	16.75	5.1	.8	2.9	1.6	1.6	5.5
6	23.55	5.70	16.95	6.7	2.0	2.2		2.0	3.3
7	24.25	10.20	16.75	6.0	3.6	2.5		2.0	3.1

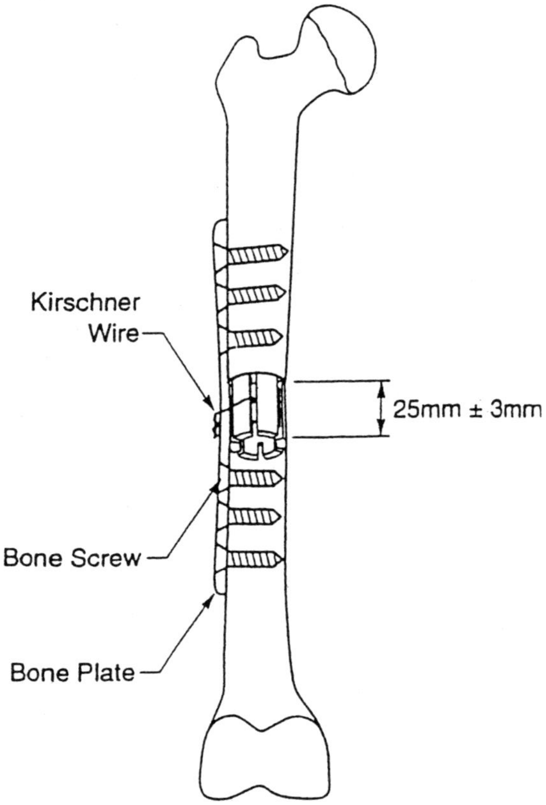

Figure 5 Surgical stabilization of a segmental replacement for guided regeneration of the cortex.

use. Some implants were broken during stabilization and discarded. Some implants broke up after implanting, probably the result of undetected cracking during stabilization. The thin-walled implants were more fragile, and fewer survived the preparation process. The dimensions of the implants actually implanted are those shown in Table 2. Dog data are included in Table 3.

2. Histology

The spinel component of the implants cannot be demineralized so the histology was based on optical microscopy, macroradiography, and microradiography of sections impregnated with a low-viscosity resin. Specimens were dehydrated with a series of ethanol solutions starting with 70% ethanol and continuing with replacements of higher concentrations until two 24-h periods of 100% ethanol were completed. The ethanol was replaced with acetone for two 24-h periods. Fifty percent of acetone and a low-viscosity resin was agitated for 48 hours, followed by 100% of the resin with agitation and a vacuum for 48 h to boil off any residual acetone. The resin was cured at 60°C overnight. A low-speed diamond saw was used to cut axial and transverse sections. The microradiographs were made from selected 300-μm sections.

3. Results

One or two wires encircling the implant were less effective in inhibiting axial motion of the implant than a single wire threaded through the holes at the root of the grooves. The thinner wire conformed well to the implant and plate, and was less likely to cause implant fracture during stabilization. In every procedure the contact plane of the implant and the contact plane of the bone plate were in alignment laterally. The medial alignment depended on the diameter of the bone and the diameter of the implant. Usually there was some mismatch medially. In the case of dog 2, with a small implant wall thickness, the mismatch was excessive and bone formed inside of the implant. Implant 7 fractured after 3 weeks and the experiment was terminated. Implants 2, 3, and 5 showed some cracking of portions of the implants, but the histology showed encapsulation of the fragments. Good bonding of implants 4 and 6 occurred (Figs. 6 and Fig. 7).

All the implants were placed in 1990 or 1991, and histological analysis was obtained for the dogs as shown in Table 3. Only one of these dogs was not sacrificed. Dog 1 was purchased by CEO and lived the active life of a farm dog until he died of abdominal tumors in December 2001. His age at the beginning of the experiment was unknown, but the pathologist concluded the death was normal for a dog over 12 years old. Both femurs were recovered. Macroradiographs and CAT scans were obtained. (Fig. 8)

Table 3 Dog Weight, Sex, and Implantation Period

Dog number	Weight	Sex	Implantation period
	(pounds)		(months)
1	60	M	137
2	45	F	9
3	55	M	$7^1/_2$
4	65	F	7
5	65	F	6
6	60	F	4
7	55	M	$^3/_4$

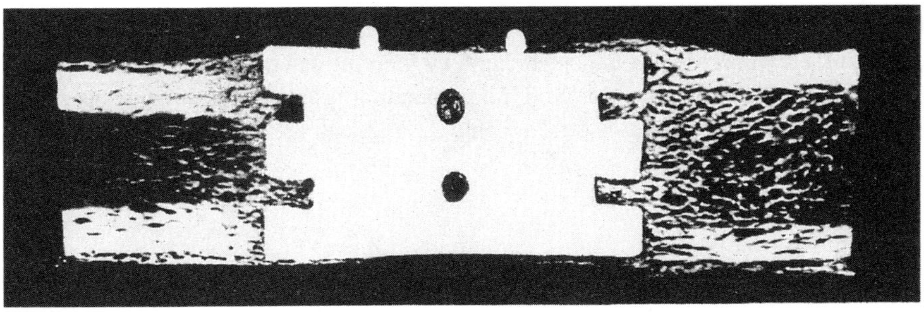

Figure 6 Microradiograph of a longitudinal section from dog 4 at 7 months showing tissue attachment.

The radiographs of the recovered femurs show enlargement of the operated limb consistent with the dimensions in Table 3. The circlage wire was clipped when the bone plate was removed in 1991, and one end projected beyond the external surface. The new bone has contours and structure consistent with the requirements of the femur of an active dog. (Force plate measurements in 2000 showed no significant difference between the two hind limbs.)

The CAT scans in Fig. 8 give a better understanding of the structural changes. Above and below the implant the cross-section of the bone shows somewhat hexagonal symmetry that registers with that of the end notches (8A,G). Trabecular bone appears in the end notches (8B,F). The regeneration grooves contain oriented trabecular bone (8C,E), and the center scan shows that the wire protrudes on one side (8D). A new cortex has formed to surround the implant. It deviates away from the implant at the center wire on the side from which it protrudes. Elsewhere the new cortex is integrated with the trabecular bone in the regeneration grooves and is close to the external surface of the implant. The regeneration grooves provide an endosteal blood supply to the new cortex.

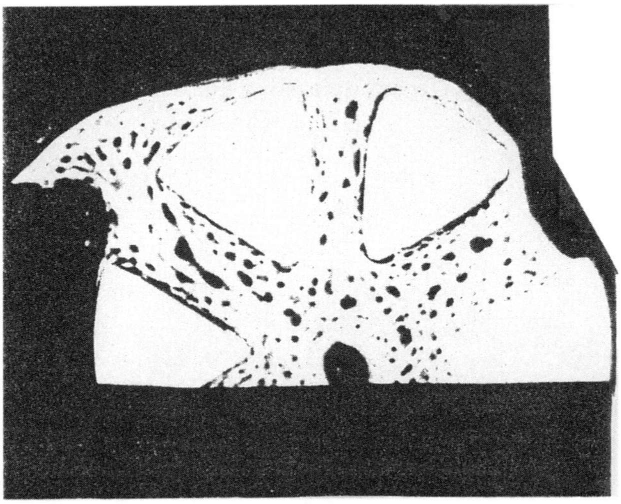

Figure 7 Microradiograph of a transverse section from dog 4 at 7 months showing trabecular bone in the holes connecting the grooves to the medullary cavity (enlarged 100%).

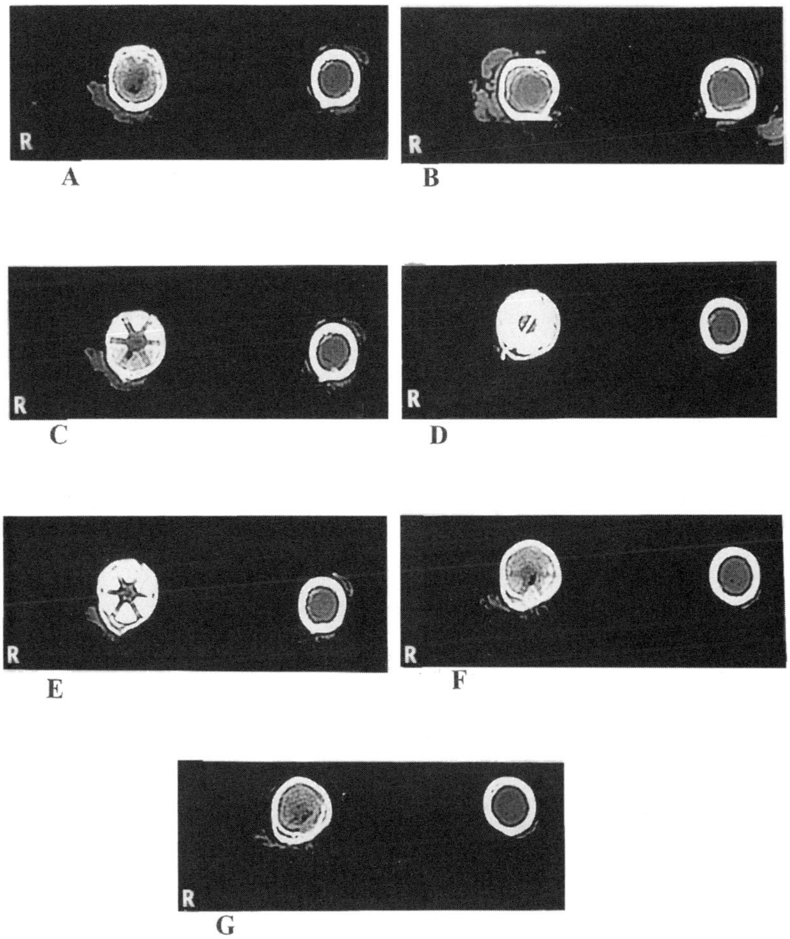

Figure 8 CAT scan of the segmental bone replacement femur. Distal to proximal. (Left) implant femur; (Right) contralateral femur.

The thickness of the new cortex is only about 50% of that in the contralateral limb (8D). The trabecular bone and implant support structure in conjunction with the new cortex is such that the combined structure is sufficient for the loads and forces imposed on the limb by an active dog. Guided regeneration is believed to be necessary for the bone replacement implant to be successful.

C. Guided Diaphysis Regeneration

1. Introduction

The bone replacement experiments, first with porous implants to achieve implant attachment and then with the longitudinal grooves to guide regeneration, described above, led to a new approach to diaphysis regeneration [25]. The porous implant experiments showed attachment to the surface pores, but there was not strong bone in all the pores to reinforce the brittle calcium phosphate/spinel ceramic. The ceramic strength was greatly reduced by the pores, and the implant failed when subjected to normal loading. In the bone replacement experiment the grooves guided regeneration of stable implants. One oversize, unstable implant regenerated in the medullary cavity, although the ostectomy was not supercritical in length. The regenerated bone was small in diameter, so it would not have been strong enough for normal loading if the bone plate were removed. Considering our evaluation of the four hypotheses, guided regeneration, with an intramedullary guide was based on (H1) nonporous implants, (H2) nonresorbable but bioactive material, (H3) an unloaded guide so that the elastic modulus was unimportant, and (H4) a tricalcium phosphate guide to accelerate bone recovery. Additional considerations were nonstabilization of the guide to prevent tissue attachment (loose in the medullary cavity), annular spacing between the guide and the endosteum to allow space for vascularization of the endosteum, and a physical barrier to cortical closure.

2. Experimental Procedure

Guide Manufacture. The raw materials for the Osteoceramic were weighed, mixed, and ground as powders by wet ball milling. The slurry was dewatered by filtering to a plastic consistency. Bubbles were removed with a weak vacuum (27 Torr), and the paste was extruded to produce the thin-walled tubes [2]. These were dried and fired to 1450°C.

Surgical Procedure. The dogs were mongrels of unknown age, seven males and five females. (The surgical procedure was conducted in accordance with the federal Animal Welfare Act. The procedures, and the justification for the research, were approved by the Iowa State University Committee on Animal Care.)

Anesthesia was induced with intravenous thiopentothal and maintained with halothane. The rear quarter was clipped and prepped with four alternate scrubs of chorohexadine and 70% alcohol. A 5-min chlorohexadine scrub followed. The leg was draped and covered with a sterile stockinet.

The skin was incised along the cranial border of the biceps femoris from the greater trocanther to the proximal patella. The skin was sutured to the stockinet with 2-0 monofilament nylon. The margins were retracted and the fascia lattus incised along the cranial border of the biceps femoris directly below the skin incision. The biceps femoris was retracted caudally and the vastus laterlis cranially to expose the femur shaft. The vastus intermedius was retracted from the cranial surface of the shaft. An eight-hole, 316L, AO/ASIF lengthening plate was contoured to the lateral femur and placed in position as a guide. The proximal and distal holes nearest the midshaft were used as a guide to drill lateral-medial bone screw pilot holes; and the plate was set aside.

A section of the adductor muscles was elevated away at midshaft, and the vastus intermedius was retracted from the cranial surface. Two transverse osteotomies, each approximately 20 mm from midshaft, were made using an oscillating saw. The midshaft bone section was measured to select an implant of the correct outside diameter and length to serve as a guide. With the bone and guide in alignment four 4.5-mm D bone screws were placed on each side of the implant. Each screw hole was drilled, measured to ensure both cortices were engaged, and tapped before placement. They were in a neutral position in the bone plate holes. The area was irrigated with saline. Marrow was removed from the ostectomy and smeared, cranially, on the implant.

Closure included 0 monofilament polyglyconate sutures to unite the fascia lata to the cranial border of the biceps, 2-0 monofilament polyglyconate for subcuterous tissue and fascia followed by 2-0 surgical steel to appose the skin. The limbs were radiographed postoperatively and at about 1- or 2- month intervals thereafter.

Three controls received identical treatment but did not include an implant or marrow smear. In general the dogs recovered normally, with partial weight bearing at 1 week and full weight bearing at 3 weeks.

Table 4 Canine, Guide Tube, Ostectomy, and Lengthening Plate Information

Dog			Ostectomy			Implant			
Number	Weight (lbs)	Sex (mm)	Length (mm)	Outside diameter cranial caudal (mm)^2	Inside diameter cranial caudal (mm)^2	Length (mm)	Inside diameter (mm)	Outside diameter (mm)	Bone plate length (mm)
246	60	F	40.0	18.1	11.1	52.6	4.2	7.3	50
256	65	F	40.1	16.3	10.5	52.0	3.7	7.3	50
322	110	M	55.4	25.1	18.7	62.4	3.4	6.7	60
408	65	F	36.8	16.3	9.4	47.6	3.6	6.0	50
406	65	F	36.0	18.2	12.9	(Control)			50
619	65	M	43.1	18.0	13.1	54.8	3.7	7.7	50
709	70	F	48.1	15.8	9.3	55.6	3.7	7.5	50
924	65	F	47.6	16.7	13.1	(Control)			50
790	60	M	43.7	15.2	10.0	(Control)			50

3. Results

Measurements of the implants and the ostectomy were taken at the time of implantation (Table 4). Additional measurements of the annular spacing were taken from postoperative radiographs and from microradiographs. The best estimates of annular spacing and the success in inducing guided regeneration are shown in Table 5. The annular spacing of dog 322 was so large that bone chips from the ostectomy were placed in the annular space to inhibit movement of the distal end. Four of the six dogs with intramedullary guides produced induced regeneration (dogs 246, 322, 408, and 709). The sequence of regeneration is shown for dog 709 in Fig. 9. In Dr. Wagner's opinion the regenerated bone became strong enough to support load for the four dogs. The plate and screws were removed from dog 408 at 9 months, from dog 709 at 7 months, and from dog 322 at 9 months. These dogs were active after recovery and additional remodeling took place. This is shown in Fig. 9. Compare D and E with F and G. The regenerated bone for dog 322 deviated medially and posterioraly from axial symmetry, possibly influenced by the annular bone chips. They were resorbed in about 8 months (Fig. 10, top). As the mineralization became more complete the regenerated bone resorbed and rebuilt to become more nearly axial, almost enclosing the implant when the experiment was terminated at 11 months (Fig. 10, bottom).

Table 5 Effect of Annular Spacing on Bone Regeneration

Dog no.	Average annular spacing (mm)	Average Endosteal diameter (mm)	Percent of endosteal diameter	Result
246	2.35	11.0	21.4	Induced regeneration
256	1.40	10.5	13.3	Bonded to implant
322	6.35	18.7	34.0	Induced but distorted
408	3.05	13.6	23.4	Induced regeneration
619	2.25	13.1	17.1	Bonded
709	2.20	9.8	22.4	Induced regeneration

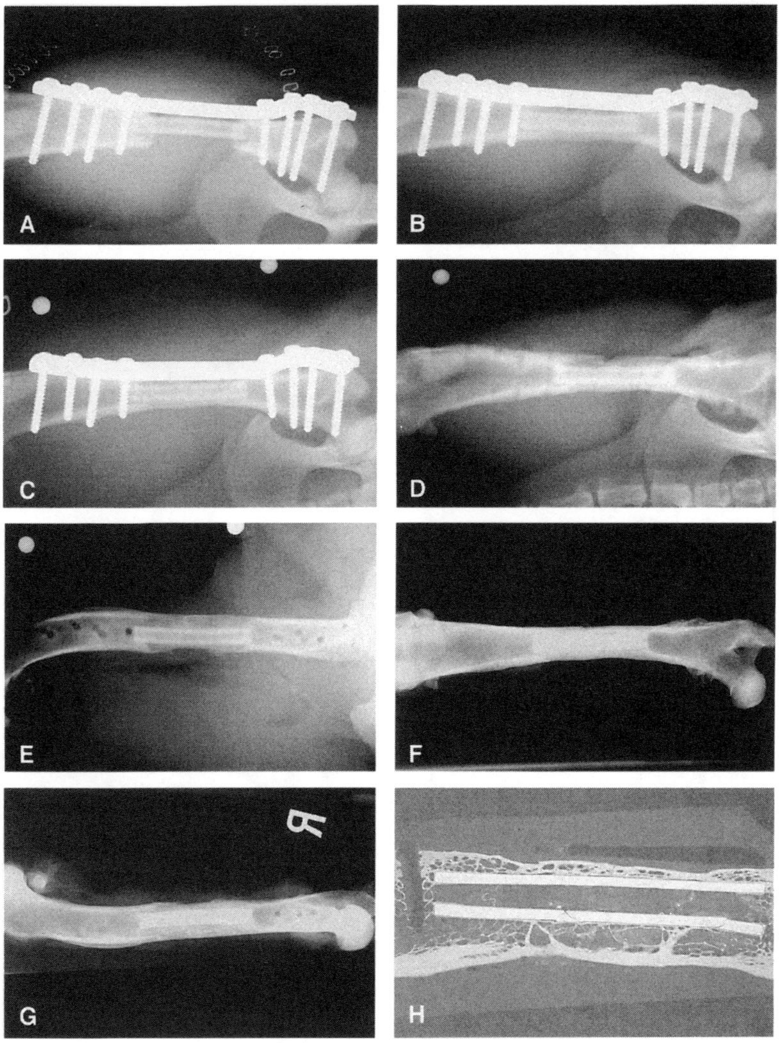

Figure 9 Dog 709 microradiographs (A–G) and microradiograph after recovery (H). (A) Postoperative; (B) 3 months; (C) 5 months; (D,E) 7 months immediately after removal of the plate and screws; (F,G) 11 months; and (H) ML section showing some transfer of load to the implant by trabecular bone.

Bone bonded to the intermedullary guides for dogs 256 and 619. This was not apparent until histological results became available. Thinking that the flexure of the bone plate was insufficient, two screws were removed from each end of the plates for dog 256 at 6 months and for dog 619 at $4\frac{1}{2}$ months (Fig. 11). The implant fractured for dog 619 at 9 months.

None of the controls developed union (Fig. 12). During the removal of the bone plate, and after sectioning the mounted specimens for microradiographs, some guides cracked. This is usually an artifact, but some also cracked during the experiment, such as the implant in dog 406 (Fig. 6). Some of the microradiographs showed cracks that were obviously artifacts because of their position in relation to the surrounding tissue.

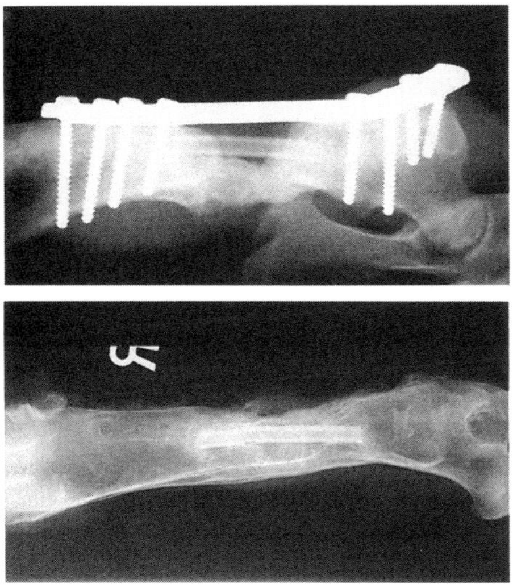

Figure 10 (Top) Dog 322 at 4 months showing deviation of regenerated bone and residue of bone chips at the distal medullary space. (Bottom) Dog 322 at 17 months showing remodeling after removal of the bone plate.

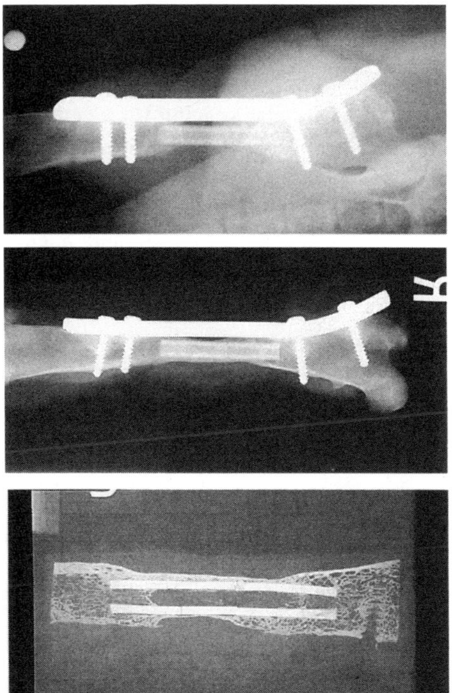

Figure 11 (Top) Dog 256 at 6 months, immediately after removing two bone screws at each end. (Middle) Dog 256 at 9 months, after removal of the bone screws showing additional regeneration. (Bottom) Dog 256 at 9 months microradiograph of longitudinal histological section.

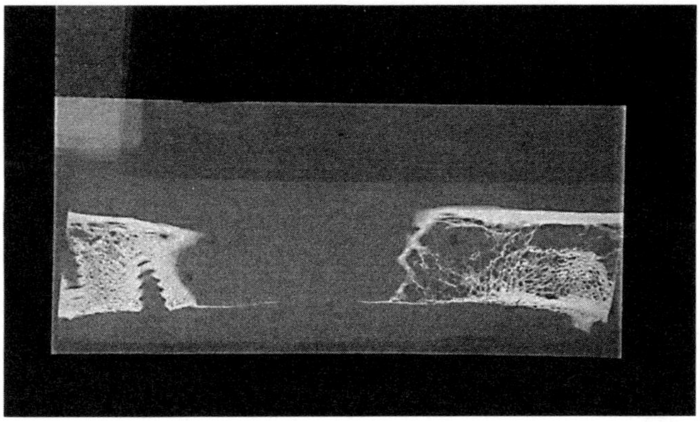

Figure 12 Dog 790, at control, at 8 months microradiograph of longitudinal section

4. Discussion

In four of the six experiments guided regeneration occurred. The bone plates were removed from three, and the regenerated bone supported normal activity of the dogs. The successful guiding functions of dogs 246, 408, and 709 had annular spacings of 21.4, 23.4, and 22.4%, respectively. Successful regeneration also occurred for dog 322, but the new bone deviated from axial orientation medially and, somewhat posteriorly, in the direction of the largest annular bone chips. The annular spacing was 34% of the endosteal diameter, and the mineralized bone that deviated from axial orientation contained shear lines (Fig. 10, top).

Two dogs did not have regenerated tissue but had tissue bonding to the ends of the guides, and the guide was carrying load. One of these, dog 256, had an annular spacing for a centered implant that was only 1.4 mm, or 13.3% of the endosteal diameter (Table 5). Removing half of the bone screws induced additional regeneration (Fig. 11).

The annular spacing for dog 619 was 17.1%, and bonding occurred at each end. When this implant was put in place the distal end was very close to the lateral bone plate and against a bone screw. The proximal end was against the medial endostéum. The implant was not centered. This appears to be the primary reason the bone bonded to the implant. The implant was supporting load at those locations. That would add to the stiffness of the bone and plate assembly and reduce the necessity for regeneration. Removing half of the bone screws did not cause bone to span the ostectomy medially. The implant shattered some time after the screws were removed and the implant became load bearing without the support of the distal screw. It appears that too little annular spacing encourages tissue bonding rather than regeneration, and that too much may allow surrounding tissue to force the regenerating bone away from the axis.

The regenerating tissue always began medially, and mineralization occurred very quickly. In each case the first radiograph after the surgery showed mineralization completely spanning the ostectomy which was more than two times the outside diameter. First radiographs were at 36, 33, 38, and 51 days for dogs 246, 322, 408, and 709, respectively. The very rigid bone plates with four screws did not allow much flexure. But bending of the bone and plate assembly would cause more strain (deflection) medially than laterally. It appears that the rigid plate was sufficiently stable for regeneration of proximal and distal bone, but that sufficient strain occurred to enhance mineralization medially.

The guides were placed as tubes with thin walls to allow normal medullary processes inside the tube. They are inherently brittle, and fracture occurred in some situations. Increasing the wall thickness would increase the strength and inhibit cracking. However, the guides are not intended to carry load and proper placement may prevent cracking if tissue bonding can be avoided.

Future experiments might be conducted with resorbable sutures around each end to control centricity. Intramedullary nails might also be considered for stabilization with the tubes centered on the nails.

The rapid regeneration achieved here indicates the calcium phosphate–containing guides may induce bone regeneration. After regeneration the osteoceramic guides would remain permanently in the medullary cavity. The consequences of this are unknown.

5. Conclusion

These preliminary experiments indicate that guided diaphysis regeneration is possible for bone replacement and for intermedullary guides. This offers the possibility to use an inorganic material to replace bone grafting in some surgical situations. The important requirements are believed to be (1) calcium phosphate chemistry; (2) stabilized proximal and distal bone; (3) annular space or grooves for vascularization of the endosteum; (4) a physical barrier to prevent periosteal closure; and (5) freedom for the implant to move when tissue attachment is not desired. The results do not conflict with these requirements. Further research is needed to test them adequately and to determine if practical guided regeneration is a good surgical procedure.

ACKNOWLEDGMENT

We thank Dr. John Keller, College of Dentistry, University of Iowa, for the micrographs of the thin sections.

REFERENCES

1. Ducheyne P, Qiu Q. Bioactive ceramics: the effect of surface reactivity on bone formation and bone cell function. Biomaterials 1999; 29:2287-2303.
2. Ducheyne P, Lemons J, eds. Biomaterials: Materials Characteristics Versus in Vivo Behavior. N.Y.: Academic Science, 1988.
3. Ducheyne P, Hench L. Comparison of the skeletal fixation of porous and bioreactive materials. Trans. 1st. Mtg. Europ. Soc. Biomat 1977; 2:S-PO.
4. de Groot K, ed. Bioceramics of Calcium Phosphate, CRC Press. 1983.
5. Ripamonti U. Osteoinduction in porous hydroxyapatite implants in heteroptic sites of different animal models. Bioimaterials 1996; 17:31-35.
6. Yamasaki H, Saki H. Osteogenic response to porous hydroxyapatite under the skin of dogs. Biomaterials 1992; 13:308-312.
7. Yuan H, Yang Z, de Bruijn J, de Groot K, Zhang X. Material-dependent bone induction by calcium phosphate ceramics: a 2.5-year study in dog. Biomaterials 2001; 22:2617-2623.
8. Albee FH. Studies in bone growth—triple calcium phosphate as a stimulus to osteogenesis. Ann. Surg 1920; 71:33-37.
9. McGee TD. Artificial bone and tooth material, 1974.
10. Dalculsi G. Biphasic calcium phosphate concept applied to artificial bone, implant coating and injectable bone substitute. Biomaterials 1988; 19:1473-1478.
11. Fulmer MT, Ison IC, Hankermeyer CR, Constantz BR, Ross J. Measurements of the solubilities and dissolution rates of several hydroxyapatites. Biomaterials 2002; 23:751-755.

12. Chu T-MG, Orton DG, Hollister SJ, Feinberg SE, Halloren JW. Mechanical and in vivo performance of hydroxyapatite implants with controlled architecture. Biomaterials 2002; 23:1283-1293.
13. Niederauer GG, McGee TD, Keller JC, Zaharias RS. Attachment of epithelial cells and fibroblasts to ceramic materials. Biomaterials 1994; 15:342-347.
14. Keller JC, Collins JG, Niederauer GG, McGee TD. In Vitro Attachment of Osteoblast-like Cells to Osteoceramic Materials. Dent. Mater 1997; 13:62-68.
15. Wolff J. Das Gestez der Transformation of Knochem. This was the first recognition that bone responds and remodels to accommodate stress. Berlin. Hischwald, 1892.
16. Burr DB, Robling AG, Turner CH. Effects of biomechanical stress on bones in animals. Bone 2002; 30:781-786.
17. Klaiwitter JJ, Hulbert SF. Application of porous ceramics for the attachment of load bearing internal orthopaedic application. J. Biomed. Mater, Res. Symp. No. 2, Part 1971; 1:161-229.
18. Simske SJ, Ayers RA, Bateman TA. Porous materials for bone engineering. Mater. Sci. For 1997; 5:151-182.
19. Graves GA, Hentrich RL, Stein HG, Bajpei PK. Resorbable ceramic implants. J. Biomed. Mater. Res. Symp. Part 2 No 1971; 1:91-115.
20. Katz JL, Pilliar RM, Berkowitch J, Christel P, Higham P, Kerpersers R, Knox GF, Serdanese A. Biomechanical stability and design. Ann. NY, Acad. Sci 1988; 523:283-286.
21. Patka P, Klein CPAT, Hollander W. Ceramics in bone replacement: reconstruction of the tibial plateau with hydroxyapatite. In:. Handbook of Bioactive Ceramics Yamamuro T, Hench LL, Wilson J, eds. Vol. II: CRC Press, 1990.
22. McGee TD, Graves A, Tweden KS, Niederauer GG. A biologically active ceramic material with enduring strength. In: Encyclopedia of Biomaterials. 1995.
23. McGee TD, Olson CE. General Requirements for a successful orthopedic implant. In: Encyclopedia of Biomaterials. Vol. 1, 1995:Part B.
24. McGee TD. Method of restructuring bone. U.S. Pat. No. 6,312,467, 2001.
25. McGee TD. Method of restructuring bone. U.S. Pat. No. 6,349,909, 2002.

10

Bioresorbable Skeletal Fixation Systems in Craniofacial Surgery

Mutaz B. Habal
The Tampa Bay Craniofacial Center/University of South Florida and University of Florida
Tampa, Florida, U.S.A.

I. INTRODUCTION

Today the use of bioresorbable biomaterials in skeletal fixation is in a process of a major evolution in the application and understanding of all fixation devices [1]. The advances made in the last decade made the utility and applications in the clinical situations possible. Skeletal fixation is an essential element in the stabilization of the different components of any bony system. This shift in the evolutionary status away from the use of metallic implants, particularly in infants and in children, toward bioresorbable material in all applications in the craniofacial skeleton and in all age groups is to be noted as an advancement in technology [2]. The metallic implants, once popular in a wide variety of uses in the 1980s, are falling to the wayside as the application and sophistication of resorbable components become more accepted by practicing surgeons, particularly for all components of the craniofacial skeleton. Notable is the increased safety to patients due to avoiding the need for a second operation to remove these implants and the avoidance of passive internal migration and growth disturbances, particularly in children [3]. These bioresorbable implants are also invisible on radiographic analysis, thus making their use in oncological clinical situations an added advantage.

II. BIOMATERIAL

The introduction and use of the polylactide and its different components as a resorbable biomaterial for sutures in the middle 1960s, followed by its extensive use for many other applications and high standard of safety throughout, led to the main impetus for basic research scientists and technology engineers to develop resorbable fixation devices for use in the craniofacial skeleton [4]. The basis of the standard units is poly(lactic acid), (PLA) in its levo and dextro forms to produce the desired form and shape needed as a copolymer of the biomaterial to be used in the craniofacial skeleton. Interest in widening applications has been mounting so as to improve the clinical applications of the bioresorbable material. Many laboratories around the world have been looking into changing the chemistry and producing different combinations to produce a compounded biomaterial that is used in skeletal fixation of the craniofacial skeleton. These different combinations being produced are made so as to vary the two different characteristics of the produced biomaterial, that is, longevity versus strength. These will be the basis of all the materials systems today marketed for use by surgeons [5].

There were many questions that had arisen regarding the validity and the side effects of the widely practiced applications of metallic implants to stabilize the craniofacial skeleton. The polylactides' base and molecular weight are the basis of all the copolymers used in the skeletal fixation systems. The key issue is to understand the longevity and mechanical strength expressed in shear pressure at different parts of the skeleton, for the skeletal fixation components that are used. The craniofacial skeleton has two units; static nonmovable units such as the orbitocranium and constantly moving units such as the jaws. Both the static components on the nonmovable parts such as the cranium and the dynamic parts that are in constant motion such as the mandible have different biology that must be understood to develop fixation systems, particularly the requirement of shear stress for the needed biomaterial [6].

III. BACKGROUND

The interest in skeletal fixation to produce the required stability for the healing process needed for the craniofacial skeleton started after World War I, although scattered reports for such stabilization systems were noted many years before the war in the literature. World War II brought importance of this need to the forefront as some of the wounded needed their face worked on after multiple injuries during their rehabilitation. It was no longer accepted to have a deformed face or a functional derangement that could be repaired by the practicing surgeon if the appropriate technologies were available. That is the start of the collaboration between the industry and clinicians to achieve the best outcome for patients [7].

The first stage in the evolution process involved just manipulation to place the craniofacial structures together; with it some techniques were developed that were still practiced by some surgeons until a few years ago. Early fixation approaches of such structures collapsed into a nonfunctional position and needed secondary and maybe tertiary repair afterward. The instability was the result of the myofunctional components of the facial skeleton. The muscle pull on the unstable bones caused the bones to shift. The craniofacial bones are membranous bones, and their healing is devoid of the presence of a callous formation as in long bones.

The second stage involved the use of an external fixator apparatus for the repair. That initially gave superior results to the previously practiced repair. However, after the external fixator in the form of an external apparatus was removed, the patients' repaired segments collapsed. The relapse was due to a similar process related to stability without the presence of a primary healing in these components that are either fractured or osteotomies.

The third stage involved the use of wire fixation, which set the stage for the use of the metallic implant systems that were the predecessors of the resorbable plating systems in use today. Their evolution was progressing slowly since World War II. Wars bring about many casualties and the rehabilitation of such patients spill over to the activities in the civilian population.

IV. SAFETY FACTORS

The application of biomaterials in the biologic system is always associated with major questions: are these biomaterials safe to use or are they harmful to the individual patient? Also, are there any by-products that are harmful to patients under any circumstance? Around the safety issue came the fundamental focus of the major function of the Food and Drug Administration (FDA) and its similar administration in Europe. These scientist panels look at all the biomaterials that are to be used [8]. The scientific study of the efficacy and the animal studies that are required before the premarket studies (PMS) are designed to collect data from clinical applications over a period of time. The data are analyzed and if the biomaterial is found to be safe and efficacious, then it is released in the marketplace, in this situation the doctors' offices and hospitals. Thus all the biomaterials referred to have passed through that process, and their final applications is the focus of the following discussions in this chapter.

V. BIORESOBABLE SYSTEMS

A. Historical Perspectives

A review of the history of skeletal fixation is helpful to the understanding of the situations in which we work today. Skeletal fixation in the craniofacial region has gone through many advances in the past few decades. Most of these advances have followed major international conflicts involving the complex treatment of large numbers of injuries [10].

Initially, fractures of the craniofacial skeleton were treated without fixation by allowing the bones to heal in open soft tissue, then performing the repair at a later time. Fractures were also treated with closed reduction after manipulation. The next development was the use of an external apparatus for fixation. This method was useful until the external fixation was removed and the repaired structures collapsed again. Those procedures were accompanied by a lack of success that was the impetus for the development of the techniques of open reduction and internal fixation in all skeletal clinical problems. That was the background for the development of the biomaterials to be used in these situations [11].

Internal fixation then came into practice, which required the use of rigid fixation. The plating system began at the turn of the century, with the use of stainless steel plates. The popularity of the applications came after the wars and packed in the mid-1970s. The use of vitalium first and titanium second as the plating systems followed the major applications of the stainless steel system, which had been accompanied by many complications. In the latter part of the last century, the use of resorbable plating systems evolved and has advanced to their present status today. Resorbable plating systems remain the state of the art for skeletal fixation in the craniofacial region, particularly in infants and children (Fig. 1).

Today's biocompatible resorbable polymers offer surgeons a new array of options for craniofacial skeletal fixation [12]. Some of the potential benefits of resorbable polymers include greater ease and accuracy of contour adaptation, clear radiographic presentation due to the absence of x-ray scatter, elimination of the need for secondary surgeries for device removal, and reduced risk of stress-shielding of the underlying bone. Known as polyesters, these copolymers have chemical, physical, material, mechanical, and biologic properties different from those of metal fixation devices. Knowledge of these differences will facilitate the utilization of resorbable implants in fixation for craniofacial trauma (Fig. 2).

Among the bioresorbable polyester craniofacial fixation devices approved for clinical use by the FDA, copolymers of lactides and glycolides are available. The first copolymer of L-lactide and glycolide (LactoSorb, W. Lorenz, Jacksonville, FL) was approved by the FDA in 1996. The lactide in LactoSorb is a homopolymer of the levo form. The ratio of the L-lactide monomer to the glycolide monomer is 82:18 in poly(L-lactide-*co*-glycolide) to take advantage of glycolide's rapid degradation time. Strength declines to approximately 70% by 6–9 weeks and resorption is complete by 12 months [14].

Approved more recently, in 1998, is a copolymer produced from a mixture of 70% L-lactide monomer and 30% D,L-lactide monomer (MacroPore, MacroPore Biosurgery, Inc., San

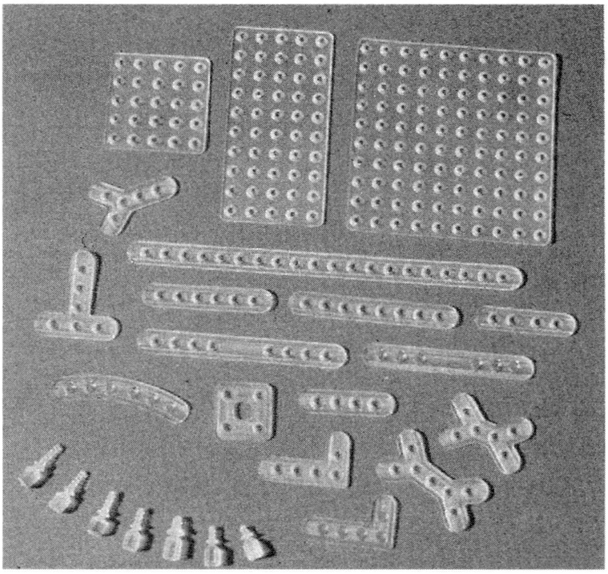

Figure 1 Components of the resorbable plating system. Plates and panels get soft with heat so they can be contoured to the configuration of the site, a major advantage.

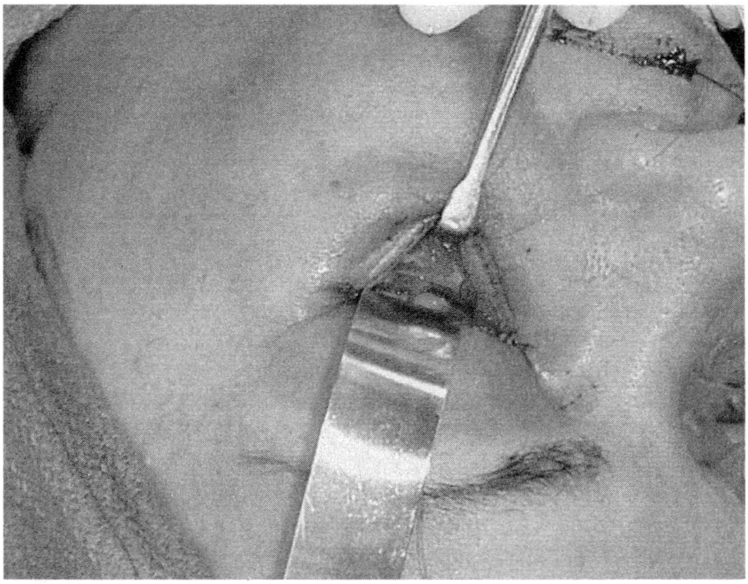

Figure 2 Repair of the orbital floor. A small panel contoured into the orbit via a transconjunctival approach to the orbit.

Diego, CA). This 70:30 ratio in poly(L-lactide-*co*-D,L-lactide) (DLLA) retains approximately 70% of its initial strength after 9 months and approximately 50% after 12 months, with resorption completed by 24–36 months. Other resorbable polyesters from Bionx, Leibinger (delta system and the new delta system), Synthes (resorbable system), KLS Martin (resorb-X), and Inion (two systems) are all FDA approved and available for surgeons to use. The differences among these systems are the ratios of the copolymers used in the compositions, which affects their longevity, a consideration of importance to surgeons. The deciding factor of which system to use is the individual surgeon's preference and ease in clinical application. In children the surgeon wants the system to resorb much faster than in the adult. A patient going into radiation therapy has a slower healing so the systems used must stay longer to allow for better bone healing. Thus surgeons have more options, based on the need of the patient and the long-term application needs [15,16].

In view of these considerations, the primary focus is on the use of poly (L-lactide) for skeletal repair and fixation due to its wide range of acceptance among practicing surgeons. The acronym DLLA will also be utilized to designate a copolymer of the two monomers, L-lactide and D,L-lactide. This combination is needed for strength to stabilize the bony components for fractures or controlled fractures as in facial osteotomy [17].

B. Biophysics

Poly(L-lactide), which has a high crystallinity, is characterized by its strength and long degradation time. Conversely, a polymer created from D,L-lactide has little strength and degrades rapidly. Combining L-lactide and D,L-lactide results in a copolymer with the intermediate characteristics of strength for 6–9 months and resorption in 24–36 months. In addition, the copolymer is optically clear and noncrystalline, resulting in minimal foreign body reactions by tissue. It should be noted that, even within a given copolymer, strength and degradation characteristics could vary according the degree of polymerization [18,19]. Therefore, the manufacturer must maintain this within the desired range. A common measure of the degree of polymerization is called intrinsic viscosity (IV), and for any given polymer the IV correlates with molecular weight. To measure IV the polymer is dissolved in a standardized amount of chloroform and then passed through a viscometer. The length of time that it takes for passage is used to calculate the IV.

At sufficiently high temperatures all materials change from hard to soft and finally to liquid. The temperature at which a material changes from hard to soft is known as the glass transition temperature (T_g). For 70:30 poly(L-lactide-*co*-D,L-lactide) the T_g is 55°C (131°F), thus allowing heat to be utilized for contouring these implants [20].

Contouring an orbital floor liner illustrates this property. After making a template of the orbital floor, the template is held against the orbital floor liner (Fig. 1A), then placed in a water bath and heated above the T_g. The floor liner becomes soft in a few seconds and simply drapes over the template when lifted from the water bath. In a few more seconds the floor liner cools below the T_g and can be removed from the template. The liner is then ready to be placed in the patient. It is useful to note that 70:30 poly(L-lactide-*co*-D,L-lactide) has shape memory, and if placed back in the water bath it will return to its original contour, thus enabling additional opportunities to raconteur it (Fig. 2). If only a portion of an implant needs to be recontoured, only that portion need be placed back in the water bath. Cyclic heating of 70:30 poly(L-lactide-*co*-D,L-lactide) to 70°C can be performed multiple times with no change in material strength (Fig. 3).

C. Biochemistry

When lactic acid undergoes polymerization, ester bonds are formed and H_2O is released. Therefore lactide copolymers are also known as polyesters. Resorption of lactide copolymers takes

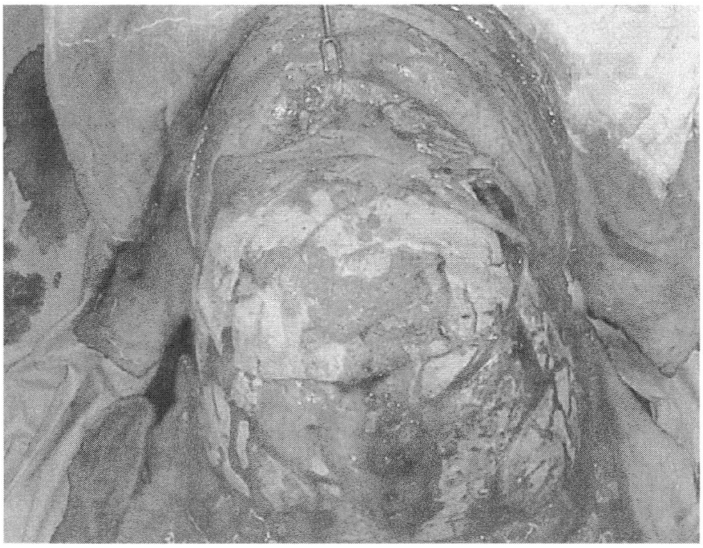

Figure 3 Total fixation of craniofacial components of a child after corrective surgery for a birth defect in the craniofacial region. All biomaterials used were bioresorbable.

place as a reversal of this process, with sorbtion of H_2O and scission of the ester linkages. This bulk hydrolysis of lactide copolymer implants continues until single lactic acid molecules are released, which are then metabolized into glucose or into CO_2 and H_2O via the Krebs tricarboxylic acid cycle.

A variety of factors are known to affect the rate of lactide copolymer resorption. A higher IV or molecular weight means there are more ester linkages that undergo scission, and this process results in a longer resorption time. A larger implant size or volume will also require more scission before implant resorption can be completed. If the polymer is packed more tightly in an orderly crystalline pattern, there is less space for H_2O access and resorption will take longer than for noncrystalline implants. Since hydrolysis occurs both on the implant surface and within its interior, implant porosity will increase surface area, facilitate H_2O access, and decrease resorption time. The molecular configuration of copolymers may alter resorption time. Greater vascularity of the implant host site as well as flexural bending from functional loading appear to be associated with an increased rate of hydrolysis.

D. Biotoxicology

The toxicology of lactides has been of minimal concern due to the relatively small volumes of implant material, slow degradation rates, and short serum half-lives. The serum half-life of the levo form is 15 min; for the D,L form it is 22 min. The normal resting blood lactate level is 1.1–2.8 mM/L. After muscular activity it will rise ten fold to 10–23 mM/L. If the assumption is made that degradation occurs over 2 months, with first-order kinetics and a half-life of 74 h, a 100-g implant would release 0.18 mM/L of lactide acid in the first minute, far less than the changes resulting from muscular activity. Two of the largest sheets of 70:30 DLLA copolymer weigh only 18 g, and degradation actually takes place over a much longer, 18–36 months, time interval. Even with first-order kinetics starting instantly, the 18 g of lactide copolymer would result in an increase in blood lactate levels of only 1.1%.

Table 1 Five Years Experience with the Bioresorbable Plating System

Category	Patients	Plates	Screws	Panel	E
Congenital	176	923	2405	121	3449
Trauma	65	103	368	35	506
Tumor	54	294	653	46	993
Cosmetic R	29	00	58	00	58
Total	324	1320	3484	202	5006

E. Biopathology

The histological responses to 70:30 DLLA copolymer have been well studied. There is an initial acute inflammatory response following implantation. By 72 h there is a narrow zone of fibrinous exudate, edematous granulation tissue, and a modest degree of fibroblast proliferation. By 7–14 days the granulation tissue has matured into a thin, cellular, fibrovascular capsule. Measurements of in vivo tissue pH adjacent to 70:30 DLLA copolymer implants have detected no change during degradation.

F. Biomechanics

The mechanical properties of 30:70 DLLA copolymer, bone, and steel are well known. The tensile strength of lactide is approximately 30% the strength of bone. With a tensile strength of 70%, lactide materials can readily be designed to accommodate the failure loads for non-weight-bearing bones. When designed as 1.0-mm thick plates, the tensile strength is approximately 190 N, or 45 lb. When metal screws are overtorqued, the threads strip the bone. When lactide screws are overtorqued, the heads shear off. The shear strength of 2.0-mm screws is approximately 85 N, or 20 lb. As the 30:70 DLLA copolymer undergoes hydrolysis, its mechanical strength will decrease. At 3 months, strength remains near 100%, decreasing to 90% at 6 months, 70% at 9 months, 50% at 12 months, and 0% by 18 months.

G. Clinical Experience

As with all fixation systems, clinical experience eventually determines the efficacy of any implant design that is most likely to produce a lasting and successful outcome. As a starting point, it is recommended that the surgeon select a copolymer design more stable than the currently employed system. An example from our experience is that the substitution of titanium with PLA systems for all repairs is imperative. The traditional metal plate with two holes on each side and 2.0-mm screws would normally have been used; however, the patient insisted on resorbable over metal fixation so four-hole plates are necessary. To assure adequate stability, the copolymer plate was contoured from a piece of 1.0-mm thick mesh with four holes on each side and attached with 2.4-mm screws (Fig. 3). While this design may be excessive in strength, it is appropriate to be conservative until more clinical experience is acquired (see Table 1) for the first 5 years of using the biomaterial implants. The screws in all the systems available today need to have the hole made first and then be tapped with a second metallic device to establish the troughs the biomaterial resorbable screw has to go in, an extra step from the routine systems used of the metallic counterpart. That may take a few minutes at the beginning of the application but it is a fast learning phase for all those who apply the system.

Table 2 Follow-up of Patients with Bioresorable Plating Systems

Follow-up (months)	1–42, arithmetic mean 22
Complications	No major complications
Average	15.4 complications per patient
Edema	+2 (age range 1 month to 94 years, sem 5.5)
Age	
Palpable	+2 (average age 11.8 years)
Extrusion	1 screw/ 1pt. Standard of care for infants and children
Infection	1 salvage accomplished

The authors' clinical experience with biomaterial use in craniofacial fixation has been summarized in Table 1 and 2. We have noted minimal complications in all the patients, and the follow-up time is enough to allow us to note that there is minimal soft tissue reaction. There have been no mishaps nor any danger to the patients, particularly in children (Fig. 3). These plates are not visible to the diagnostic techniques available today (Fig. 4). We can only see the screw holes for about a year before they fill up by the regenerating bone.

REFERENCES

1. Habal MB. Absorbable platting system—discussion of, resorbable plate fixation in pediatric craniofacial surgery. Plast Reconstruct. Surg 1997; 100:8–11.
2. Habal MB, Pieterzak WS. Key points in the fixation of the craniofacial skeleton with absorbable biomaterial. J Craniofac Surg 1999; 10:491–499.
3. Kulkarni RK, Pani KC, Neuman C, Leonard F. Polylactic acid for surgical implants. Arch Surg 1966; 93:839–43.
4. Habal MB. Absorbable, invisible and flexible plating system for the craniofacial skeleton. J. Craniofacial Surg 1997; 8:121–126.
5. Habal MB, Reddi AH. Bone grafts and bone substitutes. New York: WB Saunders, 1992.
6. Habal MB. Resorbable fixation in craniofacial. Operative Plastic Surgery. Vol. 14, 2003, in press.
7. Tharanon W, Sinn DP, Hobar PC, Sklar FH, Salomon J. Surgical outcomes using bioabsorbable plating systems in pediatric craniofacial surgery. J. Craniofacial Surg 1998; 9:441–444.
8. Bhanot S, Alex JC, Lowlicht RA, Ross DA, Sasaki CT. The efficacy of resorbable plates in head and neck reconstruction. Laryngoscope 2002; 112(5):890–898.
9. Moe KS, Weisman RA. Resorbable fixation in facial plastic and head and neck reconstructive surgery: an initial report on polylactic acid implants. Laryngoscope 2001; 111(10):1697–701.
10. Persons BL, Wong GB. Transantral endoscopic orbital floor repair using resorbable plate. J Craniofac Surg 2002; 13(3):483–487.
11. Surpure SJ, Smith KS, Sullivan SM, Francel PC. The use of a resorbable plating system for treatment of craniosynostosis. J Oral Maxillofac Surg 2001; 59(11):1271–1276.
12. Wiltfang J, Merten HA, Schultze-Mosgau S, Schrell U, Wenzel D, Kessler P. Biodegradable miniplates (LactoSorb): long-term results in infant minipigs and clinical results. J Craniofac Surg 2000; 11(3):239–245.
13. Shand JM, Heggie AA. Use of a resorbable fixation system in orthognathic surgery. Br J Oral Maxillofac Surg 2000; 38(4):335–337.
14. Wiltfang J, Merten HA, Becker HJ, Luhr HG. The resorbable miniplate system LactoSorb in a growing cranio-osteoplasty animal model. J Craniomaxillofac Surg 1999; 27(4):207–210.
15. Pensler JM. Role of resorbable plates and screws in craniofacial surgery. J Craniofac Surg 1997 Mar; 8(2):129–134.

16. Habal MB. Resorbable fixation in craniofacial. Operative Plastic Surgery. Vol. 14, 2003, in press.
17. Peltoniemi HH, Tulamo RM, Toivonen T, Pihlajamaki ???, Pohjonen T, Tormala P, Waris T. Intraosseous plating: a new method for biodegradable osteofixation in craniofacial surgery. J Craniofac Surg 1998; 9(2):171–176.
18. Peltoniemi HH, Ahovuo J, Tulamo RM, Tormala P, Waris T. Biodegradable and titanium plating in experimental craniotomies: a radiographic follow-up study. J Craniofac Surg 1997; 8(6):446–453.
19. Peltoniemi HH, Tulamo RM, Pihlajamaki HK, Kallioinen M, Pohjonen T, Tormala P, Rokkanen PU, Waris T. Consolidation of craniotomy lines after resorbable polylactide and titanium plating: a comparative experimental study in sheep. Plast Reconstr Surg 1998; 101(1):123–133.
20. Eppley BL, Sadove AM. Effects of resorbable fixation on craniofacial skeletal growth: a pilot experimental study. J Craniofacial Surg 1992; 3(4):190–196.

11

Osseointegration Principles in Orthopedics: Basic Research and Clinical Applications

Lars V. Carlsson
Institute for Surgical Sciences, Sahlgrens University Hospital, and University of Gothenburg, Gothenburg, Sweden

Warren Macdonald, C. Magnus Jacobsson, and Tomas Albrektsson
Sahlgrens University Hospital and University of Gothenburg Gothenburg, Sweden

I. INTRODUCTION

Total joint replacement has been claimed as ''the most successful surgical procedure ever.'' Whatever the technique or fixation, there is still only 90% survival at 10 years, with steeply declining survival after that time [1]. Even in an autopsy study of implants functioning successfully until death, Charnley found a fine fibrous membrane between the cement and cancellous bone in five out of six specimens [2].

Clinically, aseptic loosening is characterized by pain, restricted joint function, and loss of bone stock [3]. Radiographically, aseptic loosening presents the features of radiolucent zones at the implant–bone interface and progressive loss of bone (Fig. 1) [4]. A precursor to clinical loosening is migration of the implant, which is only accurately detectable with roentgen stereometric analysis, but thereby provides a further definition of loosening [5–9].

Charnley further described a ''destructive endosteal lesion'' of the femur around a cemented implant [10], which he postulated was due to a chronic nonsuppurative infection. But removal of aseptically loosened arthroplasty components also revealed an implant bed lined by fibrous tissue [11] which was observed as a radiolucent gap or line on radiographs [12,13]. Progression and widening of the gap was found to be correlated with failure by loosening [14] and was attributed by some to movement of the implant [15,16], but was more commonly ascribed to ''cement disease,'' a physiological reaction to cement as a material [17].

Acrylic cement is designated ''cold-curing'' because it cures or sets without the addition of heat, but the exothermic polymerization reaction can raise the bone interface temperatures to 70°C or higher, at which thermal necrosis of bone might occur [15,18–21]. Alternatively, leakage of the cytotoxic monomer into the adjacent tissues has been postulated as the cause of necrosis [22–25]. Or the elevated temperature might increase the toxicity of the monomer to cause chemical injury to the bone [26]. But in fact vital microscopic studies have shown that acrylic cement has an immediate necrotic effect on the bone and some influence on the intramedullary circulation, and that regeneration of bone tissue is also seriously impaired [27–29].

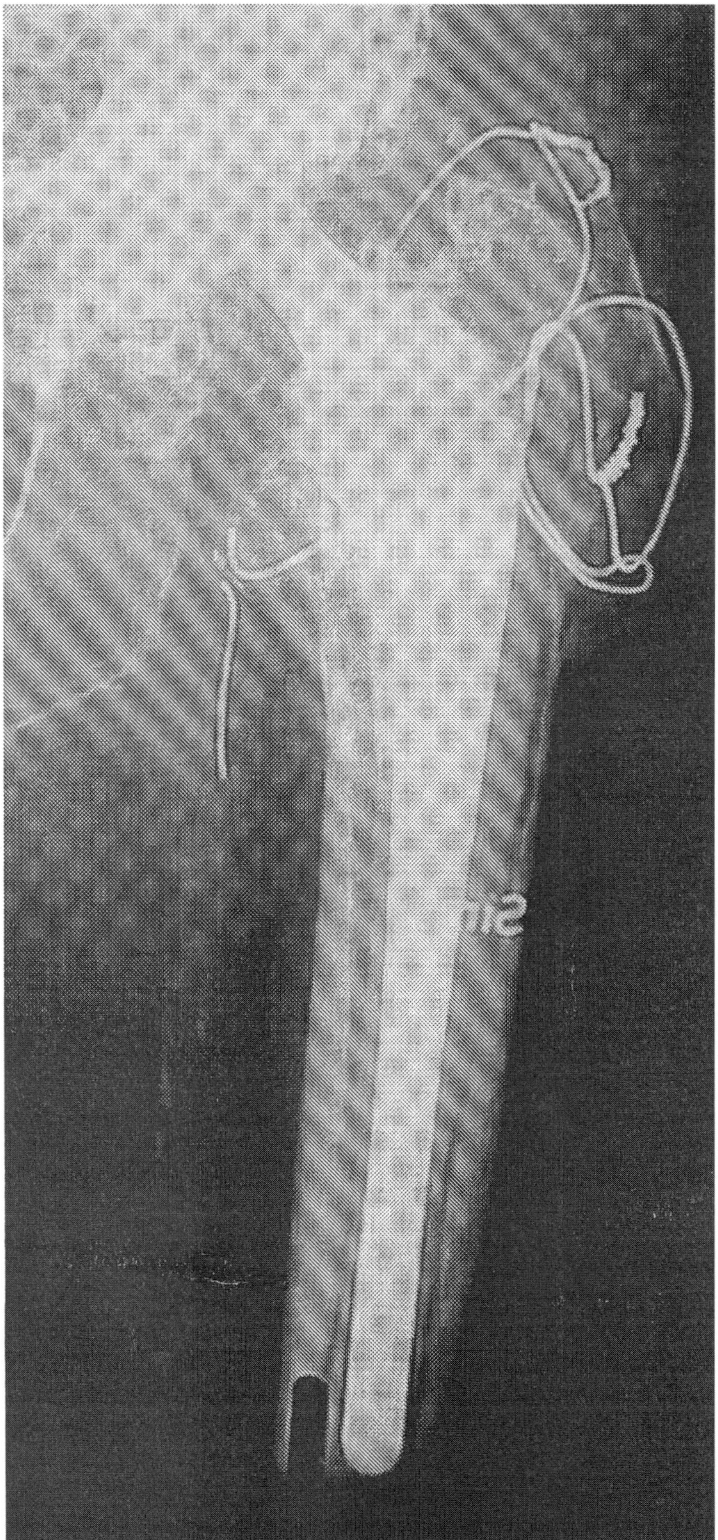

Figure 1 Radiographic appearance of linear osteolysis around a cemented stem.

Mechanical causes of failure of cemented fixation have also been advanced; the mechanical and vascular trauma of surgical preparation might cause bone necrosis adjacent to the implant mass [30,31]. Simple motion between implant and cement or cement and bone was initially thought to contribute to interfacial failure [32,33], while observations of osteolysis around radiographically stable implants [34], even in the absence of generalized cement fragmentation [35], led the Boston group to attribute failure to localized fracture of the cement mantle and fragmentation [36].

These observed hazards of cemented fixation increased the enthusiasm for cementless fixation, begun with simple impaction of implants into medullary cavities [37,38] or threaded designs [16,39]. However, whatever cementless fixation strategy was adopted, aseptic loosening with the presence of fibrous tissue interfaces was still observed [40–42]. The presence of particles of PMMA, polyethylene, or metal [43,44] drew the conclusions back to Willert and Semlitsch's earlier proposed mechanism of particle-induced osteolysis [45,46], then called "particle disease."

Particle disease has been linked with PMMA particles from the cement mantle [35,47], alone or in combination with polyethylene particles from wear of the articulating liner [15,48]. Metal particles have also been implicated, either from wear against the articulating bearing surfaces of the joint [49,50], against cement or bone [51], or from an undetermined source [52].

The mechanism of particle disease has received detailed investigation. Injection of particles at stable interfaces has been shown to cause macrophage stimulation and subsequent bone resorption [53–55], for both PMMA and polyethylene particles, and also for cobalt–chromium alloy particles [56–58]. However, although Howie's intra-articular particles of cobalt–chromium provoked macrophage proliferation and synovial degeneration, when injected in an intraosseous location the reaction was much less severe [59]. Goodman's use of larger particles (2.5 to 75 μm vs. less than 3 μm) may have diminished the response. And intraosseous-implanted wires of cobalt–chromium and c.p. titanium provoked "a discontinuous fibrous membrane that was noticeably hypocellular" [59], especially remarkable when c.p. titanium can be implanted without any fibrous response [60]. These findings of little response to particles alone have been confirmed recently [61].

Furthermore, a fibrous response to implants can have other causes: any implant in tissues will provoke a chronic inflammatory response known as the foreign body reaction [62], which is exacerbated by the presence of copious wear debris from PTFE [63] or PE [48] and can cause bone resorption. Murray and colleagues showed that surface energy and roughness of implants could cause adherent macrophages to release bone resorption mediators [64].

Indeed, most materials provoke some response from tissues by virtue of the release of ions from the surfaces or bulk material, especially metals. Metal ion release occurs through corrosive attack; less resistant materials such as stainless steel [65–68] release greater concentrations of ionic entities than the more noble Ti–6Al–4V or titanium–6% aluminium–4% vanadium alloy [69] or cobalt–chromium cast or wrought alloys [70–72]. Very resistant materials such as c.p. titanium, niobium, and tantalum release even lower levels of ionic species [73], which probably contributes to their easier acceptance by tissues. Due to the tenacious oxide coating on the surface of c.p. Ti the bone is effectively responding to a ceramic layer [74].

Several authors have also drawn attention to the differences between commonly observed aseptic loosening and the aggressive osteolytic response provoked by wear debris [4,75,76].

Periprosthetic osteolysis may be mechanically driven. Bone is a biomechanical tissue, requiring adequate stress during use to maintain bone mass. Remodeling of the proximal femur adjacent to long-standing femoral implants has been observed in cementless devices [77,78–81] and cemented implants [10,82–85]. Common findings are resorption of the medial femoral neck in cemented stems [86,87] and cementless [88], with typically 40% loss of bone mineral proximally [87] and distal progression of that bone loss [85]. In some cases the bone loss is so great that fracture of the proximal femur follows [89]. While some have attributed this to the access of wear particles proximally [81], the pattern is also consistent with finite element analysis prediction of the stress changes [90–94] and is often accompanied by distal hypertrophy—a mechanical effect and not due to particles [77,83,86].

Aspenberg and Herbertsson showed that motion between implant and bone was more important in the development of a fibrous membrane than the application of polyethylene particles alone [95]. Fluid pressure alone has also been shown to cause osteolysis in stable osseointegrated implants, even steady fluid pressure, not requiring pulsating variation [96].

Many authors, while postulating particles as the principal agents of osteolysis, have warned that the issue is multifactorial [97]. It is clear that the issue is highly complex, and to attribute the cause to one factor alone (as has been the case several times) is, in our view, overly simplistic.

II. OSSEOINTEGRATION

Osseointegration of implants was first defined as ''a direct contact between living bone and implant, on the light microscopical level'' [98]. A further definition of osseointegration was proposed in 1985: ''A structural and functional connection between ordered, living bone and the surface of a load-carrying implant'' [99]. That is to say, osseointegration is the direct opposite of and answer to orthopedic aseptic loosening.

Osseointegration for implants was first developed in clinical dentistry in the 1970s [100]. Bone cement does not function well in the craniofacial skeleton, and no reliable implants for anchorage of artificial teeth existed before the introduction of osseointegration. Excellent clinical results of 90 to 95% success were reported with osseointegrated oral implants at 5 years [101,102] and 10 years postoperatively [101]. Extraoral, skin-penetrating osseointegrated implants have been found to function equally well [103,104].

Osseointegrated implants in the craniofacial skeleton have been documented with clinical function for as long as 30 years. Implants that fail do so predominantly during the first couple of years; thereafter few failures occur [105]. This contrasts strongly with hip implant components, for which the failure rate increases with time [1].

The craniofacial experience showed that in order to establish secure osseointegration, six factors must be controlled [106]:

1. Biocompatibility of the material
2. Implant design (form and macrotexture)
3. Surface conditions of the implant (microtexture)
4. State of the host bed
5. Surgical technique used at implant insertion
6. Loading conditions applied postimplantation [98]

To achieve osseointegration of orthopedic implants it would be necessary to control these factors in the orthopedic environment, and to develop implants taking account of these conditions.

Furthermore, from the orthopedic experience of osteolysis, it is known that even if initial implant stability is achieved, the bone may retreat from or be isolated from the implant because of

1. Foreign body reaction—to the implant per se, to debris from implant component degradation or wear, or to toxic emissions from the implant
2. Damage or insult to the bone through mechanical trauma of surgery

3. Imposition of abnormal or unphysiological conditions on the bone, such as fluid pressures or motion against implant components
4. Alteration to the mechanical signals encouraging bone densification; strain reductions or "stress-shielding" of replaced or adjacent bone

A. Osseointegration in Orthopedics

1. Basic Science

The first factor investigated in the orthopedic context was the importance of implant form, or macrotexture, on fixation stability in bone [107]. Titanium screws were shown to achieve direct bone apposition after 4 to 14 months in canine and human tibiae, whereas cylinders and T plates provoked a fibrous tissue interface.

Biocompatibility of the material was investigated in a series of studies comparing commercially pure titanium (c.p. Ti) with existing orthopedic implant materials such as stainless steel, vitallium (Cobalt–chromium alloy) and titanium–6 aluminium–4 vanadium alloy [108]. In all cases the c.p. Ti components achieved higher interfacial shear strengths and greater bone–implant contact ratios than the orthopedic alloys. The interaction between c.p. Ti and bone was also characterized for cortical and cancellous bone in another series of detailed studies [109].

That close implant–bone apposition is important for the development of secure osseointegration was demonstrated in rabbit tibiae [110] and subsequently supported indirectly by findings in human acetabular implants [111–113]. Although closeness of fit has not been so carefully studied, implant stability or implant/bone movement has been shown to play a significant role in the development or prevention of osseointegration [114–117].

The effect of implant surface texture on bone incorporation has also been studied in detail [118–123] and been further demonstrated in the orthopedic environment [124–127]. In fact, it has been shown that enhancement of the bone response to implant surfaces can be achieved by increasing the roughness from an S_{qA} of less than 1 μm to a value closer to 1.5 μm [128], or by other surface treatments [129,130].

One aspect of osseointegration that was difficult to reconcile with orthopedic applications was the need for two-stage implantation, with a nonfunctional "healing-in period" between interventions of typically 3 months. This was found to be unnecessary in orthopedics; a one-stage procedure gave comparable results [131]. Immediate loading has subsequently also been proven and accepted in craniofacial applications [132–135].

Having shown the applicability of osseointegration principles in the orthopedic field, the task remained to develop an actual implant system in accordance with these principles.

3. Orthopaedic System Development

As stated above, remodeling around conventional femoral implants (cemented and cementless stems) has clearly shown that intramedullary stems are an unphysiologic design [77,91,136,137] and apply loading conditions unconducive to optimal bone response [87,138], This has led to the search for "isoelastic stems," designed to reduce stress-shielding and maintain bone mass [139–141]. The need for intramedullary stems arises from the demands of immediate and long term stability in the face of probable fixation failure.

Reliable biologic fixation offered by osseointegration permits implant design on more physiological principles, resulting in more physiological strain transference to the bone and hence encouragement of long-term stability.

The osseointegration requirement of a viable bone implant bed demands optimal bone cutting techniques in good implant support zones. Optimal bone cutting has been shown to require sharp cutting edges, controlled cutting speeds, and deep cuts under constant irrigation—not normally possible with the broaches and shapes required for intramedullary stems. But concentrating on proximal femoral geometry enabled rotational forms which could be accurately produced with sharp cutting tools and placed in viable bone support (Fig. 2).

Simplified models in computerized finite element stress analysis demonstrated that such a design would not impose unbearably high strain levels on the bone (Fig. 3), but would nevertheless apply stresses adequate to encourage and maintain bone formation. Further conventional mechanical computations and computer finite element stress analyses confirmed that component integrity could be achieved using commercially pure titanium (Fig. 4).

Mechanical tests and strain gauge studies in bone analogs and human bone supported these calculations, and human cadaver tests demonstrated more physiologic loadings with this geometry than a conventional cemented stem [142]. The mechanical integrity and durability of

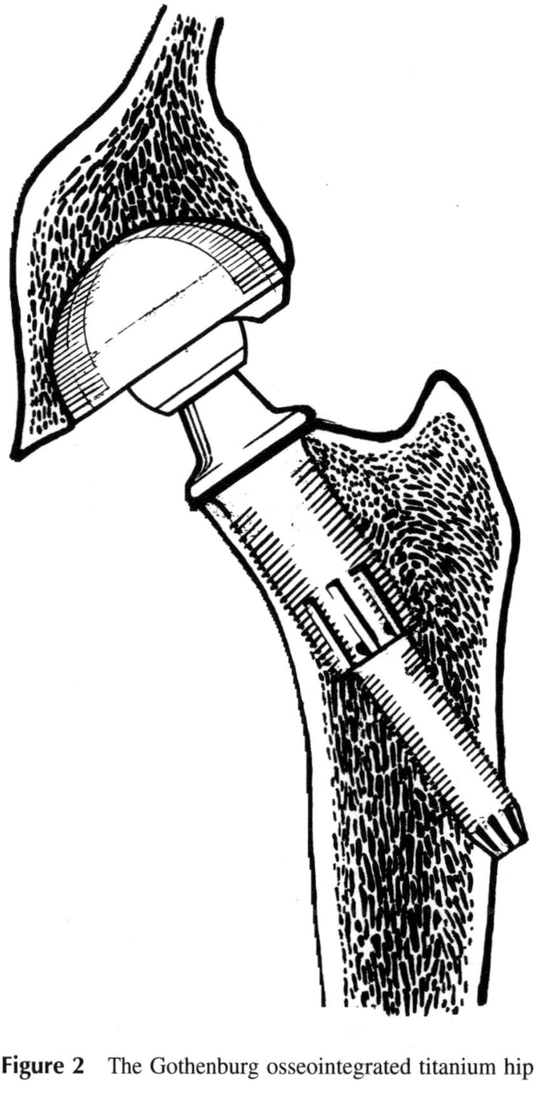

Figure 2 The Gothenburg osseointegrated titanium hip system.

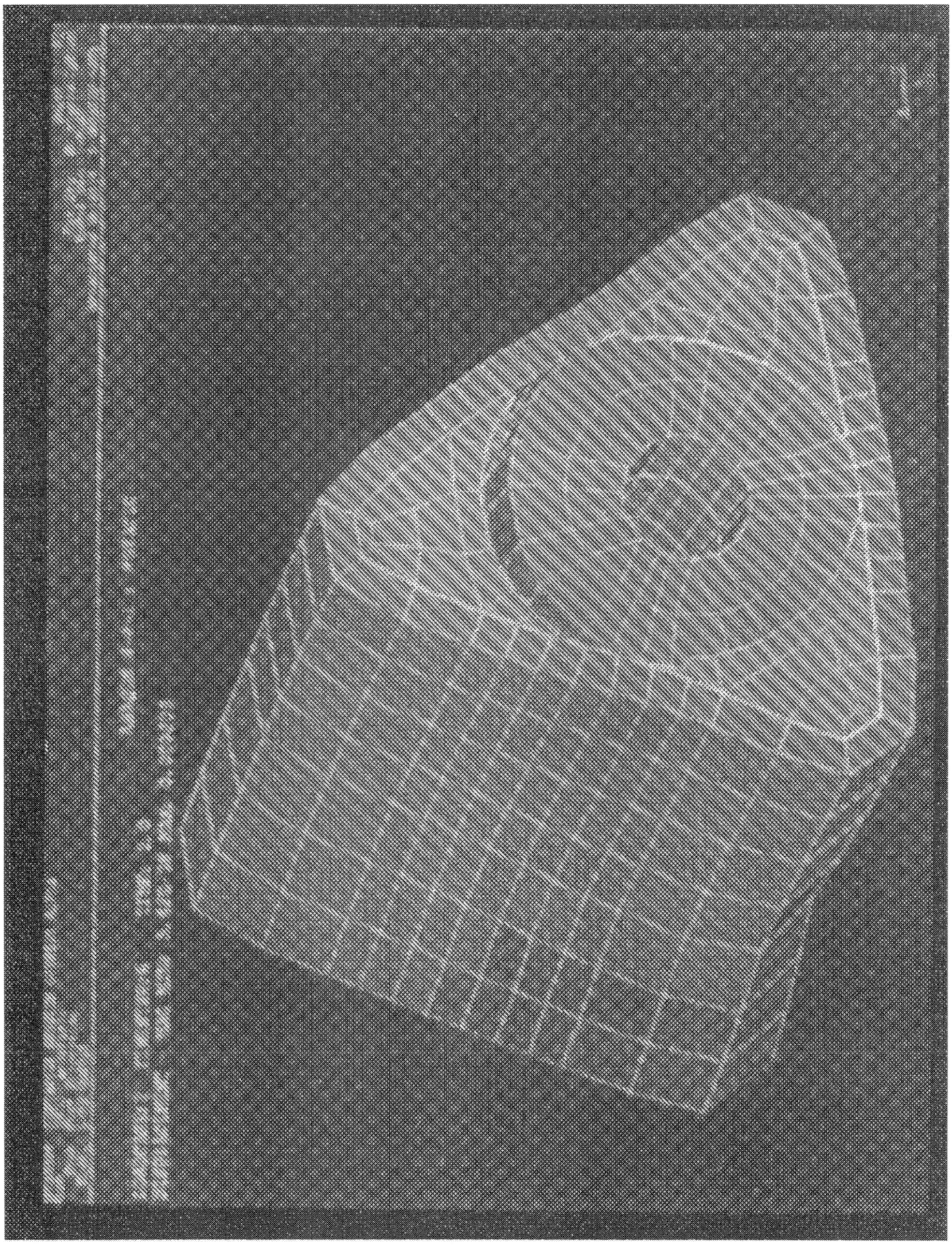

Figure 3 Finite element modeling of bone strains produced by proximal femoral osseointegrated fixation.

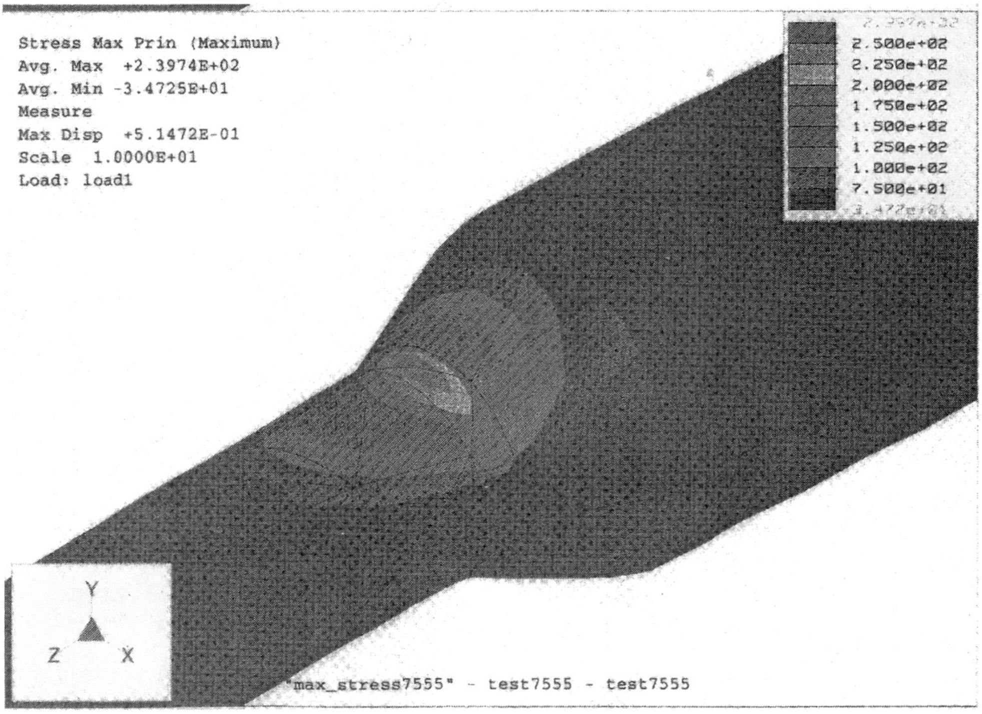

Figure 4 Finite element modeling of component strains.

the components was tested under single cycle and repetitive cyclic conditions to 10 million cycles (equivalent to at least 10 years' service in an arthroplasty patient) [143].

This material (c.p. Ti) has not been used before for the entire component of an orthopedic implant, and it was therefore necessary to validate all component interactions within the system. For the best wear resistance of the hip bearing, the bearing combination of a zirconia femoral head and an UHMWPE acetabular liner was selected. The torsional stability of zirconia heads on c.p. Ti taper trunnions was validated [144], and fretting tests of the taper fit were also undertaken [145]. These actually showed less fretting wear under cyclic loading for the zirconia/c.p. Ti combination than comparable zirconia/CoCr and CoCr/CoCr combinations.

The acetabular component was also designed and developed according to the principles of osseointegration, which therefore required cadaveric and bone analog tests of the implant stability [146]. Accuracy of surgery and proximity to bone surfaces as implanted were also important in this component, and the accuracy of the instruments developed in conjunction with the system was shown to be much better than conventional surgical preparation [147]. Retention strengths of the UHMWPE liner within the acetabular component were determined under a range of static and repetitive cyclic loading conditions [148].

3. Practical Experience

A pilot clinical trial with the GOT implant was started in 1992 at one center. Small groups of patients (4 to 5) were operated on and followed for 6 to 12 months in order to evaluate implant and instrumentation performance and surgical technique [149]. On the basis of these developments, the instrumentation and surgical procedure were amended, and a multicenter study was commenced in 1997.

The multicenter trial compared the novel GOT implant with a Harris Galante II cup and the Spectron EF femoral component. The latter cemented femoral stem and cementless cup were selected due to their excellent clinical record in the Swedish National Hip Registry [1]. The hybrid hip combination of an uncemented cup and a cemented stem were recommended by the National Institutes of Health Consensus Statement [150].

Fifty-four hips in 53 patients were included in the study, 26 women and 27 men. One patient was bilaterally operated with a GOT device on one side and the control on the other. The mean age was 59 years (44–71) at the time of surgery. The patients were randomized to receive either the GOT or Harris Galante II/Spectron arthroplasty (Fig. 5).

Clinical evaluations were performed pre- and postoperatively, after 3, 6, and 12 months, and then annually using a standard data form. Harris Hip Score [151] was calculated preoperatively and at any annual follow up. A visual analog scale for pain during activity and rest was used to describe pain in addition to the pain description according to Johnston and colleagues [152]. Migration of the implants was measured with RSA [153], which has been shown to give accurate early prediction of long-term stability in joint replacements [6,154]. Radiographs for RSA were taken immediately postoperatively and also 3, 6, and 12 months thereafter annually.

The Harris Hip Score rose from 48 to 98 and 97 for the GOT implant and control, respectively, which is comparable with other authors' findings for short-term follow-up of most hip arthroplasties. Likewise, RSA migration results were similar for the GOT group and the Spectron/Harris Galante II control group, for both the femoral components and the acetabular cups [155]. Undertaking a prospective randomized controlled clinical trial of this nature places the GOT implant system among one of the few systems to have such testing performed before wider release [156].

A further, broader multicenter clinical trial is being undertaken at present, with results to be published in 2004.

4. Conclusions

1. The principles of osseointegration can be applied to fixation of orthopedic implants, with immediate and longer-term benefits for implant survival.
2. Osseointegration enables design of implant components to more physiological performance criteria, while maintaining necessary implant endurance and performance standards.
3. Benefits are also obtained in surgical technique and instrumentation by adopting such an approach.
4. Prospective randomized clinical trials of this implant system show comparable performance to a control system chosen on the basis of the Swedish Hip Register results.

ACKNOWLEDGMENTS

The authors thank many collaborators throughout the development of the system, including Björn Albrektsson, Björn EJ Albrektsson, Lars Regnér, Tord Röstlund, and Lars Weidenhielm. Stig Wennberg's continued technical contribution with design and preparation of implants and instruments, the engineering input of Peter Elmberg and Lars-Gunnar Wässing, and the excellent graphics of Yutaka Kataoka are also gratefully acknowledged.

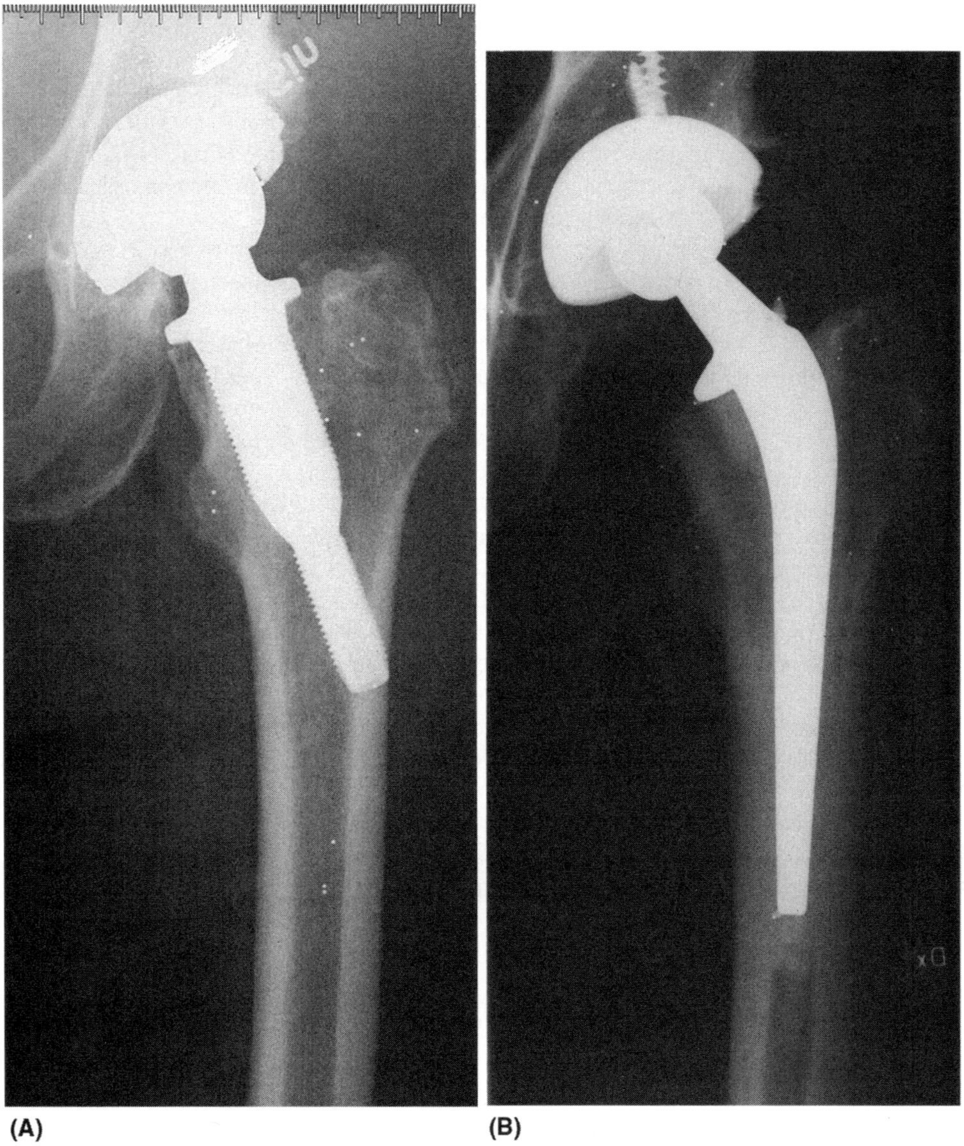

(A) (B)

Figure 5 Radiographs of clinical trial patients; (A) with GOT implant system; (B) with Spectron EF stem and Harris Galante II cup.

REFERENCES

1. Malchau H, Herberts P, Söderman P, Odén A. Prognosis of total hip replacement. Update and validation of results from the Swedish National Hip Arthroplasty Registry 1979–1998. In *67th Annual Meeting of the American Academy of Orthopaedic Surgeons*. Orlando, USA, Department of Orthopaedics, Göteborg University, Sweden, 2000.
2. Charnley J. The bonding of prosthesis to bone by cement. J Bone Joint Surg 1964; 46-B(3):518–529.

3. Kershaw C J, Atkins R M, Dodd C A F, Bulstrode C J K. Revision total hip arthroplasty for aseptic failure. A review of 276 cases. J Bone Joint Surg 1991; 73-B(4):564–568.
4. Goldring S R, Schiller A L, Roelke M, Rourke C M, O'Neill D A, Harris W H. The synovial-like membrane at the bone–cement interface in loose total hip replacements and its proposed role in bone lysis. J Bone Joint Surg 1983; 65-A:575–584.
5. Mjöberg B, Selvik G, Hansson L I, Rosenqvust R, Önnerfält R. Mechanical loosening of total hip prostheses. A radiographic and roentgen stereophotogrammetric study. J Bone Joint Surg 1986; 68-B(5):770–774.
6. Kärrholm J, Borssen B, Löwenheim G, Snorrason F. Does early micromotion of femoral stem prostheses matter? 4–7 year stereoradiographic follow-up of 84 cemented prostheses. J Bone Joint Surg 1994; 76-B(6):912–917.
7. Freeman M A R-, Bordeneuve P. Early migration and late aseptic failure of proximal femoral prostheses. J Bone Joint Surg 1994; 76-B(3):432–438.
8. Ryd L, Albrektsson B E J, Carlsson L, Dansgard F, Herberts P, Lindstrand A, Regner L, Toksvig-Larsen S. Roentgen stereophotogrammetric analysis as a predictor of mechanical loosening of knee prostheses. J Bone Joint Surg 1995; 77-B(3):377–383.
9. Walker P S, Mai S F, Cobb A G, Bentley G, Hua J. Prediction of clinical outcome of THR from migration measurements on standard radiographs. J Bone Joint Surg 1995; 77-B(5):705–714.
10. Charnley J F, Charnley F M, Hammond B T. The long-term reaction of bone to self-curing acrylic cement. J Bone Joint Surg 1968; 50-B:822–829.
11. Wilson J N, Scales J T. Loosening of total hip replacements with cement fixation: clinical findings and laboratory studies. Clin Orthop 1970; 72:145–160.
12. Insall J B, Ranawat C S, Aglietti P, Shine J. A comparison of four models of total knee replacement prostheses. J Bone Joint Surg 1976; 58-A:754–765.
13. Skolnik M D, Coventry M B, Ilstrup D M. Geometric total knee arthroplasty. J Bone Joint Surg 1976; 58-A:749–753.
14. Reckling F S, Asher M A, Dillon W L. A longitudinal study of the radiolucent line at the bone–cement interface following total joint replacement procedures. J Bone Joint Surg 1977; 59-A(3): 355–358.
15. Anderson G B J, Freeman M A R, Swanson S A V. Loosening of the cemented acetabular cup in total hip replacement. J Bone Joint Surg 1972; 54-B:590–599.
16. Ring P A. Total replacement of the hip. Clin Orthop 1970; 72:161–168.
17. Jones L C, Hungerford D S. Cement disease. Clin Orthop 1987; 225:192–206.
18. Willert H G, Ludwig J, Semlitsch M. Reaction of bone to methacrylate after hip arthroplasty. J Bone Joint Surg 1974; 56-A:1368–1382.
19. Mjöberg B. Loosening of the cemented hip prosthesis. The importance of heat injury. Acta Orthop Scand 1986; 57(Suppl 221):1–40.
20. Toksvig-Larsen S, Franzén H, Ryd L. Cement interface temperature in hip arthroplasty. Acta Orthop Scand 1991; 62:102–105.
21. Eriksson R A, Albrektsson T. The effect of heat on regeneration. An experimental study in the rabbit using the bone growth chamber. J Oral Maxillofac Surg 1984; 42:707–711.
22. Feith R. Side-effects of acrylic cement implanted into bone. Acta Orthop Scand Suppl. 1975(161): 3–136.
23. Linder L G, Harthon L, Kullberg L. Monomer leakage from polymerizing acrylic bone cement. An in vitro study on the influence of speed and duration of mixing, cement volume and surface area. Clin Orthop 1976(119):242–249.
24. Linder L, Romanus M. Acute local tissue effects of polymerizing acrylic bone cement. An intravital microscopic study in the hamster's cheek pouch on the chemically induced microvascular changes. Clin Orthop 1976(115):303–312.
25. Linder L. Tissue reaction to methyl methacrylate monomer. A comparative study in the rabbit's ear on the toxicity of methyl methacrylate monomer of varying composition. Acta Orthop Scand 1976; 47(1):3–10.
26. Stürup J, Nimb L, Kramhoft M, Steen-Jensen J. Effects of polymerization heat and monomers from acrylic cement on canine bone. Acta Orthop Scand 1994; 65:20–23.

27. Morberg P, Albrektsson T. Bone marrow pressure chamber: a permanently inserted titanium implant for intramedullary pressure measurements. J Biomed Eng 1986; 8(4):358–360.
28. Morberg P, Albrektsson T. Removal torque for bone-cement and titanium screws implanted in rabbits. Acta Orthop Scand 1991; 62(6):554–556.
29. Morberg P, Albrektsson T. Bone reactions to intramedullary insertion of methylmethacrylate. Eur J Musculoskel Res 1992; 1(1):11–17.
30. Slooff T J J H. The influence of acrylic cement. Acta Orthop Scand 1971; 42:465–481.
31. Rhinelander F W, Nelson C L, Stewart R D, Stewart B S. Experimental reaming of the proximal femur and acrylic cement implantation. Clin Orthop 1979; 141:74–89.
32. Gruen T A, McNiece G M, Amstutz H C. ''Modes of failure'' of cemented stem type femoral components. Clin Orthop 1979; 141:17–27.
33. Huddlestone H D. Femoral lysis after cemented hip arthroplasty. J Arthroplasty 1988; 3(4):285–297.
34. Harris W H, Schiller A L, Scholler J M, Freiberg R A, Scott R. Extensive localized bone resorption in the femur following total hip replacement. J Bone Joint Surg 1976; 58-A(5):612–618.
35. Jasty M J, Floyd W E, Schiller A L, Goldring S R, Harris W H. Localized osteolysis in stable, nonseptic total hip replacement. J Bone Joint Surg 1986; 68-A(6):912–919.
36. Maloney W J, Jasty M, Rosenberg A, Harris W H. Bone lysis in well-fixed cemented femoral components. J Bone Joint Surg 1990; 72-B(6):966–970.
37. Moore A T, Bohlman H R. Metal hip joint: a case report. J Bone Joint Surg 1943; 25:633.
38. Moore A T. The self-locking metal hip prosthesis. J Bone Joint Surg 1957; 39:811.
39. Ring P A. Complete replacement arthroplasty of the hip by the ring prosthesis. J Bone Joint Surg 1968; 50-B:720–731.
40. Lennox D W, McDonald D F, Riley L H J. A histologic comparison of aseptic loosening of cemented, press-fit, and biologic ingrowth prostheses. Clin Orthop 1987; 225:171–191.
41. Maloney W J, Jasty M, Harris W H, Galante J O, Callaghan J J. Endosteal erosion in association with stable uncemented femoral components. J Bone Joint Surg 1990; 72-A(7):1025–1034.
42. Maloney W J, Peters P, Engh C A, Chandler H. Severe osteolysis of the pelvis in association with acetabular replacement without cement. J Bone Joint Surg 1993; 75-A(11):1627–1635.
43. Jasty M, Maloney W J, Bragdon C R, O'Connor D, Haire T, Harris W H. The initiation of failure in cemented femoral components of hip arthroplasties. J Bone Joint Surg 1991; 73-B(4):551–558.
44. Jasty M, Bragdon C, Jiranek W, Chandler H, Maloney W, Harris W H. Etiology of osteolysis around porous-coated cementless total hip arthroplasties. Clin Orthop 1994; 308:111–126.
45. Willert H-G, Semlitsch M. Tissue reactions to plastic and metallic wear products of joint endoprostheses, in *Total Hip Prostheses* Gschwend N, Debrunner HU, eds: H Huber: Bern., 1976:205.
46. Willert H G, Semlitsch M. Reactions of the articular capsule to wear products of artificial joint prostheses. J Biomed Mater Res 1977; 11:157–164.
47. Anthony P P, Gie G A, Howie C R, Ling R S M. Localised endosteal bone lysis in relation to the femoral components of cemented total hip arthroplasties. J Bone Joint Surg 1990; 72-B(6):971–979.
48. Revell P A, Weightman B, Freeman M A R, Roberts B V. The production and biology of polyethylene wear debris. Arch. Orth. Trauma Surg 1978; 91:167–181.
49. Lombardi S V, Mallory T H, Vaughan B K, Drouillard P. Aseptic loosening in total hip arthroplasty secondary to osteolysis induced by wear debris from titanium-alloy modular femoral heads. J Bone Joint Surg 1989; 71A:1337–1342.
50. Witt J D, Swann M. Metal wear and tissue response in failed titanium alloy total hip replacements. J Bone Joint Surg 1991; 73-B:559–563.
51. Agins H J, Alcock N W, Bansal M, Salvati E A, Wilson P D, Pellici P M, Bullough P G. Metallic wear in failed titanium-alloy total hip replacements; a histological and quantitative analysis. J Bone Joint Surg 1988; 70-A:347–356.
52. Langlais F, Postel M, Berry J P, LeCharpentier Y, Weill B J. L'intolérance aux débris d'usure des prothèses. Bilan immunologique et anatomopathologique de 30 cas. International Orthopaedics (SICOT) 1980; 4:145–153.
53. Howie D W, Vernon-Roberts B, Oakeshott R, Manthey B. A rat model of resorption of bone at the cement bone interface in the presence of polyethylene wear particules. J Bone Joint Surg 1988; 70A:257–263.

54. Goodman S B, Fornasier V L, Kei J. The effects of bulk versus particulate polymethylmethacrylate on bone. Clin Orthop 1988; 232:255–262.
55. Goodman S B, Fornasier V L, Kei J. The effects of bulk versus particulate ultra-high-molecular-weight polyethylene on bone. J Arthroplasty 1988; October Suppl.:S41–S46.
56. Rae T. A study on the effects of particulate metals of orthopaedic interest on murine macrophages in vitro. J Bone Joint Surg 1975; 57-B:444–450.
57. Rae T. The toxicity of metals used in orthopaedic prostheses. An experimental study using cultured human synovial fibroblasts. J Bone Joint Surg 1981; 63-B:435–440.
58. Howie D W, Vernon-Roberts B. The synovial response to intraarticular cobalt–chrome wear particles. Clin Orthop 1988; 232:244–254.
59. Goodman S B, Fornasier V L, Lee J, Kei J. The effects of bulk versus particulate titanium and cobalt chrome alloy implanted into the rabbit tibia. J Biomed Mater Res 1990; 24:1539–1549.
60. Carlsson L V. On the development of a new concept for orthopaedic implant fixation, PhD, Thesis Göteborgs Universitet, Gothenburg, Sweden, 1989.
61. Van Der Vis H M, Marti R K, Tigchelaar W, Schuller H M, Van Noorden C J F. Benign cellular responses in rats to different wear particles in intra-articular and intramedullary environments. J Bone Joint Surg 1997; 79-B(5):837–843.
62. Coleman D L, King R N, Andrade J D. The foreign body reaction: a chronic inflammatory response. J Biomed Mater Res 1974; 8:199–211.
63. Charnley J. Tissue reactions to polytetrafluorethylene. Lancet 1963; 2:1379.
64. Murray D W, Rae T, Rushton N. The influence of the surface energy and roughness of implants on bone resorption. J Bone Joint Surg 1989; 71-B(4):632–637.
65. Williams D F, Meachim G. A combined metallurgical and histological study of tissue–prosthesis interactions in orthopaedic patients. J Biomed Mater Res 1974; 5(Symp. 1):1–9.
66. Winter G D. Tissue reactions to metallic wear and corrosion products in human patients. J Biomed Mater Res 1974; 5(Symp 1):11–26.
67. Gruen T A, Amstutz H C. A failed vitallium/stainless steel total hip replacement: a case report with histological and metallurgical examination. J Biomed Mater Res 1975; 9:465.
68. Therin M, Meunier A, Christel P. A histomorphometric comparison of the muscular tissue reaction to stainless steel, pure titanium and titanium alloy implant materials. J Mater Sci: Mater Med 1991; 2:1–8.
69. Dobbs H S, Scales J T. Behavior of commercially pure titanium and Ti-318 (Ti-6Al-4V) in orthopedic implants. in *ASTM Symposium STP 796*, Phoenix, AZ.: ASTM. 1981.
70. Emnèus H, Stenram U, Baecklund J. An x-ray spectrographic investigation of the soft tissue around titanium and cobalt alloy implants. Acta Orthop Scand 1960; 30:226.
71. Ferguson A B J, Laing P G, Hodge E S. The ionization of metal implants in living tissues. J Bone Joint Surg 1960; 42-A(1):77–90.
72. Ferguson A B J, Akahoshi Y, Laing P G, Hodge E S. Characteristics of trace ions released from embedded metal implants in the rabbit. J Bone Joint Surg 1962; 44-A(2):323–326.
73. Schider S, Bildstein H. Tantalum and niobium as potential prosthetic materials, in *Biomaterials* Winter GD, Gibbons DF, Plenk HJ, eds. 1980:13–20.
74. Albrektsson T, Brånemark P-I, Hansson H A, Kasemo B, Larsson K, Lundström I, McQueen D, Skalak R. The interface zone of inorganic implants in vivo: titanium implants in bone. Ann Biomed Eng 1983; 11:1–27.
75. Griffiths H G, Burke J, Bonfiglio T A. Granulomatous pseudotumors in total joint replacement. Skel. Radiol 1987; 16:146–152.
76. Santavirta S, Konttinen Y T, Bergroth V, Eskola A, Tallroth K, Lindholm T S. Aggressive granulomatous lesions associated with hip arthroplasty. J Bone Joint Surg 1990; 72-A(2):252–258.
77. Brown I W, Ring P A. Osteolytic changes in the upper femoral shaft following porous-coated hip replacement. J Bone Joint Surg 1985; 67-B(2):218–221.
78. Engh C A, Bobyn J D, Glassman A H. Porous-coated hip replacement. The factors governing bone ingrowth, stress shielding, and clinical results. J Bone Joint Surg [Br] 1987; 69(1):45–55.
79. Engh C A, Bobyn J D. The influence of stem size and extent of porous coating on femoral bone resorption after primary cementless hip arthroplasty. Clin Orthop 1988(231):7–28.

80. Martell J M, Pierson R H I, Jacobs J J, Rosenberg A G, Maley M, Galante J O. Primary total hip reconstruction with a titanium fiber–coated prosthesis inserted without cement. J Bone Joint Surg 1993; 75-A(4):554–571.
81. Robinson R P, Deysine G R, Green T M. Uncemented total hip arthroplasty using the CLS stem: a titanium alloy implant with a corundum blast finish. J Arthroplasty 1996; 11(3):286–292.
82. Carlsson Å S, Gentz C-F. Mechanical loosening of the femoral head prosthesis in the Charnley total hip arthroplasty. Clin Orthop 1980; 147:262–270.
83. Sarmiento A, Gruen T A. Radiographic analysis of a low-modulus titanium-alloy femoral total hip component: two- to six-year follow-up. J Bone Joint Surg 1985; 67-A(1):48–56.
84. Cohen B, Rushton N. Bone remodelling in the proximal femur after Charnley total hip arthroplasty. J Bone Joint Surg 1995; 77-B(5):815–819.
85. Kobayashi S, Eftekhar N S, Terayama K. Long term bone remodeling around the Charnley femoral prostheses. Clin Orthop 1996; 326:162–173.
86. Blacker G J, Charnley J. Changes in the upper femur after low friction arthroplasty. Clin Orthop 1978; 137:15–23.
87. McCarthy C K, Steinberg G G, Agren M, Leahy D, Wyman E, Baran D T. Quantifying bone loss from the proximal femur after total hip arthroplasty. J Bone Joint Surg 1991; 73-B(5):774–778.
88. Engh C A, McGovern T F, Bobyn J D, Harris W H. A quantitative evaluation of periprosthetic bone remodelling after primary cementless total hip arthroplasty. J Bone Joint Surg 1992; 74-A: 1009–1020.
89. Gill T J, Sledge J B, Orler R, Ganz R. Lateral insufficiency fractures of the femur caused by osteopenia and varus angulation. J Arthroplasty 1999; 14(8):982–987.
90. Weightman B, Brown I. The stress analysis of straight-stemmed femoral components for total hip replacements. J Med Eng Technol 1982; 6(1):25–27.
91. Lewis J L, Askew M J, Wixson R L, Kramer G M, Tarr R R. The influence of prosthetic stem stiffness and of a calcar collar on stresses in the proximal end of the femur with a cemented femoral component. J Bone Joint Surg 1984; 66-A(2):280–286.
92. Weinans H, Huiskes R, Grootenboer H J. Effects of material properties of femoral hip components on bone remodeling. J Orthop Res 1992; 10(6):845–853.
93. Weinans H, Huiskes R, Grootenboer H J. Effects of fit and bonding characteristics of femoral stems on adaptive bone remodeling. J Biomech Eng 1994; 116(4):393–400.
94. Vail T P, Glisson R R, Koukoubis T D, Guilak F. The effect of hip stem material modulus on surface strain in human femora. J Biomech 1998:(31).
95. Aspenberg P, Herbertsson P. Periprosthetic bone resorption. Particles versus movement. J Bone Joint Surg 1996; 78-B(4):641–646.
96. van der Vis H M, Aspenberg P, Martii R K, Tigchelaar W, Van Noorden C J F. Fluid pressure causes bone resorption in a rabbit model of prosthetic loosening. Clin Orthop 1998; 350:201–208.
97. Schmalzried T P, Kwong L M, Jasty M, Sedlacek R C, Haire T C, O'Connor D O, Bragdon C R, Kabo J M, Malcolm A J, Harris W H. The mechanism of loosening of cemented acetabular components in total hip arthroplasty. Analysis of specimens retrieved at autopsy. Clin Orthop 1992(274):60–78.
98. Albrektsson T, Brånemark P-I, Hansson H-A, Lindström J. Osseointegrated titanium implants. Requirements for ensuring a long-lasting, direct bone-to-implant anchorage in man. Acta Orthop Scand 1981; 52:155–170.
99. Brånemark P-I. Introduction to osseointegration, in *Tissue Integrated Prosthesis. Osseointegration in Clinical Dentistry*. Brånemark P-I, Zarb GA, Albrektsson T, eds: Quintessence: Berlin, 1985: 11–76.
100. Brånemark P-I, Hansson B-O, Adell R, Breine U, Lindström J, Hallén O, Öhman A. Osseointegrated implants in the treatment of the edentulous jaw. Experience from a 10-year period. Scand J Plast Reconstr Surg 1977; 16:16–21.
101. Adell R, Eriksson B, Lekholm U, Brånemark P-I, Jemt T. A long-term follow-up study of osseointegrated implants in the treatment of totally edentulous jaws. Int J Oral Maxillofac Implants 1990; 5:347–359.
102. Albrektsson T, Dahl E, Enbom L, Engevall S, Engquist B, Eriksson R A, Feldmann G, Freiberg N, Glantz P-I, Kjellman P, KJristersson L, Kvint S, Köndell P-Å, Palmquist J, Werndahl L, Åstrand

P. Osseointegrated oral implants. A Swedish multicenter study of 8139 consecutively inserted Nobelpharma implants. J Periodon 1988; 59:287–296.
103. Albrektsson T, Brånemark P-I, Jacobsson M, Tjellström A. Present clinical applications of osseointegrated percutaneous implants. Plast Reconstr Surg 1987; 79:721–730.
104. Tjellström A. Eight years' experience with titanium implants for craniofacial rehabilitation, in *Tissue Integration in Oral and Maxillofacial Construction* van Steenberghe D, ed. Amsterdam: Excerpta Medica, Elsevier Science, 1986:182–186.
105. Roos J, Sennerby L, Lekholm U, Hemt T, Gröndahl K, Albrektsson T. A qualitative and quantitative method for evaluating implant success: a 5-year retrospective analyisis of the Brånemark implant. Int J Oral Maxillofac Implants 1997; 12:504–514.
106. Albrektsson T, Albrektsson B. Osseointegration of bone implants. Acta Orthop Scand 1987; 58: 567–577.
107. Carlsson L V, Röstlund T, Albrektsson B, Albrektsson T, Brånemark P-I. Osseointegration of titanium implants. Acta Orthop Scand 1986; 57:285–289.
108. Johansson C B. On tissue reactions to metal implants, PhD Thesis, Göteborg Universitet, Sweden, 1991.
109. Sennerby L. On the bone tissue response to titanium implants, PhD Thesis, University of Gothenburg, Gothenburg, Sweden, 1991.
110. Carlsson L V, Röstlund T, Albrektsson B, Albrektsson T. Implant fixation improved by close fit. Cylindrical implant–bone interface studied in rabbits. Acta Orthop Scand 1988; 59(3):272–275.
111. Pidhorz L E, Urban R M, Jacobs J J, Sumner D R, Galante J O. A quantitative study of bone and soft tissues in cementless porous-coated acetabular components retrieved at autopsy. J Arthroplasty 1993; 8(2):213–225.
112. MacKenzie J K, Callaghan J J, Pedersen D R, Brown T D. Areas of contact and extent of gaps with implantation of oversized acetabular components in total hip arthroplasty. Clin Orthop 1994; 298:127–136.
113. Incavo S J, Ames S E, DiFazio F A, Howe J G. Cementless hemispheric acetabular components. A 4- to 8-year follow-up report. J Arthroplasty 1996; 11(3):298–303.
114. Walker P S, Schneeweis D, Murphy S, Nelson P. Strains and micromotions of press-fit femoral stem prostheses. J Biomechanics 1987; 20(7):693–702.
115. Schneider E, Kinast C, Eulenberger J, Wyder D, Eskilsson G, Perren S M. A comparative study of the initial stability of cementless hip prostheses. Clin. Orthop & Rel. Res 1989; 248:(200–209).
116. Burke D W, O'Connor D O, Zalenski E B, Jasty M, Harris W H. Micromotion of cemented and uncemented femoral components. J Bone Joint Surg 1991; 73(Br):33–37.
117. Ivanoff C-J, Sennerby L, Lekholm U. Influence of initial implant mobility on the integration of titanium implants. An experimental study in rabbits. Clin Oral Implant Res 1996; 7:120–127.
118. Carlsson L V, Röstlund T, Albrektsson B, Albrektsson T. Removal torques for polished and rough titanium implants. Int J Oral Maxillofac Implants 1988; 3(1):21–24.
119. Wennerberg A, Albrektsson T, Andersson B, Krol J J. A histomorphometric and removal torque study of screw-shaped titanium implants with three different surface topographies. Clin Oral Implant Res 1995; 6:24–30.
120. Wennerberg A, Albrektsson T, Johansson C, Andersson B. Experimental study of turned and grit-blasted screw-shaped implants with special emphasis on effects of blasting material and surface tography. Biomaterials 1996; 17:15–22.
121. Wennerberg A, Albrektsson T, Andersson B. An animal study of c.p. titanium screws with different surface topographies. J Mater Sci Mater Med 1995; 6:302–309.
122. Wennerberg A, Albrektsson T, Lausmaa J. Torque and histomorphometric evaluation of c.p. titanium screws, blasted with 25 and 75 μm sized particles of Al_2O_3. J Biomed Mater Res 1996; 30:251–260.
123. Wennerberg A, Ektessabi A, Andersson B, Albrektsson T. A 1-year follow-up of implants of differing surface roughness placed in rabbit bone. Int J Oral Maxillofac Implants 1997; 12:486–494.
124. Cohen J. Tissue reactions to metals. The influence of surface finish. J Bone Joint Surg 1961; 43-A:687–699.

125. Gotfredsen K, Hjørting-Hansen E, Jensen J S, Holmén A. Histomorphometric and removal torque analysis for TiO_2-blasted titanium implants. An experimental study on dogs. Clin Oral Implant Res 1992; 3:77–84.
126. Wong M, Eulenberger J, Schenk R, Hunziker E. Effect of surface topology on the osseointegration of implant materials in trabecular bone. J Biomed Mater Res 1995; 29:1567–1575.
127. Friedman J, An Y H, Ming J, Draughn R A, Bauer T W. Influence of biomaterial surface texture on bone ingrowth in the rabbit femur. J Orthop Res 1996; 14(3):455–464.
128. Wennerberg A. On surface roughness and implant incorporation. Göteborg. Sweden: Biomaterials/ Handicap Research; Institute of Surgical Sciences, Göteborgs Universitet, 1996.
129. Ellingsen J E. Pre-treatment of titanium implants with fluoride improves their retention in bone. J Mater Sci. Mater Med 1995; 6:749–753.
130. Sul Y-T, Johansson C B, Röser K, Albrektsson T. Qualitative and quantitative observations of bone tissue reactions to anodised implants. Biomaterials 2002; 23:1809–1817.
131. Röstlund T, Carlsson L, Albrektsson B, Albrektsson T. Osseointegrated knee prostheses: an experimental study in rabbits. Scand J Plast Reconstr Surg 1989; 23:43–46.
132. Tarnow D P, Emtiaz S, Classi A. Immediate loading of threaded implants at stage 1 surgery in edentulous arches: ten consecutive case reports with 1- to 5-year data. Int J Oral Maxillofac Implants 1997; 12(3):319–324.
133. Becker W, Becker B E, Israelson H, Lucchini J P, Handelsman M, Ammons W, Rosenberg E, Rose L, Tucker L M, Lekholm U. One-step surgical placement of Branemark implants: a prospective multicenter clinical study. Int J Oral Maxillofac Implants 1997; 12(4):454–462.
134. Friberg B, Sennerby L, Linden B, Grondahl K, Lekholm U. Stability measurements of one-stage Branemark implants during healing in mandibles. A clinical resonance frequency analysis study. Int J Oral Maxillofac Surg 1999; 28(4):266–272.
135. Cooper L F, Scurria M S, Lang L A, Guckes A D, Moriarty J D, Felton D A. Treatment of edentulism using Astra Tech implants and ball abutments to retain mandibular overdentures. Int J Oral Maxillofac Implants 1999; 14(5):646–653.
136. Huiskes R, Weinans H, Dalstra H. Adaptive bone remodelling and biomechanical design considerations for noncemented total hip arthroplasty. Orthopedics 1989; 12:1255–1267.
137. Huiskes R. The various stress patterns of press-fit, ingrown, and cemented femoral stems. Clin Orthop 1990; 261:27–38.
138. Jacobs J J, Sumner D R, Galante J O. Mechanisms of bone loss associated with total hip replacement. Orthop Clin North Am 1993; 24(4):583–590.
139. Butel J, Robb J E. The isoelastic hip prosthesis followed for 5 years. Acta Orthop Scand 1988; 59: 258–262.
140. Skinner H B. Composite technology for total hip arhtroplasty. Clin Orthop 1988; 235:224–236.
141. Cameron H U. The 3–6 year results of a modular noncemented low-bending stiffness hip implant. J Arthroplasty 1993; 8:239–243.
142. Macdonald W, Carlsson L V, Jacobsson M, Lee T Q. A Proximal femoral implant preserves physiological bone deformation: a biomechanical investigation in cadaveric bones. Proc Instn Mech Engnrs 2002; 217-H:141–48.
143. Macdonald W, Carlsson L V, Gathercole N, Jacobsson M. (in press). Fatigue testing of a proximal femoral hip component. Proc Inst Mech Eng 2003.
144. Macdonald W, Aspenberg A, Jacobsson M, Carlsson L V. Friction in orthopaedic zirconia taper assemblies. Proc Inst Mech Eng 2000; 214 pt H:685–692.
145. Ebramzadeh E, Carlsson L V, Culwell J, Jacobsson C M, Macdonald W. Fretting damage at the Morse taper interface with three combinations of biomaterials. in *44th Annual Mtg. Orthop Res Soc*, New Orleans, 1998.
146. Macdonald W, Carlsson L V, Charnley G J, Jacobsson C M. Pres-fit acetabular cup fixation: principles and testing. Proc Inst Mech Eng 1999; 213 Part H:33–39.
147. Macdonald W, Carlsson L V, Charnley G J, Jacobsson C M. Inaccuracy of acetabular reaming under surgical conditions. J Arthroplasty 1999; 14(6):730–737.
148. Macdonald W, Aspenberg A, Jacobsson M, Carlsson L V. (in press). A novel liner locking mechanism enhances retention stability. Med Eng Phys 2003.

149. Albrektsson T, Carlsson L V, Jacobsson M, Macdonald W. Gothenburg Osseointegrated hip arthroplasty. Clin Orthop Rel Res 1998; 352:81–94.
150. NIH 1994, Total Hip Replacement, National Institutes of Health Consensus Statement.
151. Harris W H. Traumatic arthritis of the hip after dislocation and acetabular fractures. An end-result study using a new method of result evaluation. J Bone Joint Surg 1969; 51-A:737–755.
152. Johnston RC, Jr FRH, Harris WH, Poss R, Muller ME, Sledge CB. Clinical and radiographic evaluation of total hip replacement. A standard system of terminology for reporting results. J Bone Joint Surg 1990; 72-A:161–168.
153. Kärrholm J. Roentgen stereophotogrammetry. Review of orthopaedic applications. Acta Orthop Scand 1989; 60:491–503.
154. Kärrholm J, Herberts P, Hultmark P, Malchau H, Nivbrant B, Thanner J. Radiostereometry of hip prostheses. Methodology and prediction of end results. Clin Orthop 1997; 344:94–110.
155. Carlsson L V, Albrektsson B E J, Jacobsson M, Macdonald W, Regnér L, Weidenhielm L. (submitted). A prospective randomized clinical trial of the Gothenburg osseointegrated hip arthroplasty; a proximal femoral implant and a press-fit cup.
156. Murray D W, Carr A J, Bulstrode C J. Which primary total hip replacement? J Bone Joint Surg 1995; 77-B(4):520–527.

12

Recent Developments in Bone Cements

Kemal Serbetci
Baskent University
Ankara, Turkey

Nesrin Hasirci
Middle East Technical University
Ankara, Turkey

Bone cements are substances used to repair the damaged or diseased areas of bones or to fix a prosthesis in the bones. Generally acrylic cements made of polymers of methylmethacrylate (MMA) are used for cement production. The scientist who first synthesized polymethylmethacrylate (PMMA) probably did not envision that this process would be a significant step in medical science. The pilots of the World War II, who carried some glassy PMMA remnants in their body as a result of shattered cockpit windshields, could not know that they already were a part of in vivo experiments for biocompatibility of polymers.

Methylmethacrylate is an ester of methacrylic acid and has been intensively studied since the second quarter of the twentieth century. In 1928, a large-scale technical synthesis for MMA was established by the company Röhm and Haas, and the production technique was patented by Bauer (1935; patent DRP 652821). By 1936, the company Kulzer (1936; patent DRP 737058) had found that a dough produced by mixing ground polymethylmethacrylate powder and its liquid monomer that hardens upon addition of benzoyl peroxide (BPO) and heating the mixture to 100°C in a stone mould. The first clinical use of these PMMA mixtures was an attempt to close cranial defects in monkeys in 1938. When these experiences became known, surgeons were anxious to try these materials in plastic surgery on humans.

The heat curing polymer Paladon 65 was soon used for closing cranial defects in humans after producing plates in the laboratory and later fitting the hardened material on the spot (Kleinschmitt, 1941). After chemists discovered that the polymerization of MMA would occur by itself at room temperature if a coinitiator is added in addition to benzoyl peroxide, the companies Degussa and Kulzer (1943, patent DRP 973590), using tertiary aromatic amines, established a protocol for the chemical production of PMMA bone cements. These studies must be considered to be the birth of PMMA bone cements.

At the end of World War II, the worldwide practical use of PMMA studies started by Otto Röhm quickly spread. PMMA bone cements were developed independently in several countries. Even though their chemical bases were identical, the advantageous handling properties of MMA polymer mixtures have remained the subject of many research projects because the cements differed considerably in this respect. Kiaer first used PMMA as a pure anchoring material by fixing acrylic glass caps on the femoral head after removing the cartilage [1–3].

Studies on the use of the plastic materials in cranial applications were started with the large-scale technical production of the polymers [4]. The fast-curing resins were also used for filling defects caused by injuries to the visceral skeleton [5]. Judet and Judet [6] were the first to introduce an arthroplastic surgical method. Soon, however, it became apparent that the PMMA (Plexiglas®) prosthesis could not be integrated in the body because of both biological and mechanical reasons. In 1958, Sir John Charnley was the first researcher who succeeded in anchoring femoral head prostheses in femur with in site autopolymerization of PMMA [7]. Charnley called the material used ''bone cement on acrylic basis.'' His studies described a totally new surgical technique [8].

Today, most of the bone cements that are used in dentistry and orthopedic surgery are made of polymethylmethacrylate, and as a group they are called acrylic-based cements. These cements have been widely used for more than 30 years [7]. In dentistry, pure PMMA is used in prosthetic applications, and the compositions that contain PMMA are used as a filler material. In orthopedic surgeries, bone cements are mainly used as a filler (for filling cavities resulting from postoperative bone loss) and for anchoring the metallic implants securely into bone medulla. The cement serves, in addition to fixation, as an interfacial phase between the high-modulus metallic implant and low-stiffness natural bone as well as to transfer and distribute static and cyclic loads due to daily activities.

I. PREPARATION OF BONE CEMENT

Most of the commercially available bone cements are composed of two parts; a pack that contains PMMA powder with an initiator, and an ampoule that contains monomeric MMA liquid, an accelerator and inhibitor.

For preparing most of the bone cements, the powder-containing pouch is cut by a sterile scissors, and contents are put in a sterile bowl (Figs 1a and b). Then the liquid ampoule is opened, and contents poured on the powder (Fig. 1c). They are mixed in a sterile container until a homogeneous dough is obtained (1–3 min). Chemicals, which provide radioopacity, may or may not be added to the mixture. The dough is applied to the desired area by the surgeon when it gets a smooth and proper texture. Curing and hardening occurs in a few minutes.

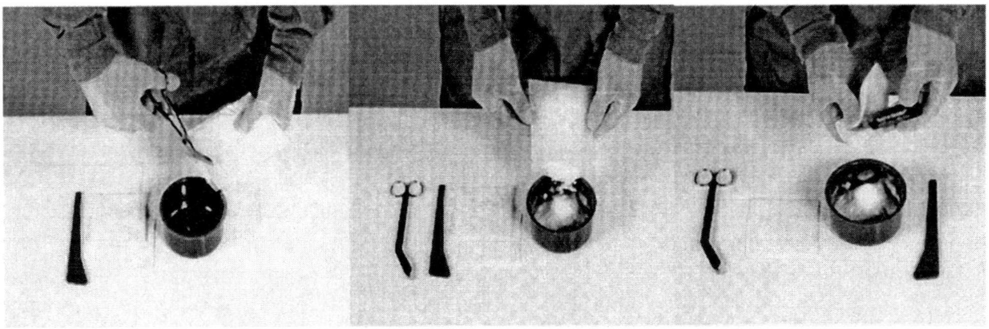

a b c

Figure 1 Preparation of cement dough. (a) Opening of the powder part; (b) introduction of powder into sterile bowl; (c) addition of liquid part.

In some operating rooms, vacuum and/or centrifugation is applied to the mixture to remove the gas bubbles (formed during mixing and polymerization) and to reduce the porosity of the cement dough. The curing process of the bone cement, known as cold curing, is the result of the free radical polymerization of MMA monomer.

The polymerization is initiated by the decomposition of initiator molecule, which is usually benzoyl peroxide. The reaction can be activated by the introduction of an accelerator, which is usually N,N-dimethyl-4-toluidine (DMPT).

During mixing of the solid and liquid components, PMMA powder partially dissolves in its monomer and gets entrapped in the newly formed polymer matrix. The mixture becomes stiff in few minutes (3–5 min). That allows enough time for the surgeon to apply the dough in situ for the primary fixation of a joint prosthesis. The surgeon manually presses or injects the cement dough into the application area in which the metal implant will be inserted. After the application, bone cement quickly hardens by an exothermic reaction. The cement location in a total hip replacement is shown in Fig. 2.

II. CHEMICAL COMPOSITION

Acrylic bone cements consist of two parts, powder and liquid components. The powder part contains essentially tiny polymer (usually PMMA) beads and an initiator such as benzoyl peroxide. Some commercial preparations include copolymers of methylmethacrylate with methacrylate (MA), styrene (S), butyl methacrylate (BMA), ethyl acrylate (EA), or ethyl methacrylate (EMA) in the powder part. Studies are being carried out on new copolymers of MMA with different comonomers like oleoyloxyethyl methacrylate (OMA) [9].

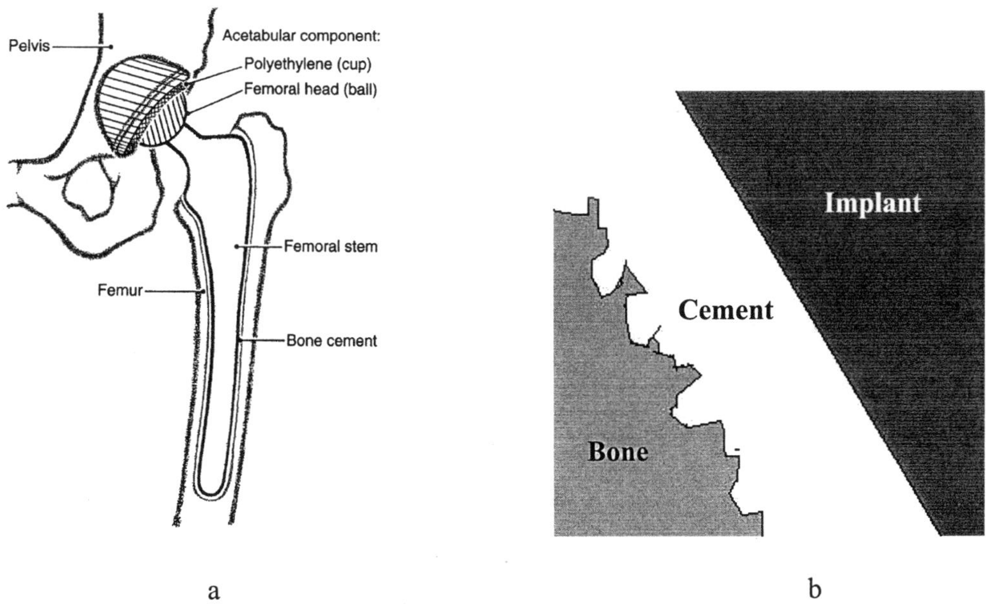

Figure 2 Total hip replacement with bone cement. (a) Schematic of implant and bone cement in hip and femur. (b) Interface between bone and implant.

Radioopacity is provided by barium sulfate ($BaSO_4$) or zirconium dioxide (ZrO_2), and they may be added to the powder component or they may be provided in a separate pack. The liquid part consists essentially of methylmethacrylate monomer. Some commercial products include butyl methacrylate, N-decyl methacrylate (DMA), or isobornyl methacrylate (IBMA) as comonomers. In order to accelerate polymerization reaction, an accelerator such as N,N-dimethyl-4-toluidine (DMPT) is also added. The liquid component also contains an inhibitor such as hydroquinone (HQ) to prevent self-polymerization of monomer before use and thus to extend the shelf-life of the cement. Some bone cements also contain a crosslinker agent, ethylene glycol dimethacrylate (EDGMA), in the liquid part. The chemical structures of MMA and PMMA are

Methylmethacrylate

$$H_2C{=}C(CH_3){-}COOCH_3$$

Polymethylmethacrylate

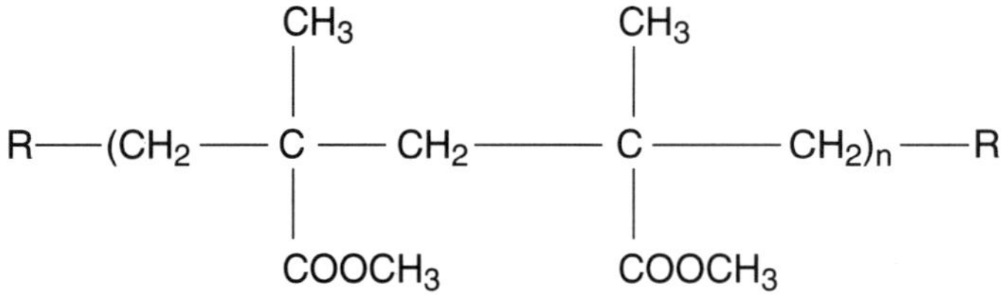

A. Polymer/Monomer Ratio

Most of the commercial bone cement formulations have a polymer/monomer ratio of 2:1 (w/v). When this ratio increases (by increasing the amount of PMMA powder or by decreasing the amount of MMA monomer), the viscosity of the cement dough increases. This causes difficulty in workability and creates problems in applying the dough deep in the bone. It also results in shorter growth time for the free radicals as well as a shorter time for the occurrence of the peak sample temperature with fast curing. On the other hand, smaller powder/liquid ratios give smaller peak concentrations of free radicals, larger peak temperatures, and higher monomer release to the surrounding tissues [10,11].

B. Initiator

Benzoyl peroxide is the most common initiator used in bone cement formulations and it is added in the powder component. There are some other initiators such as tri-n-butylborane (TBB) used in the commercial product Bonemite®. The chemical formulations of BPO and TBB are

Benzoyl peroxide

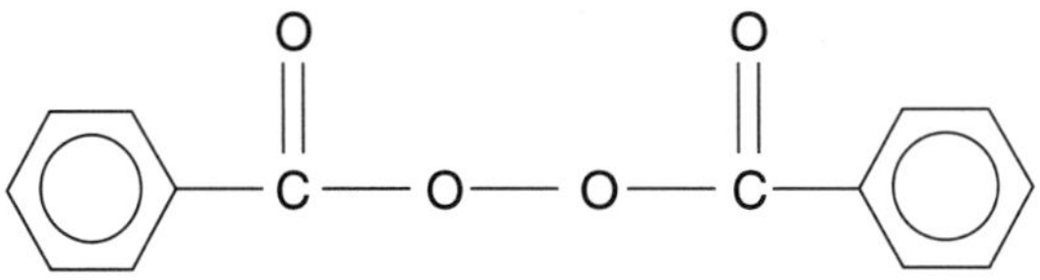

Tri-n-butylborane

$B(C_4H_9)_3$

After mixing the liquid and the powder parts, the initiator produces radicals at room temperature and starts the polymerization reaction causing an increase in temperature. Some authors suggested that the maximum temperature of TBB at curing is lower than that obtained with other conventional bone cements [12]. But at 10-year follow-up, no statistical differences were observed between the survival rates of the patients in which Bonemite or another bone cement, CMW 1, was used in total hip arthroplasty. The reaction mechanism for free radical formation by TBB is shown in Fig. 3.

C. Accelerator/Activator

Most of the bone cement formulations contain a tertiary aryl-amine, N,N-dimethyl-p-toluidine (DMPT), as an accelerator for polymerization reactions. DMPT is added in the liquid part and is water soluble to a small extent [13]. It causes the decomposition of the BPO in a reduction–oxidation process by electron transfer that produces a benzoyl radical and benzoate anion, as shown in Fig. 4.

It was shown that when the more reactive 4-N,N-(dimethylamino)phenethanol was used as accelerator, a higher radical concentration was observed in the polymerization system than with BPO, yielding cements with shorter setting times and increased strengths [14,15]. Tanzi et al. [16] studied replacing DMPT with unsaturated tertiary arylamines, such as acryloyl- (ANP) and methacryloyl- (MNP) N-phenylpiperazine. These two accelerators were expected to be less toxic. Compression tests revealed that compressive yield stress, strain at yield, and elastic modulus values were very similar to those of samples cured with DMPT and ANP, and slightly lower results were obtained with samples cured with MNP. In Boneloc®, a mixture of dihydroxypropyl-p-toluidine and N,N-dimethyl-p-toluidine was used as the accelerator system. Oldfield and Yasuda introduced 4-dimethylamino phenethyl alcohol (TDOH) as a more effective accelerator than DMPT for bone cement curing [15].

In Bonemite®, 2-5-dimethylhexane-2-5-hydroperoxide is used as the polymerization promoter [12]. Vazquez et al. studied the properties of different activators [17]. They used 4-dimethylaminobenzyl alcohol (DMOH) and 4-dimethylaminobenzyl methacrylate (DMMO) as the activators in the benzoyl peroxide–initiated redox polymerization for the preparation of acrylic bone cement based on polymethylmethacrylate beads of different particle sizes. They reported that the BPO/DMOH and BPO/DMMO redox systems had lower peak temperatures and longer setting times, and the cured materials presented higher average molecular weights

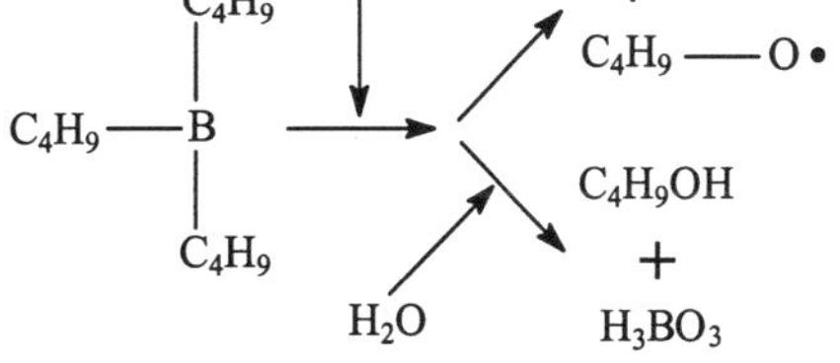

Figure 3 Free radical formation by tri-n-butylborane (TBB).

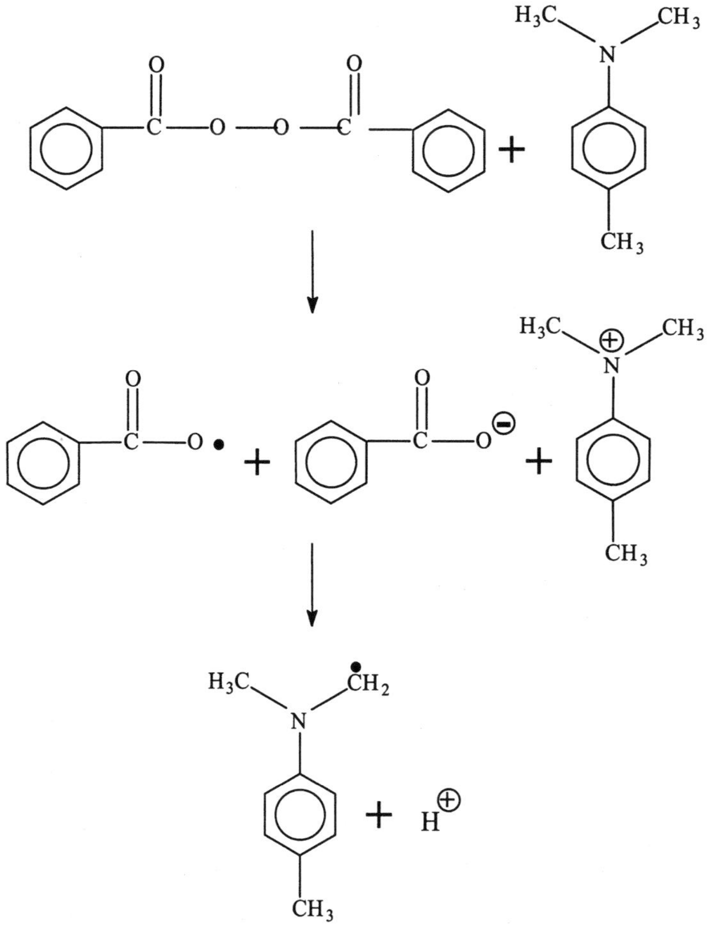

Figure 4 Reduction–oxidation process of benzoylperoxide (BPO) caused by N,N-dimethyl-p-toluidine (DMPT).

than those obtained with DMPT. In addition, they reported that these activators were three times less toxic than the classical DMPT [17]. Vazquez et al. also reported that when a long chain acid derivative bearing an aromatic tertiary amine group, 4-N,N-dimethylaminobenzyl laurate (DML), was used, lower peak temperatures and longer setting times were obtained without impairing the mechanical properties [18]. When the same group used a tertiary aromatic amine derived from oleic acid, 4-N,N-dimethylaminobenzyl oleate (DMAO), as the accelerator, they obtained a 20°C decrease in the polymerization temperature and a 7-min increase in setting time [9].

D. Inhibitor

In order to prevent self-polymerization during storage, liquid components of bone cements contain an inhibitor as a radical scavenger and usually hydroquinone is used for this purpose. It was shown that hydroquinone could be replaced by less toxic materials such as food grade di-tert-butyl-p-cresol [14].

III. MOLECULAR STRUCTURE AND POLYMERIZATION REACTION

Acrylic bone cements are polymeric materials produced by radical polymerization of methyl methacrylate (MMA), as shown in Fig. 5. The whole process starts with the formation of active unpaired electrons by dissociation of the initiator. The most commonly used initiator is benzoyl peroxide and it has an ability to split into two fragments upon dissociation of the weak peroxy bond (the single O–O bond) (Fig. 6). The initiator fragments which have unpaired electrons are called free radicals.

The unpaired electrons are energetically unstable and need to be paired and stabilized. When they find any electrons to pair up with, they do so. The carbon–carbon double bond in a vinyl monomer, like that in methylmethacrylate, has a pair of electrons which is easily attacked by the free radical to form a new chemical bond between the initiator fragment and one of the double bond carbons of the monomer molecule.

The other electron of the double bond stays on the carbon atom that is not bonded to the initiator fragment, creating a new free radical. This unpaired electron is now capable to attack the double bond of a new monomeric unit. This whole process, the breakdown of the initiator molecule to form radicals, followed by the radical's reaction with a monomer molecule is called the initiation step of the polymerization (Fig. 7).

This new radical reacts with another methylmethacrylate molecule in the same way as the initiator fragment did. Another radical is always formed when this reaction takes place over and over again. This process of adding more monomer molecules to the growing chains is called propagation (Fig. 8). As far as the radical and the monomer are present, more and more MMA molecules are added, and they build a long chain containing n monomeric units.

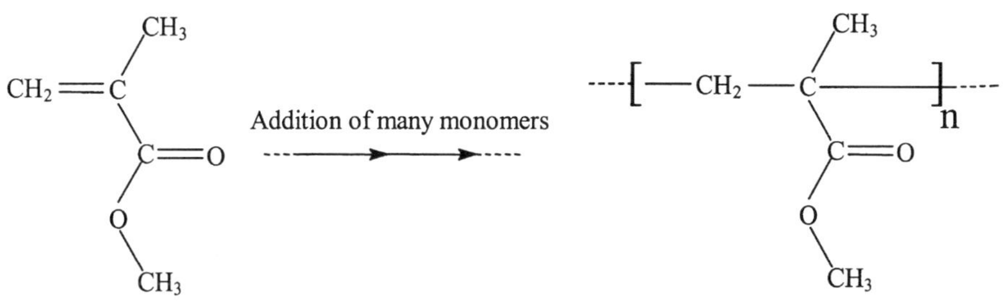

Figure 5 Polymerization reaction of polymethylmethacrylate.

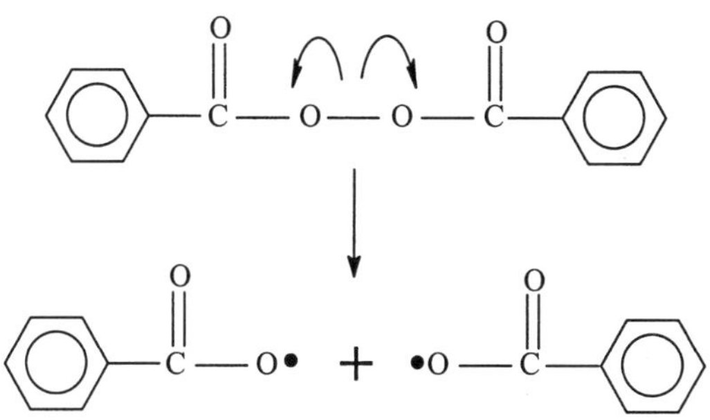

Figure 6 Initial radical formation by decomposition of benzoyl peroxide.

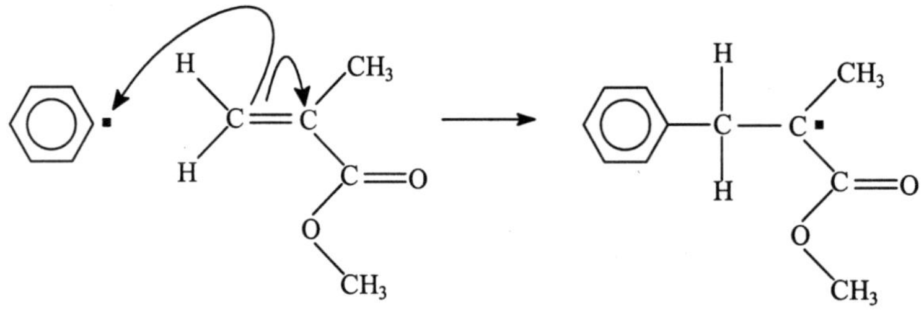

Figure 7 Initiation step of the polymerization.

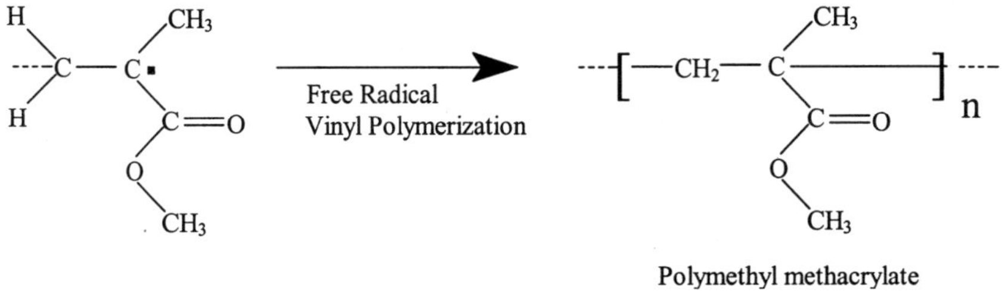

Figure 8 Chain propagation reaction.

Radicals are unstable, and when there are not enough monomeric units to combine, eventually they find a way to terminate without generating a new radical. This happens in several ways. Either disproportionation or recombination of the radical-carrying chain ends. Recombination is the simplest way, where in the two unpaired electrons join to form a pair and a new chemical bond. Termination is the third and final step of a chain-growth polymerization (Fig. 9).

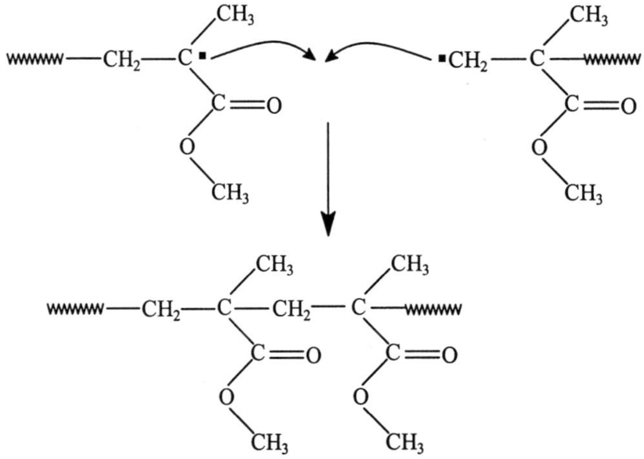

Figure 9 Termination of the polymerization reaction by recombination.

As the polymerization goes on, the viscosity of the whole mass increases, preventing the diffusion of growing chains and combination of chain ends. Therefore, rate of termination decreases. At this point, the polymerization rate and the medium temperature increase. Since thermal conductivity of the medium is poor, a peak temperature point is reached. This is called the gel effect or Trommsdorff effect. During the formation of polymer matrix, PMMA particles which are added as the powder constituent of the bone cement become entrapped in the newly formed polymer matrix.

IV. COMMERCIALLY AVAILABLE BONE CEMENTS

There are various kinds of bone cements. The chemical compositions of some commercially available and most commonly used bone cements are presented in Table 1.

V. THERMAL CHANGES DURING CURING

The dough obtained after mixing the powder and the liquid parts of the cement starts to solidify, hardens, and sets in a few minutes. Starting from the mixing of the components, the dough goes through distinct phases. Phase I, or the mixing phase, begins immediately when powder and liquid are brought together. The time that the mixture becomes homogeneous is the beginning of Phase II, or the waiting phase. The mixture is sticky at this phase. The total duration of Phases I and II is called the doughing time. At the end of Phase II, i.e., when the dough is no longer sticky, Phase III begins. The duration of this phase is called working time. Bone cement is applied to the target sites during this phase. Setting time is the duration between the beginning of Phase I and the end of Phase III. When it is just hard to manupilate the dough, Phase IV begins. The prosthesis can no longer be placed into cement dough in this phase. Complete hardening occurs at the end of this phase, and therefore it is called the hardening phase. The phases which cement dough goes through are shown in Fig. 10 [19,20].

A. Effect of Ambient Temperature

Increase in the temperature during hardening is the result of exothermic polymerization reactions. High temperatures cause quick setting and may cause damage to the surrounding tissue. The rates of the polymerization reactions and the increases in the setting temperature and setting time are sensitive to the ambient temperature. Therefore the temperature of the operating room has an important effect, and when this temperature increases, the polymerization reaction also increases and the dough hardens quickly. For example, surgical Simplex P Radiopaque Bone Cement setting time is given as 9 min at an ambient temperature of 75°F, 12 min at 70°F, and 15 min at 65°F [21].

In addition to the ambient temperature of the operating room, the temperature of the powder and liquid components of the cement and of the implant and the mixing equipment, can markedly affect the setting time and setting temperature of the cement dough. If components are stored at temperatures lower or higher than that of room temperature, sufficient time (12–24 h) must be allowed for them to reach the appropriate ambient operating room temperature before they are mixed, otherwise setting time will be correspondingly lengthened or shortened [22]. If an implant is used while still warm from the autoclave, setting time will be reduced. Mixing equipment still warm from storage or autoclaving will also induce a shorter setting time than expected.

Table 1 Some Commercially Available Bone Cements

Bone cement	Manufacturer	Composition
AKZ (Antibiotic Simplex®)	Stryker Howmedica	**Powder** (40.97 g): 29.51 g polymethylmethacrylate–styrene, 5.91 g polymethylmethacrylate, 4.00 g barium sulfate, 0.58 g benzoyl peroxide, 0.73 g erythromycin-gluco heptonate (0.5 g base), 0.24 g colistin–methane sulfonate–sodium (= 3,000,000 I.E.) **Liquid** (18.79 g / 20 mL): 18.31 g methylmethacrylate (= 19.50 mL), 0.48 g N,N-dimethyl-p-toluidine (=0.50 mL), 1.5 mg hydroquinone
C-ment 1	EMCMBV	**Powder** (40 g): 35.04 g polymethylmethacrylate containing methyl acrylate and ethyl acrylate, 0.96 g benzoyl peroxide, 4.00 g barium sulfate **Liquid** (14.40 g / 15.42 mL): 12.15 g methylmethacrylate (=12.93 mL), 1.90 g butyl methacrylate (=2.12 mL), 0.35 g N,N-dimethyl-p-toluidine (=0.37 mL), 20 ppm hydroquinone
C-ment 3	EMCMBV	**Powder** (40 g): 34.92 g polymethylmethacrylate containing methyl acrylate and ethyl acrylate, 1.08 g benzoyl peroxide, 4.00 g barium sulfate **Liquid** (16.40 g / 17.57 mL): 13.85 g methylmethacrylate (=14.73 mL), 2.16 g butyl methacrylate (=2.42 mL), 0.39 g N,N-dimethyl-p-toluidine (=0.42 mL), 20 ppm hydroquinone
Cemex Isoplastic	Tecres	**Powder** (40.00 g): 33.72 g polymethylmethacrylate (with 3% styrene), 1.08 g benzoyl peroxide, 5.20 g barium sulfate **Liquid** (13.30 g / 14.13 mL): 13.18 g methylmethacrylate (=14.00 mL), 0.12 g N, N-dimethyl-p-toluidine (=0.13 mL), 75 ppm hydroquinone
Cemex RX (LV)	Tecres	**Powder** (40 g): 35.31 g polymethylmethacrylate (with 3% styrene), 1.09 g benzoyl peroxide, 3.60 g barium sulfate **Liquid** (13.30 g / 14.13 mL): 13.18 g methylmethacrylate (=14.00 mL), 0.12 g N, N-dimethyl-p-toluidine (=0.13 mL), 75 ppm hydroquinone
Cemex Genta HV	Tecres	**Powder** (40 g): 33.11 g polymethylmethacrylate (with 3% styrene), 1.20 g benzoyl peroxide, 4.00 g barium sulfate, 1.69 g gentamicin sulfate (1.0 g base) **Liquid** (13.30 g / 14.2 mL): 13.06 g methylmethacrylate (=13.89 mL), 0.24 g N,N-dimethyl-p-toluidine (=0.255 mL), 75 ppm hydroquinone
Cemex Genta LV	Tecres	**Powder** (40 g): 33.11 g polymethylmethacrylate (with 3% styrene), 1.20 g benzoyl peroxide, 4.00 g barium sulfate, 1.69 g gentamicin sulfate (1.0 g base) **Liquid** (13.30 g / 14.2 mL): 13.06 g methylmethacrylate (=13.89 mL), 0.24 g N,N-dimethyl-p-toluidine (=0.255 mL), 75 ppm hydroquinone

Table 1 Continued

Bone cement	Manufacturer	Composition
Cerafix LV	Ceraver Osteal	**Powder** (46.50 g): 41.75 g polymethylmethacrylate, 0.45 g benzoyl peroxide, 4.30 g zirconium dioxide **Liquid** (18.69 g / 20 mL): 15.90 g methylmethacrylate (=16.91 mL), 2.41 g n-butyl methacrylate (=2.69 mL), 0.38 g N,N-dimethyl-p-toluidine (=0.4 mL), 45 ppm hydroquinone
Cerafixgenta	Ceraver Osteal	**Powder** (47.83 g): 41.75 g polymethylmethacrylate, 0.45 g benzoyl peroxide, 4.30 g zirconium dioxide, 1.33g gentamicin sulfate (=0.8 g base) **Liquid** (18.69 g / 20 mL): 15.88 g methylmethacrylate (=16.91 mL), 2.43 g n-butyl methacrylate (=2.69 mL), 0.38 g N,N-dimethyl-p-toluidine (=0.4 mL), 45 ppm hydroquinone
CMW 1 Gentamicin	DePuy, Johnson & Johnson	**Powder** (40 g): 33.89 g polymethylmethacrylate, 1.69 g gentamicin sulfate (=1.0 g base), 0.82 g benzoyl peroxide, 3.60 g barium sulfate **Liquid** (18.37 g / 19.57 mL): 18.22 g methylmethacrylate (=19.36 mL), 0.15 g N, N-dimethyl-p-toluidine (=0.16 mL), 25 ppm hydroquinone
CMW 1 radiopaque	DePuy, Johnson & Johnson	**Powder** (40 g): 35.54 g polymethylmethacrylate, 0.82 g benzoyl peroxide, 3.64 g barium sulfate **Liquid** (18.37 g / 19.57 mL): 18.22 g methylmethacrylate (=19.36 mL), 0.15 g N, N-dimethyl-p-toluidine (=0.16 mL), 25 ppm hydroquinone
CMW 2	DePuy, Johnson & Johnson	**Powder** (40 g): 34.68 g polymethylmethacrylate, 0.80 g benzoyl peroxide, 4.52 g barium sulfate **Liquid** (18.37 g / 19.57 mL): 18.22 g methylmethacrylate (=19.36 mL), 0.15 g N, N-dimethyl-p-toluidine (=0.16 mL), 25 ppm hydroquinone
CMW 2 Gentamicin	DePuy, Johnson & Johnson	**Powder** (40 g): 33.11 g polymethylmethacrylate, 1.69 g gentamicin sulfate (=1.0 g base), 0.80 g benzoyl peroxide, 4.40 g barium sulfate **Liquid** (18.37 g / 19.57 mL): 18.22 g methylmethacrylate (=19.36 mL), 0.15 g N, N-dimethyl-p-toluidine (=0.16 mL), 25 ppm hydroquinone
CMW 2000 Gentamicin	DePuy, Johnson & Johnson	**Powder** (40 g): 29.06 g polymethylmethacrylate, 5.13 g poly(methylmethacrylate–ethylmethacrylate–styrene), 1.69 g gentamicin sulfate (=1.0 g base), 3.20 g barium sulfate, 0.92 g benzoyl peroxide **Liquid** (18 g / 19.15 mL): 17.64 g methylmethacrylate (=18.76 mL), 0.36 g N,N-dimethyl-p-toluidine (=0.38 mL), 75 ppm hydroquinone
CMW 3	DePuy, Johnson & Johnson	**Powder** (40 g): 35.20 g polymethylmethacrylate, 0.80 g benzoyl peroxide, 4.00 g barium sulfate **Liquid** (17.90 g / 19.04 mL): 17.45 g methylmethacrylate (=18.56 mL), 0.45 g N, N-dimethyl-p-toluidine (=0.48 mL), 25 ppm hydroquinone

(*Continued*)

Table 1 Continued

Bone cement	Manufacturer	Composition
CMW 3 Gentamicin	DePuy, Johnson & Johnson	**Powder** (40 g): 33.55 g polymethylmethacrylate, 1.69 g gentamicin sulfate (=1.0 g base), 0.76 g benzoyl peroxide, 4.00 g barium sulfate **Liquid** (17.90 g / 19.04 mL): 17.90 g methylmethacrylate (=18.56 mL), 0.45 g N, N-dimethyl-p-toluidine (=0.48 mL), 25 ppm hydroquinone
Copal	Merck	**Powder** (42.59 g): 35.20 g poly(methylacrylate–methylmethacrylate), 4.27 g zirconium dioxide, 0.32 g benzoyl peroxide, 1.60 g gentamicin sulfate (=1.0 g base), 1.20 g clindamycin hydrochloride (=1.0 g base) **Liquid** (18.78 g / 20 mL): 18.40 g methylmethacrylate (=19.57 mL), 0.38 g N,N-dimethyl-p-toluidine (=0.43 mL), 0.4 mg chlorophyllin
Duracem 3	Sulzer	**Powder** (48.39 g): 38.55 g polymethylmethacrylate, 4.28 g poly(butyl methacrylate–methylmethacrylate), 0.40 g benzoyl peroxide, 4.76 g zirconium dioxide, 0.40 g di-cyclo-hexylphthalate **Liquid** (19.22g / 20.55 mL): 16.06 g methylmethacrylate (=17.09 mL), 2.83 g buthyl methacryalte (=3.16 mL), 0.31 g 2-[4-(dimethylamino)phenyl] ethanol, 20 ppm hydroquinone
Durus H	Macmed Orthopedics	**Powder** (42.25 g): 37.30 g polymethylmethacrylate containing about 1% ethylexyl methacrylate, 1.11 g benzoyl peroxide (titrated), 3.84 g barium sulfate **Liquid** (18.37 g / 19.54 mL): 18.19 g methylmethacrylate (=19.35 mL), 0.18 g N, N-dimethyl-p-toluidine (=0.19 mL), 60 ppm hydroquinone
Endurance	DePuy, Johnson & Johnson	**Powder** (40 g): 26.82 g polymethylmethacrylate, 8.44 g co(methylmethacrylate–styrene), 0.74 g benzoyl peroxide (titrated), 4.00 g barium sulfate **Liquid** (18.88 g / 20.08 mL): 18.50 g methylmethacrylate (=20.08 mL), 0.38 g N, N-dimethyl-p-toluidine (=0.40 mL), 0.002 g hydroquinone
Genta C-ment 1	EMCMBV	**Powder** (40.80 g): 34.97 g polymethylmethacrylate containing methylacrylate and ethyl acrylate, 0.98 g benzoyl peroxide, 4.00 g barium sulfate, 0.80 g gentamicin sulfate (=0.5 g base) **Liquid** (14.40 g / 15.42 mL): 12.15 g methylmethacrylate (=12.93 mL), 1.90 g butyl methacrylate (=2.12 mL), 0.35 g N,N-dimethyl-p-toluidine (=0.37 mL), 20 ppm hydroquinone

Table 1 Continued

Bone cement	Manufacturer	Composition
Genta C-ment 3	EMCMBV	**Powder** (40.80 g): 34.90 g polymethylmethacrylate containing methylacrylate and ethyl acrylate, 1.10 g benzoyl peroxide, 4.00 g barium sulfate, 0.80 g gentamicin sulfate (=0.5 g base) **Liquid** (16.40 g / 17.57 mL): 13.85 g methylmethacrylate (=14.73 mL), 2.16 g butyl methacrylate (=2.42 mL), 0.39 g N,N-dimethyl-p-toluidine (=0.42 mL), 20 ppm hydroquinone
Osteobond	Zimmer	**Powder** (40 g): 35.50 g poly(methylmethacrylate–styrene), 0.5 g benzoyl peroxide, 4.00 g barium sulfate **Liquid** (18.80 g / 20 mL): 18.66 g methylmethacrylate (=19.85 mL), 0.14 g N,N-dimethyl-p-toluidine (=0.15 mL), 80 ppm hydroquinone
Osteopal	Merck	**Powder** (40 g): 33.40 g poly(methylacrylate–methylmethacrylate), 0.60 g benzoyl peroxide, 6.00 g zirconium dioxide, 1 mg chlorophyllin **Liquid** (18.78 g / 20 mL): 18.40 g methylmethacrylate (=19.57 mL), 0.38 g N,N-dimethyl-p-toluidine (=0.43 mL), 0.4 mg chlorophyllin
Osteopal G	Merck	**Powder** (41.70 g): 33.14 g poly(methylacrylate–methylmethacrylate), 0.63 g benzoyl peroxide, 6.26 g zirconium dioxide, 1 mg chlorophyllin, 1.67 g gentamicin sulfate (1.0 g base) **Liquid** (18.78 g / 20 mL): 18.40 g methylmethacrylate (=19.57 mL), 0.38 g N,N-dimethyl-p-toluidine (=0.43 mL), 0.4 mg chlorophyllin
Palacos LV	Schering Plough	**Powder** (40 g): 33.40 g poly(methylacrylate–methylmethacrylate), 0.60 g benzoyl peroxide (titrated), 6.00 g zirconium dioxide, 1 mg chlorophyllin **Liquid** (18.78 g / 20 mL): 18.40 g methylmethacrylate (=19.57 mL), 0.38 g N,N-dimethyl-p-toluidine (=0.43 mL), 0.4 mg chlorophyllin
Palacos R	Schering Plough	**Powder** (40 g): 35.55 g poly(methylacrylate–methylmethacrylate), 6.13 g zirconium dioxide, 0.32 g benzoyl peroxide, 1 mg chlorophyllin **Liquid** (18.78 g / 20 mL): 18.40 g methylmethacrylate (=19.57 mL), 0.38 g N,N-dimethyl-p-toluidine (=0.43 mL), 0.4 mg chlorophyllin
Palacos R Gentamicin	Schering Plough	**Powder** (40.84 g): 33.55 g poly(methylacrylate–methylmethacrylate), 6.13 g zirconium dioxide, 0.32 g benzoyl peroxide, 1 mg chlorophyllin, 0.84 g gentamicin sulfate (0.5 g base) **Liquid** (18.78 g / 20 mL): 18.40 g methylmethacrylate (=19.57 mL), 0.38 g N,N-dimethyl-p-toluidine (=0.43 mL), 0.4 mg chlorophyllin

(Continued)

Table 1 Continued

Bone cement	Manufacturer	Composition
Palamed	Merck	**Powder** (44 g): 38.28 g poly(methylacrylate–methylmethacrylate), 5.28 g zirconium dioxide, 0.44 g benzoyl peroxide **Liquid** (18.78 g / 20 mL): 18.40 g methylmethacrylate (=19.57 mL), 0.38 g N,N-dimethyl-p-toluidine (=0.43 mL), 0.4 mg chlorophyllin
Palamed G	Merck	**Powder** (44.92 g): 38.28 g poly(methylacrylate–methylmethacrylate), 5.28 g zirconium dioxide, 0.44 g benzoyl peroxide, 0.92g gentamicin sulfate (0.55 g base) **Liquid** (18.78 g / 20 mL): 18.40 g methylmethacrylate (=19.57 mL), 0.38 g N,N-dimethyl-p-toluidine (=0.43 mL), 0.4 mg chlorophyllin
Subiton RO	Prothoplast	**Powder** (40 g): 35.03 g polymethyl acrylate (with 20% n-BUMA), 4.0 g barium sulfate, 0.97 g benzoyl peroxide **Liquid** (18.80 g / 20 mL): 18.57 g methylmethacrylate (=19.76 mL), 0.23 g N,N-dimethyl-p-toluidine (=0.24 mL)
Subiton G	Prothoplast	**Powder** (40.30 g): 34.54 g polymethyl acrylate (with 20% n-BUMA), 4.0 g barium sulfate, 0.96 g benzoyl peroxide, 0.80 g gentamicin sulfate (0.5 g base) **Liquid** (18.80 g / 20 mL): 18.57 g methylmethacrylate (=19.76 mL), 0.23 g N,N-dimethyl-p-toluidine (=0.24 mL)
Surgical Simplex P	Stryker Howmedica	**Powder** (40 g): 29.40 g poly(methylacrylate–styrene), 6.00 g polymethylacrylate, 4.0 g barium sulfate, 0.6 g benzoyl peroxide **Liquid** (18.79 g / 20 mL): 18.31 g methylmethacrylate (=19.50 mL), 0.48 g N,N-dimethyl-p-toluidine (=0.50 mL), 80 ppm hydroquinone
Zimmer	Zimmer	**Powder** (40 g): 35.70 g polymethylacrylate, 4.0 g barium sulfate, 0.30 g benzoyl peroxide **Liquid** (18.80 g / 20 mL): 18.28 g methylmethacrylate (=19.45 mL), 0.52 g N,N-dimethyl-p-toluidine (=0.55 mL), 80 ppm hydroquinone

One of the efforts to adjust the setting time and to improve the handling properties of bone cements involves prechilling the liquid monomer constituents, prior to their mixing with the powder. Lidgren et al. compared effects of ambient temperatures (4C vs. 21°C) of high viscosity, high molecular weight, antibiotic-containing polymethylmethacrylate cements mixed by hand or vacuum. If components were prechilled at 4°C, mixing became easier and handling characteristics were improved compared to the cements at 21°C. It was reported that prechilling did not cause any differences in thermal or mechanical parameters [23].

Hansen and Jensen made comparative studies of cement components stored at room temperature or chilled to 5°C and mixed either manually or under vacuum. Nine commercial bone cements were tested for various parameters such as handling characteristics, intrusion, doughing time, setting time, and exothermic temperature. It was reported that, except for the low-viscosity

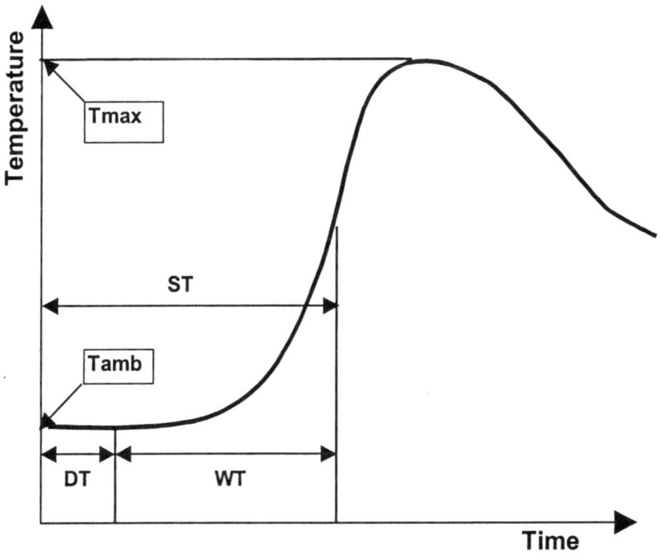

Figure 10 Change in polymerization temperature versus time (T_{max} = maximum curing temperature; T_{amb} = ambient temperature; DT = doughing time; WT = working time; ST = setting time (the time when bone cement temperature is reached to half of the difference of T_{max} and T_{amb}).

brands, it was very hard to manipulate vacuum-mixed bone cements that were kept at room temperature, as the cements became too viscous to mould test specimens. They found that prechilling and vacuum mixing prolonged the setting time and preserved a lower viscosity during the handling period [24].

Smeds et al. [25] analyzed the influence of varying viscosities and mixing temperatures (22 vs. 6°C) on macro- and microporosity, density, and compressive strength in vacuum-mixed high-, medium-, and low-viscosity cements. At 22°C, vacuum mixing significantly reduced void volume, while the reduction was most pronounced in low-viscosity cement. At 6°C, the reduction of micropores was more pronounced in high- viscosity cement. In medium- and especially low-viscosity cement, the prechilling caused an increase in the number of micropores. At 6°C, cement density was significantly reduced in low-viscosity cement. Vacuum mixing significantly increased compressive strength in all cement types. The temperature did not significantly influence compressive strength, but low- viscosity cement generally was stronger when mixed at 22°C. The authors did not recommend prechilling application to the low-viscosity cement [25].

Lewis studied the influence of the storage temperature of the cement constituents prior to mixing (21 vs. 4°C) and the mixing method (hand mixing vs. vacuum mixing) on the uniaxial tension–compression fatigue performance and porosity of Palacos R acrylic bone cement. It was reported that although the mixing method (for a given storage temperature) exerted a significant influence on the fatigue performance and porosity, the effect of storage temperature (for a given mixing method) on either of these parameters was not significant [26].

Although it is believed that high setting temperatures may damage the surrounding tissue, causing aseptic loosening and failure of the prosthesis, Marberg et al. found no statistical differences in mechanical or the histomorphological properties when inserted titanium implants in tibia of rabbits by using conventional Simplex and low temperature–curing Boneloc cements [27].

VI. MECHANICAL PROPERTIES

Mechanical properties of a material can be studied by measuring the response of the material to an applied load. Strength of a material is judged by its ability to resist stress. Stress is a quantity that describes the intensity of the internal force per unit area acting on a specific plane (area) passing through a point. When a body is subjected to external loads, an infinitely small point within a body under load can be subjected to two possible types of stresses: normal stress and shear stress (Fig. 11).

Normal stress is the intensity of the net forces acting normal (perpendicular) to an object per unit area. The simplest load applications are tensile and compression forces. In tensile force, the specimen is gradually elongated, and in compression force the specimen is gradually compressed under an applied stress. The resulting changes in length are recorded. The stress, σ, is the force applied per unit area and the strain is the fractional change in the length of the specimen. It can be measured as linear strain, ε, which is the change of length per unit length. In tensile and compression tests, the specimen is subjected to a progressively increasing tensile force until it fractures. At the beginning the test material deforms elastically. The strain is directly proportional to the stress and the specimen returns to its original form immediately on the removal of the stress. Beyond the elastic limit, applied force causes plastic deformation, in which deformation is permanent. The material either fractures or changes shape. In the elastic region, the stress and strain are linearly proportional and the ratio is defined as $E=\sigma/\varepsilon$, where E is Young's modulus.

Shear stress is the intensity of force per unit area acting tangentially to an object. In other words, it is the stress component that acts in the plane of the sectioned area. When a shear stress, which acts at an angle rather than normal to the plane, is applied, the relation between shear stress, τ, and shear strain, γ, is given as $G=\tau/\gamma$, where G is called the shear modulus. The amount of deformation in any direction depends on the magnitude and direction of the loading and on the composition of the material (Fig. 11).

Mechanical tests can be static tests when the load is applied slowly, cyclic tests when the load is applied partly or wholly for many repetitions, and impact tests when the load is applied

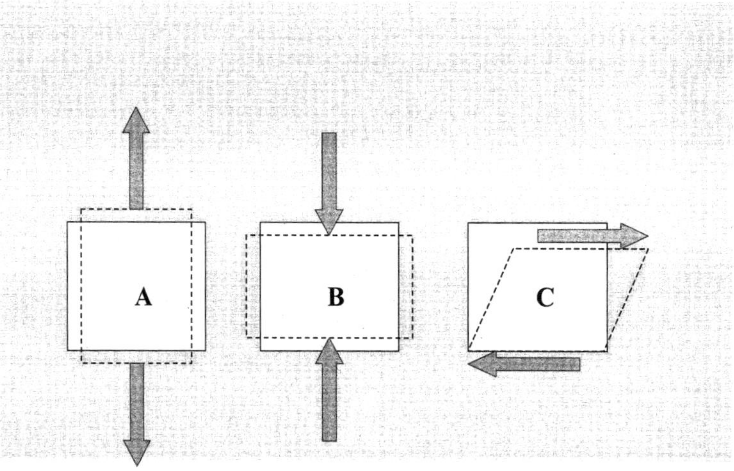

Figure 11 Basic stresses acting on a material under loading. (A) Tensile stress. (B) Compressive stress. (C) Shear stress. Dotted lines represent the final deformed shape of the material after loading.

rapidly at once. Fractures form either under constant stress (creep) or fluctuating stress (fatigue) conditions. Creep tests are carried out under certain combinations of stress and temperature. All materials exhibit an increase of strain with time when subjected to a constant stress. This phenomenon is called creep. High temperatures lead to a rapid creep, which is often accompanied by microstructural changes.

There are three common ways in which stresses may be applied: axially, torsionally, and flexurally (bending). Examples of these are seen in Fig. 12. In each type of loading, at any point of the body, normal and shear stresses exist more or less depending on the type of loading. In axial and flexural loading, normal (pure tensile or pure compressive) stress is much more than shear stress. In torsion, shear stress is much more than normal stresses. Flexural theory states that most materials will exhibit linear-plastic behavior, i.e., they will respond to an applied load by deflecting in accordance to Hooke's law, and will return to their original shape and form when the load is removed. This stress–strain relation exists only up to a certain load, after which the material will undergo some irretrievable deformation. Hooke's law states that deformation of an object under loading is proportional to the magnitude of the load and that stress is directly proportional to strain but only up to a limit, called the proportionality constant.

Fatigue test is the measurement of the failure of a material under repeated applications of a constant stress smaller than the stress which causes failure in a single application. The material initially suffers some microstructural damage. Eventually the cyclic applications of the load lead to the formation of cracks which grow larger with every application of load. A series of specimens of the material are tested to failure by application of different values of stress. Properties such as fatigue lifetime, number of cycles to failure after crack initiation, permanent set, and total deformation are also measured.

Impact test measures the brittleness of the material. In this test, a standard notch is made in a standard test specimen, which is then struck under impact conditions by a heavy weight forming the end of a pendulum. The notch serves to introduce triaxial tensile stresses into the specimen, encouraging brittle failure to occur. The weight is released from a known height and strikes the specimen on the side opposite the notch to induce tensile stresses in it. After breaking the specimen the pendulum swings on and the height to which it rises is measured. The energy absorbed in breaking the material under high-speed loading can be determined. If this value is low, the specimen is called brittle.

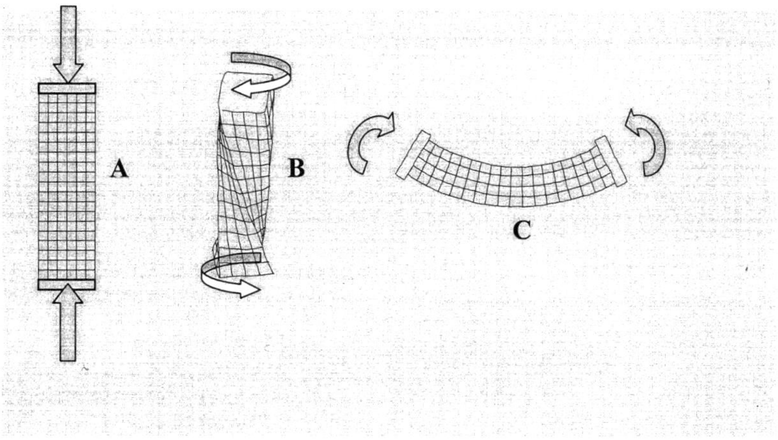

Figure 12 (A) Axial loading. (B) Torsional loading. (C) Flexural loading (bending).

If the load only slightly exceeds the proportional load, then the material may still respond elastically, but the response to the load is no longer linearly related to the applied load. When the applied load permanently changes the properties of the material, the specimen is said to be exhibiting plastic deformation. Eventually, the load to which the material is subjected will be at the material's ultimate load. At this point, the material will either fail (if it is a brittle material) or it will continue to deflect (if it is ductile) until it finally ruptures. A typical stress–strain curve is given in Fig. 13.

Fracture toughness is a quantitative way of expressing a material's resistance to brittle fracture when a crack is present. If a material has a large value of fracture toughness it will probably undergo ductile fracture. Brittle fracture is very characteristic of materials with a low fracture toughness value. There are actually four different types of fracture toughness: K_C, K_{IC}, K_{IIC}, and K_{IIIC}, each of which is used under a different mode of fracture. There are three different modes of fracture; mode I, mode II, and mode III. In bone cement fracture toughness tests, K_{IC} is used because a crack in bone cement usually causes mode I fracture. Different modes of fracture are shown in Fig. 14.

A very useful way to visualize time to failure for a specific material is with the S-N curve, which indicates stress versus cycles to failure. S-N curves use the stress amplitude, σ, plotted on the vertical axis and the logarithm of the number of cycles to failure on the horizontal axis. An important characteristic to this plot, as seen in Fig. 15, is the fatigue limit.

The significance of the fatigue limit is that if the material is loaded below this stress, then it will not fail, regardless of the number of times it is loaded. Other important terms are fatigue strength and fatigue life. The stress at which failure occurs for a given number of cycles (usually 10^6 or 10^7) is the fatigue strength. The number of cycles required for a material to fail at a certain stress is fatigue life.

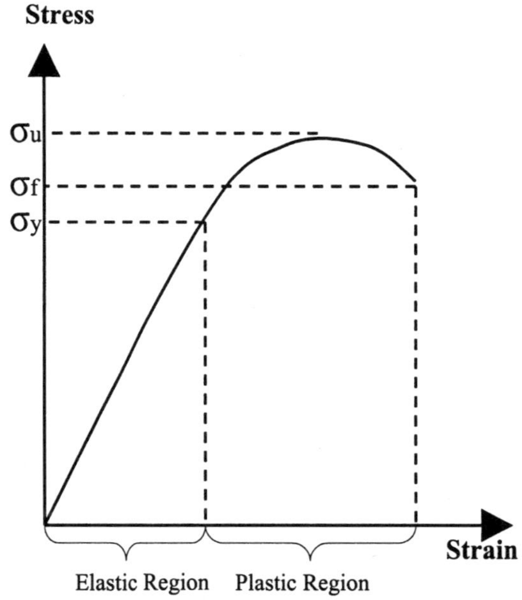

Figure 13 Typical stress-strain chart (σ_y σ_u, and σ_f represent the yield, ultimate, and fracture stresses, respectively).

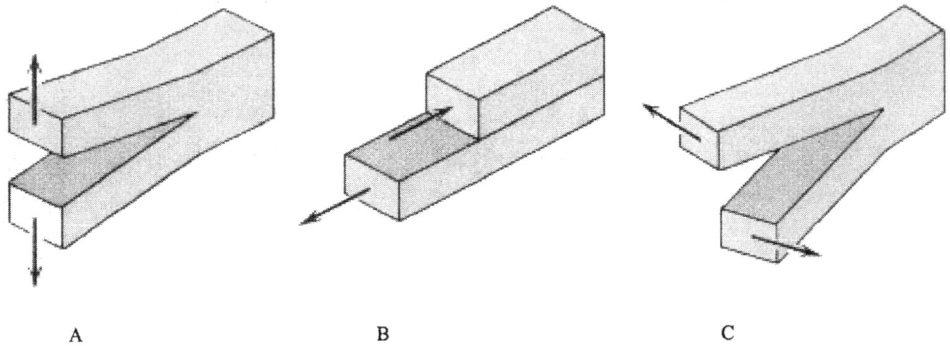

Figure 14 Different modes of fracture. (A) Mode I fracture. (B) Mode II fracture. (C) Mode III fracture.

For bone cements all these mechanical properties—tensile, compression, shear, fatigue—are important. In many cases, bone cements are used as a kind of bone substitute, although the mechanical properties of bone cements differ from that of natural human bone. But filling the surgically created cavities by bone cement is better than leaving large voids in the bone tissue. Some physical and mechanical properties of different bone cements prepared with different mixing and aging conditions obtained by numerous scientists are given in Table 2.

Mechanical strength of bone cements is very important since most of the aseptic loosening is related to the fracture of the PMMA cements. For bone cements mechanical properties are affected by various factors, and it is not easy to report strength characteristics of all new formulations because each differs from one another. Some of the factors that affect the mechanical properties are composition of cement parts (different chemical composition, existence of additives, radiopaque materials, different initiators accelerators), weight-average molecular weight of the polymer part [28], porosity [29,30], type of the sterilization method of the constituents [31], and mixing methods [32–36].

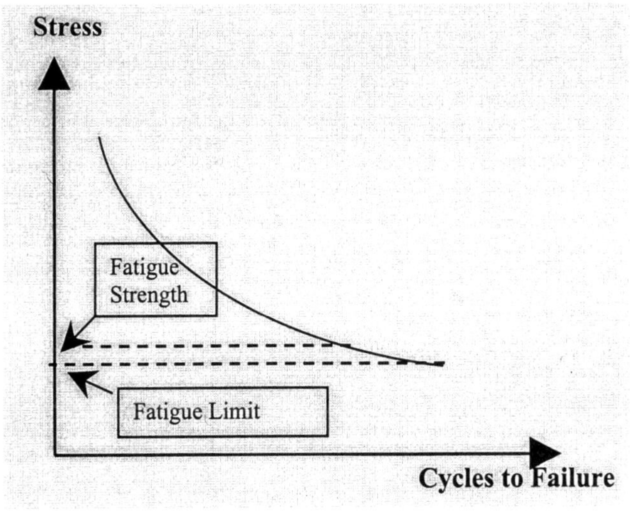

Figure 15 A demonstrative S-N chart.

Table 2 Physical and Mechanical Properties of Bone Cements

Density (g/cm^3)	1.10–1.23
Water sorption (%)	0.5
Shrinkage after setting (%)	2.75–5
Ultimate compressive strength (MPa)	72.6–114.3
Compressive elastic modulus (GPa)	1.94–3.18
Compressive strain before fracture (%)	5.0–7.5
Ultimate tensile strength (MPa)	23.6–49.2
Tensile elastic modulus (GPa)	1.58–4.12
Tensile strain before fracture (%)	0.86–2.49
Shear strength (MPa)	42.7–50.2
Fracture toughness, K_{IC} (MPa $m^{1/2}$)	1.03–2.32

It was reported that microcracks are developed in the interbead matrix just before failure and not through the prepolymerized beads. These cracks propagate, and gross mechanical failure occurs [37]. The damage accumulation failure begins before weight bearing due to cracking induced by residual stress around pores or stress raisers. These residual stresses occur during curing stages of the bone cement dough [38]. Some residual stresses are caused by the temperature differences arising after polymerization of bone cement [39].

Some attempts to prevent fracture formation and to increase the mechanical properties involve addition of rubber particles into the polymeric matrix. It was reported that addition of rubber into PMMA cement increased the fracture toughness significantly, but reduced elastic modulus and compressive strength [40].

It was also reported that penetration of bone cement into bone cavity is one important factor for the stability of implant. Penetration of a few milliliters will hold the trabecular ends firmly and enhance stiffness and resistance to compression, shear, and tension [41,42]. Deep penetration into the cancellous interstices enhance the strength of the interface between bone and the cement and affects cement–bone interfacial shear strength [43–45]. Some pressure should be exerted on the bone cement in order to provide sufficient penetration into bone trabeculae. It was shown that the human femora could resist mechanically an intramedullary pressure of at least 2 MPa. Therefore intramedullary pressures should remain below 2 MPa for safe use [46]. After cemented hip joint replacement operations, a slight increase in flexural strength and modulus of bone cement at the 2 weeks [47,48], followed by a nearly equivalent decline for 8 weeks [47] or slight increase for up to 6 months [48] were reported.

In hip arthroplasties bone cement is mainly subjected to compressive loading that is maximal at the proximal region of the stem [49]. It was measured that cement from the proximal region of the prosthesis had significantly decreased flexural strength, lower modulus, and lower molecular weight than cement used in the distal region and plug [47]. These data suggest that failures in the proximal region of the prosthesis are due to both higher loading conditions and inferior physical and mechanical properties of the bone cement in that region.

The mechanical success of an operation, such as a total hip arthroplasty, depends not only on the mechanical properties of the bone cement, but also on many other factors like surgical technique details or shape and material properties of the metallic implant [50].

VII. BONE CEMENT DELIVERY METHODS

These methods are in the scope of surgical techniques of arthroplasties. The general procedure can be summarized as follows. In an ordinary joint replacement surgery, the surgeon drills the part of the bone where the device is going to be implanted and cleans the debris in the cavity. Saline solution is spouted by a plastic syringe to drilled bone cavities in order to clean the blood and debris out of the cement application area. If this is not done well, the remaining tissue particles inhibit the cement penetration into bone trabeculae. When high-pressure spouting is applied, the cleaning operation and therefore the cement penetration is much better than by the ordinary method. The use of jet lavage is a good way to clean the bone cavity [51].

The surgeon fills the cavity with the bone cement in its soft dough state, and inserts the metallic implant into the prepared place either by hand or with an injection gun. Cement mixing in a bowl followed by finger packing does not provide adequate fill, especially in the femoral canal. Retrograde cement filling in the femur reduces risk of revision. This technique prevents air entrapment during cement application, reducing porosity. Lamination is also reduced, making it possible to achieve a completely cement-filled cavity [52].

Pressurizers are available for acetabular, femoral, and knee pressurization. They are used to maintain pressure on the bone cement until it is doughy enough to resist the force of blood pressure. In order to achieve good filling and pressurization in hip and knee arthroplasties, a small piece of bone or a cement restrictor may be used to plug the shaft. Pressurization and lavage of the cancellous bone significantly improves cement penetration.

The bone cavity should be shaped to provide an even cement mantle between the bone and prosthesis. Thickness of the cement mantle should be approximately 2–3 mm. Normally 40 g of cement is used for acetabulum and 60–80 g for femur. A stem centralizer can help guide the femoral prosthesis to a neutral position in the cement. An even cement mantle means better stress distribution and reduced risk of cement mantle failure [53,54].

The dough cures and hardens in a few minutes by an exothermic polymerization reaction and holds the metallic implant firmly in a stable position. Important points are the viscosity of the cement dough, which affects ease of application, and the porosity, which affects the mechanical strength of the cement.

A. Viscosity

Bone cements have complex viscoelastic characteristics. In a few minutes after mixing the solid and liquid parts, they go from a liquidlike consistency to a hard solid form. The changes in the flow characteristics of some commercial bone cements during the curing process are given by Dunne and Orr [55].

The viscosity is one very important parameter since it is required that the dough should attain proper texture and give enough time for the surgeon to use it with maximal workability and penetration when pressed into the cancellous bone without hardering. For good intrusion into the trabeculae, the cement should have the desired viscosity. If the viscosity is too low, blood pressure may force the cement out of the bone cavity and cause blood laminations in the cement. If the viscosity is too high, it will be hard to fill the cavity, and the dough will not penetrate into bone trabeculae well. Ideal viscosity is high enough to prevent the cement from mixing with blood or fat/bony material yet low enough to penetrate the bone adequately.

Commercial bone cements are offered as high-, medium-, or low-viscosity cements. High-viscosity bone cements typically have a doughy consistency. They have a short wetting phase and lose stickiness quickly. The working phase generally is long, and viscosity remains unchanged until the very end, when it slowly increases. Low-viscosity cements are similar to viscous oil in consistency. They have a long liquid phase, or low-viscosity wetting phase. The cement remains sticky for quite some time. Viscosity increases rapidly during the working phase, and the doughy cement becomes warm and sets quickly. For medium viscosity cements, the wetting phase is similar to that of low-viscosity cements. They lose stickiness quickly, like high-viscosity cements. During the working phase, viscosity increases slowly and continuously, as with high-viscosity cements.

High-viscosity bone cements have been shown to offer a lower incidence of revision and aseptic loosening in total hip arthroplasties. The lowest revision risk is observed when the high-viscosity cements Palacos Gentamicin and Palacos are used. Ten years of follow-up studies showed that 95% of the total hip prostheses implanted with these cements were stable. CMW offers significantly less risk reduction, demonstrating more than 85% stable implants in a 10-year follow-up. The highest risk was associated with Sulfix [56].

Malchau et al. studied the risk factors of some commercial bone cements by applying Poisson models for multivariate variables and by taking the risk ratio for Sulfix as 1. For aseptic loosening in osteoarthrosis, risk ratios with 95% confidence limits are given as 0.46–0.61 for Palacos, 0.59–0.72 for Simplex and 0.52–0.84 for CMW [57].

B. Porosity

The presence of pores in the cement in the intramedullary canal may have a positive or negative effect on the stability and the service life of the arthroplasty. On one hand, it is intuitively expected that pores would act as stress risers and initiate sites for cracks, rendering the cement susceptible to early fatigue fracture. On the other hand, Topoleski et al. [58] speculated that pores might play a role in blunting propagating cracks, thereby prolonging the life of the implant. On the whole, though, the current consensus is that every effort should be made to substantially reduce the number and size of pores.

There are two types of pores in fully polymerized bone cement: macropores (pore diameter > 1 mm) and micropores (pore diameter $\cong 0.1$–1 mm) [59]. Pore formation arises from various sources [60]: air initially surrounding the liquid monomer or powder constituents, entrapment of air during wetting of the powder by the liquid monomer, entrapment of air during mixing of the constituents, the boiling or evaporation of the volatile liquid monomer during the curing stage, and entrapment of air during the transfer of the dough to the syringe/gun.

Smeds et al. compared the properties of cements of different viscosities, mixed with vacuum or hand, and at two different temperatures. They reported that the microporosity was reduced in all cements by vacuum mixing, while the highest degree was observed in low-viscosity cement. For high-viscosity cement the micropore reduction was more pronounced at 6°C. But for medium- and low- viscosity cements, an increase in the number of micropores was reported upon prechilling [25]. Lewis also observed that vacuum mixing influenced the porosity and uniaxial tension-compression fatigue performance of the cements [26].

Interfacial porosity, the concentration of pores at the cement–metal interface in cemented femoral stems, was studied by James et al. [61]. They reported that the interfacial pores exist in almost all cases, and centrifugating the cement had no effect on the amount of these pores.

Macaulay et al. [62] studied porosities of bone cements prepared in different mixings and found that the mean percent porosity was highest in case of vacuum mixed, lowest for centrifuged, and intermediate for hand mixed cements. No significant reduction in percent porosity was reported for vacuum and centrifugational mixings.

C. Mechanisms of Anchorage

Bone cements do not form chemical bonds between the metallic implant and the natural bone. In that sense they are not glues. They fix the prosthesis in the desired area by forming a mechanical interlock between the metallic implant and the bone, and transfer the load from one to the other. Bone cement diffuses into the microscopic irregularities of the bone cavity and provides a good mechanical attachment to bone. Therefore, the strength of the cement–bone interface is related to the amount of interdigitation between the cement and bone. The apparent strength of the cement–bone interface is significantly higher when the interface is loaded in shear rather than tensile loading [63]. The specimens that have higher strengths are often associated with more trabecular bone interdigitated with cement [64]. This suggests that greater penetration should occur in cancellous bone containing larger interstitial cavities. Clinically, this is accomplished by vigorous finger packing. The cement should be pressurized as early as possible within the rasped cavity (immediately after the dough stage if possible) [65].

The attachment of bone cement to metallic implant occurs by circumferential hoop stress that is formed by contraction of the hot cement dough during the cooling process as a result of the metallic implant is firmly squeezed by hardened bone cement. Also, to increase the attachment of metallic implant to bone cement, the implants were coated with certain chemicals or the implant surface was treated with bond-forming materials. It is reported that PMMA coating on metal surfaces increased the torsional fatigue strength of the metal–cement interface [66], and hydroxyapatite coating of titanium implants increased the initial integration of the implant to the bone in dental applications [67].

VIII. MAIN SIDE EFFECTS

The main side effects of bone cements can be summarized as aseptic loosening, local temperature increase, and release of toxic molecules to the surrounding tissue.

A. Aseptic Loosening—Failure of Artificial Joints

Aseptic loosening is the leading cause of failure of cemented total hip arthroplasties. In a joint arthroplasty, a thin soft tissue gap of 0.1–3.0 mm occurs between the bone and bone cement within 1–30 years. This event, called aseptic loosening, leads to painful mechanical failure of the artificial joint. The formation of a radiolucent fibrous membrane, accumulation of inflammatory cells, and osteolysis are characteristic for aseptic loosening of polymethylmethacrylate-fixed prostheses. Since the formed membrane is produced by fibroblasts, it is likely that these connective tissue cells play a critical role in the loosening process. Some studies are summarized below.

Histologic evaluation of tissue surrounding the loosened components after joint replacement reveals the presence of a synovial-like lining adjacent to the polymethylmethacrylate cement. The tissue is heavily infiltrated with particulate cement as well as polyethylene associated with a foreign body type giant cell reaction. This tissue response which may be responsible for the bone lysis associated with loosening, is suggested by the demonstration of high prostaglandin E2 levels and enhanced bone resorbing activity in the tissue culture medium of fragments from this membrane [68,69].

The histopathology of the bone–cement interface shows that a macrophagic reaction is evident even in the presence of a stable bone–cement interface at the retrieved hip prosthesis. This is in response to particles of polymethylmethacrylate bone cement and other chemicals and supports the view that the release of particles by the cement or by the prosthetic components can precede the mechanical instability and be the primary cause of loosening. It was shown that monocytes and macrophages responding to particles of bone cement are capable of differentiating into osteoclastic cells that resorb bone. There usually exists a connective tissue layer between the cement and the bone and usually it is observed that no bone trabecula reaches the cement surface. The polymorphous features of this connective membrane are probably a result of instability and movement at the bone–cement and stem–cement interfaces. That means micromovements at the bone–cement and stem–cement interfaces can accelerate aseptic loosening [70].

In the evaluation of the microstructural characteristics of newly formed bone tissue at the interface with cement, the newly formed bone adjacent to cement in the loosened prostheses shows a maturity degree lower than that of bone adjacent to cement in stable prostheses. The lattice parameters of bone apatite do not show significant variations as compared to normal bone tissue. Bone trabeculae at the interface with loosened prostheses often show an osteoid lining characterized by a strongly demineralized lamellar and Haversian structure [71].

Fracture of the polymethylmethacrylate bone cement mantle can lead to the loosening and ultimate failure of cemented total joint prostheses. Mechanical failure of polymethylmethacrylate accelerates bone resorption at the bone–cement interface of the prostheses. Evaluation of tissue that had been retrieved from the cement–bone interface of femoral components of total hip prostheses that were loose without associated infection revealed that a critical factor associated with bone resorption was the presence of particles that were small enough (1–12 μm) to be phagocytized by macrophages. PMMA particles of less than in 12 μm size can be phagocytized by macrophages. When the mechanical failure of cement produces particles that are small enough to be phagocytized, phagocytosis of the particles results in the increased production of tumor necrosis factor by the macrophages, which may in turn lead to bone resorption and prosthetic loosening [72].

Loosening of the cemented prostheses depends not only on the failure of the implant and/or the bone cement, but also on the inflammatory response of the bone tissue against bone cement ingredients. The radiopaque additive barium sulfate was shown to increase the inflammatory response in the surrounding tissues after cemented arthroplasty. Lazarus et al. [73] showed that the inflammatory response to polymethylmethacrylate particles containing $BaSO_4$ was greater than the response to plain polymethylmethacrylate particles of similar size. Increased inflammation was measured by leukocyte counts and levels of prostaglandin E2, tumor necrosis factor, and neutral metalloprotease [73]. In comparison to pure PMMA, PMMA particles containing ZrO_2 cause a higher increase in bone resorption. But the particles containing $BaSO_4$ were associated with 50% more bone resorption than those containing ZrO_2 [74].

B. Local Temperature Increase

One of the main side effects of bone cement application is the rise of the temperature at the bone–bone cement interface during the polymerization of MMA. In bone cement formulations the powder part of the bone cement is already prepolymerized PMMA particles, and this prevents the explosive polymerization reactions. Still the highly exothermic polymerization process of the MMA, with a total polymerization heat of 544 J g^{-1}, causes an increase of the local temperature. The maximal value of the curing temperature varies from 80 to 124°C depending on the ratio of MMA to PMMA, the composition of the solid and liquid components, the size of the polymer particles, the concentrations of BPO and DMPT, and the presence of transfer agents [75]. Peak temperatures are about 25–30° higher than the threshold levels (50–60°C) and are considered a cause of thermal damage to tissues. It is reported that bone tissue degenerates when it is subjected to 47°C for a duration not less than 1 min [76].

Particle size of PMMA powder is one important factor of the peak temperature, and as the particle size decreases, the surface-to-volume ratio increases. Therefore the amount of polymer dissolved in the monomer increases, leading to higher viscosity of the dough. The increase in viscosity leads the Trommsdorff effect to occur more vigorously, and the transfer of heat becomes difficult, causing an increase in the curing temperature. This also leads to a faster setting of the dough. Higher amounts of either the initiator (BPO) or the accelerator (DMPT) increase the polymerization temperature and decrease the setting time. As the polymer/monomer ratio increases, both the polymerization temperature and the monomer release to the surrounding tissue decrease. This seems to be an advantage for temperature, but the increase in this ratio increases the viscosity of the dough so that workability and the penetration of bone cement into bone trabeculae become difficult. Therefore the optimal value of polymer/monomer ratio is given about 2:1 (w/v), and this ratio is used in most commercial bone cement formulations.

In some cases bone cements with high setting temperatures may be desirable. Some surgeons treat the giant cell tumors of bone tissue by using the technique of aggressive curettage through a large bone window followed by acrylic cement reconstruction [77]. As the bone cement self-heats, the possibility of heat necrosis in the bone tissue exists. It was mentioned that the damage to the cells due to heat may be beneficial in reducing the rate of tumor recurrence [78].

C. Release of Toxic Molecules

As are most of the organic monomeric chemicals, MMA itself is also toxic to the bone tissue. Also release of MMA monomer from the dough structure into the circulating blood causes severe drop in blood pressure. This is caused by the direct chemical effect of MMA on blood vessels. In polymerization reactions, there is always a molecular weight distribution of the obtained polymer chains. Therefore it never happens that 100% of the monomer polymerizes during curing of the bone cement. A certain amount of residual monomer remains in the hardened cement. The content of monomer in the cured cement is related to the monomer composition, glass transition temperature, polymerization temperature, and type and concentrations of accelerator and initiator systems.

The remaining monomer leads to two results: unreacted MMA monomer leaks from the cement mantle into the surrounding tissues, causing toxic effect and impairing bone remodeling [79,80]. Presence of monomer in the hardened cement acts as a plasticizer and influences the mechanical properties [80].

Residual monomer content can be detected by differential scanning calorimetry (DSC) using isothermal and dynamic modes. Because DSC tends to be less sensitive at high conversions, especially if there exists a permanent residue, gas chromatography (GC) may be used [81]. The proportion of nonconverted residual monomer remaining in the polymerized bone cement is in the range of 2–6% just after hardening, decreases to approximately 0.5% or less after 2–3 weeks, and then remains the same for years [19]. Released amounts of the residual monomer can be measured by HPLC by immersing the cements into buffer or saline solutions for some periods [19].

The studies carried out in patients undergoing total endoprosthetic replacement of the hip joint with metal prosthesis and stabilized by acrylic bone cement pointed to statistically significant changes in the blood clotting and fibrinolysis systems. It was reported that monomer or polymer release causes a tendency toward hypercoagulation and intravascular clotting. Fibrinolysis activation associated with these changes is secondary, resulting from, among others, the mechanism of the action of free fibrin monomers [82]. Bone cement affects the activity of lysosomal enzymes in peripheral blood granulocytes. It was reported that after the endoprosthesis stabilization operation, statistically significant changes occur in the activity of lysosomal marker enzymes. Lability of lysosomal membranes appears with permeation of hydrolases into supernatant [83].

The release of monomer may cause some fatality, and therefore surgeons and anesthesia experts should take precautions to protect the patient. After few minutes of the cement application, the toxic effect vanishes.

IX. BONE CEMENT MIXING METHODS

Bone cement mixing methods are categorized as manual (or hand) mixing, centrifugation, vacuum mixing, and combined mechanical mixing.

In manual mixing, the powder component is added to the liquid (which may or may not have been chilled to a temperature that is usually between -15 and 6°C) in a polymeric (usually polypropylene or polyethylene) bowl. Then these components are stirred with a polypropylene or polyethylene spatula with a speed of 1 or 2 Hz for 45–120 s.

In centrifugation mixing the hand-mixed dough is immediately poured into a syringe (from which the nozzle is detached) that is then promptly placed in a centrifuge and spun with a speed of 2300–4000 rpm for 30–180 s [84–86].

Applying vacuum to self-setting acrylic cements was first described in dentistry, but Lidgren et al. studied vacuum-mixed cements as bone cements in 1984 [87]. They analyzed the mechanical strength of high- and low-viscosity gentamicin-containing cements by using three different mixing procedures: hand, vibration, and vacuum stirring. They reported that, vacuum mixing improved the flexural and compression strength and the modulus of elasticity by 15–30%, especially for high-viscosity cement. In another study of Lidgren et al. in 1987, it was shown that vacuum mixing improves fracture strength, maximal deflection, modulus of elasticity, and hardness when compared with hand mixing. The fatigue life was ten times longer after vacuum mixing. Vacuum mixing delayed the setting time by 1 min, and also decreased the peak temperature. Radiographic analysis showed that vacuum mixing mainly reduced not only the microporosity, but also the macroporosity. The mechanical properties deteriorated slightly after 2 months in Ringer's solution, but the differences between the mixing procedures remained unchanged [88].

The vacuum system that Wixson et al. described produced set specimens of less than 1% porosity. The samples had significant improvements in the mechanical properties of tensile and compressive strength and uniaxial tensile fatigue life compared to the specimens prepared with conventional mixing [89].

Creep characteristics of vacuum-mixed acrylic bone cement were studied by Norman et al. It was found out that compressive creep testing of cylindrical specimens machined from two commercial self-polymerizing acrylic bone cements demonstrated measurable creep strains with higher creep strains for the hand-mixed cement specimens compared to vacuum-mixed ones [90].

Since 1984, a number of proprietary and experimental chambers have been used for vacuum mixing. The proprietary ones include the Simplex Enhancement Mixer (Howmedica), Stryker High Vacuum System (Stryker), MITAB (Mitab), Optivac (Mitab), Stryker Mixevac II (Stryker), and Sterivac (SD) as well as the Mitvac, Cemvac Merck, Bonelock, and Cemex systems [91]. Thus, there are no generic steps in vacuum mixing (as is the case for hand or centrifugation mixing). Generally in vacuum-mixing methods, 5 to 100 kPa of vacuum is applied for 15–150 s to the dough during mixing stage. In some of them air within the powder is evacuated before mixing. The mixing apparatus, the quantity in the vacuum, the vacuum application time, and the mixing frequency differ among these methods.

A number of combined mixing devices have been used by various workers. One involves adding the powder to the liquid constituents in a stainless bowl that is then placed on a vibrating plane (50 vibrations/s) while stirring the mixture with a polypropylene spatula [92]. Another device is a proprietary machine (Hauschild, Germany) that simultaneously mixes and centrifugates (12 s, 2950 rpm) the cement mixture [93].

It was confirmed that porosity of bone cement decreased by applying vacuum and/or centrifugation. The decrease in porosity leads to improvement in fatigue performance. In two separate studies, Lewis found that the increased viscosity of bone cement did not improve the fatigue performance as much as vacuum mixing, and that the most important factor in fatigue performance of bone cement was the weight-average molecular weight (MWw), rather than mixing method [94,95].

Hansen et al. [96,97] and Fritsch et al. [98] claimed that vacuum mixing was not suitable for all commercial bone cements and this method provided good improvement in fatigue performance of a few commercial products. Gilbert et al. showed that vacuum mixing increased the shrinkage of bone cement and this shrinkage under certain constrained conditions may result in the development of porosity at the implant–bone cement interface [99]. Naturally, this would weaken the mechanical attachment of bone cement to implant.

The effect of mixing methods on monomer liberation was studied by Bettencourt et al. [100]. Residual monomer content was determined by high performance liquid chromatography. It was reported that monomer release in phosphate buffer is significantly reduced when vacuum is applied.

Following the mixing of the powder and liquid constituents using any of the aforementioned processes, the usual procedure is to immediately inject the cement dough into a cartridge or tube with a device, usually a cement gun made of polyvinylchloride (PVC) or polytetrafluoroethylene (PTFE), and pressurize the mold into the desired cavity where final setting occurs.

X. ADDITIVES

A. Radio-opacity

Polymethylmethacrylate is not a radiopaque material, i.e., it is almost impossible to determine the borders of the cement applied during the surgery by ordinary x-ray imaging. Until 1972, polymethylmethacrylate did not contain any radiopaque material and was, therefore, radiolucent. Since 1972, radiopaque materials—barium sulfate ($BaSO_4$) or zirconium dioxide (ZrO_2)—have been added into the bone cement in order to provide radio-opacity. Radiopaque material addition of about 10% (w/w) to the powder part provides sufficient radio-opacity to determine the borders of the bone cement by ordinary x-ray imaging. Otherwise, the areas which bone cement occupies can be determined by using magnetic resonance imaging (MRI) [101].

Addition of radiopaque materials may have some disadvantages. It was observed that, osteolysis, i.e., bone resorption around bone cement application area, was more severe when radiopaque agents were used [102]. This situation was more evident in $BaSO_4$ cases than in ZrO_2 cases [74,103]. On the other hand, pollution of zirconium dioxide by radioactive elements is possible. Various bone cements containing zirconium oxide (Palacos®, Implast®, Sulfix-6®) show a certain degree of radioactivity. These x-ray contrast media remain in the body for decades as components of the bone cement. Although no specific study has been reported, this situation seems to increase cancer incidence [104].

In addition to two former molecules, the possibility to confer radiopacity by introducing an x-ray opaque iodine containing methacrylate in the liquid phase of the bone cement has been studied and 2,5-diiodo-8-quinolyl methacrylate (IHQM) was proposed as a new radiopaque agent [105]. It was given that the incorporation IHQM produced a decrease in the value of peak temperatures and a slight increase in the setting time [105]. When static and dynamic mechanical properties of acrylic bone cements that include $BaSO_4$, ZrO_2, and IHQM were examined, it was shown that in relation to the radiolucent cement, the addition of zirconium dioxide improved the tensile strength significantly, the fracture toughness, and the fatigue crack propagation resistance. In contrast, the addition of barium sulfate produced a decrease in the tensile strength, but did not affect the fracture toughness and improved the crack propagation resistance. When the iodine-containing monomer IHQM was used, although the tensile strength and the fracture toughness increased, the fatigue crack propagation resistance remained as low as it was for the radiolucent cement [106,107].

Organo-bismuth compounds such as triphenyl bismuth were also studied as radiopaque agents by Deb et al. It was found that addition of triphenyl bismuth in the bone cement matrix up to 25% by weight of the polymer did not affect the polymerization temperature and setting time. If the addition was done via dissolution in monomer phase, an increase in strain and reduction in brittleness were observed [108].

B. Antibiotics

Surgical operating rooms have sterile conditions, but even under these conditions some bacteria can pass through all of the protective barriers and contaminate the open body tissues during the surgery. In order to prevent postoperative infections, some small quantities of antibiotics can be added into the bone cement. Some scientists suggested that the destruction of microorganisms in the tissues is more likely to be achieved by topical and intravenous administration of antibiotics during the operation than by incorporation of antibiotics in the cement [109].

Adding small quantities of antibiotics into bone cement has been a topic studied by many research groups. Two important points are the release of antibiotics from the cement matrix and the effect of antibiotics on the properties of the cement. Separate experiments indicate that each antibiotic can diffuse out from polymerized cement, and the concentration of each antibiotic is consistently above the minimal inhibitory concentration of 96% of the isolates [110]. For example, when the prophylactic effect of gentamicin-containing bone cement on postoperative infections in total hip arthroplasties was compared with that of systemically given antibiotics, it was observed that incidence of postoperative infections in the patients with gentamicin-containing bone cement was less than the group which was treated with systemic antibiotic therapy. The difference was statistically significant [111]. It was also reported that the presence of small amounts of antibiotics in acrylic cement had no influence on the compressive and diametral tension strengths of the cement. Mechanical tests demonstrated that adding a small amount of antibiotic did not reduce the strength of the cement below acceptable standards, and did not change the handling characteristics [112,113]. But contrary results were also reported. It was claimed that addition of antibiotics caused a reduction in the fatigue life of bone cement [114] and that large amounts of antibiotics decreased the compressive and tensile strengths of bone cements [115].

Otsuka et al. reported that Cephalexin release from bioactive glass containing bone cements was observed for more than 2 weeks and in some cases over 4 weeks, and the cements set in 5 min [116,117]. When Indomethacin was used as a model drug and loaded to a self-setting bioactive glass bone cement and implanted in the subcutaneous tissue of the rats, maximal release in a few days was reported with excellent biocompatibility [118].

In vitro gentamicin sulfate release from CMW 1 acrylic bone cements was studied by Cabanillas et al. by using o-phthaldialdehyde as a derivatizing reagent, and it was reported that the method is easy, quick, and reliable for the detection of antibiotics [119].

Frutos et al. [120] studied release of gentamicin sulfate from bone cements prepared from CMW 1 with addition 2-hydroxyethyl methacrylate monomer to liquid and poly(N-vinyl-2-pyrrolidone) to the solid components of CMW cement. A three-step model for release kinetics which include initial burst, diffusion-controlled, and release from the formed cracks and pores, was proposed. Sivakumar et al. [121] observed zero-order pattern for release of gentamicin from coralline hydroxyapatite granules.

Bone cements can be used as local drug carriers, and the release depends on the properties of the carrier. If the release is too low, the concentrations of antibiotics will not be inhibitory [122]. Bone cements can also be used for release of antineoplastic agents. Methotrexate release from calcium phosphate systems was studied by Lebugle et al. [123]. In the in vivo studies, which involved implantation in the external femoral condyle of rabbits, it was reported that the circulating concentration of methotrexate was below toxic level and no local necrosis was observed.

C. Fibers

Polymethylmethacrylate was first used for orthopedic implant fixation in total hip prostheses in the 1960s and since then it has been used for stabilization of knee, shoulder, elbow, and other prostheses. However, in the first applications many of the prostheses were removed because fractures of the surrounding PMMA cement were significant [7]. Therefore many studies have been carried out to improve the mechanical properties of bone cement. Early researchers tried to improve the mechanical properties by using additives, and they carried out various mechanical tests such as tensile, compressive, three-point bending, fracture toughness, and fatigue tests.

One of the effords for improving the properties of bone cement involves dispersing small quantities (typically 1–2 vol%) of enforcing materials such as carbon [124], graphite [125], aramid [126], bone particle [127,128], polyethylene [129], titanium [130], ultra high molecular weight polyethylene [131], or PMMA [132] fibers in the cement matrix. Although most of the results regarding the properties of these filamentary composite materials have been encouraging, the biocompatibility issues regarding some of these fibers are not known yet [133]. Thus none of these reinforced cements have been approved as yet by the FDA for clinical use.

The first attempt to use additives in bone cement was presented at the Fifth Biomaterials Congress by Knoell in 1975. He proposed that adding small amounts of graphite fiber would enhance the mechanical properties of bone cements [125]. In 1979 Wright and Trent [133] studied aramid fiber reinforcement of bone cement and conducted both tensile and fracture toughness tests with aramid fiber–containing cements and observed enhancement of mechanical properties. By adding up to 7% aramid fibers having average lengths of 1.3 cm the tensile strength of bone cement could be increased from 30.8 to 42.8 MPa, and the fracture toughness from 1.53 to 2.85 MPa $m^{1/2}$.

Saha and Pal studied carbon and aramid fiber reinforcement of bone cement. In their 1982 study they examined the stress relaxation, creep behavior, and other mechanical properties of bone cement that had been reinforced by 6-mm long and 8-μm thick chopped carbon fibers. They observed that increased deformation due to creep in 24 h was reduced from 70 to 45% by carbon fiber reinforcement, and stress relaxation in carbon fiber–reinforced specimens was greater compared to unreinforced bone cement [134]. They also prepared bone cement specimens containing 1% carbon fibers with 2 and 4% aramid fibers. Although all fiber-reinforced specimens exhibited good mechanical results, they found that adding 4% aramid fibers into the bone cement formulation provided best improvement in compressive strength, and adding 2% aramid fibers into the bone cement formulation provided best improvement in compressive elastic modulus [135].

Pourdeyhimi et al. [136] studied Kevlar 29® fiber reinforcement of bone cement and observed improved fracture toughness even at very low fiber contents. By adding up to 7% Kevlar 29® fibers, the flexural strength increased from 67.4 to 82.6 MPa, flexural elastic modulus increased from 1.28 to 1.3 GPa, fracture toughness increased from 1.47 to 2.61 MPa $m^{1/2}$.

Wagner et al. [137] studied the use of high-performance polyethylene fibers and activated the surfaces by oxygen plasma for a better adhesion in order to reinforce bone cement. But no significant reinforcement was observed

Gilbert et al. [138,139] studied a self-reinforced composite PMMA (SRC-PMMA). They described a composite material consisting of high strength, high ductility, and made of PMMA fibers embedded in a PMMA matrix. The results of their study showed that the tensile strength, tensile modulus, and tensile strain to failure were significantly greater for the SRC-PMMA compared to the commercial PMMA cements.

Kim and Yasuda studied plasma surface treatment of fillers, and they provided chemical attachment of those fillers to PMMA matrix that would lead to better mechanical results than the previously reported results [140].

Fibers of PMMA were developed by Wright et al. in 2002 by extrusion of PMMA melt. In the study, the effects of melt viscosity on the resultant properties of fibers were examined and processing parameters were optimized. Fibers with ultimate tensile strength values in the range of 60 to 225 MPa, moduli values from 1.5 to 3.5 GPa, and strain to failure from 10 to 40% were obtained. It is proposed that the fibers could be used in bone cement formulations to improve their mechanical properties [141].

Murakami et al. developed rubber-toughened PMMA powder, and they reported that the cement containing rubber demonstrated significantly high values in elongation and fracture toughness compared to the ones prepared without rubber [142].

XI. MODIFICATIONS OF ACRYLIC-BASED BONE CEMENT COMPOSITIONS

Polymethylmethacrylate has been used as a material for prosthetic fixation since the 1960s. Since then it has been the basic molecule of acrylic-based bone cements. Alternatively, scientists synthesized copolymers of MMA to improve the mechanical properties of bone cement. Simplex-P® (Howmedica) includes poly(MMA-*co*-styrene) in powder part, and Sulfix-6® (Sulzer Brothers, Switzerland) includes poly(MMA-*co*-butyl methacrylate) in the powder and MMA and BMA monomers in the liquid part. The curing temperatures of these two do not differ much from the other commercial bone cements.

One method for improving the properties of bone cements involves the development of new formulations. Large numbers of formulations have been introduced and these have constituents that differ marginally or markedly from those in the current generation of commercial formulations. For example, a composition called Bioactive Bone Cement was presented [143,144]. These cements consist of silane-treated $CaO–MgO–SiO_2–P_2O_5–CaF_2$ glass powder as the filling particles and bisphenol-α-glycidyl methacrylate–based resin as the organic matrix.

Some scientists started to work on bioerodible bone cement composites [145], wherein as the polymer degrades, bone ingrowth would replace the volume and eventually the polymer composite would be completely replaced by new bone. Domb et al. [146] synthesized some degradable polymeric structures to use as bioerodible bone cement. They produced polypropylene fumarate (PPF), crosslinked it to methylmethacrylate monomer, and added tricalcium phosphate (TCP) and calcium carbonate to bone cement dough. The compressive yield strength of the resultant bone cement was less than one-tenth of the ordinary bone cement. Gresser et al. [147] formulated a resorbable bone cement from N-vinyl-2-pyrrolidinone (VP), the unsaturated polyester polypropylene fumarate (PPF), and the inorganic filler tribasic calcium phosphate or hydroxyapatite. Lewandrowski et al. [148] investigated an injectable form of a resorbable bone cement based on in situ crosslinking of the unsaturated polyester poly(propylene glycol-*co*-

fumaric acid) (PPF). They filled the material with calcium gluconate/hydroxyapatite (CG/HA) and cured to a hard cement that was degradable by hydrolysis. They evaluated the osteoconductive properties of the injectable cement and observed good ingrowth of new bone into the cement in vivo.

In the 1990s the search for new formulations of bone cements for the purpose of reducing the curing temperature and the adverse biological effects became important. In 1992 Trap et al. [149] tried a new formulation of bone cement: a methylmethacrylate/n-decyl methacrylate/isobornyl methacrylate (MMA/DMA/IBMA) polymer called Boneloc®. They observed a considerable decrease of both released and residual MMA. Among the acrylic-based bone cements, the commercial preparation which had the least curing temperature was Boneloc (Polymers Reconstructive A/S, Farum, Denmark) [79,150–153]. In Boneloc the liquid part included 49.28% methylmethacrylate, 29.57% n-decyl methacrylate, and 19.71% isobornyl methacrylate. The powder part included 88.91% poly(MMA/DMA). Some studies involving Boneloc are summarized below.

In 1993 Nimb et al. [154] observed improved cortical histology at canine tibial diaphysis after Boneloc application compared to inert bone wax and conventional PMMA applications. They observed that the new formulation of the cement was characterized by a reduced exothermic temperature at curing and reduced leakage of chemicals to the adjacent bone, so it was claimed that new cement was better than PMMA bone cement in terms of both bone necrosis and repair as well as bone remodeling.

In 1995 Wykman and Sanderjoo [155] investigated the thermal properties of Boneloc during total hip replacement using a thermocouple connected to a digital thermometer. They measured the peak temperature at bone–bone cement interface as 43°C, and they found no difference regarding the setting time of the cement. But in the same year Suominen [156] reported that Boneloc cement failure occurred in four of eight total hip replacements in 3 years. Similarly, Havelin et al. [150] reported that Boneloc cement failure rate was six times greater than other bone cements for acetabular components in total hip replacements performed on 760 patients. In 2000, Walczak et al. [157] reported that aseptic loosening showed 79% survival in 4 years, and failure of the bone cement was evident radiologically in 55% of the patients. These values are significantly worse than the ones that were performed with other commercial products. Therefore, the product was withdrawn from clinical use.

To summarize, the curing temperature of Boneloc bone cement was decreased by changing the chemical composition of the polymer, but this composition did not have acceptable fatigue strength. Strain-controlled fatigue tests of Boneloc produced good results, but stress-controlled fatigue test results of Boneloc were all worse than the other commercial cements [79,157].

Pascual et al. [158] analyzed the effect of size and size distribution of polymethylmethacrylate beads on the classical kinetic parameters peak temperature and setting time for acrylic bone cement formulations prepared with PMMA particles with diameters in the range 10–60 μm and a relatively wide size distribution. In addition, they studied the combined effects of the concentration of the free radical initiator benzoyl peroxide and the activator N,N-dimethyl-4-toluidine for different particle sizes. They compared their data with those of commercially available formulations like CMW and Rostal. Their results indicated that the use of PMMA particles with average diameter of 50–60 μm and a relatively wide size distribution (10–140 μm diameter) significantly changes the curing parameters (peak temperature and setting time) of the cement formulations in comparison with the classical behavior of the commercial systems CMW and Rostal, without any noticeable loss in the mechanical properties. Pascual and coworkers found that as the PMMA average bead size decreases, the peak curing temperature increases and the average curing time decreases; as the concentrations of BPO or DMT decreases curing temperature decreases and setting time increases.

Vila et al. [159,160] investigated the effect of adding an elastomeric second phase, acrylonitrile–butadiene–styrene (ABS), into the liquid part of bone cement. They studied the effect on the fracture toughness and fatigue crack propagation behavior of the modified polymethylmethacrylate bone cement. Their results showed that with an increase in ABS content ductility and fracture toughness both increased, although both strength and stiffness decreased. The fracture toughness increased up to 60% when the amount of ABS reached 20% (v/v). They observed a decrease in the crack propagation at between one and two orders of magnitude. Storage in saline solution at 37°C also caused a decrease of about two orders of magnitude in the crack propagation rate for the plain and modified cements prepared in air or under vacuum. In the same study they found that porosity did not have any noticeable effect on the fatigue crack propagation behavior of the cement.

Jayabalan et al. prepared random and block copolymers of poly(propylene fumarate-*co*-ethylene glycol) with addition of HA. They reported that these cements have low setting temperatures but low compressive strength [161].

In order to make the bone cement adhere to the bone and metallic prostheses, the cement composed of 4-methacryloyloxyethyl trimellitate anhydride (4-META) and methylmethacrylate as monomers and tri-n-butyl borane (TBB) as an initiator was evaluated in a few studies. The results showed clear adherence of bone cement to both bone and metal [162–165]. Yang et al. used azobisisobutyronitrile (AIBN) as one of the initiators of bone cements [166].

In order to increase the biocompatibility of the cements, compositions similar to protein structure were developed by Muh et al. [167]. Amino acid–based dimethacrylate monomer was prepared by adding hydroxyethyl methacrylate to lysine diisocyanate. Low volume shrinkage and high thermal stability with good mechanical properties were observed. With addition of hydroxyapatite the Young's modulus increased from 3740 to 5250 MPa. Vazquez et al. [168] derived acrylic bone cements with the use of tertiary aromatic amine derived from oleic acid and acrylic monomer derived from the same acid. 4-N,N-Dimethyl aminobenzyl oleate was used as activator. Cements demonstrated very low polymerization exotherm values ranging between 55–62°C. Also increase in strain to failure and decrease in Young's modulus compared to Palacos was observed. Partial substitution of MMA by oleyloxyethyl methacrylate caused significant increase in tensile strength and elastic modulus.

A. Calcium Phosphate Bone Cements

In bone structure, the main inorganic compounds are $Ca_3(PO_4)_2$, $CaCO_3$, $CaF_2.3Ca_3(PO_4)_2$, and $CaCO_3.3Ca_3(PO_4)_2$, which form complex crystals similar to those of the apatite group. Therefore it is believed that addition of these kinds of substances would increase biocompatibility of cements giving better anchorage to the bone.

Calcium phosphate bone cements (CPCs) contain different crystal structures [such as alpha-tricalciumphosphate, beta-tricalciumphosphate, or hydroxyapatite, $Ca_{10}(PO_4)_6(OH)_2$] of calcium phosphate granules dispersed in polymeric matrix. Calcium hydroxy apatite, commonly called hydroxyapatite (HA), can also be defined in inorganic chemistry as calcium phosphate fluoride hydroxide chloride. The general formula is $Ca_{10}(PO_4)_6(F,OH,Cl)_2$. Biological apatites exist in calcified tissues such as enamel, dentin, and bone, and they differ from pure HA in stoichiometry, composition, and crystallinity. Some minor elements (e.g., Sr^{2+}, Pb^{2+}, Ba^{2+}, etc.) may also exist in apatite structure.

Hydroxyapatide is proven to be a biocompatible material and it is osteoconductive [169–172], i.e., hydroxyapatite strongly integrates with bone [169]. So a large number of studies were carried out on addition of hydroxyapatite into bone cement formulations in order to improve mechanical strength as well as to enhance biocompatibility of the cement. It was observed that existence of inorganic materials decreases the mechanical properties of bone cements. Therefore, the use of self-curing calcium phosphate cements is limited to non-load-bearing applications. Some of them are resorbable and only used in bone graft applications [173]. There are also studies to strengthen the commercial CPCs by adding various chemicals. It was reported that addition of polycations poly(ethyleneimine) and poly(allylamine hydrochloride) exhibited compressive strength six times greater than the pure alpha-BSM trademark material [174].

Two paste systems containing HA were developed by Saito et al. [175]. One paste contains HA granules, the monomer of bisphenol A–glycidyl methacrylate and 2,2-bis-(4-methacryloyl ethoxyphenyl)propane and triethyleneglycol dimethacrylate in 35:35:30 w/w ratios and benzoyl peroxide as initiator. The other paste includes HA powder, acrylic monomers and diethanol-p-toluidine as accelerator. It was reported that tensile and bending strength values of these cements were higher than PMMA cements.

Giunti et al. observed enhanced mechanical properties as well as a decrease in curing temperature for hydroxyapatite-added bone cements [170]. Castaldini and Cavallini also observed a decrease in curing temperature for the cements that contain HA in the powder part [171].

Park et al. examined properties of bone particle–impregnated PMMA and reported that the elastic modulus increased from 2.82 to 3.50 GPa by adding up to 35% bone particles into bone cement formulation [127].

Dandurand et al. proposed that HA impregnation into the acrylic bone cement decreased the mechanical strength [172]. They linked the PMMA powder and HA by using thermally stimulated current and obtained better mechanical characteristics than the ordinary PMMA bone cement. Mongiorgi et al. reported that the curing temperature of the bone cement decreased from 100 to 58 °C by adding 66 % α-tricalcium phosphate (α-TCP) [176].

Deb et al. [177] studied water absorption characteristics of modified hydroxyapatite–reinforced poly(ethyl methacrylate-n-butyl methacrylate) (PEMA-nBMA) bone cements. They found that the introduction of HA reduced the water uptake of the bone cement, with a further decrease if the HA filler was surface treated with a silane coupling agent.

Yang et al. [178] prepared a kind of organic/inorganic hybrid material, polymethylmethacrylate-SiO_2 (SiO_2 content 72% w/w) by incorporating PMMA covalently into an SiO_2 glass network via the sol-gel approach. They examined the effect of the addition of tricalcium phosphate (TCP), hydroxyethyl methacrylate (HEMA), and ethylene glycol dimethacrylate (EGDMA) on the properties of the sol-gel hybrid bone cement. They found that the addition of TCP increased the Young's modulus and the polymerization time; a decrease was observed for the tensile, bending, and compressive strengths and the polymerization temperature.

Kwon et al. studied biocompatibility and mechanical properties of 10 and 30% HA-impregnated bone cement formulations. Their mechanical test results showed significant decrease of the flexural and diametral tensile strength linear with the increased amount of HA, but a significant increase of the attachment of bone cement to bone tissue [179].

Morita et al. added 4-methacryloxyethyl trimellitate anhydride (4-META) into MMA monomer as an adhesion promoting agent. They obtained an improvement in mechanical properties without disturbing the biocompatibility of the HA-containing bone cement [180].

Yang et al. used dynamic mechanical analysis (DMA) to characterize the properties of acrylic bone cement with the addition of tricalcium phosphate, hydroxyethyl methacrylate, and ethylene glycol dimethacrylate [181]. They confirmed that TCP was incompatible with bone cement and they found that by adding HEMA and EGDMA to bone cement that contained TCP, the incompatibility between acrylic bone cement and TCP could be ameliorated.

Vallo et al. [182] examined the influence of HA filler proportion on the flow characteristics, mechanical properties, and residual monomer content of the resultant composite. It was found that degree of polymerization was not affected by the addition of HA, and porosity and the pore size of the hardened bone cement increased with increasing amount of the particulate filler. Their results showed that the workability of the modified formulation limited to a maximum of 15% w/w incorporation of the ceramic filler. In their study, up to 15% HA in the powder component of the bone cement increased the flexural modulus from 2 to 2.5 GPa. For the samples having 2.5% HA, compressive strength increased up to 4% higher than that of the unreinforced cement, but began to decrease when HA percentage increased more. Bone cement formulations containing proportions of HA higher than 2.5% displayed similar values of compressive strength compared to the unreinforced cement. Similarly, the samples having 2.5% HA displayed a maximum average value of 1.62 MPa $m^{1/2}$, which is statistically higher than the fracture toughness of the unreinforced cement, 1.33 MPa $m^{1/2}$, but began to decrease when HA percentage increased more. It was found that the compressive yield strength of the modified bone cement having 15% HA in the powder component was slightly lower than the unreinforced cement, and fracture toughness of the modified bone cement having 15% HA in the powder component was slightly higher than the unreinforced cement [182].

Curing time and cure duration of HA-containing bone cement was investigated monitoring the propagation of ultrasonic pulses through a sample of curing bone cement, and it was reported that cure time was not significantly affected by the addition of HA particles [183]. Hydroxyapatite-containing PMMA bone cement formulations were developed by Serbetci et al., and it was reported that addition of hydroxyapatite decreased the polymerization temperature and increased the compressive strength [184].

Physicochemical, biological, and handling properties of an injectable calcium phosphate bone cement were examined by Ooms et al. by implanting the paste into femoral trabecular bone of goats [185]. X-ray diffraction studies showed that cement had set as an apatite with formation of new bone and without any inflammatory reaction of fibrous encapsulation.

In apatite cements some certain porosity is needed to assure that new bone tissue grows in it. Beruta et al. prepared porous bone substitute compounds by mixing aqueous dispersions of α-tricalcium phosphate grains and calcium-deficient hydroxyapatite powders with PMMA [186]. The inorganic powders placed in the polymeric matrix acted as local microsponges, and new bone formation was observed in the pores when implanted in rabbits.

Nilsson et al. prepared partially resorbable bone cements containing an apatite phase with a resorbable phase by mixing α-tricalcium phosphate, calcium sulfate hemihydrate and aqueous solutions of sodium hydrophosphate [187]. Lu et al. studied the biodegradation mechanism of HA and β-tricalcium phosphate structures implanted into cavities in rabbit femoral and tibiae condyles [188]. They concluded that biodegradation was rapid in β-tricalcium phosphate and was very weak in hydroxyapatite structures.

Arcis et al. studied mechanical properties of light-curing polymers by adding micro- or nanoparticles of hydoxyapatite modified with citric, malic, acrylic, or methacrylic acid [189]. They reported that addition of hydoxyapatite increased Young's modulus and surface hardness of the material, but caused a decrease in flexural strength. Surface modification of hydroxyapatite, especially for micro samples, favored mechanical properties. The effect of citric acid on the setting properties of injectable calcium phosphate bone cement was studied by Sarda et al. [190]. Addition of citric acid retarded dissolution precipitation reactions and increased compressive strength.

Injectable calcium phosphate bone cements were implanted in bone defects at the distal end of rabbit femora by Khairoun et al. [191]. Examinations with scanning and transmission electron microscopies showed good biocompatibility with direct contact of the cement and bone.

Crystalline carbonate and sodium carbonate substituted hydroxyapatite powders were prepared by Suchanek et al. [192]. It was reported that chemical composition and crystallinity of these powders were similar to those of mineral constituents of hard tissues. Therefore it was proposed that these materials can be used for fabrication of bone-resembling implants.

B. Glass Based (Bioactive) Bone Cements

Bioactive materials are ones that achieve a specific biological response at the interface of the biomaterial which forms a bond between the tissues and the material. Generally the formed layer is biologically active hydroxycarbonate apatite, which is chemically and structurally equal to the mineral phase of the bone. The thickness and the strength of the bonding depends on the chemical composition [193–195].

Clinical applications of bioactive glasses and glass ceramics are concentrated on non-load-bearing applications such as middle ear surgery to replace damaged ossicles [196], in vertabral surgery [197], or in dentistry.

The reactions taking place in the interface occurs in various stages. The first is ion exchange and in this step Na^+ or K^+ ions exchange with H^+ ions. The second step is dissolution of silica network forming soluble $Si(OH)_4$. The third step is deposition of a SiO_2-rich layer on the surface containing alkali or alkaline earth cations. The fourth step is migration of Ca^{2+} and $PO_4{}^{3-}$ groups to the surface and formation of Ca-P containing film. And the fifth step is crystallization of this film by incorparation of OH^-, $CO_3{}^{2-}$, or F^- ions forming hydroxyl, carbonate, or fluorapatite layers [198,199]. It was reported that the depth of the calcium- and phosphate-enriched surface reaches 20 nm in 2 min and reaches 200 nm in an hour [200].

Some bioactive glasses include materials such as Bioglass®; bioactive glass ceramics such as Ceravital®; AW glass ceramics, dense hydroxyapatite such as Durapatite® and Calcitate®; or bioactive composites such as polyethylene–Bioglass, polysulfone–Bioglass, and polyethylene–hydroxyapatite Hapex® mixtures [201,202].

Many bioactive silica glasses contain five Ca for each P as molar ratio and 45% (w/w) SiO_2. It was reported that glasses do not bond to bone if the Ca/P ratio is substantially lower than 5:1 [203]. Gross et al. reported that addition of Al_2O_3, Ta_2O_5, TiO_2, Sb_2O_3, and ZrO_2 into bioactive silica glass ceramics (Ceravital) inhibit bone formation [204,205].

Some mica-containing bioactive multiphase glass ceramics were found to form bonds with bone even though they have Al_2O_3 in their structures [206,207]. For various glass compositions prepared from the powders of $(NH_4)_2HPO_4$, $Ca(OH_2)$, and Na_2CO_3, the setting times were found to be in the range of 7 min to over 3 h [208].

Tamura et al. [209] studied glass or glass ceramic powder containing bone cements of bisphenol A–glycidyl methacrylate and implanted them into the tibial metaphyses of rabbits. They reported that the ones containing glass ceramic powder were stronger than the others. Kobayashi et al. [210] compared bone-bonding strengths of bone cements containing apatite and wollastonite (AWC), glass ceramic, hydroxyapatite, and beta-tricalcium phosphate with organic matrix of Bis-GMA. X-ray microanalysis showed that AWC had contacted to the bone with Ca-P rich layer when implanted to rabbits and demonstrated higher failure loads than the others.

Bioactive beads consisting of $MgO–CaO–SiO_2–P_2O_5–CaF_2$ glass at various sizes were added into PMMA polymeric matrix at different proportions and packed into the intramedular canals of rat tibiae to examine osteoconductivity. It was reported that as the glass bead size decreased and the content increased, compressive strength and elastic modulus values as well as affinity index were also increased [211,212]. Addition of phosphoric ester monomer to the cement as an adhesion-promoting agent increased the bending strength and decreased Young's modulus [213].

Glass ionomer cements, which are used in dentistry, set rapidly and have good mechanical strength, but it was reported that the presence of even a small quantity of polyacrylic acid inhibits apatite formation in the body environment [214].

Apatite and wollastonite containing glass ceramic (AW-GC) powder was developed and mixed with bisphenol A–glycidyl dimethacrylate–based resin. It was reported that AW-GC particles were completely absorbed and replaced by new bone when implanted into tibial metaphyses of rabbits [215].

Partially crystallized glass beads added (70% w/w) into a PMMA matrix increased the bending strength and achieved very high osteoconductivity, and these cements were proposed for weight-bearing conditions [216].

Bioactive glasses are also used for coating the metallic implants to achieve better bonding between metal–glass and glass–bone interfaces. Some bioactive glass materials bond to soft tissues as well as to hard bony structures [217,218].

C. Two-Solution Acrylic Bone Cement

In order to obtain homogeneous dough structures, two-solution systems such as solutions of polymethylmethacrylate powder predissolved in methylmethacrylate have been developed. These cements follow the same addition polymerization reactions as commercial ones. But it was reported that porosity is eliminated by mixing and delivering via a closed system, decreasing the dependence of material properties on the surgical technique. In this study, two solutions of constant polymer-to-monomer ratio (80 g of PMMA:100 mL of MMA) were prepared and benzoyl peroxide initiator was added to the first solution, while N,N-dimethyl-p-toluidine activator was added to the second. It was reported that these two-solution cement compositions are comparable to Simplex P bone cement in polymerization exotherm and setting time and demonstrate flexural mechanical properties and flexural strengths and moduli higher than commercial Simplex P. Residual monomer content was significantly affected by both the individual concentrations of BPO and DMPT, and their molar ratios and had significantly higher residual monomer contents versus Simplex P; however, this can be attributed to their higher initial monomer concentration rather than a lower degree of conversion [219,220].

XII. CONCLUSION

Bone cements are substances based on acrylic polymers and are used in orthopedic and dental applications. Bone cement research is concentrated on enhancing mechanical properties and improving biological compatibility. As mentioned previously some studies attempt to decrease porosity by adjusting the ambient temperature or mixing techniques or to increase strength and compatibility by altering chemical composition by using acrylates other than PMMA or by adding various chemicals (such as co-polymers of acrylates, fibers, antibiotics or inorganic components such as HA and bioglass). There is a very delicate balance in the combination of all chemicals; therefore, an intense care should be taken and all the properties should be checked before any application, since each additive through improving one property may cause a weaking in other properties.

REFERENCES

1. Kiaer S. Preliminary report on arthroplasty by use of acrylic head. Stockholm: Cliniquièm Congrès International de Chirurgie Orthopèdique, 1951.

2. Haboush EJ. A new operation for arthroplasty of the hip based on biomechanics, photoelasticity, fast setting dental acrylic and other considerations. Bull Hosp It Dis NY 1953; 14:242.
3. Henrichsen E, Jansen K, Krogh-Poulson W. Experimental investigation of the tissue reaction to acrylic plastics. Acta Orthop Scand 1953; 22:141–146.
4. Worringer E, Thomalske G. Über die Plastiche Deckung von Schädelknochendefekten mit autopolymerisierender Kunstharzmasse. Eine neue Schnellmethode. Arch Psychiatr Nervenkr 1953; 191: 100–113.
5. Rau H. Plastische Deckung deformierter Schädeldefecte. Arch Chir 1963; 304:926–929.
6. Judet J, Judet R. The use of an artificial femoral head for arthroplasty of the hip joint. J Bone Surg Br 1956; 32:166.
7. Charnley J. Anchorage of the femoral head prosthesis of the shaft of the femur. J Bone Joint Surg Br 1960; 42:28–30.
8. Charnley J. Acrylic Cement in Orthopaedic Surgery. Baltimore: Williams and Wilkins, 1970.
9. Vazquez B, Deb S, Bonfield W, Roman JS. Characterization of new acrylic bone cements prepared with oleic acid derivatives. J Biomed Mater Res 2002; 63(2):88–97.
10. Pascual B, Gurruchaga M, Ginebra MP, Gil FJ, Planell JA. Goni. Influence of the modification of P/L ratio on a new formulation of acrylic bone cement. Biomaterials 1999; 20(5):465–474.
11. Turner RC, Atkins PE, Ackley MA, Park JB. Molecular and macroscopic properties of PMMA bone cement: free-radical generation and temperature change versus mixing ratio. J Biomed Mater Res 1981 May; 15(3):425–432.
12. Morita S, Kawachi S, Yamamoto H, Shinomiya K, Nakabayashi N, Ishihara K. Total hip arthroplasty using bone cement containing tri-n-butylborane as the initiator. J Biomed Mater Res 1999; 48(5): 759–763.
13. Linder L. Tissue reaction to methyl methacrylate monomer; a comparative study in the rabbit's ear on the toxicity of methyl methacrylate monomer of varying composition. Acta Orthop Scand; 1976 Feb; 47(1):3–10.
14. Brauer GM, Steinberger DR, Stansbury JW. Dependence of curing time peak temperature and mechanical properties on the composition of bone cement. J Biomed Mater Res; 1986 Jul-Aug; 20(6):839–852.
15. Oldfield FF, Yasuda HK. ESR study of MMA polymerization by a peroxide/amine system: bone cement formation. J Biomed Mater Res; 1999 Mar 15; 44(4):436–445.
16. Tanzi MC, Sket I, Gatti AM, Monari E. Physical characterization of acrylic bone cement cured with new accelerator systems. Clin Mater 1991; 8(1–2):131–136.
17. Vazquez B, Elvira C, Levenfeld B, Pascual B, Goni I, Gurruchaga M, Ginebra MP, Gil FX, Planell JA, Liso PA, Rebuelta M, San Roman J. Application of tertiary amines with reduced toxicity to the curing process of acrylic bone cements. J Biomed Mater Res 1997; 34(1):129–136.
18. Vazquez B, San Roman J, Deb S, Bonfield W. Application of long chain amine activator in conventional acrylic bone cement. J Biomed Mater Res 1998; 43(2):131–139.
19. Kühn KD. Bone Cements: Up-to-Date Comparison of Physical and Chemical Properties of Commercial Materials. Berlin: Springer-Verlag, 2000.
20. Albright JA, Brand RA. Orthopaedic Basic Science: Appleton & Lange, 1987.
21. Official Website of Stryker Howmedica Osteonics www.howost.com.
22. Baleani M, Cristofolini L, Toni A. Temperature and ageing condition effects on the characterization of acrylic bone cement. Proceedings of the Institute of Mechanical Engineers Part H. J. Eng. Med 2001; 215:113–118.
23. Lidgren L, Bodelind B, Moller J. Bone cement improved by vacuum mixing and chilling. Acta Orthop Scand 1987; 58(1):27–32.
24. Hansen D, Jensen JS. Prechilling and vacuum mixing not suitable for all bone cements. Handling characteristics and exotherms of bone cements. J Arthroplasty 1990; 5(4):287–290.
25. Smeds S, Goertzen D, Ivarsson I. Influence of temperature and vacuum mixing on bone cement properties. Clin Orthop 1997(334):326–334.
26. Lewis G. Effect of mixing method and storage temperature of cement constituents on the fatigue and porosity of acrylic bone cement. J Biomed Mater Res 1999; 48(2):143–149.

27. Morberg P, Johansson CB, Malchau H. No biological advantage with a low temperature curing versus a conventional bone cement on experimental mechanical and histomorphometrical study in the rabbit tibia. J Mater Sci Mater Med 1999; 10:329–331.
28. Lewis GJ. Relative roles of cement molecular weight and mixing method on the fatigue performance of acrylic bone cement: Simplex P versus Osteopal. J Biomed Mater Res 2000; 53(1):119–130.
29. Jasty M, Davies JP, O'Connor DO, Burke DW, Harrigan TP, Harris WH. Porosity of various preparations of acrylic bone cements. Clin Orthop; 1990 Oct; 259:122–129.
30. Murphy BP, Prendergast PJ. The relationship between stress porosity, and nonlinear damage accumulation in acrylic bone cement. J Biomed Mater Res; 2002 Mar 15; 59(4):646–654.
31. Lewis G, Mladsi S. Effect of sterilization method on properties of Palacos R acrylic bone cement. Biomaterials 1998 Jan-Feb; 19(1–3):117–124.
32. Wixson RL, Lautenschlager EP, Novak MA. Vacuum mixing of acrylic bone cement. J Arthroplasty 1987; 2(2):141–149.
33. Lidgren L, Bodelind B, Moller J. Bone cement improved by vacuum mixing and chilling. Acta Orthop Scand 1987 Feb; 58(1):27–32.
34. Davies JP, O'Connor DO, Greer JA, Harris WH. Comparison of the mechanical properties of Simplex P Zimmer Regular and LVC bone cements. J Biomed Mater Res; 1987 Jun; 21(6):719–730.
35. Chin HC, Stauffer RN, Chao EY. The effect of centrifugation on the mechanical properties of cement. An in vitro total hip arthroplasty model. J Bone Joint Surg Am; 1990 Mar; 72(3):363–368.
36. Lewis G. Effect of two variables on the fatigue performance of acrylic bone cement: mixing method and viscosity. Biomed Mater Eng 1999; 9(4):197–207.
37. Topoleski LD, Ducheyne P, Cuckler JM. Microstructural pathway of fracture in poly(methyl methacrylate) bone cement. Biomaterials; 1993 Dec; 14(15):1165–1172.
38. Lennon AB, Prendergast PJ. Residual stress due to curing can initiate damage in porous bone cement: experimental and theoretical evidence. J Biomech; 2002 Mar; 35(3):311–321.
39. Zor M, Kucuk M, Aksoy S. Residual stress effects on fracture energies of cement bone and cement implant interfaces. Biomaterials; 2002 Apr; 23(7):1595–1601.
40. Puckett AD, Roberts B, Bu L, Mays JW. Improved orthopaedic bone cement formulations based on rubber toughening. Crit Rev Biomed Eng 2000; 28:457–446.
41. Krause W, Krug W, Miller JE. Cement bone interface–effect of cement technique and surface preparation. Trans Orthop Res. Soc 1980; 5:76.
42. Halawa M, Lee AJ, Ling RS, Vangala SS. The shear strength of trabecular bone from the femur and some factors affecting the shear strength of the cement–bone interface. Arch Orthop Trauma Surg 1978; 92:19–30.
43. Markolf KL, Amstutz HC. Penetration and flow of acrylic bone cement. Clin Orthop 1976 Nov-Dec; 121:99–102.
44. McCarthy TC, Wells MK, Gorman HA. Bone–methyl methacrylate interfacial shear strength: an experimental study of dogs. Am J Vet Res 1977 Jan; 38(1):75–79.
45. Krause WR, Krug W, Miller J. Strength of the cement–bone interface. Clin Orthop 1982 Mar; 163: 290–299.
46. Sakkers RJ, Valkema R, de Wijn JR, Lentjes EG, van Blitterswijk CA, Rozing PM. The intramedullary hydraulic pressure tolerance of the human femur. Clin Orthop; 1995 Feb; 311:183–189.
47. Bargar WL, Brown SA, Paul HA, Voegli T, Hseih Y, Sharkey N. In vivo versus in vitro polymerization of acrylic bone cement: effect on material properties. J Orthop Res 1986; 4(1):86–89.
48. Looney MA, Park JB. Molecular and mechanical property changes during aging of bone cement in vitro and in vivo. J Biomed Mater Res; 1986 May-Jun; 20(5):555–563.
49. Yettram AL, Wright KW. Biomechanics of the femoral component of total hip prostheses with particular reference to the stress in the bone cement. J Biomed Eng; 1979 Oct; 1(4):281–285.
50. Chang PB, Mann KA, Bartel DL. Cemented femoral stem performance. Effects of proximal bonding geometry, and neck length. Clin Orthop; 1998 Oct; 355:57–69.
51. Breusch SJ, Norman TL, Schneider U, Reitzel T, Blaha JD, Lukoschek M. Lavage technique in total hip arthroplasty: jet lavage produces better cement penetration than syringe lavage in the proximal femur. J Arthroplasty 2000; 15(7):921–7.

52. Scott S, McCaskie AW, Calder SJ, Wildin C, Gregg PJ. Current cementing techniques in hip hemi-arthroplasty. Injury 2001; 32(6):461–464.
53. Ramaniraka NA, Rakotomanana LR, Leyvraz PF. The fixation of the cemented femoral component. Effects of stem stiffness cement thickness and roughness of the cement–bone surface. J Bone Joint Surg Br; 2000 Mar; 82(2):297–303.
54. Dowd JE, Cha CW, Trakru S, Kim SY, Yang IH, Rubash HE. Failure of total hip arthroplasty with a precoated prosthesis. 4- to 11-year results. Clin Orthop; 1998 Oct; 355:123–136.
55. Dunne NJ, Orr JF. Flow characteristics of curing polymethyl methacrylate bone cement. J Eng Med 1998; 212:199–207.
56. www.bonecement.com.
57. Malchau H. Prognosis of total hip replacement. Scientific Exhibition presented at the 67th Annual Meeting of the AOS, Orlando, FL, March 15–19, 2000.
58. Topoleski LDT, Ducheyne P, Cuckler JM. Microstructural pathway of fracture in poly(methyl methacrylate) bone cement. Biomaterials 1993; 14:1165–1172.
59. Wang JS, Franzén H, Jonsson E, Lidgren L. Porosity of bone cement reduced by mixing and collecting under vacuum. Acta Orthop Scand 1993; 64:143–146.
60. Friis EA, Stromberg LJ, Cooke FW, McQueen DA. Fracture toughness of vacuum mixed PMMA bone cement. 19thAnnu. Mtg. Soc. Biomater, Birmingham, AL, 1993.
61. James SP, Schmalzried TP, McGarry FJ, Harris WH. Extensive porosity at the cement–femoral prosthesis interface: a preliminary study. J Biomed Mater Res 1993; 27:71–78.
62. Macaulay W, DiGiovanni CW, Restrepo A, Saleh KJ, Walsh H, Crossett LS, Peterson MG, Li S, Salvati EA. Differences in bone-cement porosity by vacuum mixing centrifugation and hand-mixing. J Arthroplastry 2002; 17(5):569–575.
63. Mann KA, Werner FW, Ayers DC. Mechanical strength of the cement–bone interface is greater in shear than in tension. J Biomech 1999; 32(11):1251–1254.
64. Mann KA, Ayers DC, Werner FW, Nicoletta RJ, Fortino MD. Tensile strength of the cement–bone interface depends on the amount of bone interdigitated with PMMA cement. J Biomech 1997; 30(4): 339–346.
65. Markolf KL, Amstutz HC. Penetration and flow of acrylic bone cement. Clin Orthop 1976; 121: 99–102.
66. Davies JP, Singer G, Harris WH. The effct of a thin coating of polymethylmethacrylate on the torsional fatigue strength of the cement–metal interface. J Appl Biomater 1992; 3:45–49.
67. Jones JD, Saigusa M, VanSickels JE, Tiner BD, Gardner WA. Clinical evaluation of Hydroxyapatite-coated titanium plasma-sprayed and titanium plasma-sprayed cylinder dental implants. Oral Surg Oral Med Oral Pathol Oral Radiol Endod 1997; 84:137–141.
68. Frondoza CG, Tanner KT, Jones LC, Hungerford DS. Polymethylmethacrylate particles enhance DNA and protein synthesis of human fibroblasts in vitro. J Biomed Mater Res 1993; 27(5):611–617.
69. Goldring SR, Jasty M, Roelke MS, Rourke CM, Bringhurst FR, Harris WH. Formation of a synovial-like membrane at the bone–cement interface. Its role in bone resorption and implant loosening after total hip replacement. Arthritis Rheum 1986; 29(7):836–842.
70. Pazzaglia UE. Pathology of the bone–cement interface in loosening of total hip replacement. Arch Orthop Trauma Surg 1990; 109(2):83–88.
71. Savarino L, Stea S, Ciapetti G, Paganetto G, Donati ME, Alvergna P, Pizzoferrato A. Microstructural investigation of bone–cement interface. J Biomed Mater Res 1995; 29(6):701–705.
72. Horowitz SM, Doty SB, Lane JM, Burstein AH. Studies of the mechanism by which the mechanical failure of polymethylmethacrylate leads to bone resorption. J Bone Joint Surg Am 1993; 75(6): 802–813.
73. Lazarus MD, Cuckler JM, Schumacher HR, Jr, Ducheyne P, Baker DG. Comparison of the inflammatory response to particulate polymethylmethacrylate debris with and without barium sulfate. J Orthop Res 1994; 12(4):532–541.
74. Sabokbar A, Fujikawa Y, Murray DW, Athanasou NA. Radio-opaque agents in bone cement increase bone resorption. J Bone Joint Surg Br 1997; 79(1):129–134.
75. Huiskes R. Some fundamental aspects of human joint replacement: heat generation and conduction analyses of acrylic bone cements in situ. Acta Orthop Scand 1980; 185:43.

76. Eriksson AR, Albrektsson T. Temperature threshold levels for heat-induced bone tissue injury: a vital microscopic study in the rabbit. J Prosthet Dent 1983; 50:101–107.
77. Bini SA, Gill K, Johnston JO. Giant cell tumor of bone. Curettage and cement reconstruction. Clin Orthop 1995; 321:245–250.
78. Krishnan EC, Nelson C, Neff JR. Thermodynamic considerations of acrylic cement implant at the site of giant cell tumors of the bone. Med Phys 1986; 13(2):233–239.
79. Kindt-Larsen T, Smith DB, Jensen JS. Innovations in acrylic bone cement and application equipment. J Appl Biomater 1995; 6:75–83.
80. Lewis G. Properties of acrylic bone cement: state of the art review. J Biomed Mater Res 1997; 38(2):155–182.
81. Vallo CI. Residual monomer content in bone cements based on poly(methyl methacrylate). Polym Int 2000; 49:831–838.
82. Gaszynski W, Lewicki R, Pruszynska-Bednarek A. The effect of bone cement used for stabilization of endoprostheses on blood clotting system activity. Anaesth Resusc Intensive Ther 1976; 4(4): 237–244.
83. Gaszynski W, Pruszynska-Bednarek A, Majka A, Siwik J. The effect of bone cement used for endoprosthesis stabilization on the activity of lysosomal enzymes in peripheral blood granulocytes. Anaesth Resusc Intensive Ther 1976; 4(3):187–194.
84. Jasty M, Davies JP, O'connor DO, Burke DW, Harrigan TP, Harris WH. Porosity of various preparations of acrylic bone cements. Clin Orthop Rel Res 1990; 259:122–129.
85. Hansen D, Jensen JS. Mixing does not improve mechanical properties of all bone cements. Acta Orthop Scand 1992; 63:13–18.
86. Demian HW, Wey AC, Shalaby SW. Bone cement with exceptionally uniform dispertion of radiopacifier. 21st Annu. Mtg. Soc. Biomater, San Fransisco, CA, 1995.
87. Lidgren L, Drar H, Moller J. Strength of polymethylmethacrylate increased by vacuum mixing. Acta Orthop Scand 1984; 55(5):536–541.
88. Lidgren L, Bodelind B, Moller J. Bone cement improved by vacuum mixing and chilling. Acta Orthop Scand 1987; 58(1):27–32.
89. Wixson RL, Lautenschlager EP, Novak MA. Vacuum mixing of acrylic bone cement. J Arthroplasty 1987; 2(2):141–149.
90. Norman TL, Kish V, Blaha JD, Gruen TA, Hustosky K. Creep characteristics of hand- and vacuum-mixed acrylic bone cement at elevated stress levels. J Biomed Mater Res 1995; 29(4):495–501.
91. Wang JS, Franzén H, Toksvig-Larsen TA. Comparison of seven bone cement mixing systems. Acta Orthop Scand 1994; 65(suppl. 260):62.
92. Lidgren L, Drar H, Möller J. Strength of polymethylmethacrylate increased by vacuum mixing. Acta Orthop Scand 1984; 55:536–541.
93. Trieu HH, Paxon RD, Carrol ME, Bert JM. A comperative study of bone cement preparation using a new centrifugation mixing technique. 20th Annu. Mtg. Soc. Biomater. Boston, 1994:416.
94. Lewis G. Effect of two variables on the fatigue performance of acrylic bone cement: mixing method and viscosity. Biomed Mater Eng 1999; 9(4):197–207.
95. Lewis G. Relative roles of cement molecular weight and mixing method on the fatigue performance of acrylic bone cement: Simplex P versus Osteopal. J Biomed Mater Res 2000; 53(1):119–130.
96. Hansen D, Jensen JS. Prechilling and vacuum mixing not suitable for all bone cements. Handling characteristics and exotherms of bone cements. J Arthroplasty 1990; 5(4):287–290.
97. Hansen D, Jensen JS. Mixing does not improve mechanical properties of all bone cements. Manual and centrifugation-vacuum mixing compared for 10 cement brands. Acta Orthop Scand 1992; 63(1): 13–18.
98. Fritsch E, Rupp S, Kaltenkirchen N. Does vacuum-mixing improve the fatigue properties of high-viscosity poly(methyl-methacrylate) (PMMA) bone cement? Comparison between two different evacuation methods. Arch Orthop Trauma Surg 1996; 115(3–4):131–135.
99. Gilbert JL, Hasenwinkel JM, Wixson RL, Lautenschlager EP. A theoretical and experimental analysis of polymerization shrinkage of bone cement: a potential major source of porosity. J Biomed Mater Res 2000; 52(1):210–218.

100. Bettencourt A, Calado A, Amaral J, Vale FM, Rico JMT, Monteiro J, Castro M. The influence of vacuum mixing on methylmethacrylate liberation from acrylic bone cement powder. Int J Pharmaceutics 2000; 219:89–93.
101. Fehrman DA, McBeath AA, DeSmet AA, Tuite MJ. Imaging barium-free bone cement. Am J Orthop 1996; 25(2):172–174.
102. Ingham E, Green TR, Stone MH, Kowalski R, Watkins N, Fisher J. Production of TNF-alpha and bone resorbing activity by macrophages in response to different types of bone cement particles. Biomaterials; 2000; 21(10):1005–1013.
103. Imhurst JA, Brooks RA, Rushton N. The effects of particulate bone cements at the bone-implant interface. J Bone Joint Surg Br; 2001; 83(4):588–592.
104. Hopf W, Hopf CG, Globel B. About radioactivity in some PMMA bone cements. Acta Orthop Belg 1990; 56(2):443–444.
105. Vazquez B, Ginebra MP, Gil FJ, Planell JA, Lopez Bravo A, San Roman J. Radiopaque acrylic cements prepared with a new acrylic derivative of yodo-quinoline. Biomaterials 1999; 20: 2047–2053.
106. Ginebra MP, Aparicio C, Albuixech L, Fernández-Barragán E, Gil FJ, Planell JA, Morejón L, Vázquez B, San Román J. Improvement of the mechanical properties of acrylic bone cements by substitution of the radio-opaque agent. J Mater Sci Mater Med 1999; 10:733–737.
107. Ginebra MP, Albuixech L, Fernandez-Barragan E, Aparicio C, Gil FJ, San RJ, Vazquez B, Planell JA. Mechanical performance of acrylic bone cements containing different radiopacifying agents. Biomaterials; 2002; 23(8):1873–1882.
108. Deb S, Abdulghani S, Behiri JC. Radiopacity in bone cements using an organo-bismuth compound. Biomaterials 2002; 23:3387–3393.
109. Hill J, Klenerman L, Trustey S, Blowers R. Diffusion of antibiotics from acrylic bonecement in vitro. J Bone Joint Surg Br 1977; 59(2):197–199.
110. Rosenthal AL, Rovell JM, Girard AE. Polyacrylic bone cement containing erythromycin and colistin. I. In vitro bacteriological activity and diffusion properties of erythromycin colistin and erythromycin/colistin combination. Int Med Res 1976; 4(5):296–304.
111. Josefsson G, Lindberg L, Wiklander B. Systemic antibiotics and gentamicin-containing bone cement in the prophylaxis of postoperative infections in total hip arthroplasty. Clin Orthop 1981; 159: 194–200.
112. Marks KE, Nelson CL, Lautenschlager EP. Antibiotic-impregnated acrylic bone cement. J Bone Joint Surg Am 1976; 58(3):358–364.
113. Murray WR. Use of antibiotic-containing bone cement. Clin Orthop 1984; 190:89–95.
114. Klekamp J, Dawson JM, Haas DW, DeBoer D, Christie M. The use of vancomycin and tobramycin in acrylic bone cement: biomechanical effects and elution kinetics for use in joint arthroplasty. J Arthroplasty 1999; 14(3):339–346.
115. Lautenschlager EP, Jacobs JJ, Marshall GW, Meyer PR, Jr. Mechanical properties of bone cements containing large doses of antibiotic powders. J Biomed Mater Res; 1976; 10(6):929–938.
116. Otsuka M, Fujita H, Nakamura T, Kokubo T. Effects of ceramic component on cephalexin release from bioactive bone cement consisting of bis-GMT/TEGDMA resin and bioavtive glass ceramics. Biomed Mater Eng 2001; 11:11–22.
117. Otsuka M, Matsuda Y, Kokubo T, Yoshihara S, Nakamura T, Yamamuro T. A novel skeletal drug delivery system using self-setting bioactive cement. IV. Cephalexin release from cement containing polymer coated bulk powder. Biomed Mater Eng 1993; 3:229–236.
118. Otsuka M, Nakahigashi Y, Matsuda Y, Kokubo T, Yoshiharas S, Fujita H, Nakamura T. The in vitro and in vivo indomethacin release from self-setting bioactive glass bone cement. Biomed Mater Eng 1997; 7:291–302.
119. Cabanillas PF, Pena ED, Barrales-Rienda JM, Frutos G. Validation and in vitro characterization of antibiotic-loaded bone cement release. Int J Pharmaceutics 2000; 209:15–26.
120. Frutos P, Diez-Pena E, Frutos G, Barrales-Rienda JM. Release of gentamicin sulphate from a modified commercial bone cement. Effect of (2-hydroxyethyl methacrylate) comonomer and poly(N-vinyl-2-pyrrolidone) additive on release mechanism and kinetics. Biomaterials 2002; 23:3787–3797.

121. Sivakumar M, Manjubala I, Rao KP. Preparation, characterization and in-vitro release of gentamicin from coralline hydroxyapatite–chitosan composite microspheres. Carbohydrate Polymers 2002; 49: 281–288.
122. Walenkamp GH. Gentamicin PMMA beads and other local antibiotic carriers in two-stage revision of total knee infection: a review. J Chemother 2001; 1:66–72.
123. Lebugle A, Rodrigues A, Bonnevialle P, Voigt JJ, Canal P, Rodriguez F. Study of implantable calcium phosphate systems for the slow release of methotrexate. Biomaterials 2002; 23:3517–3522.
124. Friis EA, Cooke FW, Yasuda HK. Fracture toughness of surface-treated carbon fiber reinforced composite bone cement. Fifth World Biomater. Congr. Toronto. Canada, 1996:913.
125. Knoell A, Maxwell H, Bechtol C. Graphite fiber reinforced bone cement. Ann Biomed Eng 1975; 3:225–229.
126. Pourdeyhimi B, Wagner HD, Schwartz P. A comparison of mechanical properties of discontinuous Kevlar 29 fiber reinforced bone and dental cements. J Mater Sci 1986; 21:4468–4474.
127. Park HC, Liu YK, Lakes RS. The material properties of bone-particle impregnated PMMA. J Biomech Eng 1986; 108:141–148.
128. Liu YK, Park JB, Njus GO, Stienstra D. Bone particle impregnated bone cement: an in vitro study. J Biomed Mater Res 1987; 21:247–261.
129. Friis EA, Cooke FW, McQueen DA. Toughening of PMMA bone cement by addition of flexible fibers with varying interface strengths. San Fransisco: 21st Annu. Mtg. Soc. Biomater, 1995:95.
130. Topoleski LDT, Ducheyne P, Cuckler JM. The fracture toughness of titanium-fiber-reinforced bone cement. J Biomed Mater Res 1992; 26:1595–1617.
131. Pourdeyhimi B, Wagner HD. Elastic and ultimate properties of acrylic bone cement reinforced with ultra-high-molecular-weight polyethylene fibers. J Biomed Mater Res 1989; 23:63–80.
132. Gilbert JL, Ney DS, Lautenschlager EP. Self-reinforced PMMA composites: strength, fatigue and fracture toughness evaluation. San Fransisco: 20th Annu. Mtg. Soc. Biomater, 1994:141.
133. Wright TM, Trent PS. Mechanical properties of aramid fiber reinforced acrylic bone cement. J Mater Sci 1979; 14:503–505.
134. Pal S, Saha S. Stress relaxation and creep behaviour of normal and carbon fiber reinforced acrylic bone cements. Biomaterials 1982; 3:93–95.
135. Saha S, Pal S. Improvement of mechanical properties of acrylic bone cement by fiber reinforcement. J Biomech 1984; 17:467–478.
136. Pourdeyhimi B, Wagner HD, Schwartz P. A comparison of mechanical properties of discontinuous Kevlar 29 fiber reinforced bone and dental cements. J Mater Sci 1986; 21:4468–4474.
137. Wagner HD, Cohn D. Use of high performance polyethylene fibers as a reinforcing phase in poly(-methylmethacrylate) bone cement. Biomaterials 1989; 10:139–141.
138. Gilbert JL, Ney DS, Lautenschlager EP. Self-Reinforced composite poly(methyl methacrylate): static and fatigue properties. Biomaterials 1994; 16:1043–1055.
139. Buckley CA, Gilbert JL, Lautenschlager EP. Thermomechanical processing of PMMA into high strength fibers. J App Polym Sci 1992; 44:1321–1330.
140. Kim HY, Yasuda HK. Improvement of fatigue properties of poly(methyl methacrylate) bone cement by means of plasma surface treatment of fillers. J Biomed Mater Res 1999; 48(2):135–142.
141. Wright DD, Lautenschlager EP, Gilbert JL. The effect of processing conditions on the properties of poly(methyl methacrylate) fibers. J Biomed Mater Res 2002; 63:152–160.
142. Murakami A, Behiri JC, Bonfield W. Rubber-modified bone cement. J. Biomater. Sci 1988; 23: 2029–2036.
143. Kobayashi M, Nakamura T, Tamura J, Lida H, Fujita H, Kokubo T, Kikutani T. Mechanical and biological properties of bioactive bone cement containing silica glass powder. J Biomed Mater Res 1997; 37(1):68–80.
144. Tamura J, Kitsugi T, Lida H, Fujita H, Nakamura T, Kokubo T, Yoshihara S. Bone bonding ability of bioactive bone cements. Clin Orthop 1997; 343:183–191.
145. Wise DL, Wentrowrth RL, Sanderson JE, Crooker SC. Evaluation of repair materials for evulsive combat-type maxillofacial injuries. in Biopolymeric Controlled Release Systems Wise DL, ed. Boca Raton. FL: CRC press, 1984.

146. Domb AJ, Laurencin CT, Israeli O, Gerhart TN, Langer R. The formation of propylene fumarate oligomers for use in bioerodible bone cement composites. J Polym Sci A 1990; 28:973–985.
147. Gresser JD, Hsu SH, Nagaoka H, Lyons CM, Nieratko DP, Wise DL, Barabino GA, Trantolo DJ. Analysis of a vinyl pyrrolidone/poly(propylene fumarate) resorbable bone cement. J Biomed Mater Res 1995; 29(10):1241–1247.
148. Lewandrowski KU, Gresser JD, Wise DL, White RL, Trantolo DJ. Osteoconductivity of an injectable and bioresorbable poly(propylene glycol-*co*-fumaric acid) bone cement. Biomaterials 2000; 21(3): 293–298.
149. Trap B, Wolff P, Jensen JS. Acrylic bone cements: residuals and extractability of methacrylate monomers and aromatic amines. J Appl Biomater 1992; 3:51–57.
150. Havelin LI, Espehaug B, Vollset SE, Engesater LB. The effect of the type of the cement on early revision of Charnley hip prosthesis. J Bone J Surg 1995; 77(A:1543–1550.
151. Nimb L, Sturup J, Jensen JS. Improved cortical histology after cementation with a new MMA-DMA-IBMA bone cement: an animal study. J Biomed Mater Res 1993; 27(5):565–574.
152. Wykman AG, Sandersjoo GA. Low polymerization temperature with Boneloc. In vivo measurements in 11 hip replacements. Acta Orthop. Scand 1995; 66(3):218–219.
153. Suominen S. Early failure with Boneloc cement. 4/8 femoral stems loose within 3 years. Acta Orthop. Scand 1995; 66(1):13.
154. Nimb L, Sturup J, Jensen JS. Improved cortical histology after cementation with a new MMA-DMA-IBMA bone cement: an animal study. J Biomed Mater Res 1993; 27(5):565–574.
155. Wykman AG, Sandersjoo GA. Low polymerization temperature with Boneloc. In vivo measurements in 11 hip replacements. Acta Orthop Scand 1995; 66(3):218–229.
156. Suominen S. Early failure with Boneloc cement. 4/8 Femoral stems loose within 3 years. Acta Orthop Scand 1995; 66(1):13.
157. Walczak JP, D'Arcy JC, Ross KR, James SE, Bonnici AV, Koka SR, Morris RW. Low-friction arthroplasty with Boneloc bone cement: outcome at 2 to 4 years. J arthroplasty 2000; 15(2):205–214.
158. Pascual B, Vázquez B, Gurruchaga M, Goni I, Ginebra MP, Gil FJ, Planell JA, Levenfeld B. San Román J. New aspects of the effect of size and size distribution on the setting parameters and mechanical properties of acrylic bone cements. Biomaterials 1996; 17:509–516.
159. Vila MM, Ginebra MP, Gil FJ, Planell JA. Effect of porosity and environment on the mechanical behavior of acrylic bone cement modified with acrylonitrile-butadiene-styrene particles: part I. Fracture toughness. J Biomed Mater Res 1999; 48(2):121–127.
160. Vila MM, Ginebra MP, Gil FJ, Planell JA. Effect of porosity and environment on the mechanical behavior of acrylic bone cement modified with acrylonitrile-butadiene-styrene particles: part II. Fatigue crack propagation. J Biomed Mater Res 1999; 48(2):128–134.
161. Jayabalan M, Thomas V, Sreelatha PK. Studies on poly(propylene fumarate-*co*-ethylene glycol) based bone cement. Biomed Mater Eng 2000; 10:57–71.
162. Ishihara K, Nakabayashi N. Adhesive bone cement both to bone and metals: 4-META in MMA initiated with tri-n-butyl borane. J Biomed Mater Res; 1989; 23(12):1475–1482.
163. Werning JW, Maniglia AJ, Anderson JM. Biomechanical assessment of a new adhesive bone cement for otologic surgery. Am J Otol; 1995; 16(3):269–276.
164. Maniglia AJ, Nakabayashi N, Paparella MM, Werning JW. A new adhesive bonding material for the cementation of implantable devices in otologic surgery. Am J Otol; 1997; 18(3):322–327.
165. Sakai T, Morita S, Shinomiya K, Watanabe A, Nakabayashi N, Ishihara K. Prevention of fibrous layer formation between bone and adhesive bone cement: in vivo evaluation of bone impregnation with 4-META/MMA-TBB cement. J Biomed Mater Res; 2000; 52(1):24–29.
166. Yang JM, You JW, Chen HL, Shih CH. Calorimetric characterization of the formation of acrylic type bone cements. J Biomed Mater Res 1996; 33(2):83–88.
167. Muh E, Zimmermann J, Kneser U, Marquardt J, Mulhaupt R, Stark B. Lysineurethanedimethacrylate — a novel generation of amino acid based monomers for bone cements and tissue repair. Biomaterials 2002; 23:2846–2854.
168. Vazquez B, Deb S, Bonfield W, Roman JS. Characterization of new acrylic bone cements prepared with oleic acid derivatives. J Biomed Mater Res 2002; 63:88–97.

169. Dai KR, Liu YK, Park JB, Clark CR, Nishiyama K, Zheng ZK. Bone-particle–impregnated bone cement: an in vivo weight-bearing study. J Biomed Mater Res 1991; 25(2):141–156.
170. Giunti A, Moroni A, Olmi R, Vicenzi G. Composite acrylic cement with added hydroxyapatite: a study of the polymerization temperature. Ital J Orthop Traumatol 1983; 9(3):369–375.
171. Castaldini A, Cavallini A. Setting properties of bone cement with added synthetic hydroxyapatite. Biomaterials 1985; 6(1):55–60.
172. Dandurand J, Delpech V, Lebugle A, Lamore A, Lacabanne C. Study of the mineral-organic linkage in an apatitic reinforced bone cement. J Biomed Mater Res 1990; 24(10):1377–1384.
173. Burnie J, Gilchrist T, Duff S, Drake C. Controlled release glass (CRG) for biomedical uses. Biometerials 1981; 2:244–245.
174. Mickiewicz RA, Mayes AM, Knaack D. Polymer–calciumphosphate cement composites for bone substitutes. J Biomed Mater Res 2002; 61:581–592.
175. Saito M, Maruoka A, Mori T, Sugano N, Hino K. Experimental studies on a new bioactive bone cement: hydroxy apatite composite resin. Biomaterials 1994; 15:156–160.
176. Mongiorgi R, Valdre G, Giardino R, Maggi G, Prati C, Bertocchi G. Thermodynamical aspects of the polymerization reaction of PPMA cement mixed with phosphatic mineral phases. Boll Soc Ital Biol Sper 1993; 69(6):365–372.
177. Deb S, Braden M, Bonfield W. Water absorption characteristics of modified hydroxyapatite bone cements. Biomaterials 1995; 16(14):1095–1100.
178. Yang JM, Lu CS, Hsu YG, Shih CH. Mechanical properties of acrylic bone cement containing PMMA-SiO_2 hybrid sol-gel material. J Biomed Mater Res 1997; 38(2):143–154.
179. Kwon SY, Kim YS, Woo YK, Kim SS, Park JB. Hydroxyapatite impregnated bone cement: in vitro and in vivo studies. Biomed Mater Eng 1997; 7(2):129–140.
180. Morita S, Furuya K, Ishihara K, Nakabayashi N. Performance of adhesive bone cement containing hydroxyapatite particles. Biomaterials 1998; 19(17):1601–1606.
181. Yang JM, Li HM, Yang MC, Shih CH. Characterization of acrylic bone cement using dynamic mechanical analysis. J Biomed Mater Res 1999; 48(1):52–60.
182. Vallo CI, Montemartini PE, Fanovich MA, López JMP, Cuadrado TR. Polymethylmethacrylate-based bone cement modified with hydroxyapatite. Biomed Mater Res 1999; 48:150–158.
183. Viano AM, Auwarter JA, Rho JY, Hoffmeister BK. Ultrasonic characterization of the curing process of hydroxyapatite-modified bone cement. J Biomed Mater Sci 2001; 56:593–559.
184. Serbetci K, Korkusuz F, Hasirci N. Mechanical and thermal properties of hydroxyapatite-impregnated bone cements. Turkish. J Med Sci 2000; 30:543–549.
185. Ooms EM, Wolke JG, van der Waerden JP, Jansen JA. Trabecular bone response to injectable calcium phosphate (Ca-P) cement. J Biomed Mater Res 2002; 61:9–18.
186. Beruto DT, Botter R, Fini M. The effect of water in inorganic microsponges of calcium phosphate on the porosity and permeability of composites made with polymethylmethacrylate. Biomaterials 2002; 23:2509–2517.
187. Nilsson M, Fernandez E, Sarda S, Lidgren L, Planell JA. Characterization of a novel calcium phosphate/sulphate bone cement. J Biomed Mater Res 2002; 61:600–607.
188. Lu J, Descamps M, Dejou J, Koubi G, Hardouin P, Lemaitre J, Proust JP. The biodegradation mechanism of calcium phosphate biomaterials in bone. J Biomed Mater Res 2002; 63:408–412.
189. Arcis RW, Lopez-Macipe A, Toledano M, Osorio E, Rodriguez-Clemente R, Murtra J, Fanovich MA, Pascual CD. Mechanical properties of visible light-cured resins reinforced with hydroxyapatite for dental restoration. Dental Materials 2002; 18:49–57.
190. Sarda S, Fernandez E, Nilsson M, Barcells M, Planell JA. Kinetic study of citric acid influence on calcium phosphate bone cements as water-reducing agent. J Biomed Mater Res 2002; 61:653–659.
191. Khairoun J, Magne D, Gauthier O, Bouler JM, Aguado E, Daculsi G, Weiss P. In vitro characterization and in vivo properties of a carbonated apatite bone cement. J Biomed Mater Res 2002; 60: 633–642.
192. Suchanek W, Shuk P, Byrappa K, Riman RE, TenHuisen KS, Janas VF. Mechanochemical-hydrothermal synthesis of carbonated apatite powders at room temperature. Biomaterials 2002; 23:699–710.
193. Gross U, Strunz V. The interface of various glasses and glass-ceramics with bony implantation Bed. J Biomed Mater Res 1985; 19:251.

194. Yoshii S, Kakutani Y, Yamamuro T, Nakamura T, Kitsugi T, Oka M, Kokubo T, Takagi M. Strength of bonding between A-W glass ceramic and the surface of bone cortex. J Biomed Mater Res 1988; 22:327.
195. Kitsugi T, Yamamuro T, Kokubo T. Bonding behavior of a glass-ceramic containing apatite and wollstonite in segmental replacement of the rabbit tibia under load-bearing conditions. J Bone Joint Surg 1989; 71A:264.
196. Reck R, Storkel S, Meyer A. Bioactive glass-ceramics in middle ear surgery: a 8-year review, in Bioceramics: Materials Characteristics Versus in Vivo Behavior Ducheyne P, Lemons J, eds. Vol. 523: Annals of the New York Academy of Science, 1988.
197. Yamamuro T. Replacement of the spine with bioactive glass-ceramic prosthesis. in Handbook of Bioactive Ceramics Yamamuro T, Hench LL, Wilson J, eds. Vol. 1. Boca Raton, FL: CRC press, 1990.
198. Ogino M, Ohuchi F, Hench LL. Compositional dependence of the formation of calcium phosphate films on bioglass. J Biomed Mater Res 1980; 14:55–64.
199. Kim CY, Clark AE, Hench LL. Early stages of calcium-phosphate layer formation in bioglasses. J Non-Cryst Solids 1989; 113:195–202.
200. Clark AE, Pantano CG, Hench LL. Auger spectroscopic analysis of bioglass corrosion films. J Amer Ceram Soc 1976; 59:37–39.
201. Hench LL. Bioactive ceramics in Bioceramics: Materials Characteristics Versus in Vivo Behavior Ducheyne P, Lemons J, eds. Vol. 523: Annals of the New York Academy of Science, 1988.
202. Gross U, Kinne R, Schmitz HJ, Strunz V. The response of bone to surface active glass/glass ceramics. Vol. 4: CRC Critical Reviews in Biocompatibility, 1988:2.
203. Hench LL, Paschall HA. Direct chemical bonding between bio-active glass-ceramic materials and bone. J Biomed Mater Res 1973; 4:25–42.
204. Gross UM, Strunz V. The anchoring of glass ceramics of different solubility in the femur of the rat. J Biomed Mater Res 1980; 14:607–618.
205. Gross U, Roggendorf W, Schmitz HJ, Strunz V. Testing procedures for surface reactive biomaterials. in Biological and Biomechanical Performance of Biomaterials Christel P, Meunier A, Lee AJC, eds. Amsterdam: Elsevier, 1986.
206. Hohland W, Vogel W, Naumann K, Gummel J. Interface reactions between machinable bioactive glass-ceramics and bone. J Biomed Mater Res 1985; 19:303–312.
207. Vogel W, Hohland W, Naumann K, Vogel J, Carl G, Gotz H, Wange P. Glass-Ceramics for medicine and dentistry. in Handbook of Bioactive Ceramics Yamamuro T, Hench LL, Wilson J, eds. Vol. 1. Boca Raton, FL: CRC press, 1990.
208. Chen WS, Monroe EA. Phosphate glass cement bone graft. Biomaterials 1991; 12:561–564.
209. Tamura J, Kitsugi T, Iida H, Fujita H, Nakamura T, Kokubo T, Yoshihara S. Bone bonding ability of bioactive bone cements. Clin Orthop 1997; 343:183–191.
210. Kobayashi M, Nakamura T, Okada Y, Fukumoto A, Furukawa T, Kato H, Kokubo T, Kikutani T. Bioactive bone cement: comparison of apatite and wollastonite containing glass ceramic hydroxyapatite and beta-tricalcium phosphate fillers on bone-bonding strength. J Biomed Mater Res 1998; 42: 223–237.
211. Shinzato S, Nakamura T, Kokubo T, Kitamura Y. A new bioactive bone cement: effect of glass bead filler content on mechanical and biological properties. J Biomed Mater Res 2001; 54:491–500.
212. Shinzato S, Nakamura T, Kokubo T, Kitamura Y. Bioactive bone cement: effect of filler size on mechanical properties and osteoconductivity. J Biomed Mater Res 2001; 56:452–458.
213. Shinzato S, Nakamura T, Tamura J, Kokubo T, Kitamura Y. Bioactive bone cements: effects of phosphoric ester monomer on mechanical properties and osteoconductivity. J Biomed Mater Res 2001; 56:571–577.
214. Kamitakahara M, Kawashita M, Kokubo T, Nakamura T. Effect of polyacrylic acid on the apatite formation of a bioactive ceramic in a simulated body fluid. Biomaterials 2001; 22:3191–3196.
215. Okada Y, Kobayashi M, Neo M, Kokubo T, Kitamura Y, Nakamura T. Ultrastructure of the interface between bioactive composite and bone: comparison of apatite and wollastonite containing glass-ceramic filler with hydroxyapatite and beta-tricalcium phosphate fillers. J Biomed Mater Res 2001; 57:101–107.

216. S. Shinzato, T. Nakamura, K Ando, Kokubo T, Kitamura Y. Mechanical properties and osteoconductivity of new bioactive composites consisting of partially crystallized glass beads and poly(methyl methacrylate). J Biomed Mater Res 2002; 60:556–563.
217. Wilson J, Pigott GH, Schoen FJ, Hench LL. Toxicology and Biocompatibility of Bioglass. J Biomed Mater Res 1981; 15:805–817.
218. Wilson J, Nolletti D. Bonding of soft tissues to bioglass. in Handbook of Bioactive Ceramics Yamamuro T, Hench LL, Wilson J, eds. Boca Raton, FL: CRC press, 1990.
219. Hasenwinkel JM, Lautenschlager EP, Wixson RL, Gilbert JL. A novel high-viscosity, two-solution acrylic bone cement: effect of chemical composition on properties. J Biomed Mater Res; 1999 Oct; 47(1):36–45.
220. Hasenwinkel JM, Lautenschlager EP, Wixson RL, Gilbert JL. Effect of initiation chemistry on the fracture toughness fatigue strength and residual monomer content of a novel high-viscosity, two-solution acrylic bone cement. J Biomed Mater Res; 2002 Mar 5; 59(3):411–421.

13

Three-Dimensionally Engineered Hydroxyapatite Ceramics with Interconnected Pores as a Bone Substitute and Tissue Engineering Scaffold

Akira Myoui, Noriyuki Tamai, Masataka Nishikawa, and Hideki Yoshikawa
Osaka University Graduate School of Medicine
Osaka, Japan

Nobuhito Araki
Osaka Medical Center for Cancer and Cardiovascular Diseases
Osaka, Japan

Takanobu Nakase
Osaka National Hospital
Osaka, Japan

Shosuke Akita
Hoshigaoka Kouseinenkin Hospital
Osaka, Japan

I. BONE GRAFTS AND BONE SUBSTITUTES

Orthopedics has long been associated with the development of tissue engineering techniques and biomaterials through the use of bone grafting techniques, including autograft, allograft, and xenograft, as well as through bone lengthening techniques with distraction osteogenesis; fixation devices made of metal, ceramic, or biodegradable materials; and organic or nonorganic bone substitutes. However, in the current clinical scene, when bone grafts are required autogenous bone grafting has been the gold standard because of its obvious advantages in osteogenic capacity, osteoconduction, mechanical properties, and the lack of adverse immunological response. In fact, autograft accounts for nearly 70% of surgical procedures involving bone grafts or bone substitutes in Japan according to a survey by the Japanese Orthopedic Association. [1] On the other hand, autogenous bone grafting techniques have some limitations, [2] such as the requirement of additional surgery for harvesting and the availability of grafts of the correct size and shape. Furthermore, autograft techniques all carry the risk of donor site morbidity, [3–5] which can include long-lasting pain, fracture, nerve damage, and infection, none of which are experienced when other techniques are employed.

Although allogeneic bone is widely used in the United States, its use is quite limited in Japan, accounting for as little as 3% of procedures. [1] This is probably due to the lack of a well organized tissue bank system such as the Musculoskeletal Transplant Foundation in the United States as well as religious or philosophical difficulties with utilizing tissue from other people or corpses. In addition, allograft carries the risk of a host immune response, which can sometimes result in complete resorption of the transplanted grafts or transmission of occult disease. [6]

Among the various biomaterials, calcium hydroxyapatite (HA) is considered biocompatible, nonimmunogenic, and osteoconductive, and HA-based materials have been employed in the field of bone repair for over 20 years. [7–12] Numerous products for clinical use as bone substitutes have been granted approval by the U.S. Food and Drug Administration and the Ministry of Health, Labour, and Welfare (OPSR) of Japan. In Japan, synthetic HA ceramics, including both solid and porous forms, have been widely used because of their availability and the variety of designs and sizes. Porous HA ceramics in particular were expected to encourage host bone ingrowth in the porous structure and to be incorporated into host bone. However, recent clinical studies have revealed that the pores are not replaced by new bone for a substantial period of time, probably due to limited interconnection of pores. Ayers et al. reported that bone ingrowth had penetrated into porous HA implants no further than 300 μm from the interface at 4 months after implantation in humans. [13]

II. PORE SIZE, POROSITY, AND INTERCONNECTIVITY

Several researchers have explored the optimal pore size for in vivo osteoconduction and 150 to 500 μm in diameter was reported to be optimal. [14,15] These studies were not focused on the size of interpore connections but on the size of the pore itself. However, both the rate and distribution of osteogenesis around and throughout the implant are affected by the size and number of interconnecting channels, and pore size and porosity may not be the primary determinants for osteoconductivity.

Several reports claim that natural coral-derived HA, which also has an interconnected porous structure, shows good osteoconductivity. [10,16] Although this product holds the top share in the bone substitute market in the United States, there are some limitations, such as negative impact on the environment, fragility, and the possibility of disease transmission. [17] Therefore, when a three-dimensional structure that considers interconnecting porosity is included in the design of synthetic porous HA ceramics, the result could be an ideal bone substitute that may replace autogenous bone grafting. Increasing the porosity or pore size theoretically increases the interconnectivity of the pores, but also weakens the strength of this graft material. [15] Thus, engineering an ideal porous HA ceramic that confers mechanical support and promotes bone ingrowth has, until now, proven difficult for biomaterial specialists.

III. SYNTHETIC FULLY INTERCONNECTED POROUS HYDROXYAPATITE CERAMICS

The conventional method used to manufacture synthetic porous HA ceramics is by sintering an HA slurry mixed with organic polymer beads. [18] The polymer beads melt and vaporize during the sintering process, eventually leaving pores in the ceramic material. However, as shown in

Fig. 1, the pores resulting from this method are irregular in size and shape and not fully interconnected with one another.

Together with Toshiba Ceramics, MMT, and the National Institute for Materials Science Biomaterials Center, we recently developed a fully interconnected porous HA ceramic (IP-CHA) featuring 75% porosity, 150-μm average pore size, and 40-μm average interpore connections by adopting the foam-gel technique. [19] This approach involves a crosslinking polymerization step that gelatinizes the foamlike CHA slurry in a rapid manner, thus promoting the formation of an interconnected porous structure. Briefly, the new method is as follows:

1. *Slurry preparation*: slurry was prepared by mixing hydroxyapatite (60 wt%) with a crosslinking substrate (40 wt% polyethyleneimine).
2. *Foaming and gelatinization*: the slurry was mixed with a foaming agent (1 wt% polyoxyethylene lauryl ether) and stirred until the mixture had a foamy appearance. Pore size was controlled by regulating the stirring time.
3. *Gelatinization*: to gelatinize the foamed slurry, another water-soluble crosslinking agent (poly functional epoxy compound) was added, and the mixture was cast by pouring into a mold. The porous structure stabilized in less than 30 min. The foamy HA gel was removed from the mold, dried, and sintered at 1200°C.

A. Macro- and Microstructure and Mechanical Properties of IP-CHA

Scanning electron microscopy (SEM) analysis revealed that most of the IP-CHA pores were spherical, similar in size, approximately 100 to 200 μm in diameter, and showed uniform connec-

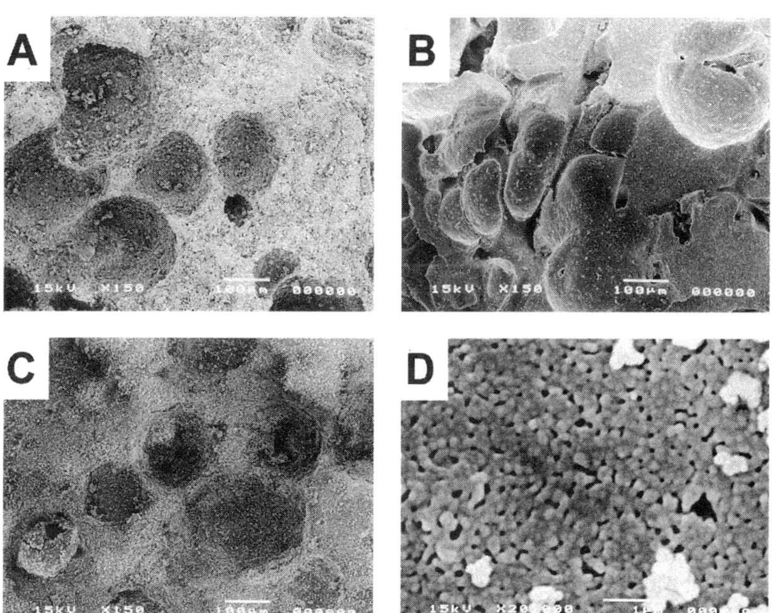

Figure 1 SEM photographs of macro/microstructures of synthetic porous HA ceramics currently used in Japan. The materials were manufactured by (A) Asahi Optical Co., Ltd. (manuf. A), (B, D) Mitsubishi Material Corp. (manuf. M) and (C) Sumitomo Osaka Cement Co., Ltd. (manuf. S). Magnification: (A,B,C) 150 ×; (D) 20,000 ×.

tions with one another (Fig. 2). On the other hand, the pores of commercial synthetic porous HAs available in Japan (Fig. 1A-C), manufactured by Asahi Optical (manuf. A), Mitsubishi Material (manuf. M), and Sumitomo Osaka Cement (manuf. S), respectively, were separated by irregular, relatively thick walls, and interpore connections were sparsely distributed. The wall surface of IP-CHA was very smooth and HA particles were lined closely to one another and bound tightly (Fig. 2), while the surfaces of the other HAs were very rough and the particles were loosely attached one another (Fig. 1D).

In IP-CHA, the majority of the interpore connections ranged from 10 to 80 μm in diameter, with a maximum peak at about 40 μm, which would theoretically be permissive to cell migration or tissue invasion from pore to pore. [20] Interpore connections larger than 10 μm accounted for as much as 91% of the total porosity in IP-CHA. The calculated available porosity, the proportional volume of pores in the material that were connected by interpore connections larger than 10 μm in diameter, was 73.4% (total porosity) $\times$ 0.91 $=$ 67.1% in IP-CHA. However, the diameter and the total number of the interpore connections were much less in the other synthetic HAs. The calculated available porosities of the commercial synthetic porous HAs ranged from 2.5% to 36.7%, which is much less than that of IP-CHA. The compression strength of IP-CHA is 12 MPa and that of the commercially available synthetic porous HAs ranges from 4.8 to 60 MPa, while the compressive strength of cancellous bone is 1–12 MPa. [21] Taking the porosity of IP-CHA into consideration, its initial compressive strength is above average. This is probably because the HA particles in the framework of IP-CHA are tightly bound to

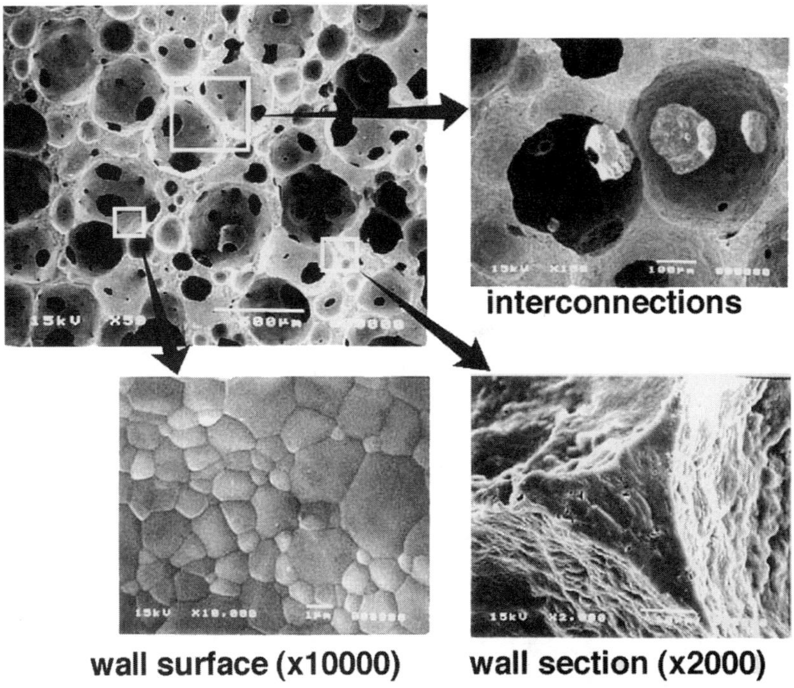

Figure 2 SEM photographs of macro/microstructures of IP-CHA. Spherical pores (100–200 mm in diameter) were divided by thin walls and interconnected. Wall surface is very smooth, and HA particles were tightly bound.

one another, as is seen in solid form HA ceramics. The structural characteristics and compression strength of IP-CHA and porous HAs available in Japan are summarized in Table 1.

B. Bone Ingrowth in Vivo

Macroporosity is known to influence the biological performance of calcium phosphate in vivo. Holmes et al. reported that pores of approximately 100 μm in diameter could provide a framework for bone growth into the pore volume, which then becomes vascularized easily. [22] Most of the pores of IP-CHA are large enough to meet such criteria and, more importantly, the pores are fully interconnected and more likely to allow bone ingrowth. We histologically analyzed bone ingrowth in cylindrical blocks (6 mm in diameter) of IP-CHA and the other synthetic porous HAs using the rabbit femoral condyle model. Within 6 weeks after transplantation of IP-CHA, mature bone tissue ingrowth was seen in all the pores throughout the block (Fig. 3A). In the pores, bone formation through interpore connections with osteoblastic rimming, bone marrow, and vessels were observed. At 6 months, the pores facing host cortical bone were occupied with dense bone matrix, while the pores surrounded by fatty marrow were mostly replaced by bone marrow tissue accompanied by a thin layer of bone matrix on the internal surface of the pores (Fig. 3B).

Figure 4A shows the histology of the central area of the tested porous HA blocks at 3 and 6 weeks after implantation. The penetration of bone tissue into the central pores, as deep as 3 mm from the surface, was seen in most animals with IP-CHA within 3 weeks and in all animals within 6 weeks. However, tissue penetration in the other synthetic porous HAs was limited, and the degree of bone tissue penetration was proportional to the calculated available porosity of the HA used. These results suggest that calculated available porosity is one of the most important factors in determining the capacity of bone ingrowth in the porous scaffold. The summarized data of in vivo bone ingrowth are shown in Fig. 4B.

We also examined the sequential change in the compression strength of the IP-CHA implanted in rabbit femoral condyle. The initial compressive strength of IP-CHA was approximately 10 to 12 MPa. The implanted IP-CHA steadily increased its compressive strength with time until 9 weeks after implantation, finally reaching a value of about 30 MPa. Such an increase in strength resulting from bone tissue penetration enhances the mechanical stability of the implant at the defect site.

C. Clinical Application in Orthopedic Surgery

As we reported previously, [11,12] porous HA is a good material to replace the bone defect in treating benign bone tumors because of its biocompatibility, osteoconduction, and convenience,

Table 1 Structural Characteristics and Compressive Strengths of Synthetic Porous Hydroxyapatite Ceramics

	Overall porosity (%)	Calculated available porosity (%)	Average pore diameter (μ m)	Range of pore diameter (μ m)	Inter connection diameter (μ m)	Compressive strength (Mpa)
IP-CHA	73.4	67.1	191	30–400	39	12.0
Manuf. A	50.5	28.5	242	80–500	22	32.3
Manuf. M	64.5	36.7	268	40–500	15	8.0
Manuf. S	47.6	2.5	176	30–400	<1	60.5

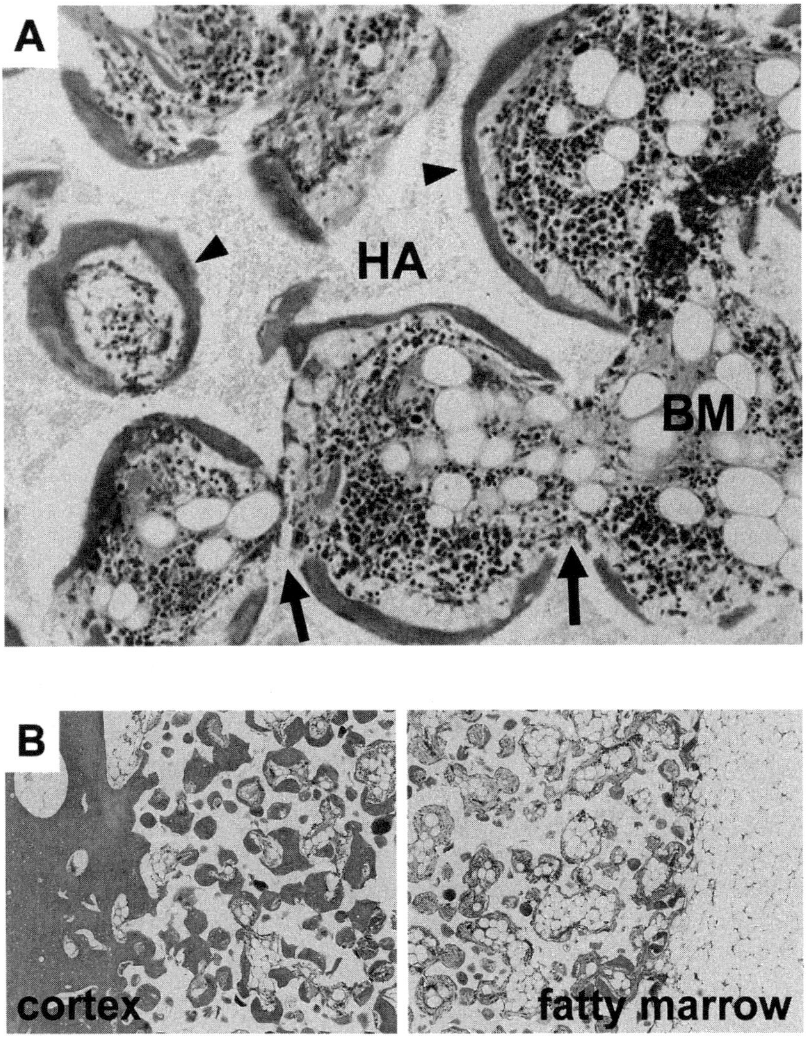

Figure 3 HE-stained histological sections of IP-CHA implanted in rabbit femoral condyle (A) 6 weeks and (B) 6 months after implantation. (A) New bone matrix (arrowheads) and bone marrow tissue (BM) penetrated into the central portion of the cylindrical IP-CHA block (HA) (6 mm in diameter) through interconnections (arrows). (B) At 6 months, pores facing the cortex were filled with dense bone matrix, and pores surrounded by fatty marrow were filled with fatty marrow and had thin layers of bone.

and it eliminates the need for autograft harvesting procedures. However, as a late complication, pathological fractures of the implanted sites have been reported. [12] This is probably due to poor bone ingrowth in the material as a result of poor incorporation of the material into host bone. If bone ingrowth throughout the porous HA implant is achieved more quickly, such late complications could be avoided.

We applied IP-CHA as a bone substitute for the treatment of 25 patients with benign bone tumors at Osaka University Hospital and affiliated hospitals. The average age of the patients was 47 years (range 22 to 75). Histological diagnoses were enchondroma in ten, giant cell tumor of bone in three, and others in twelve patients. The tumors were located in the upper extremities

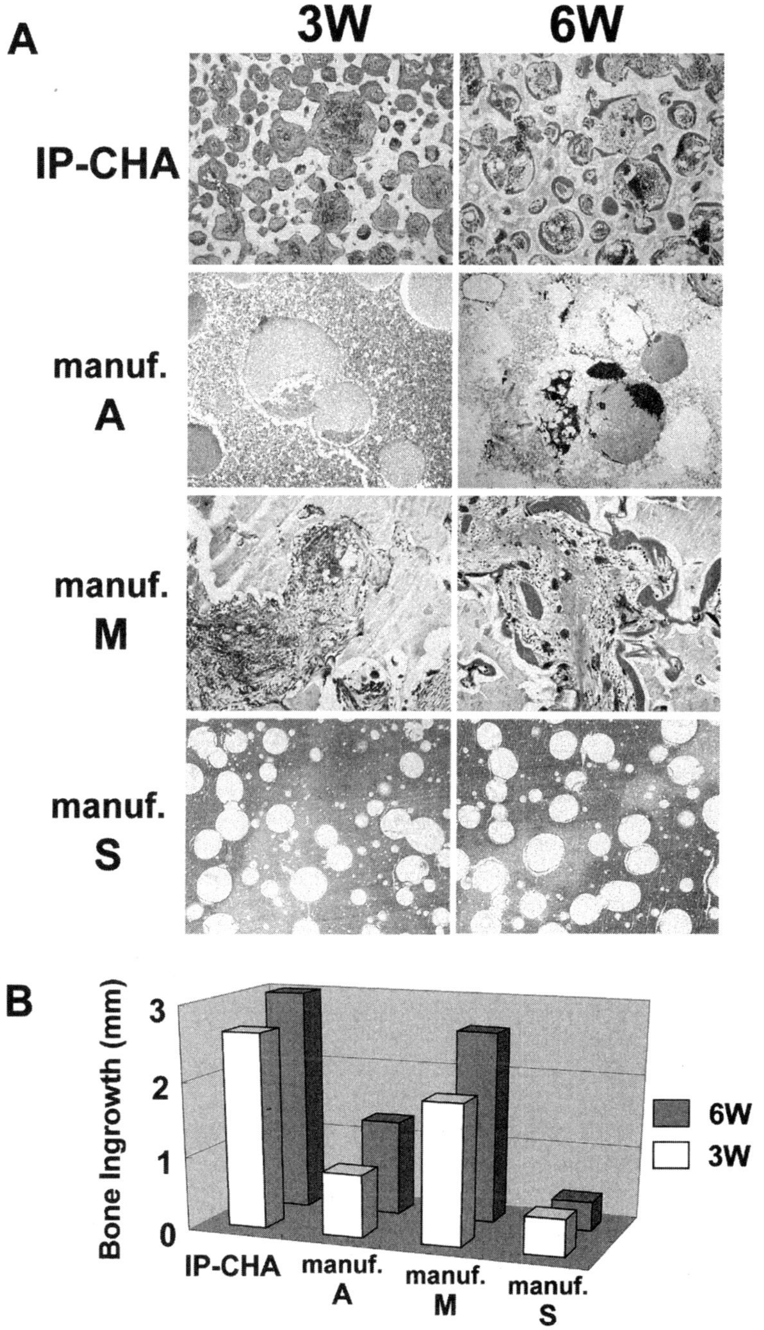

Figure 4 (A) Sequential photomicrographs of central areas of synthetic porous HAs implanted in rabbit femoral condyle at 3 and 6 weeks after implantation. Penetration of bone tissue into central pores was seen in most animals with IP-CHA by 3 weeks and in all animals by 6 weeks. However, tissue penetration into other synthetic porous HAs was limited. (B) Average depth of new bone penetration from the surface of synthetic porous HAs at 3 and 6 weeks after implantation.

in thirteen, lower extremities in nine and pelvis in three. Mean follow-up period was 16 months (range 3 to 30 months). After adequate removal of the tumors, IP-CHA blocks and/or granules of 2 to 5 mm in diameter were used to fill the bony defects. Particular care was taken not to leave any space and, at the same time, not to apply excess pressure to break the material or its porous structure. In some cases, fibrin glue was used to keep the IP-CHA in position. If necessary, the blocks were carved using surgical air-tome with diamond bur. An internal fixation device was used in one case with an aneurysmal bone cyst of the femoral neck. No patients who were implanted with IP-CHA exhibited any signs of host inflammatory reaction, rejection, infection, or abnormal results in blood tests. Two patients with giant cell tumor of bone developed local recurrences.

Radiographic examinations were periodically carried out and revealed that the radiolucent line between the implanted IP-CHA and host bone tended to decrease with time after operation and eventually disappeared. The radiographic density at the implanted site increased with time, and the IP-CHA granules appeared to fuse with one another, eventually forming a dense radiopaque shadow (Fig. 5A, B). These radiological findings were interpreted as bone regeneration around the IP-CHA implants; in other words, incorporation into host bone.

Radiological evaluation criteria for the degree of incorporation of the implant into host bone were as follows:

- Grade 0. osteolysis (increase of radiolucency around the implant)
- Grade 1. no change
- Grade 2. minor incorporation (minor increase of radiographic density of the implanted area and disappearance of radiolucent line between the implant and host bone)
- Grade 3. major incorporation (major increase of radiographic density and/or disappearance of discrimination between IP-CHA granules)

As shown in Fig. 5D, IP-CHA incorporation in host bone on radiographs was observed in 72% of the cases by 2 months and in all cases by 3 months. At 6 months after operation, all cases showed major incorporation, which is much more rapid than seen in our previous experience with other commercial porous HAs. [12] Neither pathological fracture nor deformity was observed at the implanted site based on radiographic examinations during the follow-up period.

Some of the cases showed interesting radiographic changes, as shown in Fig. 5B. The radiographic density of the implanted site greatly increased within 3 months after implantation, while in the peripheral areas the expanded cortex exhibited self-correction with time. Moreover, at 12 months a radiolucent band seemed to appear in the bone axis, suggesting the regeneration of bone marrow cavity. These findings also indicate good incorporation of IP-CHA into host bone and, possibly, biodegradability by being involved in the host bone remodeling process at the implanted site.

We also applied IP-CHA to treat nonunion of fracture sites. Fig. 5C illustrates the radiographs of a representative case. After the removal of interposed fibrous tissue between bone fragments, IP-CHA granules were implanted in the defect together with fresh autogenous bone chips. The nonunion was successfully treated within 6 months following surgery with abundant callus formation and increased radiodensity of the gap between host bone fragments. We have treated only three cases of nonunion in this manner, but complete union with incorporation of the IP-CHA granules was accomplished in all cases. In these cases, the amount of necessary autogenous bone was reduced by using IP-CHA as a bone substitute.

However, it should be noted that our favorable results in clinical use are based only on radiographic findings because we do not currently have an effective noninvasive method to evaluate bone ingrowth in the implants. Biopsy of the IP-CHA implant site was performed with informed consent at the time of revision surgery for the treatment of recurrent tumor 10 months

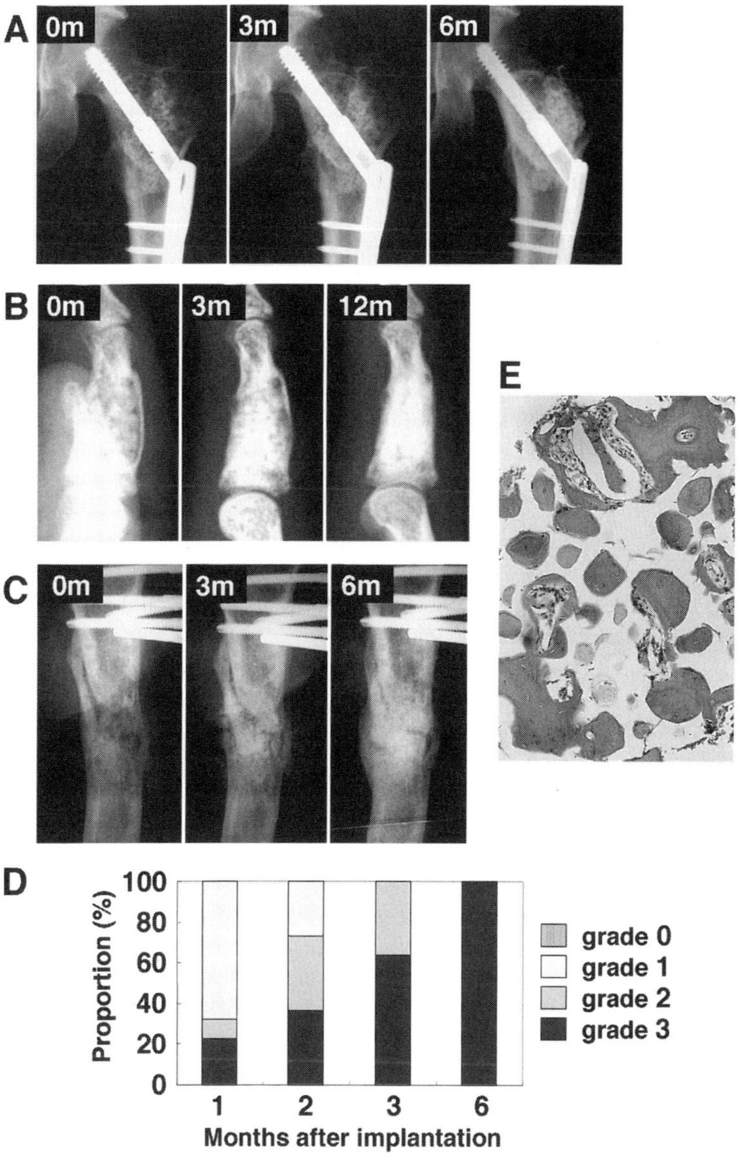

Figure 5 Results of the clinical application of IP-CHA in treatment of patients with (A,B,D,E) benign bone tumors and (C) nonunion. (A) Radiographs of an aneurysmal bone cyst in a 35-year-old male immediately after, and at 3 and 6 months after surgery. Ten grams of IP-CHA granules (2 to 5 mm in diameter) were implanted and reinforced by hip screw/plate. Radiopacity increased and discrimination between granules disappeared with time. (B) A case of an enchondroma of mid-phalanx, 28-year-old male. As radiodensity increased, expansive deformity of the affected bone self-corrected. At 12 months, a radiolucent band seemed to appear in the axis of the phalanx, suggesting regeneration of the bone marrow cavity. (C) Nonunion of the proximal diaphysis of the femur, 19-year-old male. IP-CHA granules were implanted together with autogenous cancellous bone chips harvested from iliac crest. Note the abundant external callus formation and disappearance of the gap at 6 months. (D) Radiographic evaluation of the incorporation of IP-CHA implanted to bone defects in patients with benign bone tumors. Vertical axis represents the proportion of patients in each grade to all patients. See text for grading criteria. (E) Histological section of IP-CHA recovered from a patient with recurrent giant cell tumor of bone 10 months after initial surgery. Most pores were replaced with newly formed bone in this section.

after initial implantation. Histological examination revealed that most of the pores were replaced with newly formed bone (Fig. 5E). Although we do not have much data regarding the depth of new bone penetration into the IP-CHA implant in human bone, such histological observations are encouraging.

D. Scaffold for Bone Tissue Engineering

The main principle of current tissue engineering strategy for bone replacement is to use a combination of osteoconductive porous scaffold and osteoinductive molecules with appropriate carriers or osteogenic cells. [23,24] Osteoinductive factors such as bone morphogenetic protein-2 (BMP-2), [23] osteogenic protein-1 (OP-1, also known as BMP-7), and fibroblast growth factor-2 (FGF-2) mixed with natural or synthetic carriers such as collagen as well as poly-lactide–based synthetic polymers can be implanted along with porous scaffolds. The scaffolds can be prefabricated with osteoblasts or mesenchymal osteoprogenitor cells, usually from autogenous bone marrow cell aspiration, cultivated for several weeks, and then implanted into bony defects. [24] The latter approach is exclusively for larger bone defects or to treat poor bone healing conditions.

The requirements for a scaffold in bone tissue engineering are (1) biocompatibility (should not induce host inflammatory reaction), (2) osteoconductivity (material itself should support migration of new bone onto its surface), (3) interconnected-porous structure (should allow bone ingrowth and nutrient supply), (4) appropriate mechanical strength (should tolerate mechanical load until integration into host bone is completed), and (5) biodegradability (either by simple hydrolysis at an appropriate rate coupled with new bone formation or by active resorption by osteoclasts as a part of normal bone remodeling). HA is a well-characterized material with biocompatibility and osteoconductivity, and the pores of IP-CHA have a well-organized 3D interconnective structure. Compression strength of IP-CHA is equal to or higher than cancellous bone (1–12 MPa) and coralline HA (2–4 MPa), [25] which has been used both clinically and in several studies on cell-based bone tissue engineering. Although data regarding biodegradability are still limited, IP-CHA seems to meet most of the requirements as a scaffold for bone tissue engineering. Moreover, such a synthetic scaffold can be prefabricated into specific sizes and shapes to match bone defects, and even into composite with solid form of HA in order to reinforce its initial mechanical strength (Fig. 6).

We investigated whether IP-CHA can be utilized as a scaffold for cell-based bone tissue engineering using a rat subcutaneous model. In this study, we used bone marrow-derived mesenchymal stem cells as a source of bone producing cells. [26] Bone marrow cells were collected from the femur of 7-week-old Fischer 344 rat and were cultivated in minimal essential medium supplemented with 15% fetal bovine serum. Cells were trypsinized and a cell suspension of 10^6 cells/mL was prepared. IP-CHA discs ($R = 5$ mm, $h = 2$ mm) were soaked in the cell suspension overnight and further cultured in the same medium with β-glycerophosphate, ascorbic acid, and 10^{-8} M dexamethasone for 14 days. The discs were then implanted into the subcutaneous tissue of rats and harvested 2 or 4 weeks after implantation. Histological examination revealed that newly formed bone replaced approximately half of the pores within 4 weeks, and the remaining pores were filled with fibrous connective tissue accompanied by blood vessels (Fig. 7). Alkaline phosphatase activity, a marker for osteoblastic differentiation, of the harvested implants was increased with time, suggesting a steady increase in new bone formation in the pores. Therefore, IP-CHA seems to provide a good scaffold for cell-based bone tissue engineering.

E. Vascular Prefabrication

Vascular network invasion into porous implants is another important aspect of using such materials as bone substitutes for large bone defects or in the construction of tissue-engineered bone,

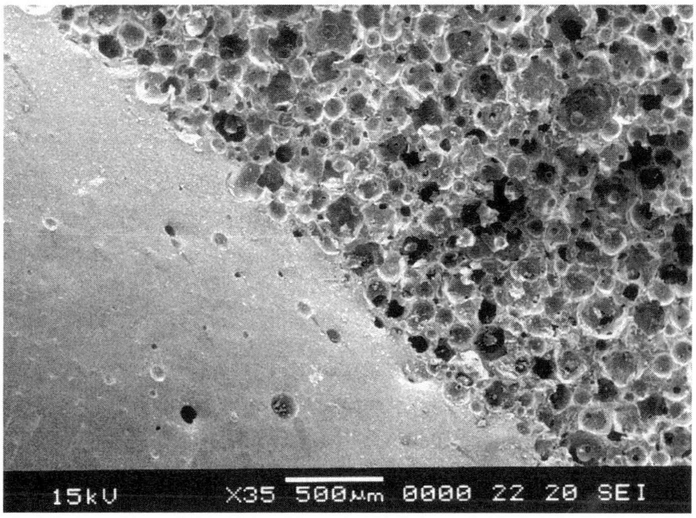

Figure 6 SEM photograph of the solid/interconnected porous HA composite. Magnification: 35 ×.

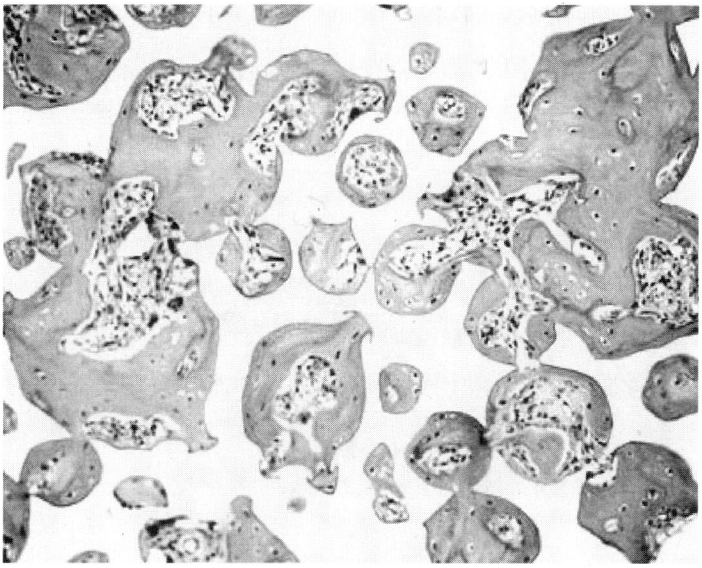

Figure 7 Histologic section of the IP-CHA and bone marrow-derived culture-expanded osteoprogenitor cells recovered from rat subcutaneous tissue 4 weeks after transplantation. Pores were filled with newly formed bone and some parts of the bone surface were lined with plump osteoblasts.

because cells cannot survive farther than a few hundred micrometers from a nutrient supply. The rate of new bone ingrowth into the porous material depends on vascular invasion from the surface of the implant, which is not fast enough in large implants to transport nutrients to cells transplanted in pores of the implant. Therefore, we examined if prefabrication of IP-CHA with a vascular bundle enhances vascular network invasion into the pores via interpore connections. As shown in Fig. 8, when an IP-CHA cylindrical block was prefabricated with rat superficial inferior epigastric vessels, vascular invasion in the pores increased both in number and size compared to the control, resulting in more abundant fibrous connective tissue formation. Our findings suggest that inserting a vascular bundle into such interconnecting porous implants at the site of implantation supports vascular network invasion, which may eventually enhance bone ingrowth in the implants. This technique may be also useful for constructing cell-based tissue-engineered bone, since transplanted cells cannot survive without a nutrient supply. This kind of approach to support blood supply generation in porous biomaterials has also been reported by others using subcutaneous tissue and muscle flaps. [27,28]

IV. CONCLUSIONS

The interconnected porosity encourages bone ingrowth in the material and eventually leads to good incorporation of the material into host bone. The foam-gel technique is an innovative method that generates a three-dimensional fully interconnected porous structure in synthetic HA ceramics. To date, comparable structure has only been achieved by using natural coral. Our study indicated that IP-CHA, a novel synthetic HA ceramic with an interconnected porous structure manufactured using this technique, showed a several-fold increase in initial compressive strength compared with coralline-based materials. Moreover, IP-CHA exhibited excellent bone ingrowth in an animal model and favorable performance in clinical use, although the latter was evaluated only radiographically and with a short-term follow-up. We believe that IP-CHA is an excellent bone substitute for filling bone defects and should be considered as an alternative to autogenous bone. In addition, IP-CHA seems likely to serve as a good scaffold for cell-based or cytokine-based tissue-engineered bone.

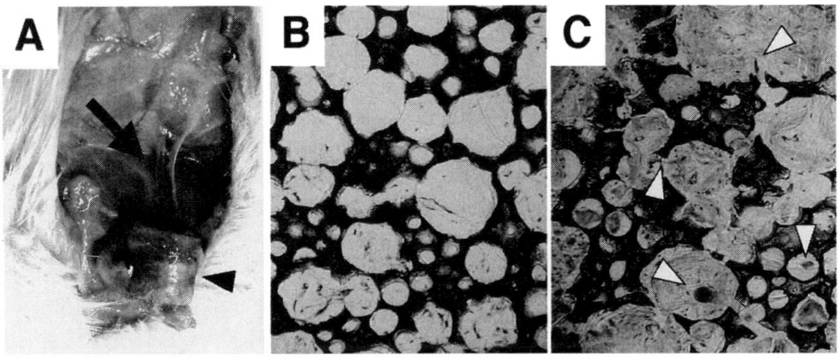

Figure 8 Prefabrication of IP-CHA with vascular network. (A) Rat superficial inferior epigastric vessels (arrow) were exposed, ligated, cut, and inserted into IP-CHA cylindrical block (arrowhead) and then examined 6 weeks later. Compared with (B) the control, (C) vascular invasion (open arrowhead) in the pores of IP-CHA with vascular bundle was greater both in number and size of vessels, with more abundant fibrous connective tissue formation.

ACKNOWLEDGMENTS

The authors thank Dr. Hajime Ohgushi for invaluable advice regarding bone tissue engineering techniques using bone marrow cells. We also thank Toshiba Ceramics Co. and MMT Co. for supplying materials and technical assistance, and Miss Kanae Asai for her expert histological work.

This work was supported in part by grants from New Energy and Industrial Technology Development Organization (NEDO) and the Ministry of Health, Labour, and Welfare, Japan.

REFERENCES

1. The Japanese Orthopedic Association, Tissue Transplantation and Regeneration Committee. Current status of tissue transplantation in orthopedics [in Japanese]. J Jpn Orthop Assoc 2002; 76:255–261.
2. Prolo DJ, Rodrigo JJ. Contemporary bone graft physiology and surgery. Clin Orthop 1985; 200: 322–342.
3. Arrington ED, Smith WJ, Chambers HG, Bucknell AL, Davino NA. Complications of iliac crest bone graft harvesting. Clin Orthop 1996; 329:300–309.
4. Banwart JC, Asher MA, Hassanein RS. Iliac crest bone graft harvest donor site morbidity. A statistical evaluation. Spine 1995; 20:1055–1060.
5. Younger EM, Chapman MW. Morbidity at bone graft donor sites. J Orthop Trauma 1989; 3:192–195.
6. Horowitz MC, Friedlaender GE. Immunologic aspects of bone transplantation. A rationale for future studies. Orthop Clin North Am 1987; 18:227–233.
7. Sartoris DJ, Gershuni DH, Akeson WH, Holmes RE, Resnick D. Coralline hydroxyapatite bone graft substitutes: preliminary report of radiographic evaluation. Radiology 1986; 159:133–137.
8. Holmes RE, Bucholz RW, Mooney V. Porous hydroxyapatite as a bone graft substitute in diaphyseal defects: a histometric study. J Orthop Res 1987; 5:114–121.
9. Bucholz RW, Carlton A, Holmes RE. Hydroxyapatite and tricalcium phosphate bone graft substitutes. Orthop Clin North Am 1987; 18:323–334.
10. Bucholz RW, Carlton A, Holmes R. Interporous hydroxyapatite as a bone graft substitute in tibial plateau fractures. Clin Orthop 1989; 240:53–62.
11. Uchida A, Araki N, Shinto Y, Yoshikawa H, Kurisaki E, Ono K. The use of calcium hydroxyapatite ceramic in bone tumour surgery. J Bone Joint Surg [Br] 1990; 72:298–302.
12. Yoshikawa H, Uchida A. Clinical application of calcium hydroxyapatite ceramic in bone tumor surgery. In: Wise DL, ed. Biomaterials and Bioengineering Handbook. New York: Marcel Dekker: 433–455.
13. Ayers RA, Simske SJ, Nunes CR, Wolford LM. Long-term bone ingrowth and residual microhardness of porous block hydroxyapatite implants in humans. J Oral Maxillofac Surg 1998; 56:1297–1301.
14. Hulbert SF, Morrison SJ, Klawitter JJ. Tissue reaction to three ceramics of porous and non-porous structures. J Biomed Mater Res 1972; 6:347–374.
15. Flatley TJ, Lynch KL, Benson M. Tissue response to implants of calcium phosphate ceramic in the rabbit spine. Clin Orthop 1983; 179:246–252.
16. Baramki HG, Steffen T, Lander P, Chang M, Marchesi D. The efficacy of interconnected porous hydroxyapatite in achieving posterolateral lumbar fusion in sheep. Spine 2000; 25:1053–1060.
17. Vuola J, Taurio R, Goransson H, Asko-Seljavaara S. Compressive strength of calcium carbonate and hydroxyapatite implants after bone marrow–induced osteogenesis. Biomaterials 1998; 19:223–227.
18. Uchida A, Nade SM, McCartney ER, Ching W. The use of ceramics for bone replacement. A comparative study of three different porous ceramics. J Bone Joint Surg [Br] 1984; 66:269–275.
19. Tamai N, Myoui A, Tomita T, Nakase T, Tanaka J, Ochi T, Yoshikawa H. Novel hydroxyapatite ceramics with an interconnective porous structure exhibit superior osteoconduction in vivo. J Biomed Mater Res 2002; 59:110–117.

20. Steinkamp JA, Hansen KM, Crissman HA. Flow microfluorometric and light-scatter measurement of nuclear and cytoplasmic size in mammalian cells. J Histochem Cytochem 1976; 24:292–297.
21. Martin RB, Chapman MW, Sharkey NA, Zissimos SL, Bay B, Shors EC. Bone ingrowth and mechanical properties of coralline hydroxyapatite 1 yr after implantation. Biomaterials 1993; 14:341–348.
22. Holmes RE, Wardrop RW, Wolford LM. Hydroxylapatite as a bone graft substitute in orthognathic surgery: histologic and histometric findings. J Oral Maxillofac Surg 1988; 46:661–671.
23. Petite H, Viateau V, Bensaid W, Meunier A, de Pollak C, Bourguignon M, Oudina K, Sedel L, Guillemin G. Tissue-engineered bone regeneration. Nat Biotechnol 2000; 18:959–963.
24. Ohgushi H, Caplan AI. Stem cell technology and bioceramics: from cell to gene engineering. J Biomed Mater Res 1999; 48:913–927.
25. Holmes R, Mooney V, Bucholz R, Tencer A. A coralline hydroxyapatite bone graft substitute. Preliminary report. Clin Orthop 1984; 188:252–262.
26. Ohgushi H, Dohi Y, Katuda T, Tamai S, Tabata S, Suwa Y. In vitro bone formation by rat marrow cell culture. J Biomed Mater Res 1996; 32:333–340.
27. Bernard SL, Picha GJ. The use of coralline hydroxyapatite in a ''biocomposite'' free flap. Plast Reconstr Surg 1991; 87:96–105.
28. Casabona F, Martin I, Muraglia A, Berrino P, Santi P, Cancedda R, Quarto R. Prefabricated engineered bone flaps: an experimental model of tissue reconstruction in plastic surgery. Plast Reconstr Surg 1998; 101:577–581.

14

The Histological and Immunological Aspects of the Interfacial Membranes of Cemented Total Hip and Knee Arthroplasties

Najat Al-Saffar
Imperial College London
London, England

Jochanan H. Boss
Bnai-Zion Medical Center
Haifa, Israel

I. INTRODUCTION

The worldwide prestige enjoyed by cemented total joint replacements is well deserved. To our professional predecessors recalling the pitiful fate of chair-ridden patients with end stage joint disease in the 1950s, the phenomenal results attained by replacement of a destroyed articulation by an artificial joint manufactured of alloplastic components spells a success story. Interpositional arthroplasty, placing one substance or another between resected joint surfaces, goes back a long time, but it was not until Smith-Petersen promoted the cast cobalt–chrome–molybdenum alloy to serve as artificial joint's constituents that a breakthrough was first recorded in the late 1940s [1]. All in all, the clinical outcome was however still poor. In an attempt at achieving durable pain-free joint motion, Moore and MacIntosh innovated the hemiarthroplasties [2,3]. Notwithstanding striking functional improvements in patients' subjective and objective status, a metallic endoprosthesis replacing a femoral head or a tibial plateau did not bring about the anticipated long-term relief because the diseased acetabulum or femoral condyles were left in situ, thus continuing to be a source of pain and impaired performance. Sir John Charnley ushered in a genuine headway with his innovation of an artificial hip joint incorporating a stemmed, stainless steel femoral head articulating with a polymeric acetabular socket, polymethylmethacrylate fixating the components to the bone [4]. The kindred total knee arthroplasty, designed by Gunston, comprised a cemented metallic femoral component and a cemented polymeric tibial insert [5].

Since Charnley's groundbreaking scientific enterprise and application of his perceptions in the practice of joint replacements, novel designs of the components, fixation techniques (cemented or otherwise), and materials were introduced. But for a small number of metal-on-metal or ceramic-on-ceramic artificial joints, modern arthroplasties customarily utilize an ultrahigh molecular weight polyethylene component for one of the articulating surfaces. For reasons beyond the scope of this chapter, stainless steel–based weight-bearing surfaces are no longer in vogue, and cobalt-, titanium-, or ceramics-based parts are generally orthopedic surgeons' first choice. Cobalt–chromium–molybdenum alloy has superb wear resistance. Titanium–aluminum–vanadium alloy has a close-to-bone modulus of elasticity, excellent biocompatibility and osseointegrative properties. For hip replacements, femoral heads manufactured of aluminum- or zirconium-oxide ceramics have optimal frictional and wear characteristics when articulating with polyethylene [6].

A. Survival Rate and Advantages of Cementation Systems

Albeit pain relief, motion stabilization, and correction of the joint deformity attained in a majority of arthritic patients by the replacement of a destroyed joint by an artificial one, a host of quandaries faces surgeons, bioengineers, and materials scientists in their pursuit of an optimal operative procedure, implant design, fixation technique, instrumentation, alloplastic parts' wear resistance, and compatibility with the osseous environment [7]. The mechanical anchorage of alloplastic components is realized by cement–bone interdigitation. Unfortunately, it is often lost in the wake of formation of an ever-expanding soft tissular interfacial membrane in between the acrylic polymethylmethacrylate (PMMA) cement mantle and the bone. In view of the inherently inevitable quota of prosthetic failures, first and foremost due the aseptic loosening of one component or another, cementless arthroplasties have been anticipated to remedy many ills of cemented joint replacements. No matter how strong initially, any material without a capacity for self repair will ultimately fail given sufficient repetitive loading. Hence, breakdown products, be they polymeric, metallic, or ceramic in nature, accumulate within the joint cavity, pseudocapsule, and at the implant–bone interface of the artificial joints [8]. Having reviewed the scientific basis of component fixation, Freeman and Tennant contend that in view of equivalence of the clinical results, an advantage for cemented versus cementless fixation of total hip arthroplasties cannot be premised [9].

At the hands of skilled orthopedic surgeons using up-to-date artificial joints, the prosthetic failure rates are low, whether the components are cemented or otherwise. Seventy-five percent of cemented femoral stems have been found to be well fixated at the 22-year follow-up of patients younger than 40 years at the time of their operation, a laudable result when this young population's active lifestyle is taken into account [10]. Based on revision rates of aseptic loosening of cemented femoral stems as the endpoint, the survival of hip arthroplasties in one published series of cases was 100, 99, and 98% at the 5-, 10-, and 15-year postoperative intervals, respectively; the survival of the cemented acetabular cups in this series was 99, 98, and 84% at the 10-, 12-, and 15-year postoperative examinations, respectively. In addition to host- and surgeon-related factors, the design of the components also plays a decisive role in the success or failure of total joint replacements. Ten to 15 years after its implantation, femoral osteolysis was detected in 8% of patients with a Charnley low-friction arthroplasty, the acetabular socket and femoral stem of which had to be revised in 8 and 5% of the cases, respectively. In comparison, femoral osteolysis was discovered in just 2% of patients with a Wrightington Frusto-Conical artificial hip joint, and just 3 and 4% of the femoral stems and acetabular cups had to be revised, respectively. The results affirm an edge of the Wrightington over the Charnley system [11]. An impressive long-term outcome of cemented total knee arthroplasties has also been reported. In a recently published series of patients with posterior cruciate ligament-retaining knee arthroplasties, the survivorship was 96 and 82% at the 15- and 23-year postoperative intervals, respectively [12]. Ten to 15 years after implantation, no case of aseptic loosening of cemented Kinematic artificial knee joint has been recorded in 102 patients who were operated on for osteoarthritis, rheumatoid arthritis, osteonecrosis, or hemophilic arthropathy [13]. Usually, cemented knee arthroplasties with relatively conforming articular surfaces manifest excellent long-term survival. In the experience of the Mayo Clinic's orthopedists, the condylar knee prostheses have a 21-year success rate of 91%, the posterior stabilized prostheses a 16-year success rate of 94%, and the constrained condylar knee prostheses a 7-year success rate of 98% [14]. In one convincing comparative study of patients with total knee arthroplasties examined at the sixth postoperative year, the incidence of aseptic loosening of the cemented femoral stems has been as low as 0.6%, while aseptic loosening of press-fit, noncemented femoral stems has been met with in as many as 9.8% of the patients [15]. It should be kept in mind that arthroplasties fail by mechanisms other than wear debris–induced aseptic loosening, e.g., an altered stress-strain fields-initiated bone remodeling, interfacial motion, improper orientation of components, or inadequate cementation techniques [16–19]. Their histopathological aspects are not dealt with herein.

B. Impact of Wear Debris on Survival of Cemented Arthroplasties

Having surveyed a token part of the myriad publications embodying experts' perceptions, the cemented total arthroplasties attest to an outstanding achievement as far as an individual patient's fate is concerned. Yet, in view of the over one million arthroplasties performed globally every year, it is not surprising that the number of revisions of failed prostheses is legion. Initially, the PMMA was blamed for most ills of artificial joints fixated to bone with the acrylic cement, so much so that the epithet ''cement disease'' was pinned to components' aseptic loosening [19]. This ideation was soon realized to be false. Unjustifiably, so-called polyethylene disease was then equated with aseptic loosening, again condemning one material for the fiasco of an entire prosthesis [20]. Surgeons' frustration with the need to revise their patients' cemented artificial joints rationalized the quest for cementless fixation techniques. These, however, have not alleviated the loosening predicament: noncemented prosthetic parts fail like their cemented counterparts. The loosening is most often due to an increased resorption of the bone supporting the alloplastic components. The augmented osteolysis is secondary to an exaggerated osteoclastic activity that is stimulated by intermediates secreted by the constituent cells of a granulomatous membrane arising at the bone–implant interface. Macrophages are stimulated by ingested submicron- to micron-sized wear particles, deposited at the bone–implant interface, to synthesize and secrete the mediatory substances. Given that formation of interfacial membranes is independent of the fixation methods, at least to a certain degree, reflecting in fact the inflammatory granulomatous response to breakdown products, be they polymeric, metallic, or otherwise, prosthetic failure may be viewed as a ''small particles disease.'' with proper design and surface topography of the implant and also load patterns directing the genre of cellular activities, osteogenesis prevails under favourable settings; while osteolysis, fibroblastic proliferation, inflammation, and granulomatous responses monopolize the scene under adverse conditions [6].

C. Response to Wear Particles and Formation of the Interfacial Membrane

Acrylic cement is not a glue. It does not bond an implant to the bone. Rather, PMMA is an adhesive, anchoring an alloplastic material to the underlying bone via mechanical interlocking without chemically binding the one to the other. The PMMA has an entirely mechanical function, which is contingent on the formation of bone–cement interdigitations. The fixation strength depends on the depth of the cement projections penetrating into the bone. Interdigitation with bony interstices increases with increasing pressure applied to the curing methylmethacrylate and the size of the intertrabecular spaces of the bone. Improved PMMA quality, distal plugging of the bone cavity, pressurized insertion of the cement, and pressure lavage, among others, increase the bone–cement interfacial shear strength [21–23]. Alongside well-fixated components, the intraoperatively achieved interlock of the bony trabeculae and cement persists, but it is lost when loosening ensues. Stable cemented bio-materials are accepted by their bony environs without an undue inflammatory reaction, and their surface is, at least partially, osseointegrated. A thin, fibrous interfacial membrane—a millimeter or so in width—may be present at the interface of stable components without impeding their function. This contrasts with the situation aside an unstable, loose cemented component which is separated from its bony bed by a thick interfacial membrane. In addition to the width of this soft tissular layer, the extent and intensity of inflammatory granulomatous responses at the implant–bone interface determine the biomechanical and biochemical (i.e., mediators' secretion) consequences of the interfacial membrane [24–26].

The late failure of the cemented acetabular sockets of total hip arthroplasties is representative of the chain of events leading from a stable to an unstable setting: loosening begins circumferentially at the socket's intra-articular margins, and the small particles-driven granulomatous membrane interposes itself between the bone and the cement, advancing with time toward the implant's dome [27,28].

D. Osseointegration Versus Formation of Interfacial Membrane

Osseointegration specifies lack of a soft tissular membrane nearby a foreign material within the osseous environs. It is operationally defined, at the light microscopical level, by the proximate contact of the bony tissues with the surface of the implant (Figs 1 and 2). This definition holds true even though, at the ultrastructural level, optimally biocompatible and properly integrated implants, e.g., titanium alloy–based screws, do not directly abut on the bone but are separated from it by a 100- to 200-nm-wide amorphous proteoglycan film and by an about 100-nm-wide lamina limitans–like layer [28–30]. In this respect, bioactive ceramics are unique in so far as their hydroxyapatite-like phase of nanocrystals grows in the direction of the crystals of the implant across the interface, whereby a genuine chemical bond between the biomaterial and the bone is created [31].

Leukocytes, in general, and monokaryonic and polykaryonic macrophages in particular, are short-lived cells. Their persistence at interfaces many years postoperatively, therefore, indicates ongoing control of the local scene by chemical and physical properties of the implants or their breakdown products. A filmy interfacial membrane, typical of a well-functioning implant, is a compelling criterion of the bio-material1s biocompatibility when associated with a weak foreign body response. As a rule, the thickness of an inflamed interfacial membrane increases with the in vitro measured cytotoxicity of the biomaterial. Deductively, the absence of an interfacial membrane implies a pre-eminent harmonious coexistence of the biomaterial and the bone. Thus, osseointegration specifies optimal biological (that is, biocompatible material) and biomechanical (that is, motionless interface) settings [32–34].

E. Impact of Surface Topography on Fixation Strength

The surface affects functions through the variations in the cement adhesion and abrasion properties. The smoother the surface, the lower is the fixation strength of the cement–metal interface. Conversely, rough surfaces encourage high fixation strengths. On the other hand, smooth surfaces are less abrasive in the face of interfacial motion. Vice versa, rough surfaces are more abrasive. To recount surgeons' dilemma, rough-surfaced implants benefit from the enhanced cement attachment such that the probability of interfacial motion is low, but concurrently higher rates of debris generation ensue once motion does occur. While a smooth implant surface is conducive to interfacial motion, the rates of debris generation are low incidental to motion [35]. Moreover,

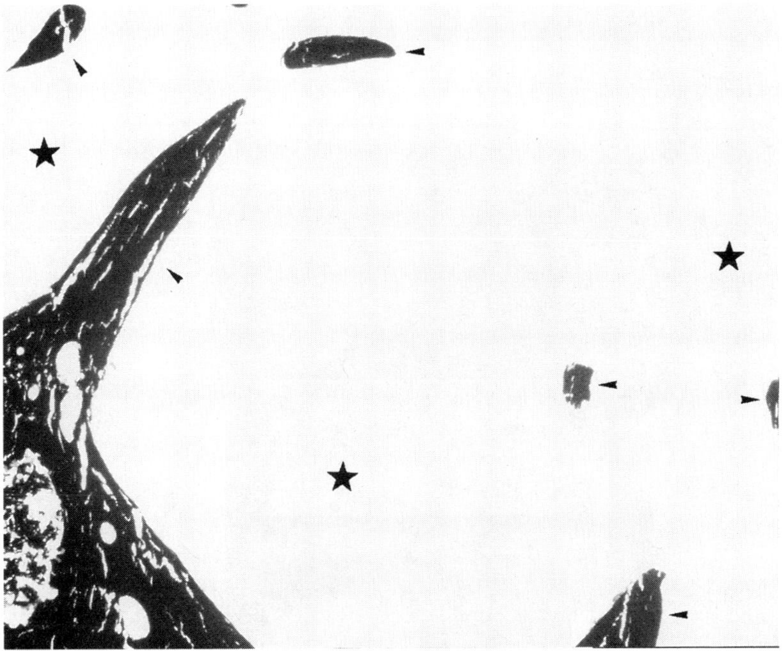

Figure 1 Appropriately osseointegrated acrylic mantle of cemented femoral stem of a total hip arthroplasty revised for aseptic loosening of the acetabular socket 10 years after implantation. In addition, there is excellent interlock of the bony projections (arrows heads) with the bulk of polymethylmethacrylate (asterisk). Section of an undecalcified specimen stained by von Kossa method, 25×.

interfacial motion interferes with osseointegration. Twenty diurnal cycles of interfacial movements of a 0.5-mm amplitude suffice to bolster up fibrous tissue proliferation rather than bony ongrowth onto titanium implants in rabbit tibiae [36].

F. Common Causes of Failure of Prosthetic Joints

Dicta professing to foretell that all prostheses are condemned to eventually fail if the patient lives long enough or that the ''degenerative changes of the bone–cement interface'' are inevitable [37,38] are overly pessimistic and therapeutically nihilistic. Charnley has disputed the widely held belief that bone and PMMA cannot peacefully coexist side by side [4]. In support of his deductions, PMMA particles do not impair healing of rat tibial cortical defects. In contrast, materials such as polyethylene and resorbable calcium phosphate incite an inflammatory response and inhibit bony repair [39].

Proficient cementation offers implants better stability than an optimal press-fit technique [40]. In as much as loss of the bone–implant interlock is a leading cause of failure of cemented arthroplasties [9], understanding the nature of the interfacial processes is pivotal for researchers probing the perplexities of effective versus mal-functioning artificial joints. Awareness of the role of biological events in prosthetic failure does not deflect researchers from assessing the participation of mechanical issues in the loosening process, say, alignment, cortical reinforcement, canal/fill ratio, stress-shielding, strain patterns, cement fracture, fatigue cracks in the cement, fatigue of the cement–bone unit, cement creep, cement–metal interface debonding,

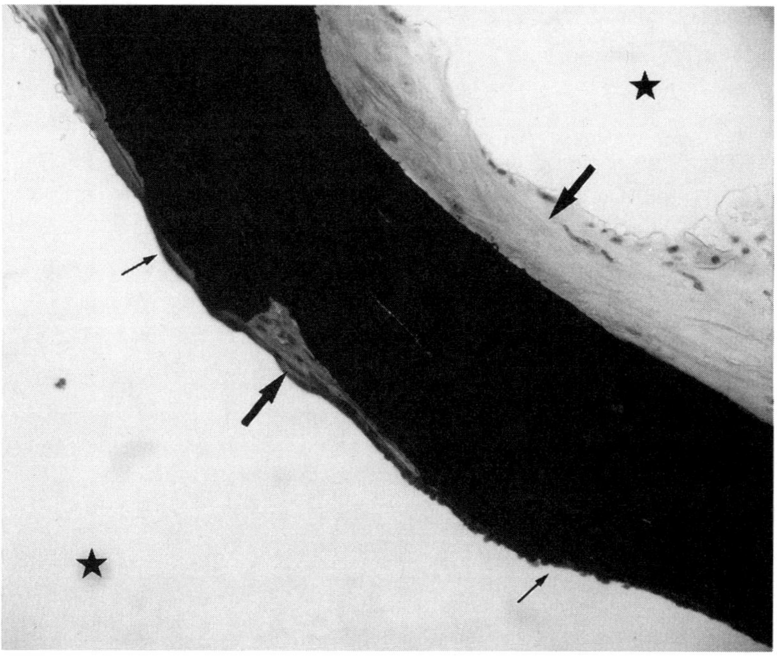

Figure 2 A bony beam within the polymethylmethacrylate bulk (asterisks). The surface of this PMMA projection is segmentally osseointegrated (short arrows) and segmentally separated from the PMMA by a thin layer of hypocellular fibrous tissue (long arrows). Section of an undecalcified specimen stained by von Kossa method, 100×.

gross polyethylene upsets, and microfracture of the osseous trabeculae buttressing the construct. Cracks in the acrylic mantle, primarily initiated at voids, act as stress raisers, resulting in fatigue failure of the cement, an important albeit infrequent mode of loosening of the stem of total hip arthroplasties. Both a thin (less than 2 to 3 mm wide) acrylic coat and a low canal/fill ratio raise the stress within the cement, thus increasing its fracture proneness. Thin PMMA mantles being discontinuous as a rule, the ensuing bone–metal contact points weaken the cement coat [41–45].

To explain complex immunopathological processes involved in responses to wear debris, the periprosthetic tissues have been the focus of extensive analyses by different approaches, including evaluation of retrieved interfacial membranes, in vitro interactions of cells cultured with biomaterials, and functional and histological assessment of artificial joints implanted in experimental animals. In this context, this chapter reviews some advances in the research of diverse aspects of the so-called wear debris disease.

1. Cellular and molecular mechanisms of bone resorption mediated by macrophages and their fused variants, i.e., the multinucleated giant cells.
2. The expression and the specific role of proinflammatory cytokines, growth factors, and transcription factors in the local inflammatory granulomatous reactions.
3. The response of the osseous microvasculature to the presence of an implant and the activation of endothelial cells by biomaterials.
4. The role of mast cells–derived cytokines and mediators in bone resorption.

Lessons learned from the experimentation promise to aid our understanding of the effectiveness of orthopedic implants in long-term use in patients with view to developing a new generations of biomaterials. Cemented or cementless techniques are used to replace diverse joints. This chapter spotlights cemented total hip and knee arthroplasties, not just because they make up a majority of total joint replacements, but with the hope that the knowledge acquired by exploring quandaries raised by these time-honored procedures may serve as the ''gold standard'' against which what happens at the interface of innovative artificial joints could be measured.

II. CYTOLOGICAL AND HISTOLOGICAL ASPECTS OF THE IMPLANT–BONE INTERFACE

A. Characteristic Features of Implant's Particulate Wear Debris in the Periprosthetic Tissues

Willert and associates published a series of pioneering research articles; the first, brought out in 1974, was dedicated to a ''reaction of bone to methacrylate'' [46]. The authors discussed in more general terms the ''material breakdown and tissue reaction in clinical joints'' after they realized that both polymeric (cement and polyethylene) and metallic wear particles amassing in the pseudocapsule and interface of artificial joints provoked formation of a foreign body granulation tissue. Ongoing growth of the granulomatous interfacial membrane was purportedly responsible for the loosening of the prosthetic parts [47]. Macroscopically, the bone-facing cement displayed a highly wrinkled and papillary relief. Microscopically, flattened beads at the cement's front were assembled into an irregular, rough, and open, porous-type exterior. The acrylic surface conformed with the contour of the adjacent bony tissue. External mechanical influences, volumetric changes of the cement, and the effect of the abutting bone seemed to lead to this reciprocal adaptation, resulting in a ''bone–cement composite.'' Under high loads, particles broke off where beads were either isolated or incompletely attached to the surface of the PMMA [48].

The amounts of polyethylene debris polluting the periprosthetic tissues vary from one artificial hip joint design to another. With metal ball and polyethylene cup combinations, the annual linear polymeric wear ranges from 100 to 300 μm. Cups paired with ceramic balls generate less debris, the linear wear varying from 50 to 150 μm per year. Metal-on-metal and ceramic-on-ceramic pairings generate small amounts of particles, the linear wear rate ranging from 2 to 20 μm per year [49]. In fact, scanning electron microscopical appraisal of ceramic femoral heads retrieved after over 10 years of service discloses adequate surface roughness and sphericity, and ceramic particles have not been located in at least some of the sampled periprosthetic tissues [50]. Articulating metallic surfaces, even though disclosing myriad delicate scratches, yield an annual wear volume of about 5 μm^3, a low value when compared to the quantities of polymeric debris accumulating each year. The macrophagic reaction to metallic particles is said to be mild; at any rate it is less violent than that to cement particles tallied in histological sections of specimens retrieved from the same patients. Compared to generation of particles from cobalt–chrome alloy–based femoral components, titanium alloy–based parts release a great deal more debris [51]. Titanium alloy is especially prone to abrasion by acrylic third body wear [52]. It is doubtful that a less vigorous inflammatory response to the titanium than to the other kinds of detritus affects a compensatory influence, wear debris of titanium alloy–based femoral heads being infamous for initiating prosthetic failure [53,54]. Titanium and its alloys do not behave as harmlessly toward their osseous environment as hitherto envisioned [6,55]. In vitro phagocytosed, submicron-sized titanium particles exert adverse effects on the cytoskeletal actin arrangement of the osteoblasts. In consequence of the ensuing decreased cellular activities, the cells' spreading and adhesion force are reduced, as is the gene expression for collagen type I and fibronectin [56].

At the macroscopical examination of freshly retrieved specimens, differences are discernible between different patients' interfacial membranes and between different segments of the same patient's specimen. The color of the membranes varies from yellow to red to gray to ''dirty'' black. Some membranes are firm, elastic or rubbery, while others are pulpy, soft to gelatinous; sometimes they are friable and granular to the touch. Bits of acrylic cement remain embedded in the membranes after their peeling off from their surroundings [20].

Goldring and his coworkers introduced the epithet of ''synovial-like membrane'' to the soft tissular layer separating the bone and cement from one another [57,58]. This term is a misnomer. Interfacial membranes do not resemble a synovial capsule. If anything, they histopathologically recall the features of chronic synovitis. Santavirta et al. have highlighted the ''aggressive'' nature of the interfacial membranes in view of their central role in periprosthetic osteolysis and prosthetic failure [59]. The interfacial membranes evince a more-or-less three-tiered construct: an inner synovial-like, a middle inflammatory granulomatous, and an outer mainly fibrotically sclerosed coat [60]. The interfacial membrane dimensions and intensity of their inflammatory granulomatous–cicatrizing changes account for the biomechanical and biochemical consequences. Theoretically, bland fibrous films, less than a couple of millimetres across, may protect the bone–implant unit by smoothing and evenly distributing the stresses. On the other end of the spectrum are those membranes which range from a few millimetres to two centimetres or more in width and teem with macrophages and lymphocytes. They not only facilitate interfacial macro-motion and interfere with optimal stress transfer, but perhaps more importantly they are the site of production of proinflammatory cytokines, bone-resorbing factors, and growth factors. Repetitive biopsies have established that the first signs of an interfacial membrane are evident 10 weeks postoperatively. For the interpretation of the periprosthetic events it is consequential to realize that the characteristics of membranes nearby a nonloosened and a loose component may be morphologically indistinguishable from each other even though quantitative differences between them are overt [61]. Aside from the metallic, polymeric, and/or ceramic foreign body–induced granulomas usually encountered in most regions of the sampled operation specimens, the interfacial membranes may elsewhere exhibit histological aspects of a nonspecific, well-vascularized granulation tissue (Fig. 3).

B. Cellular Invasion and Activation at the Bone–Implant Interface

The macrophages infiltrating the interfacial membranes of cemented artificial joints exhibit various morphological features. Directly facing the acrylic mantle, they are often palisading, unistratified or pseudostratified, closely to spaciously spaced, and range from elongated to polyhedral to cuboidal. They are intermixed with some lymphocytes and a few polykaryonic macrophages (''giant cells''). Segmentally, this cellular barrier is necrotic, replaced by a hyalinized fibrous tissue, has undergone fibrinoid change, or is covered by fibrinous precipitates (Fig. 4). This synovial-like film rests on a fibrous tissue, which varies in width from one case to another and is infiltrated by lymphocytes, macrophages, and plasma cells (but unless bacterial or fungal infections have occurred, there are no neutrophils). This upper layer blends with the underlying, prosthetic debris–driven granulation tissue of the middle layer.

As a rule, large collections of closely packed, roundish to polyhedral, foamy to finely granular, mono-, bi- and polykaryonic macrophages dominate the scene in the middle layer. Ultrastructural observations evince that the foamy or granular look of these cells (Fig. 5) is bestowed on them by the myriad ingested, submicron- to micron-sized polyethylene particles. The particulate polymer in the macrophage's phagocytic vacuoles takes up oil red O and, under

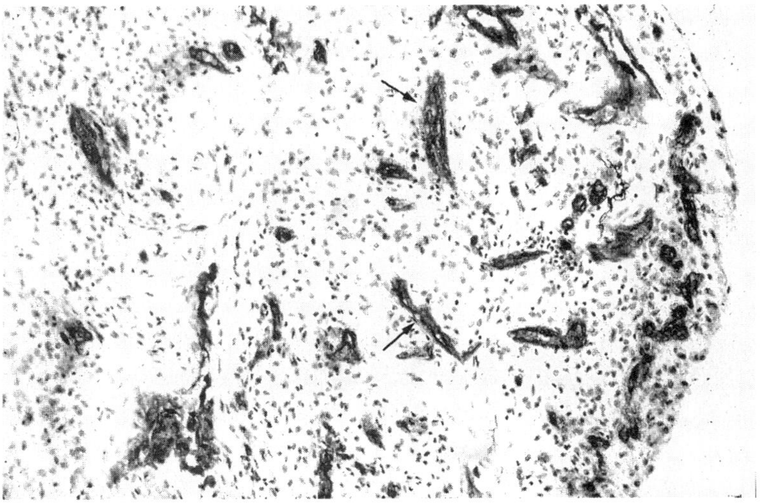

Figure 3 Segment of an interfacial membrane disclosing the histological features of a nonspecific, well-vascularized granulation tissue. Note the many crisscrossing blood vessels (arrows). Immunostaining with the lectin Ulex Europaeus agglutinin 1, 200×.

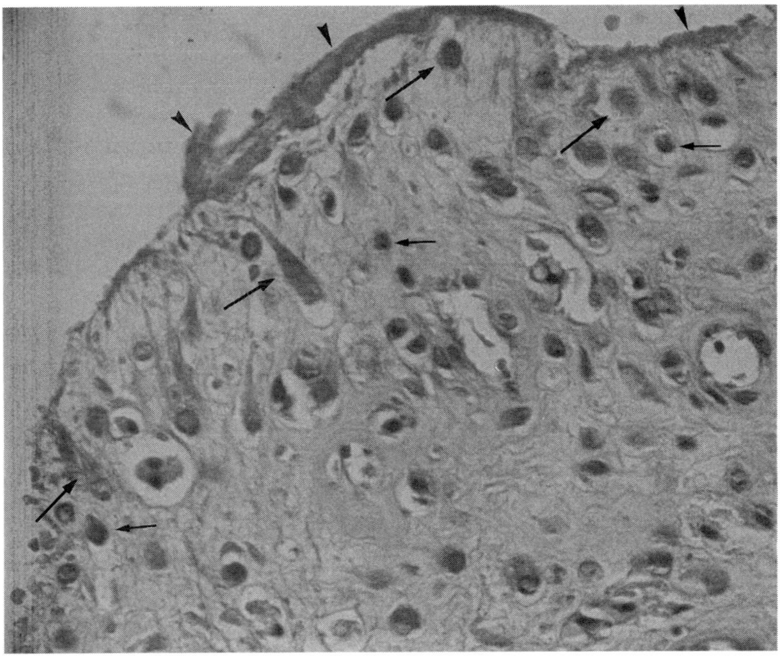

Figure 4 The ''synovial-like'' features of the cement-facing aspect of an interfacial membrane disclosing elongated, polyhedral or cuboidal, spaciously spaced, plump macrophages (large arrows) which are intermixed with sparse lymphocytes (small arrows). A fibrinous precipitate segmentally covers the surface (arrow heads). Hematoxylin and eosin, 400×.

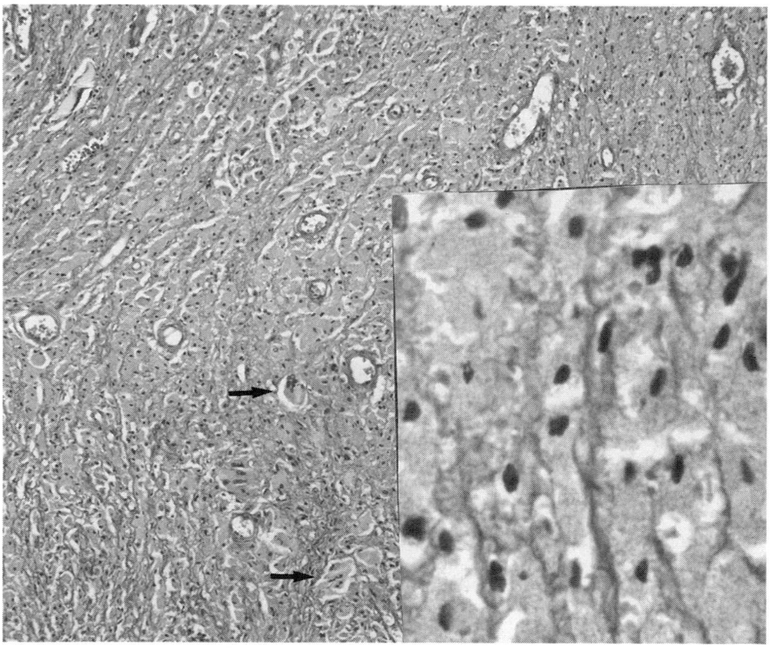

Figure 5 Large sheet of foamy to finely granular, mostly mononuclear, polyhedral cells in the midzone of the interfacial membrane of a failed artificial hip joint. There are some binucleated cells as well as polykaryonic macrophages, so-called giant cells (arrows). The insert exemplifies, at higher magnification, the finely granular quality of these submicron- to micron-sized polyethylene particle–laden cells. Masson trichrome stain, 100x and (insert) 630×.

polarized light, yields punctate and variously sized fibrous birefringence [62]. (Fibers are defined as threadlike structures which are at least three times as long as they are broad.) Distinctive tribological conditions at one site or another are responsible for different modes of wear debris generation. Splintering and delamination of material from tibial inserts are overwhelming when contrasted to those occurring at acetabular cups. Unless a catastrophic event complicates the postoperative course, say, intra-articular interposition of an acrylic fragment or impingement, polyethylene splinter–provoked giant cell granulomas are less common in the interfacial membranes of hip arthroplasties than in those of knee joint replacements [63–65]. Bizarrely shaped globular or oblong polyethylene fragments, ranging from tens to thousands of microns in length, are incorporated in the interfacial membranes and pseudocapsules of many artificial joints. Such fragments are found within then phagocytic vacuoles of the giant cells or are encircled by a multitude of mono- and polykaryonic macrophages (Fig. 6). The rate of degradation of the components of artificial joints is enormous. Thus, 4×10^7 to 4×10^{10} particles are cast off into the joint cavity annually. There is consensus that the billions of 100- to 500-nm-sized polymeric particles retained per gram of periprosthetic tissue are the most critical species of prosthetic debris as far as the fate of the arthroplasties is concerned [66]. Occasionally, the polyethylene debris induces a flourishing interfacial membrane at one site while elsewhere the cement is directly attached to the bone [67].

The wide distribution in the sizes of particulate polymers in the periprosthetic tissues of failed prostheses is truly remarkable. Sometimes, massive polyethylene fibers and flakes abound in the interfacial membranes, especially those of artificial knee joints; they are less frequent in

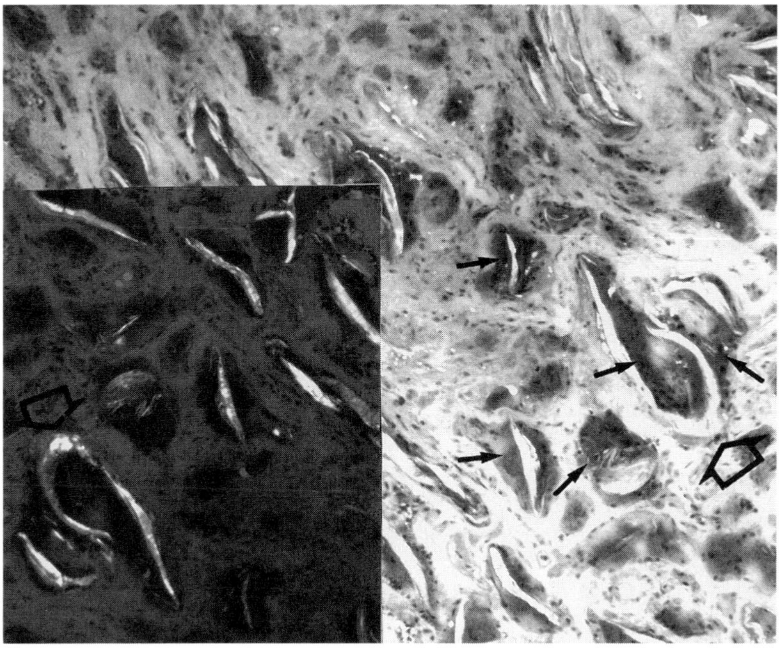

Figure 6 Giant celled granulomatous reaction to extra large, oblong, and bizarrely shaped polyethylene splinters in the interfacial membrane of an artificial knee joint which failed 12 years postoperatively. Other than the oversized foreign body giant cells (arrows), few mononuclear macrophages, lymphocytes, and fibrocytes occupy the fibrous background. The insert displays the same area but photographed under polarized light to demonstrate the birefringence of polyethylene. Note the duplicate of the polyethylene filament in the microphotographs taken under regular and polarized light (open arrow). Section of an undecalcified specimen stained by von Kossa method, 100×.

the interfacial membrane of hip arthroplasties. Different tribological settings, conformity of bearing surfaces, configurations of the articulating surfaces, and mechanisms of wear bring about different modes of load transfer as also stress and strain effects. These dissimilarities are also reflected in the electron microscopical measurements of the ingested polymeric debris. In the periprosthetic tissues of hip arthroplasties, 90 and 75% of the intracellular polyethlene particles are less than 1.0 and 0.5 μm in size, respectively. Seventy and 45% of the ingested polyethylene particles in the periprosthetic tissues of artificial knee joints are less than 1.0 and 0.5 μm in size, respectively. Joint tribology- and geometry-related peculiarities determine the modes of wear debris generation at various anatomical locales, modifying not only the histological scene but also the cellular responses and, therefore, the pathways to bone resorption [64,68]. Usually, widely spread, confluating sheets of large, polyhedral, finely granular, oil red O–positive, birefringent macrophages are the dominating elements in periprosthetic tissues obtained at revision operations of loose hip arthroplasties. These distinctive macrophages occur in the interfacial membranes of stable artificial joints as well, albeit in smaller numbers. In addition, there are numerous granulomatous foci with polykaryonic, at times gigantic multinucleated, macrophages, particularly in the interfacial membranes of artificial knee joints. Some of these giant cells abut on huge noningestable slabs, while others partly or totally phagocytose the bizarrely shaped, rugged polyethylene shards, which may be up to several hundred microns across [69].

The pathogenic primacy of ''cement disease'' [16] in the loosening process of artificial joints having lost its popularity, most authors currently subscribe to the supremacy of the concentration, shape, and size of polyethylene particles deposited at the bone–implant interface in prosthetic failure [70]. The faults of routine histological assessment of particulates are conceded because dimensions are underestimated by reason of the evaluation of 4-μm-thick sections and, additionally, structures smaller than 0.5 μm^2 are below the resolution power of light optical microscopes. It may, nonetheless, be of interest to briefly refer to statistics with respect to the intracellular lading of polymeric debris in pseudocapsules of a cementless total hip arthroplasty. The mean load of the polyethylene particles has been measured to be 13 ± 8, 25 ± 16, and 49 ± 42 μm^2 in monokaryonic (mean areal size 190 ± 65 μm^2), small polykaryonic (mean areal size 605 ± 200 μm^2), and large polykaryonic macrophages (mean areal size 1420 ± 490 μm^2), respectively. Even though giant cells may ingest particles as large as ~ 250 μm^2, most particulates larger than ~ 45 μm^2 amass in the intercellular space. The mean particle size has been found to be 2.34 ± 3.20, 3.02 ± 3.46, and 4.70 ± 11.25 μm^2 in the monokaryonic, small polykaryonic, and large polykaryonic macrophages, respectively. Finally, the relative loading (percent of cell body occupied by the foreign bodies) is 6.7 ± 3.6, 4.1 ± 2.4, and $3.5 \pm 2.9\%$ in the monokaryonic, small polykaryonic, and large polykaryonic macrophages, respectively [71].

Cement chunks are sparsely strewn in the majority of interfacial membranes. They abound in the periprosthetic tissue of patients in whom the acrylic mantle has crumbled for one reason or another, say, a poor cementation technique or cement fatigue. They manifest as variously sized, optically empty, polycyclic spaces which usually contain myriad tiny, black, birefringent barium sulfate or zirconium oxide grains. (The lipid-soluble PMMA in routinely processed samples is extracted by the solvents used for the preparation of the histological sections; the $BaSO_4$ or ZrO_2 is preserved.) The space–tissue boundaries exhibit collagen fibers or a several-cells-thick covering of mono- and polykaryonic macrophages, which are encircled by the concentrically oriented collagen fibers (Fig. 7). In addition, $BaSO_4$- or ZrO_2-laden macrophages invest the formerly cement chunk–containing spaces. Micron-sized PMMA bead–ingesting macrophages exhibit a bubbly cytoplasm dirtied by the tiny, black $BaSO_4$ or ZrO_2 grains (Fig. 8). The radiopaque additives to the cement most likely share in bringing about aseptic loosening because their ingestion reinforces phagocyte's potential to produce bone-resorbing intermediates. More often than not, and in contrast to past impressions, cement chunk–provoked granulomatous reactions are minor in comparison to the profuse response induced by both the polyethylene or metal particulates. Skin patch tests to N,N-dimethylparatoluidine (the accelerator used in bone cement) are positive in 10% or so of patients with cemented arthroplasties. In these allergic patients, an immune reaction to the N,N-dimethylparatoluidine may be blamed, at least partly, for onset of prosthetic failure within 2 years of implantation [6,69,72–74].

Within the periprosthetic tissues, metallic debris appears as black, irregularly shaped; barely visible to 5-μm-sized particles in round to polyhedral; eosinophilic; granular; mono-, bi-, and (rarely) polykaryonic macrophages (Fig. 9). Metallic debris is detectable by light microscopy in over 80% of periprosthetic tissues retrieved from patients with failed artificial joints. Viewed with scanning electron microscopy, these particles vary from round to oval to rectangular, many having sharp edges and corners. They may be scattered extracellularly between the collagen fibers. Macrophages concurrently ingesting metallic and polymeric particles exhibit black grains on a foamy background, unless the cells are overloaded by metallic dust to such a degree that all cellular details are obscured [51]. As a rule, there are just small amounts of metallic debris in the interfacial membranes of artificial joints which have served for a short term [75]. Metallosis, i.e., the macroscopically discernible grayish discoloration of periprosthetic tissues by stored metallic particles, is rare in patients with cemented arthroplasties. It is principally associated with implants disclosing abraded surfaces. Incidentally, metallosis is most often met

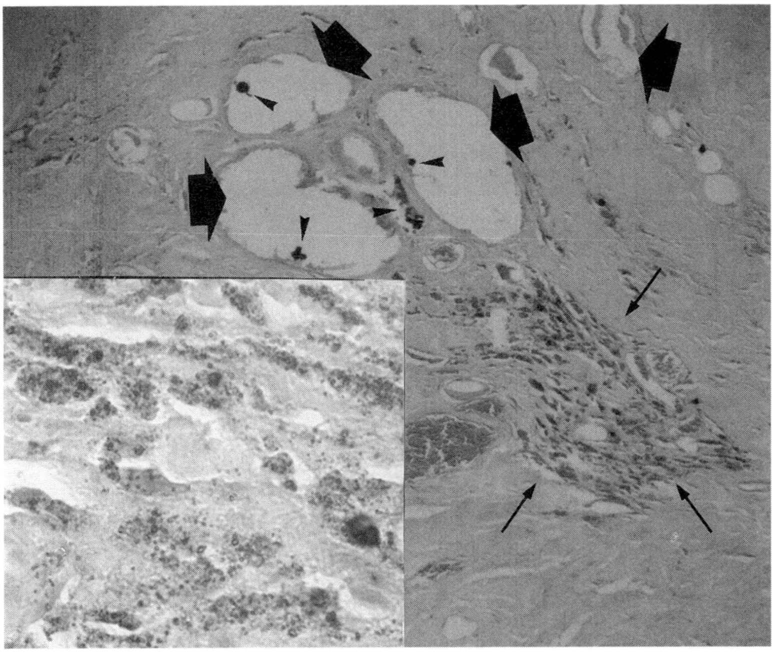

Figure 7 Cement chunks in the periprosthetic tissue of a patient whose total knee arthroplasty failed during the 13th postoperative year. They manifest as variously sized, optically empty, polycyclic spaces (thick arrows) which contain remnants of the clumped, black barium sulfate grains (arrow heads). Note also the hyalinized make-up of the surrounding fibrous tissue. Subjacent to these cement chunks, an aggregate of $BaSO_4$-laden macrophages is present (thin arrows), and its cells with their phagocytosed, finely dispersed $BaSO_4$ grains are illustrated (in the insert) at a higher magnification. Hematoxylin and eosin, 100x and (insert) 630×.

with at revision operations of noncemented total hip arthroplasties, usually those with titanium- or titanium alloy–based parts. The metallic particles originate from the articulating as well as anchoring surfaces [51,76,77].

Metallic dust–laden macrophages typically accumulate around blood vessels (Fig. 10). Most particles have areas smaller than 1 μm^2. Occasionally larger (about 10 to 300 μm long), jagged metallic fragments are randomly dispersed in the fibrous and granulation tissues. Otherwise, they are enclosed in giant cells of discrete granulomas or scattered in the extracellular space. Much metallic detritus is also present in some sizable foci of coagulation necrosis (Fig. 11). Metallic debris abounds in the periprosthetic tissues of some long-term-serving artificial joints [51,78]. As sometimes occurs in patients with cemented titanium alloy–based total hip arthroplasties, particulate metals amassing early in the postoperative period are blamed for premature prosthetic failure [54]. Extensive but focal necrosis of the fibrous and granulation tissues accompanies some cases of severe metallosis of the interfacial membranes of short-term surviving cemented arthroplasties. This type of necrosis has also been described in patients with worn necks of cemented titanium alloy stems of artificial hip joints failing within less than 10 years of the implantation. Histologically, copiously phagocytosed, nanometer-sized titanium alloy particles may fully mask the polyethylene debris taken up by the same macrophages [79].

Metallic debris–laden cells mingle with helper and cytotoxic T-lymphocytes, scant plasma cells, some mast cells, and varying numbers of eosinophils. On the other hand, B-lymphocytes

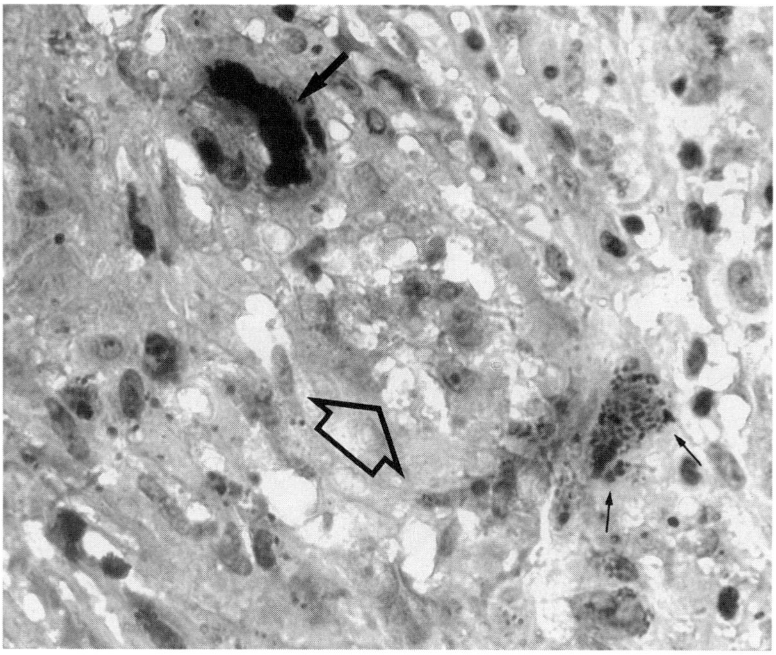

Figure 8 Micron-sized PMMA bead–ingesting macrophages exhibiting a bubbly cytoplasmic body (open arrow), the polymethylmethacrylate having been dissolved during section preparation. Several $BaSO_4$ grain–laden macrophages (small arrows) are present next to the cement bead–ingesting cells. Also present are a few residual bony shreds (large arrow), which are enclosed by macrophages. Microphotograph taken from the same specimen as that illustrated in Fig. 6. Section of an undecalcified specimen stained by von Kossa method, 630×.

are rather rare. This infiltrate's constitution indicts cell-mediated immunity as a key player in the inflammatory reaction. The expression of E selectin on endothelial cells of blood vessels in the vicinity of metallic deposits and the occurrence of necrotizing arteritis or epithelioid cell granulomas in a back-ground of an eosinophil-rich infiltrate imply, but do not affirm, participation of immune processes in the initiation or maintenance of the inflammatory reaction [80–82]. The augmented proliferation of lymphocytes cultured in a cobalt or chromium ion–enriched media favors the role of hypersensitivity in the loosening process of some patients' formerly stable artificial joints. Almost one-third of patients tested 1 year postoperatively manifest skin hypersensitivity to nickel, cobalt, chromium, or titanium. There is no documentation, however, to support the every so often heard assertion that eczematous and urticarial dermatoses constitute an allergic reaction to constituents of patients' implants. The cutaneous lesions do not rapidly remit following implant removal. But it must be acknowledged that this does not invalidate the argument that metal hypersensitivity contributes to periprosthetic inflammation, because the responsible constituent(s) may be continuously released from metallic residues left behind in the bone or soft tissues [83–86]. The infiltrate's composition is nonpartisan, eosinophils, for instance, making up a substantial portion of the cell population sometimes incited by polyethylene fiber prostheses used for the reconstruction of lost structures [69].

The interfacial membrane's midzone merges imperceptibly with the outer, densely textured fibrous layer, in which are strewn rare debris-induced granulomas and chronic inflammatory infiltrates. Newly formed cartilaginous islands and bony plates within the outer layer are, at

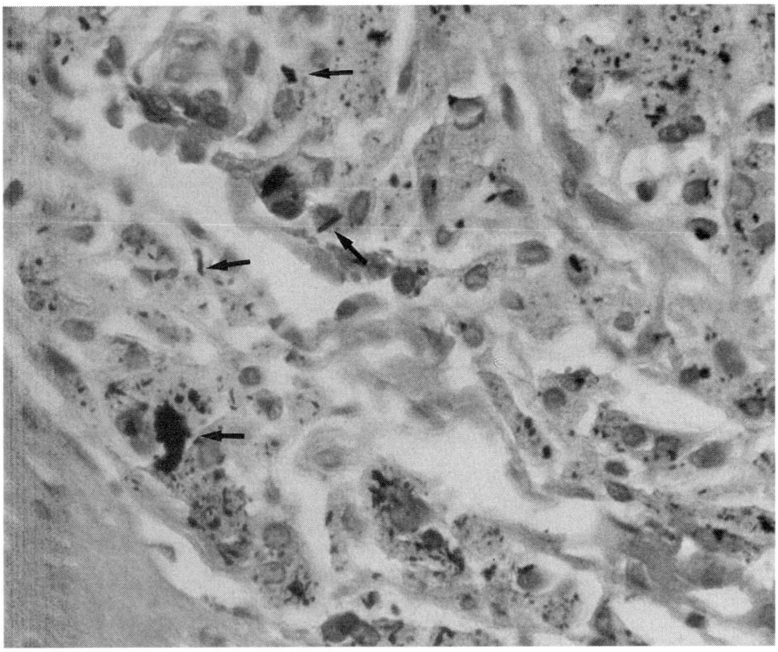

Figure 9 Metallic wear debris manifests as black, irregularly shaped, and slightly varying in particle size, which are mostly present in eosinophilic, round to polyhedral, monokaryonic or polykaryonic macrophages. Even though the micron-sized particles predominate, several jagged and up to ~300-μm-long metallic fragments (arrows) are usually scattered in the intercellular matrix, as in this interfacial membrane of a patient whose artificial hip joint was revised 16 years postoperatively. Section of an undecalcified specimen stained by von Kossa method, 630×.

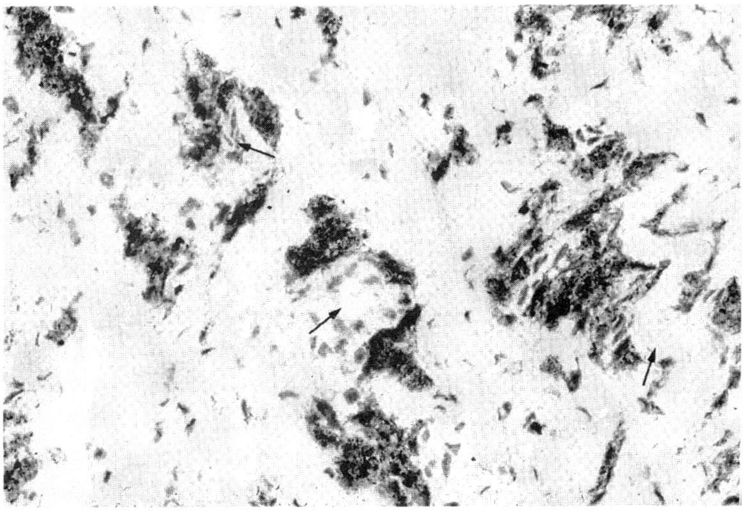

Figure 10 Metallic particle–laden macrophages favoring disposition around the blood vessels (arrows) of this interfacial membrane. Immunohistochemical stain with CD68, 300×.

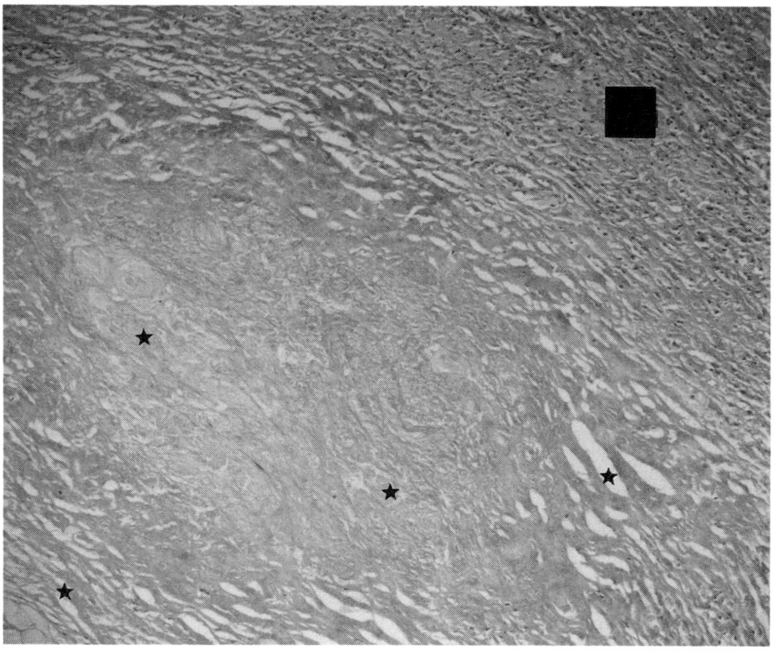

Figure 11 Massive coagulation necrosis (asterisks) of the fibrous and granulation tissues (square) of the interfacial membrane of a patient whose titanium alloy stem of a cemented total hip joint replacement loosened 9 years postoperatively. Albeit not recognizable at this magnification, the macrophages of the granulation tissue were overloaded with metallic and polyethylene particles. Masson trichrome stain, 100×.

least partially, continuous with the peri-implant osseous shell, which makes up the foundation of the artificial joint. This shell comprises woven- and lamellar-fibered, sclerotic (compacta-type structured) and spongy bone, exhibiting signs of ongoing remodeling, including alkaline phosphatase-rich osteoblasts along the newly formed bone (Fig. 12). Osteoclasts typically amass on the cement-facing side of the bone, whereas plentiful osteoblasts accrue on the opposite, marrow-facing side of the bone. In patients whose interfacial membrane is overloaded with breakdown products, polyethylene and metallic debris–laden macrophages, and every so often cement-induced granulomas as well, are strewn within the fibrous and chronically inflamed intertrabecular spaces of the bony shell. These newly formed, inflamed and foreign body–induced granulomatous fibrous tissues may partially replace the bone such that there are osseous remnants or just chips in the deeper endosteal portion of the cortex (Fig. 13). Not infrequently, the remodeled acetabular bone is separated from the cemented socket of an artificial hip joint by a layer of a newly formed cartilage (Fig. 14). Blackish discoloration of a bony area may be evident at the inspection of specimens of severe metallosis with inundation of the intertrabecular fibrous tissue by metallic particle–laden cells. The sporadically found granular hemosiderin-laden macrophages suggest residues of past bleedings. Phagocytes with a bluish cytoplasmic hue in Prussian blue–stained sections possibly contain submicroscopic corrosion detritus emanating from one or another metallic part of the prostheses [87–89].

There are functionally crucial differences in histomorphometrical parameters of the bone around well-fixated acetabular cups when compared with those of the bone adjacent to the granulomatous synovial-like membrane of loosened hip prostheses. The static data evidence increased osteoclastic surface, osteoclast–related eroded surface, osteoid volume, osteoid width,

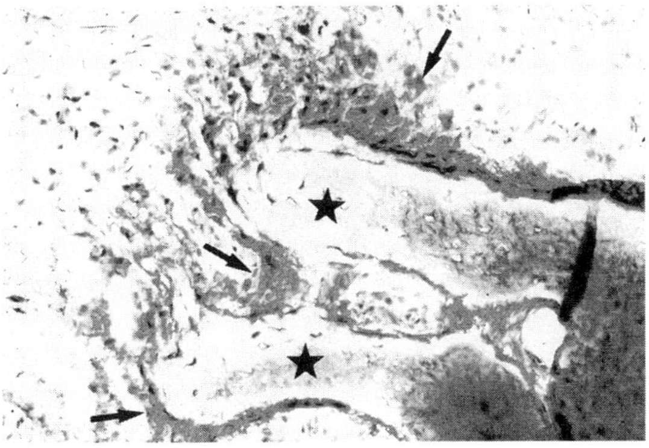

Figure 12 Many alkaline phosphatase-positive osteoblasts (arrows) covering the surfaces of the bone (asterisks) under an interfacial membrane of a cemented tibial component, indicative of vigorous new bone formation and bone remodeling. Histochemical stain for alkaline phosphatase using naphthol AS-BI phosphate as a substrate and Fast Red BR salt on a cryostat section of the interfacial membrane, 280×.

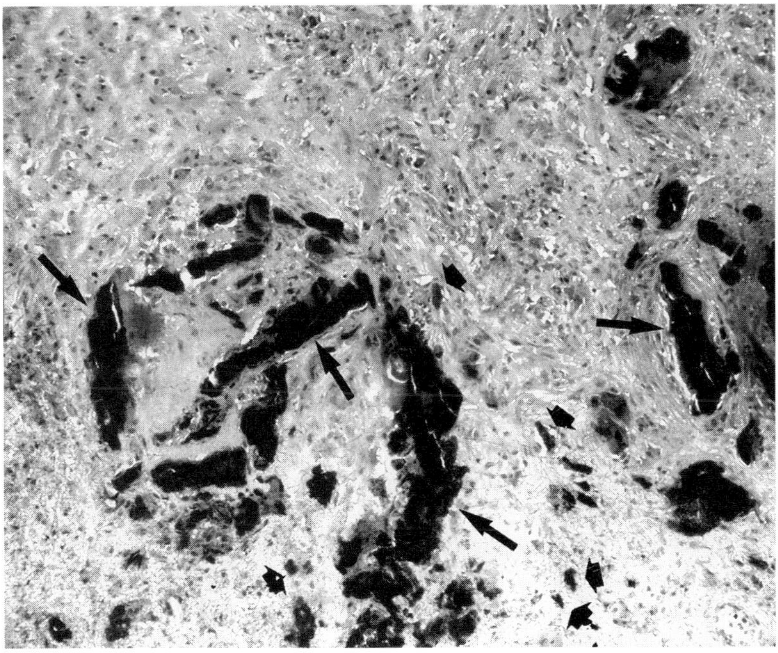

Figure 13 A mildly inflamed fibrous tissue with polyethylene and metallic debris–laden macrophages permeates the bone underlying the interfacial membrane of a cobalt–chrome alloy–based femoral stem of a cemented artificial hip joint which was revised 14 years postoperatively. The intertrabecular spaces of the bony shell are enlarged due to extensive osteolysis. The endosteal aspect of the cortex comprises jagged osseous beams (long arrows), representing either broken up trabeculae (which had buttressed the implant) or tattered osseous residues succeeding severe osteoclastic osteolysis. Note also the osseous gravel (short arrows) representing leftover bone after advanced bone resorption. Section of an undecalcified specimen stained by von Kossa method, 100×.

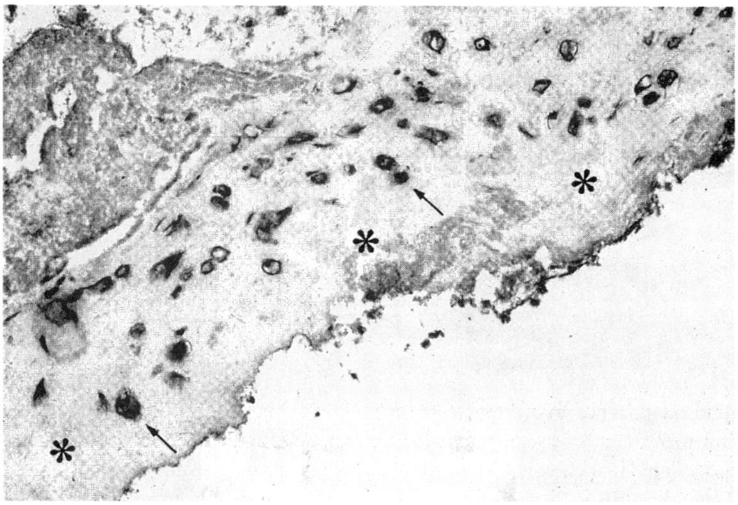

Figure 14 Implant-abutting covering of cartilage (asterisks) nearby the cemented acetabular socket of a total hip joint replacement. Remodeled bone underlying the cartilage. Note the metalloproteinase-1 within the chondrocytes (arrows). Immunohistochemical stain with rabbit antibody to collagenase (MMP-1), 340×.

and low-mineralized bone matrix in the cases of loosened acetabular sockets. The dynamic data disclose an enhanced mineral apposition rate, mineralizing surface, and bone formation rate near loose components. Osteocytic canalicular processes—the intercellular network activated in the response of the bone to mechanical stress—are abundant in the osteoid and low-mineralized bone at these sites. Osteoclastic bone resorption, defective bone formation, or both are coupled with the macrophage-mediated foreign body type granulomas in the synovial-like interfacial membrane of the loosened artificial hip joints. This high-turnover periprosthetic remodeling leads to formation of a bone with poor biomechanical properties. Most likely it results from the effects of foreign body–induced cellular responses combined with inappropriate cyclic mechanical loading, and thus it may contribute to hip prosthesis loosening [90,91].

C. Wear Debris–Induced Osteolysis

Debris-driven osteolysis is the principal cause of failure of cemented artificial joints. The chain of events leading to component loosening is initiated by dispersal of particles in the "effective joint space," that is, the sum total of all the periprosthetic territories accessible to the joint fluid. Schmalzried et al. have brought up the concept of the effective joint space to emphasize the potential of the joint fluid and the detritus floating within it to penetrate between the bone and implant, even though the components are well fixated. Having taken up and been stimulated by particles, macrophages release proinflammatory mediators, chemoattractants, growth factors and enzymes: Interleukin-1α (IL-1α), interleukin-1β (IL-1β), interleukin-6 (IL-6), interleukin-8 (IL-8), interleukin-10 (IL-10), prostaglandin E_2 (PGE_2), tumor necrosis factor-α (TNF-α), granulocyte macrophage colony stimulating factor (GM-CSF), monocyte chemoattractant protein-1, monocyte inflammatory protein-1α, platelet-derived growth factor A, platelet-derived growth factor B, stromelysin, collagenase, and gelatinase. Their net effect augments matrical breakdown, intensifies inflammatory responses, and also recruits and activates ever more fibroblasts, angioblasts, osteoclasts, and macrophages which erode bone on the one hand and boost the growth of the interfacial membrane on the other. Quantitative and qualitative interactions of these mediatory substances determine where and how much bone is lost. Macrophages resorb bone by so-called surface resorption, in contrast to the osteoclast-mediated lacunar resorption. The increased osteoclast differentiation of macrophages exposed in vitro to PMMA is associated with increased lacunar bone resorption. To retrace, the particles-driven vicious circle is kicked off by activation of the macrophages and succeeded by the incremental production of intermediates, osteolysis, component loosening, implant instability, micromotion and then macro-motion, generation of ever more wear debris, additional recruitment and stimulation of macrophages, and proliferation of fibrous tissue. It terminates with enlargement of the interfacial membrane to a degree incompatible with sufficient function of the prosthesis. The negative osseous balance cooperates with the positive soft tissular balance to finalize the arthroplastic failure [20,92–113].

In a cemented, tibial hemiarthroplasty model in rabbits, Goodman et al. have demonstrated that there are greater amounts of N-acetyl-β-D-glucosaminidase and β-glucuronidase in periprosthetic membranes of loose than in those of stable components. The increased content of these lysosomal enzymes parallels the density of the inflammatory cell population in the interfacial membranes. For this reason, cellular influx rather than cellular synthesis is probably more decisive to the overall concentration of the enzymes [114].

Understanding of bone resorption has dramatically advanced in the past few years with definition of the roles of RANKL, RANK, TRANCE, and OPG in regulating osteoclast formation. RANKL, the osteoprotegerin ligand, is essential for osteoclastogenesis. It can substitute for the osteoblasts, whose activities are physiologically required for the differentiation of osteoclasts. RANK is osteoprotegerin's receptor. The osteoclast differentiation factor TRANCE is inhibited by osteoprotegerin (OPG), which acts by competing for the binding of RANKL to its receptor on preosteoblasts [115–119]. In addition to the previously assessed promoters of osteoclastogenesis, e.g., the monocyte colony stimulating factor, TNF-α, IL-1β, IL-6, and soluble IL-6 receptor, which are richly produced by cells of the interfacial membranes, the wear debris–laden macrophages apparently also express mRNAs encoding for RANKL, RANK, and OPG. Deserving of attention in the context of osteolysis are that the macrophages, i.e., precursors of osteoclasts, abound in the interfacial membranes and evidently express RANK, which is needed for the maturation of the osteoclasts. Albeit an abundance of bone-resorption stimulating factors, binding of the RANKL to its receptor is prerequisite for the formation of osteoclasts so as to overcome the inhibitory effects of the OPG in the periprosthetic tissues [120].

A critical analysis of the microscopical features of the periprosthetic tissues readily validates the deduction of Shoji et al. that there is nothing distinctive to the interfacial membrane which portrays the morphological expression of the body's defence mechanism against living or inanimate "invaders" [121]. The noncemented components of loosened artificial joints are surrounded by comparable layer of soft tissue, denying the cement any esoterically exclusive capacity in the formation of a synovial-like interfacial membrane. In vitro, osteoblasts attach to acrylic cement and synthesize proteins. As demonstrated in an experimental loosening model, implant instability rather than its fixation mode, cemented versus noncemented, correlates with the intensity of the inflammatory response and amounts of released mediators. Bone directly abuts on PMMA implants which are not subject to a loading force. Autoptically, osseointegration of the acrylic mantle may be observed in individuals whose artificial hip joints functioned well prior to their demise and revealed extensive radiolucency. These radiolucent lines represent regions of osteoporosis, cortical cancellization, cortical thinning, or bony minitrabeculae, the cement mantle itself being apposed by a shell of newly formed bone (the so-called neocortex), which is supported by a network of osseous trabeculae arising from the endosteal aspect of the cortex. Orthopedists are prevented from seeing the neocortex on x-rays because the densities of the barium sulfate–impregnated cement and newly formed bone are similar. What is clinically considered to express ''prosthetic stability'' is histologically represented by a spectrum of tissular reactions ranging from direct bone–cement contact to an interfacial membrane that is up to 1.5 mm in thickness. This spectrum may reflect varying degrees of microinstability. Polymeric and metallic detritus is often present in small, biologically negligible amounts in the interfacial membranes of stable arthroplasties [20,56,122–126].

There is no biological necessity for every implant to be sequestered from its tissular bed by foreign body granulation tissue. With attainment of a balanced host–implant relationship, biocompatible materials in bulk formed under perfect circumstances are not walled off by granulation tissue but rather are enveloped by a thin layer of bland fibrous tissue (Fig. 2) or appose their viable surroundings directly, without intermediary of any kind of tissue. Perceptions of the synovial-like interfacial membrane are based on microscopical evaluations of samples retrieved at revision operations of loose arthroplasties. Inferences ought to be construed with caution when what failed and not what succeeded is the ultimate object accessible for our studies. For example, PMMA in bulk form does not harm its environs. Osteoblasts adhere to and proliferate on its exterior and amply synthesize collagen. However, diverse cells, including macrophages, exposed in vitro to pulverized acrylic cement are lethally damaged [6].

D. Periprosthetic Bone Necrosis

Bony trabeculae of the implant bed and cement bulges amply interlock with each other during the immediate postoperative period. Investigators have from time to time drawn attention to a layer of necrotic osseous and soft tissues alongside the cement. This tissular death is claimed to result from the reaming, heat produced in the course of the exothermic polymerization of the methylmethacrylate, and effects of the toxic (monomeric) methylmethacrylate or N,N-dimethyl-paratoluidine oozing from the curing cement. In the wake of resorption of the necrotic bone and its substitution by a soft tissue, the bone–cement interlock and, hence, component stability are lost [127–130]. Allegedly, bone rapidly undergoes necrosis after cementation. It is highly questionable whether heat-induced bone necrosis occurs under ordinary circumstances, because protein-denaturating temperatures are obtained only with excess cement. Based on theoretical and experimental data, temperatures of about 58°C are expected at the cement–bone interface. Coagulative necrosis of the bone occurs at 72°C, which is not reached at the surface of curing methylmethacrylate spread in a less than 10-mm-thick film. Finite element numerical analysis predicts that permanent thermal necrosis of bone is not likely to take place with cement collars of a centimetre or less [6,131]. Methylmethacrylate1s LD_{50} to human cells is 700 ppm. The highest measured levels in the periprosthetic tissues of hip arthroplasties are less than 260 ppm. However, it has not been excluded that condensed monomer may focally achieve toxic concentrations [132].

Samples obtained autoptically during the second postoperative week exhibit ample interlock of the bony trabeculae and cement, fingerlike acrylic projections penetrating into the fatty-hematopoietic marrow, and cement directly abutting upon the viable osseous and soft tissues in the absence of a substantial inflammatory infiltrate [128]. These observations confirm Radin et al.'s earlier findings in ewes with artificial hip joints: the authors saw no signs of thermal injury in the immediate postoperative period, the cortical and trabecular bone and also the hematopoietic marrow alongside the cement appeared histologically to be healthy. In fact, it takes some time prior to sprouting of fibrous tissue at the interface, focal and mild inflammatory infiltration, cortical necrosis, and bone remodeling [38].

Biopsies from the boundary regions of stable components reveal a spectrum of tissular reactions, ranging from direct bone–cement contact to an up to 1.5-mm-thick fibrous layer.

Necrotic bone, purportedly incurred at the time of cementation, is not seen adjacent to polymethylmethacrylate because—so it is maintained—it has been resorbed and replaced by viable bone. When the opportunity lends itself to assess specimens of clinically sufficient artificial joints retrieved postmortem, the acrylic mantle is found to directly contact mineralized bone and osteoid seams or to be insulated from the bone by a thin layer of bland, fibrous tissue. Osteoid partakes in the formation of the cement-contacting face of the osseointegrated cement. No osteoblasts are seen in relation to this osteoid. Having been laid down at earlier times, the osseous matrix evidently fails to mineralize, persisting as such at stable interfaces for long periods, plausibly for several years [127,133]. Osteoid abutting on cement is a peculiarity reflecting a variation of the physiological mineralization process, a consequence of toxic effects of the implant's dissolved constituents, an inhibition of calcification by ions leaching from components, a bioincompatibility-dependent osteocyte-mediated process, or a local passive phenomenon [27,134].

In addition to the isolated segments of osteonecrosis adjacent to the cement, myriad, variously sized and shaped necrotic bony fragments are randomly strewn in the interfacial membranes of long-term-serving total arthroplasties. The presence of osteoclasts, Howship's lacunae, newly formed appositional osseous matrix, and osteoblasts alongside these fragments suggests that the bone of the implant's bed breaks down continuously, its spicules being incorporated in the interfacial tissues, where they undergo remodeling. Some cement extensions, osseous trabeculae, or both may be loaded to failure during patient's activities. Which of the interdigitating intrusions fails first depends on the mass of the cement on the one hand and density of the bone on the other. Exceeding the strength of the buttressing intrusions, interfacial motions fracture the cement, bone, or both. Loss of function of the artificial joints under these circumstances is attributable to mechanical failure rather than to the more common prosthetic debris–driven biological failure [20].

E. Cement–Bone Integration

The cement mantles of well-fixated tibial inserts of many patients' total knee arthroplasties rest on shelves of newly formed fibrocartilage. Here as well as along-side surfaces with cement–bone interlock persisting for the duration of the implants' service, the PMMA is accepted by the cartilaginous and osseous environs without development of a functionally unruly inflammatory granulomatous reaction. That fibrous or fibrocartilaginous tissue is present at the interface of the tibial component is compatible with secure fixation, even if a radiolucent line is röentgenologically evident. The radiolucent line corresponds histologically to fibrocartilaginous tissue, which is supported by a relatively massive bony construct, the trabeculae of which are thicker and not as regular as those of the normal spongiotic bone. Bony bulges protrude to various depths into the cartilage. The bony construct is akin to a new subchondral osseous plate. It possibly arose in the wake of joining together of the cut trabeculae at the plane of bone section. Formation of fibrocartilage is related to the bone's requirement for a relatively compliant cushion to accommodate the load-bearing stress and is not connected to a purported untoward effect of the cement [135–137]. The inoffensive circumstances are also reflected by the normal fatty-hematopoietic cells occupying the adjacent intertrabecular spaces [133].

Thin, bland interfacial membranes consist of a densely textured collagenous tissue, the fibers of which are oriented parallel to the surface of the cement. Closely arrayed Sharpey-like fibers emanate from the bone and connect with collagenous fibers of the interfacial membrane. Projections of the acrylic mantle bulge into the interfacial membrane. Being contiguous with adjacent cement, they differ from the small, roundish, polycyclic cement chunks scattered in the interfacial membrane. Both the cement projections and chunks are bounded by macrophages, intermixed with lymphocytes, and concentrically encircled by fibrous tissue which merges with that of the interfacial membrane. The macrophages contain tiny acrylic beads and finely dispersed barium sulfate granules, implying ongoing breakdown of the PMMA, albeit a tardy one [138]. In the patient with a stable acetabular socket but a disintegrated femoral cement mantle, localized cortical osteolysis may be provoked almost exclusively by a granulomatous reaction to acrylic particles. Metallic debris appears to play a causative role in other patient's femoral osteolysis. The levels of metals in osteolytic cavities are considerably higher than their levels in either the interfacial membrane or joint pseudocapsule [139]. The prosthetic debris-burdened granulation tissue, occupying localized osteolytic cavities, not infrequently extends into the intertrabecular spaces of the adjacent bone, sometimes it perforates the cortex, and rarely spreads into the muscles [74,124,140].

In the course of the learning process, as much if not more is mastered by trial and success rather than by trial and error [141]. Correlations of biomechanical and clinical data with morphological observations indicate that cement osseointegration is achievable, provided that interfacial motions are at a minimum, secure fixation of the components persists, and a stable situation prevails. The remodeled trabecular bone intimately interdigitating with the cement shapes up a neocortex, which lasts for over 15 years after the implantation of artificial hip joints. The axial micromotion at the interface of these well-fixated components is 40 μm or less [142]. The cement is thrown into papillary undulations where it is in contact with the cancellous bone. Subsequent to the loosening process, the smooth cement exterior reflects motion-induced surface polishing. Acrylic and osseous morsels dirtying the surface attest to past fracture of the cement and bone [20].

The presence of lamellar-fibered bone near osseointegrated cement confirms environmental "quietness"; woven-fibered bone would be expected were ossification overly speedy. In biopsies taken at revision operations for femoral stem loosening, interfacial territories close to stable acetabular cups manifest ample cement–bone interlock: direct contact of the cement with the underlying bone and tracts of bland fibrous tissue succeed each other. In addition, lengthy segments of calcified bone and short sections of nonmineralized osseous matrix (osteoid) alternate with each other aside the osseointegrated cement [127,135,143]. There is direct cement-to-bone contact along the femoral component of cemented artificial hip joints in cases revised for acetabular loosening. In histological sections of undecalcified samples of patients preoperatively given tetracycline, the segmentally atypical fluorescence patterns of the bone in direct contact with or close to the cement mantle imply that the osseous matrix has not been fully mineralized [144].

Biocompatibility expresses a biomaterial1s ability to perform with appropriate reactions in applications planned for a particular objective. The appropriateness of the host1s tissular response determines the biocompatibility of a material [6,145]. The biocompatibility of metals depends on the electrochemical interactions which they undergo with their environs as well as the release of ions. The biocompatibility of a metal correlates with its corrosion resistance. As far as metal in bulk form is concerned, some mono- and polykaryonic macrophages, a few lymphocytes, some mast cells, and rare neutrophils are scattered along its surface [146]. In contrast, metallic breakdown products elicit a severe, intractable inflammatory macrophagic reaction. In the interfacial membranes of loosened cemented arthroplasties, mono- and polykaryonic macrophages packed with submicron- to 3-μm-sized, roundish to rod-shaped to jagged metallic particles abound in the chronically inflamed fibrous tissue. The quantity of particulate metal is larger in periprosthetic tissues of patients with titanium alloy–based than in those of patients with either a cobalt–chrome–or stainless steel–based device. The data in the literature concerning the dimensions of the metallic particles vary to some degree from author to author, depending on several factors, prime among which is the measurement technique; while the metal species makes virtually no difference. Salvati et al. have calculated the mean size of isolated particles to range from 0.86 to 1.06 μm in the short dimension and from 1.57 to 1.79 μm in the long dimension. As measured by computer-assisted image analysis of routinely prepared sections, the mean size of the particles ranges from 0.34 to 0.40 μm and from 1.57 to 1.79 μm in the short and long dimensions, respectively. Large metallic fragments, up to several microns in length, are scattered in small numbers both intra- and extracellularly and also in granulomas [76,133,139].

F. T-Lymphocytic and B-Lymphocytic Responses

The recruitment and retention of T-lymphocytes in the periprosthetic tissues concern scientists and surgeons alike because of the cells' potential role in hyper-sensitivity reactions to prosthetic materials, primarily metals, chemicals used in the manufacture of polyethylene, and PMMA components [72,82,85]. These lymphocytes, which are partly arranged in a perivascular pattern (Fig. 15), seem to drive antigen-specific immune responses, as evidenced by the presence of predominating T-cell clones—indicative of the clonal expansion of the T-cells—and a restricted T-cell receptor expression. As has been alluded to above, some patients' skin patch test to N,N-dimethylparatoluidine is positive. This has been presumed by some investigators to be incidental to repetitive loosening events of cemented arthroplasties. Prior exposure to dental or occupational cement may have a detrimental effect on the survival of cemented prostheses if presensitization were responsible for an exaggerated cellular reaction at the implant site. Proliferation and transformation of T-lymphocytes at the bone–cement interface and also in the peripheral blood have been reported. The authors point out that there is no corresponding B-lymphocytic response [59,72,147–150].

The occurrence of B-lymphocytes in the inflammatory infiltrates at the cement–bone interface has been rarely perceived. Local follicular proliferations of B-cells in the interfacial membranes (Fig. 16) are associated with bone resorption in patients with early clinical failure

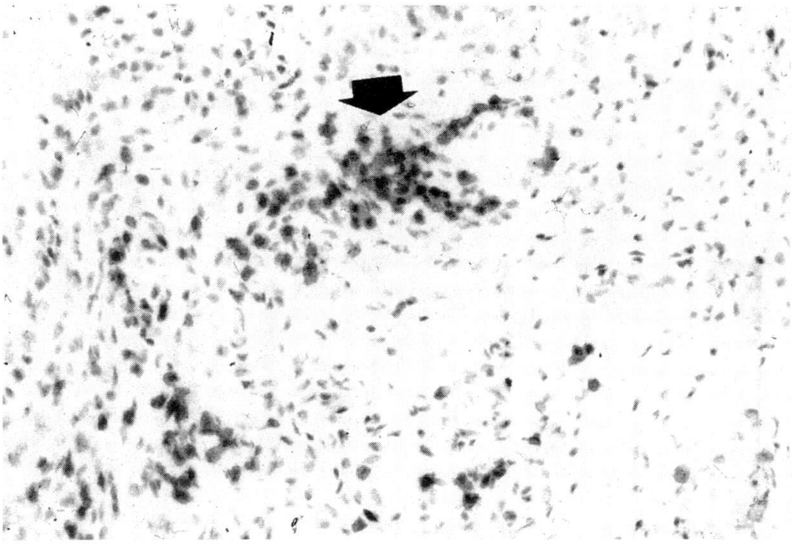

Figure 15 Perivascular infiltrates of the T-lymphocytes (arrow) are often present in the interfacial membranes of loosened, cemented components. Immunohistochemical stain with monoclonal antibody CD3 (Dako), 300×.

of cemented metal-on-metal total hip arthroplasties [151]. In the few cases thus far analyzed, the lymphocytic follicles have been established to be made up of both kappa and lambda light chain–immunoreactive lymphocytes, indicating a polyclonal proliferation of the constituent cells. A selective recruitment of B-lymphocytes cannot be attributed to a reaction against the cement only and is evidently related to the deposits of metallic debris as well.

Hypersensitivity to the acrylic cement or its constituents has been claimed to aggravate the foreign body reaction to wear debris. Inflammatory infiltrates around cement particles are more extensive and the lymphocytic production of cytokines in vitro is higher in the presence of PMMA in patients undergoing revision arthroplasties for aseptic loosening than for mechanical failure or sepsis [152]. Macrophages dominate the inflammatory cell population of acetabular interfacial membranes of loosened, cemented total hip arthroplasties. Close to 20 and 60% of these cells are B7-1 and B7-2 positive, antigen-presenting cells, respectively. Activated CD28- and CTLA-4-positive, mostly CD4 T-lymphocytes occur in abundance in these tissues. The B7-CD28 costimulatory pathway is active at the bone–cement interface in the face of a low CTLA-4 expression, favoring distinct connection between aseptic loosening and a chronic delayed hypersensitivity reaction, which may be activated by autologous protein-coated wear debris [153].

The immunocompatibility of PMMA has not been explored as extensively as its biocompatibility. The proportion of apoptotic phytohemoagglutinin-stimulated lymphocytes increases following their in vitro exposure for 4 h to the extracts of three of ten different cement types. All these ten cements, however, inhibit lymphocytic proliferation in vitro and decrease the lymphocyte's IL-2 receptor expression but only two of the ten cements affect the expression of the early activation antigen. If so, then the release of a product(s) from the cement hampers the function of lymphocytes activated by one stimulus or another [154]. It has earlier been reported that the augmented transformation of lymphocytes cultured in polymethylmethacrylate-enriched media correlates with positive patch tests in patients with a loose artificial hip joint. These patients' blood contains more T-cells, B-cells, and activated (i.e., IL-2 receptor-positive) than the

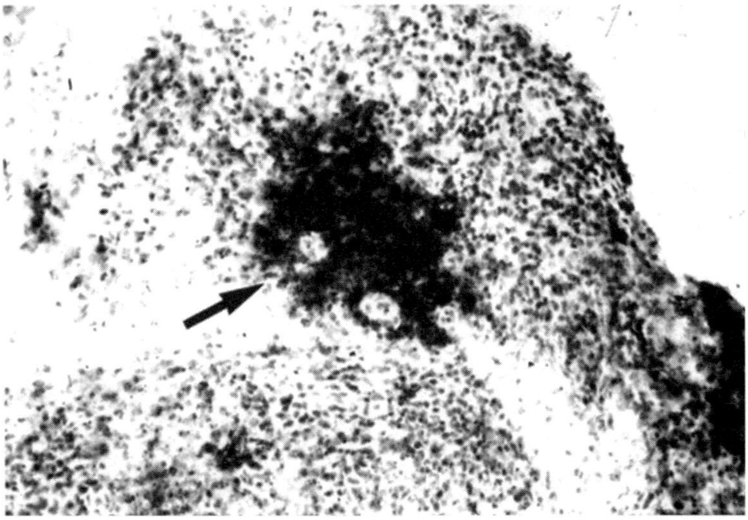

Figure 16 An infiltration of B-lymphocytes with the formation of a lymphoid follicle (arrow) in the periprosthetic tissues of a tibial component of a failed metal-on-metal total knee replacement. Immunohistochemical stain with monoclonal antibody CD22 (Dako), 300×.

blood of patients with stable implants. Although indicative of a cell-mediated immune reaction to the cement or one of its products, the findings do not correlate with the presence or absence of an aggressive interfacial membrane [148]. Normal and immunodeficient mice evince identical reaction patterns to subcutaneously implanted cement particles, casting doubt on a role of cellular or humoral immune mechanisms in the body's response to PMMA [155]. To summarize, the antigenicity of acrylic cement and the potential role of anti-PMMA immune response in prosthetic failure remain uncertain.

In the course of investigating the role of immune-mediated prosthetic failure, the complement-activating capacity of polyethylene has been explored. The results of in vitro hemolytic assays and double antibody immunoperoxidase staining technique indicate that the polyethylene particles activate the alternative complement pathway. The occurrence of factors Bb, C3a, iC3b, and SC5-9 have been demonstrated immunohistologically in samples of the pseudocapsules of failed arthroplasties obtained at revision operations. Additionally, particulate polyethylene boosts the adsorption of activated complement components. Such a localized activation of the complement cascade is possibly operative in the recruitment, proliferation, and activation of the macrophages during the early events of osteolysis [156].

G. Interfacial Micromotion and the Generation of Wear Particles

Interfacial micromotions are unavoidable in view of the differences between the elastic moduli of the implant and bone. Migration of wear particles throughout the effective joint space is reinforced by the pumping action of these micromotions. Accumulating at cement–bone interfaces, the wear particles recruit and activate the macrophages. The interfacial micromotions further stimulate the macrophages, by which the inflammatory reaction is bolstered. Additionally, they impair the regeneration of the bone [157–160].

Excessive interfacial shear stresses (due to component instability) augment the prosthetic detritus by adding breakdown products generated at the surfaces of the intraosseous components to those generated at the articulating surfaces. Poly-methylmethacrylate is not pathogenic. The literature's testimony with respect to its bioincompatibility is pathetic. Leukocytic infiltrates, bleedings, and clustered foamy macrophages occur at the interface shortly after implantation, such changes being encountered in the early postoperative period after many an inoffensive surgical intervention [160,161].

Given that the granulomatous response to breakdown products of alloplastic components of artificial joints plays the crucial role in prosthetic failure, a comment with respect to the modes of wear debris generation is in place. The main source of the wear particles are the primary bearing surfaces, whether related to motion in between the two articulating surfaces or to an interposed third body, e.g., an acrylic fragment. Less significant, albeit by no means trivial, sources are a surface of an intraosseous part, a primary bearing surface rubbing against a secondary surface (e.g., a cement-contacting femoral head which penetrated through the polyethylene socket), and two secondary nonbearing surfaces rubbing against one another (e.g., a polyethylene surface moving against a cement surface) [162,163]. Fractographic studies evince that fractures of the femoral cement mantles are commonplace. Significantly, excessive generation of wear particles occurs at these fracture surfaces [144]. Lastly, contact of the asperities on the mating surfaces at the fracture site of PMMA may produce large amounts of cement particles [164].

H. Cellular Responses to Metallic Particles and Metal Ions

The metals and their alloys from which artificial joints are fabricated dissolve in body fluids. To what degree is tissular necrosis due to toxic effects of the metallic ions released from the prostheses themselves or from their wear particles? Human synovial fibroblasts exposed in vitro to 50 and 200 nmol/mm^3 of cobalt or nickel chloride, respectively, undergo degeneration; no such effects are evident following incubation of the cells with chromic chloride, ammonium molybdate, or ferric chloride at concentrations up to 500 nmole/mm^3. The toxic effects of $CoCl_2$ on cells have been demonstrated in cultures of macrophages. The in vitro exposure to $CoCl_2$ injures metallic particles–laden macrophages isolated from interfacial membranes as evidenced by an inhibition of the cell activation and reduction of the cell proliferation (Fig. 17). Fibroblasts exposed to pure cobalt or vanadium particles under similar culture conditions are poisoned, but they are not so affected by pure nickel, chromium, molybdenum, titanium, or aluminum particles. This ostensible discrepancy is most likely related to the high solubility of cobalt and vanadium. Nonetheless, osteogenesis does ensue in the vicinity of metallic debris–containing macrophages' infiltration the interfacial membrane (Fig. 18).

It is not without an untoward repercussion that cobalt–chromium alloy–based surfaces articulating against themselves release notably high quantities of cobalt ions. These most likely are harmful to the tissue, whether through direct toxicity or immunization of the host [166]. Past experience supports this perception in as much as delayed hypersensitivity reaction to cobalt, but not to nickel or chrome, has been observed in patients with McKee hip arthroplasties failing within 1 to 3 years of operation. Sizable necrosis of the bone and pseudocapsule has been witnessed on assessing these patient's periprosthetic retrieved specimens. The high prevalence of metal sensitivity in the patients with McKee arthroplasties is in contrast to the rarity with which immune responses to metals have been noticed in patients with Charnley prostheses [167].

Metallic dust–laden macrophages are encountered in a majority, if not in all, of retrieved periprosthetic tissues subjected to a diligent search. As early as 1973, Charosky and his coauthors claimed that metallic particles were omnipresent, the greatest amounts being encountered in

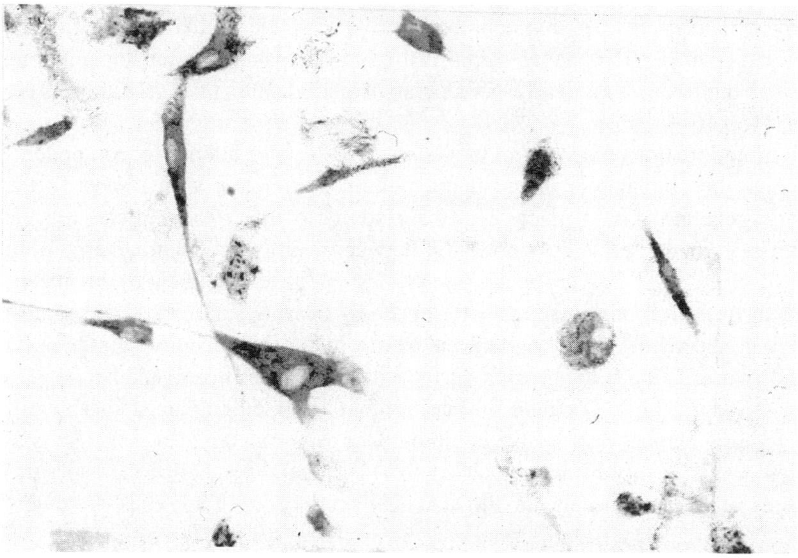

Figure 17 Inhibition of cell activation and reduction of cell proliferation of metallic debris-laden adherent macrophages, isolated from the tissues at the bone–cement interface by enzymatic digestion, cultured in vitro in the presence of cobalt chloride for 48 h. Immunohistochemical stain with monoclonal antibody EBM11 to the CD68 macrophage associated antigen (Dako), 400×.

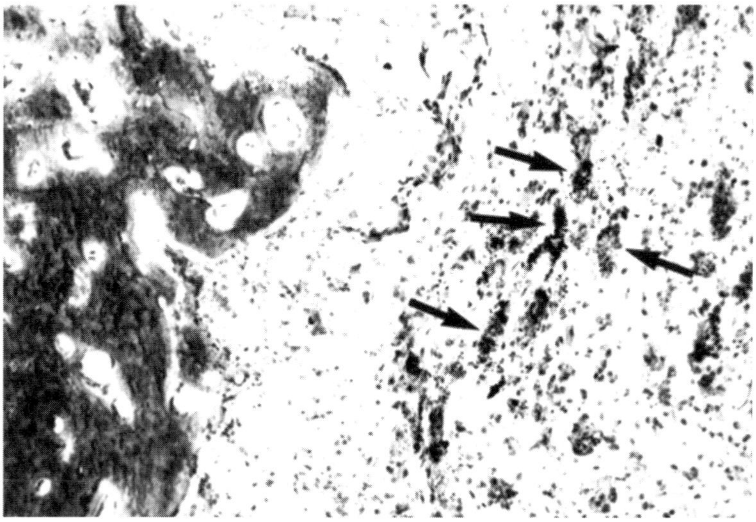

Figure 18 Newly formed osseous trabeculae (asterisks) surrounded by metallic particle–laden CD68 immunoreactive macrophages (arrows) within the bony shell under an interfacial membrane of a cemented femoral stem. Immunohistochemical stain with monoclonal CD68, 400×.

interfacial membranes of infected implants. Metallic particles have been premised to be birefringent under polarized light. The published illustrations speak for the presence of blackish, metallic dust in the polyethylene-burdened macrophages [168]. While metallic fragments may disclose so-called pseudobirefringence, in as much as sharp edges brighten under polarized light, true birefringence is not a property of metallic particles. In some patients with late-onset prosthetic failure, the metallic debris seems to prevail over other species of wear debris [77]. Yet, its relative contribution to the loosening process remains contentious.

The concentration, size, and topography of the metallic particles profoundly affect cell growth and cytokine production in vitro culture tests. Even though there is agreement that cobalt–chromium is more toxic than titanium and its alloys, the differential effect of diverse wear particles on the level of cytokine release is contentious. Be this as it may, particulate metals reduce the rate of uptake of ^{3}H-thymidine, indicating compromised growth and proliferation of cell in vitro [169–170]. Compared to fresh prepared samples, the in vitro toxicity of ''aged'' stainless steel, cobalt, and chrome particles is reduced. But monocytes exposed to aged metals produce extra large amounts of IL-6 and PGE_2. Causing the release of high levels of mediators of osteolysis, the long-lasting tissular storage of the metallic particles is especially detrimental to the integrity of the periprosthetic bone [171]. Having loosened after a service period of just a few years, cemented titanium alloy–based femoral stems with Al_2O_3 ceramic heads may disclose severe deformation of their neck, accompanied by extensive tissular necrosis overburdened by metallic debris [79]. Indeed, the linkage of metallosis and necrosis is convincing. Incidental necrosis and prosthetic failure have been appreciated years ago [172]. Like other authors, we have witnessed the presence of polymeric debris, in smaller or larger amounts, wherever metallic detritus collects [78,79].

Apoptotic cells are more numerous in tissues burdened by metallic debris than in those primarily containing polymeric or ceramic detritus. It is suggested that apoptosis is of central importance in the loosening of implants releasing metal ions. In the past, its role in the loss of fixation has probably been underestimated because apoptotic cells do not provoke an inflammatory response. Metals released in toxic concentrations lead to cell degeneration and necrosis, followed by discharge of intracellular enzymes and the ingested metallic particles. Repeating itself, this cycle terminates in necrosis. For poorly understood reasons, this necrotic tissue is sluggishly substituted by fibrous tissue. Excessive amounts of particulate metal amass under certain circumstances, say, hard ceramic femoral heads rubbing against the softer titanium femoral stems or bacterial metabolic product–expedited corrosion of the implants. Metallic and polymeric particle–driven responses are similar to each other in many respects, both recruiting macrophages and triggering the release of mediators which induce osteolysis [173–178]. In vitro exposure of macrophages to clinically relevant concentrations of cobalt, chromium, or titanium up-regulates the expression of bone-resorbing cytokines and inhibits the production of bone-forming cytokines [179–180].

The levels of metals gauged in the tissues reflect the constituent elements in the alloys making up particular artificial joints. But these ratios are skewed in cases with very low concentrations, implying that corrosion products accumulate in the tissues in addition to the wear particles. The amounts of cobalt, chromium, nickel, and molybdenum vary from one patient to another as well as from one site to another in one and the same sample. The total content of these metallic species ranges from about 3 to 250 μg per gram of dry tissue. Titanium is stored in considerably larger amounts at certain sites and under certain circumstances, say, around screw holes of cementless acetabular sockets. The mean titanium level is about 1620 μg per gram of dry tissue [139].

Macrophages are more or less indiscriminate in their response to the ''small particles.'' It is the number of such particles that decisively establishes the vigor of the macrophagic activation. This may account, among others, for the qualitative monotony of the biochemical and morphological scene nearby different alloplastic components. Yet the macrophagic response is controlled by the composition of the particles: cells exposed in vitro to submicron-sized particles of titanium–aluminum–vanadium release larger amounts of bone resorbing intermediates (IL-1α, IL-1β, IL-6, and PGE_2) than macrophages stimulated by commercially pure titanium particles, polyethylene particles retrieved from interfacial membranes of failed artificial joints, or in vitro fabricated ultrahigh molecular weight polyethylene particles. Levels of IL-1β in synovial fluids are higher in patients whose failed artificial joints contain a titanium alloy–based component than in those of patients with loosened cobalt–chrome–based prostheses [107,177].

The loss of material from metallic parts occurs mainly in the form of corrosion products [181]. But a small albeit not negligible portion of the loss is due to friction-induced generation of particles at articulations, motion at metallic–polymeric interfaces, and impingement phenomena [182]. Compared to the abounding polymeric detritus, there are as a rule just relatively small amounts of particulate metal in the periprosthetic tissues obtained at revision operations. Metallosis is less frequently encountered in association with components fabricated from cobalt alloy or stain-less steel than in association with titanium- or titanium alloy–based components. In addition, it is more often incidental to motion-induced fretting-corrosion of unstable, uncemented rather than cemented femoral stems. Wear-through of polyethylene inserts may lead to metal-on-metal contact of bearing surfaces such that metallic detritus is generated in large amounts. Juxtaposed to a cemented part, metallosis may be related to abrasion under harsh mechanical settings. Particles may also profusely emanate from femoral stems abrading against a cement fragment, metallic surfaces abrading against intra-articular third bodies, a screw rubbing against an acetabular socket backing, a cable used for the fixation of a trochanteric osteotomy, a titanium alloy taper of a stem-head junction, or a broken head of a screw embedded within acetabular cups. Under these and kindred adversities, the macrophagic reaction to metallic debris often dominates over that to polymeric debris. Periprosthetic tissues of hip arthroplasties with a ceramic femoral head serving for over 15 years may be replete with metallic dust–laden macrophages but nearly devoid of macrophages filled with polymeric particles [76,128,183–188].

The McKee-Farrar and Sivash artificial hip joints with a cobalt–chrome metal-on-metal articulation are suboptimal by current standards. Their bearing surfaces, nevertheless, reveal low wear at operations performed for revision of loose femoral stems during the second or third postoperative decade. A combined linear femoral and acetabular wear of 4 μm or less per year is manifold smaller than that typically seen with metal–polyethylene articulations. The dimensions of the interfacial membranes and the intensity of the metallic particle–induced foreign body inflammation are small when compared with those usually found at revision of artificial hip joints with a polyethylene acetabular bearing. Limited deposits of foreign bodies in these cases might be unexpected, taking into account that wear of the articular surfaces, impingement of the femoral neck on the acetabular rim, burnishing of the femoral stems, and corrosion produce particulate metal. Innovators might do well to keep in mind that the long-term results of the McKee-Farrar prosthesis are comparable with those of Charnley's low-friction arthroplasties, the 20-year survival probability (after exclusion of revisions for infections and dislocation) being 77 and 73% and the incidence of osteolytic lesions being 7 and 45%, respectively. It is intriguing that no instance of metallosis has been observed in a series of 24 aseptically loosened McKee-Farrar prostheses [180]. In addition to a limited particle burden, a tolerable inflammatory response to particulate metals may be responsible for this, at first sight unforeseen, virtue of these metal-on-metal arthroplasties [180].

The volumetric wear of metal-on-metal articulations is usually less than that of metal-on-polyethylene bearings. Myriad metallic particles are generated though in the wake of third body abrasion, impingement of the femoral component against the rim of the acetabular cup, and increased adhesive or abrasive wear. There is an increased potential for ionic and particulate release compared with polyethylene-on-metal couples. In fact, serum levels of chromium and cobalt are greatly elevated in patients with long-term surviving McKee-Farrar arthroplasties. On the average, close to 6 mm^3 of metallic debris is produced yearly, and the mean number of particles is 1.7 billion per gram of periprosthetic tissue. Artificial joints' specific details, i.e., the size of their femoral ball or contact with edges of recesses machined into a bearing zone, alter the amounts of the generated debris [189].

Phagocytosis of submicron- to micron-sized particles of any metallic species stimulates the macrophages, but the magnitude of the cellular activation differs from one element to another. More IL-1β is produced by macrophages ingesting particles released from titanium alloys than from cobalt–chrome alloys. In addition to the macrophages, polymorphonuclear leukocytes are also stimulated in vitro and in vivo by phagocytosed titanium or cobalt–chromium particles. In the subcutaneous pouch model of rats, neutrophils degranulate in the presence of particulate metals, freeing β-glucuronidase and myeloperoxidase. The data suggest that neutrophil-mediated acute inflammatory reactions to metallic debris play a role in the chain of events leading to the later, macrophage-dominated chronic inflammatory response. Titanium and titanium alloy particles prompt the fibroblasts to synthesize IL-6, basic fibroblast growth factor, PGE_2, stromelysin, as well as tissue inhibitor of metalloproteinases. The augmented activities of fibroblasts, macrophages, and neutrophils combine with the diminished activities of the osteoblasts to increase osteolysis and decrease osteogenesis [190–193].

I. Metallic Particles Versus Polyethylene Particles

Proliferation of MG63 osteoblast-like cells exposed in vitro to polyethylene particles is promoted and, concurrently, the production of PGE_2 and transforming growth factor-β1 is increased. Including everything, the change of the profile of the cytokines favors osteolysis and inhibits osteogenesis [194]. Interfacial membranes of well-fixated and also of some loose components retrieved autoptically or at revision operations performed after the prostheses have served successfully for 15 or more years and sometimes have displayed distinctive features: these membranes consist of a mildly inflamed, densely textured, and focally hyalinized fibrous tissue with scant cement and polyethylene debris–induced granulomas but widespread collections of metallic dust–overloaded cells. Overall, unless unusually stern biomechanical circumstances, say, impingement, have existed during the service of the prosthesis, metallic particles are scant in the periprosthetic tissues of short-term-surviving joint replacements when compared to the plentiful present polymeric debris. It is as if the metallic particle–induced macrophagic response vicariously supplants the usually dominating polymeric particle–mediated macrophagic reaction in the case of long-term-surviving artificial joints. In other cases of prosthetic joints which also have successfully functioned for prolonged time periods, the bone–implant interface discloses an osseointegrated cement mantle, a thin bland fibrous layer, or a polymeric debris–driven granulomatous membrane [76,133,195].

J. Localized Versus Linear Osteolysis

Given that PMMA is not enzymatically broken down, the membranes cannot grow except at the expense of the endosteal siding of the bone. Characterized by the formation of an intraosseous cavity filled with soft tissue, localized osteolysis differs macroscopically from the coatlike interfacial membrane. In so far as inflammatory granulomatous reactions and mediator production are concerned, essentially no differences between the interfacial membranes and tissues curetted from osteolytic cavities are evident. Localized osteolysis, related to a poor cementation technique with sites of less than 1-mm-thick acrylic mantle, is of more than just academic interest since fractures easily spread through hollowed cortical bone. Acetabular bone loss is larger in patients with well-fixated cementless cups showing expansile or localized osteolysis (roentgenographically characterized by a sharply demarcated radiolucent space with a rounded or scalloped appearance and extending away from the surface of the implant) than that in patients with cemented cups associated with linear patterned osteolysis (roentgenographically characterized by radiolucent spaces appearing directly adjacent to the implant and having a relatively uniform width of more than 1 mm). The prevalence of femoral osteolysis is higher in patients with cementless than in those with cemented acetabular sockets. These differences are related to the ease of dissemination of the wear particles, which in turn is a function of the dimensions of the effective joint space as well as to the extent and size of the generated debris particles [196–199].

It is instructive from the historical outlook to keep in mind the often-witnessed massive osteolysis with the former use of the ill-famed Teflon sockets which, splintering with ease, polluted the tissues with extra-large amounts of polymeric detritus [200]. Nowadays, with the utilization of contemporary biomaterials, the femoral and iliacal bones are less frequently affected by extensive localized osteolysis, whether the adjoining component is securely fixated or otherwise. Both the distribution and extent of the localized osteolysis, rather than its morphological features at the microscopical level, typify this complication [104,201,202]. Examination of the tissues removed from the site of pathological fractures through the attenuated cortex shows necrotic bone fragments incorporated in the debris-induced granulation tissue and luxuriant osteoclastic activity along the underlying bone [203].

K. Wear Particles Determine the Pattern of Periprosthetic Osteolysis

Cement mantle defects provide routes through which wear particles from the defects cause localized osteolytic lesions if the escaping wear debris provokes giant-celled granulation tissue from which bone-resorbing factors are released into the bony environs. Friable granulation tissue, newly formed bone, and turbid fluid fill the osteolytic cavities [204]. A deficient cement mantle between an acetabular cup and the bone is an important mechanism of formation of polyethylene detritus. The direct contact of the polymer with the bone has been evidenced to be associated with abrasion in one-third of cases with a segmentally absent acrylic mantle around a loose acetabular component [205].

Aseptic loosening of artificial joints is associated with either linear osteolysis or focal bone erosion, presenting roentgenologically as radiolucent lines and cystic bone lesions, respectively [206]. Linear osteolysis is more likely to develop around cemented components, whereas expansive, localized, osteolytic lesions are mostly associated with porous-coated cementless acetabular sockets. In contrast to the effective joint space at cement surfaces, which makes up potential access channels for the joint fluid sustaining dissemination of micron-sized particles, an osseointegrated ingrown bone confines the debris-laden fluid to one locale or another, where it occasions localized osteolysis [207]. Gauging samples obtained at postmortem, Harris has ascertained that, as late as the 17th postoperative year, linear osteolysis around cemented components progresses slower than localized osteolysis [208]. Although both patterns contribute to the loosening of components due to loss of the bony support, focal and expansive lesions may lead to a major destruction of bone stock, notably when occurring at multiple sites. Twenty to 25 years postoperatively, osteolysis by itself accounts for 35% of cases of acetabular loosening but as a main and contributory factor it is responsible for loosening of the cemented femoral components of 85% of the patients [209].

L. Biomaterial-Dependent Histological Features of the Bone–Implant Interfaces

It goes without saying that a biomaterial-dependent feature of the interfacial membranes relates to the artificial joint's makeup, say, stainless steel, cobalt and its alloys, titanium and its alloys, PMMA, polyethylene, polyacetal, or ceramics. In other points, the microscopical scene at the implant–bone junction is monotonous in so far as the tissular reaction to diverse species of prosthetic debris expresses a prototypical foreign body response. The uniform and common motif of these tissular reactions may have been unanticipated but for preconceived notions inherited from our predecessors' command of the principles of general pathology [84,110]. The histological features observed in samples taken at revision operation of cementless and cemented knee components are also similar to each other with but a few exceptions such as a more widely hyalinized fibrous tissue in membranes abutting on an acrylic mantle [210]. The periprosthetic tissues of well-fixated and loosened total hip arthroplasties are microscopically similar one to another in so far as both evince the inner synovial-like coat overlying a layer rich in macrophages and foreign body giant cells which aggregate in a fibrous matrix. The reaction patterns in tissues retrieved from osteolytic regions near a hydroxyapatite-coated or a cemented femoral stem are also similar to each other, including the intensity of the inflammatory and granulomatous responses, with the self-evident absence of poly-methylmethacrylate-induced reactions in cases of cementless arthroplasties and presence of Ca-P particle–laden macrophages in cases of hydroxyapatite-coated artificial joints [205]. The limited armament that the body can mobilize in its defence against inanimate ''invaders'' accounts for the histological monotony of the interfacial membranes of intraosseous implants [211]. Histologists cannot deduce, without recourse to physicochemical analysis, the makeup of debris by appraising wear particles' shape, size, color, and physique or estimating the relationships between resident and immigrating cells partaking in the inflammatory granulomatous reaction at the interface.

The above-portrayed histological characteristics of the periprosthetic tissues embody generalizations based on analyses of autoptically or operatively retrieved specimens of generic total artificial joints. Orthopedic surgeons, bioengineers, and materials scientists came up with variations on the theme of total joint replacement. The modifications, ranging from cementless technique to structurally reconstructed parts (e.g., porous-coated) to compositionally redesigned components (e.g., hydroxyapatite-coated), are not the subject of this chapter. As far as cemented fixation is concerned, trivial differences have been recorded between the histological findings of the interfacial membranes around one variant artificial joint and another. Suffice it to remind the reader that innovations may bring about unforeseen complications. Macroscopic corrosion at the head–neck junction of modular femoral components is most likely at fault for the overloading with metallic particles (perhaps toxic) with its attendant extensive tissular necrosis [212]. Because of enhanced creep resistance, improved wear properties, and increased strength, use of carbon fiber–reinforced polyethylene-based tibial inserts has been advocated, which one expects to prevail over the effects of the harsh biomechanical situation at the femoral-tibial articulation, where contact stress in noncongruent designs exceeds the yield strength of polyethylene. Unsurprisingly, in addition to the granulomatous reaction to the common species of prosthetic detritus, the retrieved specimens also reveal a carbon particle–induced foreign body granulation tissue. This is worrisome because the carbon particulate–induced granulomatous response inhibits healing processes [213,214]. The results of cemented total hip arthroplasties with ''modern'' titanium alloy femoral stems and cobalt–chrome alloy heads are reportedly poor, 8% of the hips having been judged clinically fair or poor and 11% of the femoral stems to be radiologically loose at the second to eighth postoperative follow-up year. Stem cement debonding is the presenting initial setback in 85% of these cases. This high failure rate—especially when compared with the failure rate of similarly designed cobalt–chrome alloy stems—has persuaded some surgeons to abandon employing titanium alloy femoral stems for cemented total hip replacements. In these authors' practice, metallic particle–overloaded macrophages abound in patients' and experimental animals' tissues neighboring titanium- and titanium alloy–based devices [215,216].

Shadowy perspectives are accentuated in synopses of treatment modalities issuing from pathologists' pens. As a rule, pathological evaluations are carried out on what failed rather than on what succeeded. Particulars of the synovial-like membrane, periprosthetic osteolysis, and aseptic loosening of artificial joints are over-represented in the literature. This misrepresents the actual clinical situation. The outcomes of cemented total arthroplasties exemplify a *succès fou* of reconstructive surgery. Serious postimplantation complications and rates of revision have declined in recent years despite an increasing number of patients at risk. Being an indication of 70% of all revision operations, aseptic loosening of cemented artificial joints is still an urgent predicament. In Sweden, its incidence at the 10-year post-operative interval has declined to less than 3% during the past 15 years. In addition to a proper choice of the implant for each individual patient, accurate surgical and specifically cementing techniques are essential to a satisfactory long-term service of arthroplasties [217]. The results of illustrative in vitro experiments, carried out at the beginning of the modern arthroplasty era, have acquainted practitioners with the importance of a clean and dry cancellous bone surface at the time of implantation, adherent blood being responsible for weakening the tensile and shear bond strengths to a critical degree [218]. The component's mechanical properties as well may contribute to artificial joints fate. By promoting very efficient stem-to-cement-to-bone force transmission, the normalized, proportionalized, cemented femoral stem lessens hoop stresses, thus reducing the incidence of interface failure. In addition to highlighting the meaning of an ideal biomechanical situation, the low loosening rate of these stems (1.6% at the 15-year postoperative interval) attests to the commendable intraosseous behavior of polymethylmethacrylate [219]. The findings of a follow-up study of over 63,000 patients convincingly elucidates the part played by a proper design in deciding the fate of arthroplasties: complete polyethylene failure, i.e., a fracture or total wear through the acetabular insert, have befallen ~2.5 and ~0.3% of patients with a total hip replacement possessing an all-polyethylene or a metal-backed acetabular cup, respectively [220]. How surgeons implant an artificial joint and their patients' bone quality may turn the scales: femoral stems implanted with a 2- to 5-mm-thick proximal cement mantle evince a better outcome than those implanted with thicker or thinner cement mantles. The results are better in patients with femoral stems filling more than half of the medullary canal than in those whose stems fill half or less of the canal. Loosening, cement fracture, and cement–stem or cement–bone detachment more likely ensue in patients whose stems are oriented in more than 5 degrees of varus than in those whose stems have been implanted in a neutral or valgus orientation. Finally, the best results are observed when patients' proximal medial cancellous bone of the femur is about 2 mm thick when compared with the outcome in patients with a thicker cancellous bone [221].

Disagreement between the champions of cemented versus cementless total arthroplasties will not go out of existence in the foreseeable future. Adherents of cementless systems have to persuade their interlocutors that the outcome data do not favor cementation of femoral stems of hip arthroplasties. Currently, cemented implantation of femoral stems is associated with excellent to good clinical results in 97%, thigh pain in 3%, implant subsidence in 0%, and endosteal cavitation in 6% of the patients. This unfavorably compares with excellent to good clinical results in 88%, thigh pain in 40%, implant subsidence in 22%, and endosteal cavitation in 12% of the patients with noncemented femoral stems [222]. These data are based on the experience of orthopedic surgeons whose patients were operated upon in the 1980s and whose artificial joints are, thus, viewed as old fashioned by present-day criteria. Nevertheless, 13 years postoperatively, the 94% survival rate of cemented total artificial knee joints implanted between 1976 and 1989 connotes attainment of efficacious surgery [223]. This result stands out when collated, to cite an instance, with a 33% failure rate at less than 5 years follow-up evaluation of patients with a unicompartmental porous-coated anatomic cemented artificial knee joints implanted between 1985 and 1992 [224]. Roentgen stereophotogrammetric analysis, which has ushered in a novel option of matching diverse fixation modes with each other, evinces that secure fixation of uncemented femoral components is rarely achieved such that ''the use of cement still seems justified [225,226]. Be that as it may, the clinical, morphological, and biochemical information gathered during decades of proficiency with cemented total arthroplasties ought to serve as the gold standard by which the results of other treatment modalities, primarily cementless total joint replacements, should be measured.

Micromovement of the femoral stem within a strong intact cement mantle is compatible with prolonged survival of total hip arthroplasties provided that wear debris generation is kept to a minimum [227]. The cemented component's fixation strength is the product of the mechanical interlock achieved and the extent to which bone–cement interdigitation persists for the period of the implant's intended service. The implant–cement interfacial strength is governed by the roughness of the surface of the component and the topography of the surface of the adjacent bone [35,228–230]. Long-lasting fixation strength depends on patients' age, level of activity, and bone quality on the one hand and the relative volume of the cement involved in mechanical interlock with the bone, the wear of the cement and polyethylene surfaces, and the shedding of particles from the metallic surfaces on the other. Rough surfaces secure greater fixation strength but also yield larger amounts of motion-induced wear debris [35,231]. The actual volume of interdigitation is smaller than the theoretically accessible volume because of the interfacial void content, polymerization shrinkage, viscosity of curing methylmethacrylate, improperly applied pressure. and miscellaneous surgeon-related factors [232]. Amounts, dimensions, and shapes of the polyethylene particles deposited interfacially seal the fate of the majority of artificial joints failing during the first postoperative decade. Generation of wear debris is affected by multiple factors, prime among which are patient-related variables, lubrication, diameter of the femoral head, third body abrasion, as well as manufacture-related quality statistics, e.g., base resins, fabrication methods, fusion defects, surface finish, and sterilization technique [233].

III. The Diversity of Cellular Responses at the Bone–Implant Interface

Much of our understanding of the complex biological processes involved in periprosthetic bone erosion is derived from in-depth analysis of samples retrieved during revision surgery of failed implants or removed at autopsy. These include the interfacial membrane, synovial membrane, synovial fluid, bone, and bone marrow. Interfacial membranes are formed as a result of a massive influx of cells, responding to the presence of foreign bodies. With subsidence of the acute inflammatory reaction, the earliest recruited cells are macrophages and fibroblasts, followed by endothelial cells, T-lymphocytes, and mast cells.

Mono- and polykaryonic macrophages are the dominating cellular component at the bone–implant interface of most revised artificial joints (Fig. 19). However, in a certain proportion of cases, the T-lymphocytes are, at least segmentally, the most prominent part of the cellular infiltration (Fig. 20). Ingestion of undigestible particles, i.e., phagocytosis in the absence of disintegration of uptaken biomaterials, results in the cells' increased metabolic rate, augmented motility, irreversible transformation, and release of intermediates. It is well known that the macrophages produce over 100 proinflammatory mediators, cytokines, cytokine receptors, growth factors, and proteases. A subset of the cells may be persistently stimulated. Interestingly, new bone formation ensues in the midst of metallic debris–laden macrophages.

Following phagocytosis, the macrophages acquire a substantially increased capacity for cell fusion with the formation of multinucleated giant cells (MNGC). Cell fusion is a complex

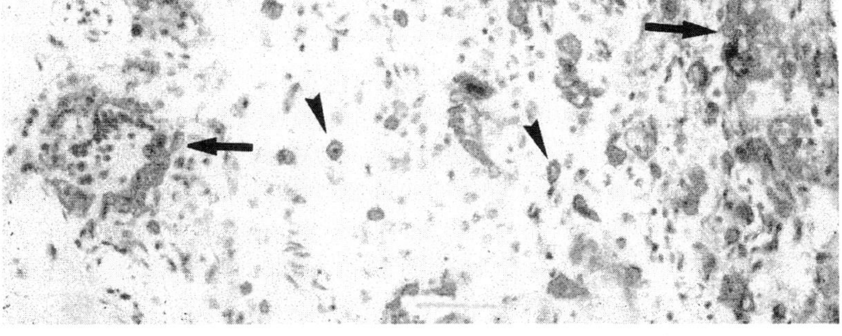

Figure 19 Abundant superficially congregated and some, more deeply, scattered CD13-immunoreactive, monokaryonic (arrow heads) and polykaryonic (arrows) macrophages within the inflammatory infiltrate of the cement mantle–facing aspect of an interfacial membrane. Immunohistochemical stain with the monoclonal antibody CD13 (Dako), 250×.

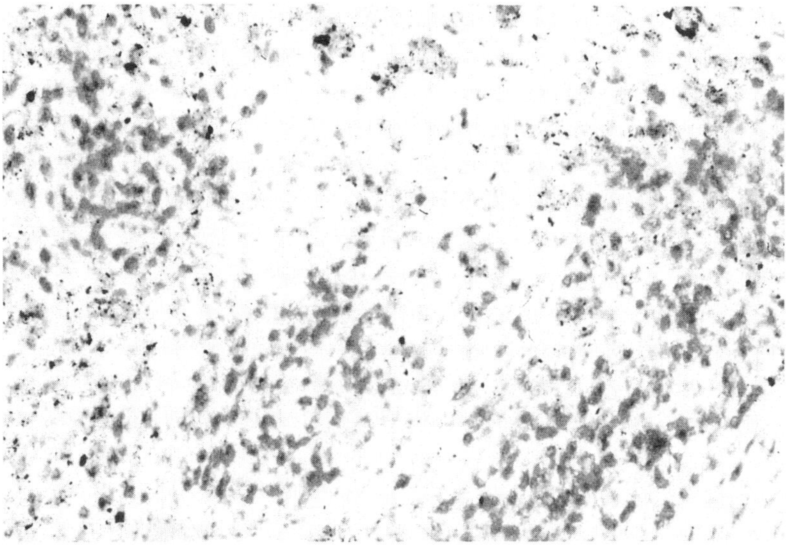

Figure 20 Within the interfacial membrane of this aseptically loosened, cemented cobalt–chrome alloy–based total hip arthroplasty, the inflammatory infiltrate consists primarily of T-lymphocytes. Immunohistochemical stain with monoclonal antibody CD3 (Dako), 300×.

process, involving the expression and interaction of a number of membrane-associated molecules. It is thought to impact the functional attributes of the developing giant cells. The foreign body MNGCs share certain morphologic, phenotypic, and functional characteristics with authentic osteoclasts [234–236]. The MNGCs express the osteoclast lysosomal enzyme tartrate-resistant acid phosphatase (TRAP), integrin (vitronectin receptor $\alpha V\beta 3$, CD51/CD61), and calcitonin receptor. The last, a lineage-specific marker, is expressed by committed mature osteoclasts. The expression of the calcitonin receptor is enhanced following the attachment of the cells to mineralized bone surfaces [237].

In view of the central role of the osteoclasts in periprosthetic bone erosion, the molecular mechanisms and mediators involved in the conversion of wear debris–containing macrophages into MNGCs has gained wide interest. The results of in vitro culture models imply the possible differential regulation of macrophage fusion by hematopoietic growth factors, cytokines, lymphokines, and selective adhesion molecules as well as by chemical stimuli. Hematopoietic growth factors, namely, GM-CSF and macrophage colony stimulating factor (M-CSF), are implicated in the regulation of the differentiation of macrophages and early precursors of MNGCs and osteoclasts [238–240]. In addition, induction of macrophage-derived MNGCs has been evidenced in response to phorbol myristate acetate, interferon-γ, and concanavalin A–stimulated lymphocyte-conditioned media [241,242]. Using either mouse bone marrow– or pulmonary alveoli–derived mononuclear cells, McInnes and Rennick have demonstrated that interleukin-4 (IL-4) promotes an aggregation of cells and development of MNGCs [243]. Later experiments by McNally and Anderson have, for the first time, revealed a clear distinction of the specific roles of diverse cytokines in the evolution of different morphological variants of the MNGCs. Thus, IFN-γ, IL-3, and GM-CSF have been identified in association with emerging Langhans type of MNGCs, i.e., relatively small giant cells with less than 50 nuclei arrayed in a circular fashion at the cytoplasmic periphery, whereas IL-4 consistently induces large foreign body syncytia, i.e., cells with over 100, randomly distributed nuclei [244].

Recent progress made in the cognition of the intercellular signalling pathways affecting generation of MNGCs is impressive. But the use of in vitro systems does not allow a precise identification of the principal factor(s) responsible for MNGCs-mediated periprosthetic osteolysis. This knowledge is particularly important for the development of novel strategies for the prevention of the generation and survival of these cells and control of their bone-resorptive activity at the site of artificial joints. To be stressed, in this context, is the expression and specific localization pattern of GM-CSF by particle-laden macrophages in the lining and sublining layers at sites close to biomaterial surfaces and deposited wear debris [236]. Macrophages at the interface also manifest simultaneous expression of the intercellular adhesion molecule-1 (ICAM-1/CD54) and complement receptor C3biR (CR3, CD11b/CD18), implying a mechanism of homotypic adhesion [84]. The activity of the complement receptor C3biR supports that of the lymphocyte function antigen LFA-1 (CD1) in the cell–cell fusion of particle-containing macrophages as well as their adhesion to the surfaces of biomaterials [245,246]. The adhesion of activated macrophages and MNGCs to biomaterial's surfaces through these processes plays a role in their degradation as a result of the cell's uncontrolled release of oxidative products, such as hypochlorous acid and derivatives of nitric oxide [247].

Of particular interest in the context of orthopedic reconstructive surgery is the development of MNGCs that fulfil the criteria of osteoclasts in terms of their antigenic properties and functional characteristics as bone-resorbing cells. Conversion of fused macrophages into mature osteoclast-like giant cells and their terminal differentiation appear to be determined by the expression and acquisition of antigen phenotypic markers of osteoclasts. The latter include the calcitonin receptor, the integrin vitronectin receptor (CD51/CD61), an osteoclast lineage specific marker, and other markers shared with macrophages, such as TRAP, CD9, CD13, CD68, and ICAM-1/CD54 [248–252]. Other cells, say, macrophages, osteocytes, and fibro-blasts, degrade the organic osseous matrix by a range of lysosomal enzymes that are formed in response to the effects of bone-resorptive hormones and cytokines [253].

The osteoclasts are unequalled in their capacity for lacunar resorption of the mineralized bony matrix under physiological conditions. After fusion of committed precursor cells of the monocyte–macrophage lineage, the differentiating MNGCs acquire and express several enzymes, cytokines, and cell membrane–associated receptors. In addition to the above-mentioned TRAP and vitronectin receptor, matrix metalloproteinase-9 (MMP-9) and transforming growth factor-α (TGF-α) are also expressed by the osteoclasts [254–256]. Osteoclasts undergo a rapid phenotypic change and cellular polarization during bone resorption by creating two membrane domains. The sealing zone is one such domain. It constitutes the specialized cell–extracellular matrix adhesion structure that mediates a tight attachment of the cell to the bone matrix. The second domain, the ruffled border, makes up the area of complex folding of the cell membrane which is closely apposed to the resorption lacuna within the surface of the mineralized bone. It plays a central role in the processes of demineralization and dissolution of the osseous matrix [257,258]. The resorption is facilitated by the low pH generated via an active secretion of protons by the V-type proton pump in the ruffled border and intracellular vacuoles [259].

Lack of an absolutely specific, phenotypic osteoclastic marker makes it difficult to ascertain the functional similarity between MNGCs within the interfacial membranes and osteoclasts at the bone–implant interface. Nevertheless, immunological studies of cytokines and their receptor profiles and also of other potential mediators support a common biological role for these cells in line with the expression of common bone resorptive markers, i.e., IL-1, TGF-α, M-CSF, MMP-9, ICAM-1, CD13, as well as classical markers, i.e., TRAP, calcitonin receptor and vitronectin receptor [84,234,256–262].

In addition to the intermediates-induced effects of MNGCs and macrophages on osteolysis, prosthetic failure is also attributed to compromised osteogenesis. Diminished bone formation may be due to impeded proliferation, impaired differentiation, or decreased functional competence of the osteoblasts. Wear particles and metal ions are toxic to osteoblasts in vitro. In their pioneering research, Vermes et al. have challenged cultures of the MG-63 osteoblastic cell line with micron-sized particles of different provenance (commercially pure titanium particles, chromium orthophosphate, medical-grade ultrahigh molecular weight polyethylene particles, and polystyrene). These studies have revealed, for the first time, the osteoblasts' capacity for phagocytosis of particles. It appears that the cells' gene expression is thereby altered through the action of various nuclear transcription factors. Particle-stimulated osteoblasts' synthesis of type I collagen is reduced due to suppression of procollagen $\alpha/1$ gene expression. Further, the cell's activation of protein tyrosine kinases is associated with the activation of the nuclear transcription factor (NFkB) which is central to the production and release of a wide range of inflammatory cytokines and PGE_2 [263,264]. Lastly, prosthetic particle–induced caspase-dependent apoptosis is demonstrated at a high level with premature osteoblasts. The contact of this cell fraction with wear particles markedly impacts the expression of gene coding for inflammatory cytokines and those controlling the architecture of the nuclei [265].

These findings collectively unravel a critical role of particle-exposed cells in osteolysis through activation of osteoclasts on the one hand and in osteogenesis through suppression of osteoblasts on the other. If this is indeed the case, a novel approach in the control of prosthetic failure envisages therapies that promote the survival and activities of osteoblasts to achieve optimal new bone formation.

A. The Role of Cytokines in Periprosthetic Osteolysis

Cytokines comprise a group of low molecular weight proteins that are mainly produced by cells pertaining to the body's defence mechanism and have profound impact in the regulation of immune reactions, inflammation, and tissue remodeling. Released as soluble mediators, these factors act in a paracrine or autocrine fashion to coordinate those cell–cell interactions that are essential for the development and growth of cells within the skeletal tissues. Autocrine regulation of cellular functions is achieved by the binding of a cytokine produced by a particular cell to a specific counter-receptor expressed by the same cell. The paracrine mechanism of activity comprises the binding of that cytokine to specific counter-receptors expressed on adjacent cells. The production of the cytokines is highly regulated and is provoked by a variety of biological, chemical, and mechanical stimuli. Even though pleiotropic and overlapping in their biological properties, most cytokines appear to have selective actions. This is apparent, for example, in the specific regulatory role of M-CSF in the generation and differentiation of precursors of osteoclasts and the bone-resorbing capacity of their mature counterparts [266]. Expression of cytokines is essential for the maintenance of the body's competent defence mechanisms. Yet, up-regulated and uncontrolled production of cytokines is implicated in the pathogenesis of certain illnesses, including autoimmune disorders and neoplasms. But on the bright side the recent identification of a large number of naturally occurring cytokine inhibitors and cytokine receptor antagonists opens up novel options for the control of many diseases. In particular, blocking the TNF-α with a soluble TNF-α receptor fusion protein or monoclonal antibodies is emerging as a viable therapeutic tool in the control of inflammatory joint disorders.

The intimate relationship between bone cells and immune cells in terms of the responsiveness of the former to osteotropic cytokines and endocrine factors formed by the latter is based on the regulatory role of bone marrow cells (marrow stromal cells, fibroblasts, and monocytes) in bone formation, resorption, and remodeling [267,268]. The above-discussed cytokines have been extensively studied in relation to periprosthetic pathological bone erosion and in particular the level of their production in the bone-implant interface membranes retrieved from patients undergoing joint revision of failed prostheses. Likewise, special attention is due to the part played by the macrophages because their activation by the ingested implant's wear debris is associated with the release of mediators of inflammation, fibrosis, and osteolysis.

B. Interleukin-1 and Tumor Necrosis Factor-α

Interleukin-1 and TNF-α are the first macrophage-derived cytokines to have been recognized as in vivo and in vitro osteolysis-promoting intermediaries [269–271]. Both exert a potent effect on the generation of osteoclast precursors and development of their different maturational stages [272]. The IL-1–induced hypercalcemia occurring in an in vivo model of osteoclast formation is reminiscent of the osseous environment in patients with rheumatoid arthritis, in which osteoclast-induced osteopenia is an everyday x-ray finding [273,274]. Interleukin-1 and TNF-α perform in a paracrine fashion through binding to specific receptors on adjacent cells, a function that is central to the regulation of the cellular responses. These cytokine-producing cells also express receptors facilitating an autocrine function: those cytokines that bind to cells from which they have been released autoregulate the cell's activation, growth, and signaling. As to this chain of events, it is similar to that operative in patients with some chronic inflammatory or autoimmune diseases [275,276].

Assessment of expression of cytokines in the interfacial membranes reveals a substantially raised level of their production. Horowitz et al. have discovered that polymethylmethacrylate provokes the macrophages to boost their synthesis of IL-1, TNF-α, and PGE_2. Their studies comprise specimens of interfacial membranes surrounding cemented femoral stems of total hip arthroplasties as well as of samples of in vitro cultured cell line J774 macrophages challenged with 1- to 12-μm-sized Simplex or Palacos cement particles. Notwithstanding the phagocytosis-mediated increase in the release of bone-resorbing cytokines, ^{3}H-thymidine incorporation is inhibited, suggestive of toxic effects of the cement particles [277]. Larger amounts of IL-1β than IL-1α are released at the interface of cemented acetabular sockets of total hip replacements and at the interfaces of cemented versus cementless failed artificial joints [278,279]. Conditioned media of cultured peripheral blood monocytes exposed to 1- to 2-μm-sized nonpolymerized methacrylate particles or pulverized polymethylmethacrylate contain IL-1, TNF-α and PGE_2, which directly affect bone resorption, as verified in the calcium release assay using pre-labeled murine limb bone assay [280].

The release of potent osteoclastogenic cytokines, such as IL-1 and TNF-α, by wear particle–activated macrophages has attracted the attention of clinical researchers and materials scientists. Using an in vivo murine bone marrow model, Merkel et al. have shown that PMMA particles induce c-src, a protein that is specifically expressed when the marrow macrophage progenitors commit to the osteoclast lineage. Since TNF-α is the only osteoclastogenic factor known to be c-src inductive, the authors have analyzed the role of this cytokine as the possible major cytokine triggering the particle-induced osteoclastogenesis. They have established in a gene deletion experiment that the deletion of the expression of TNF-α receptors p55 and p75 or p55 alone results in failure of the production of the c-src protein by bone marrow macrophages and in protection from osteolysis following the implantation of the cement particles in the calvariae of wild-type mice [281].

More recently, TNF-α cooperating with the nuclear factor-κ B (NF-κB) has been implicated in the biological process responsible for aseptic loosening of artificial joints. In this study, the authors employed a mouse cell culture and a mouse in vivo model to assess the inflammatory responses to titanium particles, induction of TNF-α, and signaling process leading to activation and translocation of the NF-κB transcription factor. Binding of TNF-α to the TNF-α receptors type I (p55) or type II (p75) initiates a rapid nuclear translocation of the NF-κB by way of degradation of the inhibitory factor-κB (I-κB), the intracellular inhibitor of NF-κB. The titanium particle–induced decrease of the cellular I-κB and translocation of NF-κB to the nucleus preceded by 30 min the production of TNF-α by the J774 cell line macrophages. Extensive bone resorption and fibrous encapsulation of the titanium particles implanted in the animal's calvariae accompany the excessive TNF-α production within 1 week of the operation. These findings have been reinforced by the results of testing mice that genetically overproduce TNF-α (hTNF-α-Tg), mice defective in TNF-α signaling TNF-α (TNF-R −/−), and animals that are NF-κB transcription factor deficient (NF-κB1 −/−): The hTNF-α-Tg mice have a grossly exaggerated inflammatory response, the TNF-R(−/−) mice show scant inflammation and bone resorption, and the NF-κB1 −/− mice disclose an inflammatory response but no bone resorption. To conclude, phagocytosis of particles stimulates the macrophages, resulting in the activation of nuclear transcription factor-kappaB and production of tumor necrosis factor-α. The latter directly incites fibroblastic proliferation, eliciting an interfacial fibrosis and recruits as well as activates osteoclasts to resorb the adjacent bone [95].

This chain of events holds true for cement particle–stimulated macrophages as well: exposure of murine osteoclast precursors to PMMA particles prompts the nuclear translocation and activation of NF-κB. A soluble inhibitor of TNF-α dampens the particle–directed NF-κB activation. This response is abolished in TNF-α–deficient osteoclast precursors. Thus, PMMA particle–induced activation of NF-κB is a secondary event resulting from the enhanced TNF-α expression. The nuclear transcription factor NF-κB mediates TNF-α signaling, and this transcription complex is necessary for osteoclastogenesis [282].

C. Interleukin-6

Interleukin-6 is produced by cells of the body's defense mechanism as well as by the osteoblasts and osteoclasts [283,284]. It has been recognized to be a potent bone-resorbing factor in view of its ability to induce the formation of osteoclast-like cells and the expression of vitronectin receptor. Interleukin-6, an osteoclast functional antigen, plays a regulatory role in homeostasis of calcium and in osteoclastic bone resorption mediated by osteotropic hormone, such as parathyroid hormone–related protein [285]. For example, in multiple myeloma–related focal osteolysis, the up-regulated production of IL-6 by the marrow stromal cells is associated with local stimulation of osteoclasts at the site of neoplastic cell aggregates next to the bone [286]. The close proximity of the bone surfaces to the cytokine-producing, particle-laden macrophages suggests that similar mechanisms of local osteoclast activation or formation of MNGCs are mediated by IL-6 in the interfacial membranes as well.

Production of high levels of IL-6 by macrophages is stimulated by ingestion of wear particles. Expression of IL-6 has been demonstrated in sections of interfacial membranes and macrophage cell lines as well as peripheral blood monocytes in vitro exposed to prosthetic particles. Diverse particle types differently affect macropahgic activities. Relative to particles derived from cobalt–chrome or polyethylene devices, those of titanium–aluminum–vanadium origin are more powerful inducers of IL-6. It has been recognized, however, that toxicity of cobalt–chrome is a limiting factor in assessing the level of elaborated mediators because of the reduced cellular activity and increased cell death, especially when cultured cells are exposed to high concentrations of this alloy [108,169,287–289].

High levels of IL-6 have been uncovered in the synovial fluid and the synovial membrane of patients with erosive rheumatoid arthritis as also in the synovial fluid of patients with prosthetic failure [279,290,291]. Nivbrant and his coworkers' observations are of interest in the context of the cytokines' role in osteolysis: the synovial fluid from patients with well-functioning or loose artificial hip joints contain higher amounts of IL-6, IL-1β, and TNF-α than that of patients with osteoarthritis. It should be emphasized that the increased levels of the cytokines are linked to the loosening process and bone resorption and are independent of the duration of implantation or the patients' age or gender [292].

Conditioned media of macrophages exposed to PMMA particles contain an abundance of TNF-α that stimulates cultured osteoblasts to release IL-6, GM-CSF, and PGE_2. Extrapolating to the clinical setting, Horowitz and Purdon suggest that following on mechanical failure or fragmentation of the cement mantle, PMMA particle–primed macrophages release TNF-α that, in turn, impacts the expression of IL-6, GM-CSF, and PGE_2 by nearby osteoblasts. The authors implicate these cytokines as triggering factors in the recruitment of macrophages and osteoclast precursors into the interfacial area, where they affect bone resorption and implant failure [293].

Contact of the particulates with and attachment to the cell surfaces institute the sequence of events that triggers the alterations in specific membrane receptors and transfer of external signals through the receptor's cytoplasmic domain. The transfer of the external signals is succeeded by the synthesis of protein, phosphorylation of cytoplasmic protein, activation of transcription factors [among them the NF-κB and nuclear factor interleukin-6 (NF-IL-6)], and modulation of gene expression [294].

The levels of IL-6 and TNF-α in the interfacial membrane serve as a measure of macrophagic activation. Obtained by enzymatic digestion of granulomas at the bone-implant interface, ~0.7-μm-sized titanium particles amplify 40-fold and seven-fold the release of TNF-α and IL-6 respectively by adherent peripheral blood monocytes. That cells treated with cytochalasin B ingest 95% less particles than untreated monocytes but persevere in their release of cytokines suggests that particle–induced activation is independent of ingestion of particles. There is evidence to suggest that cell surface–bound complement receptor proteins CD11b and CD18, and predictably other receptor proteins as well, are involved in recognition of prosthetic debris. Binding of particles to membrane surface ligands apparently suffices to activate cytokine production in the absence of phagocytosis. This mechanism's inhibition by actinomycin D (which blocks RNA synthesis) cycloheximide (which impairs protein synthesis) implies that the activation of cytokine production by binding particles to the cells' surface ligands is dependent on transcriptional and translational processes. By these means, the external stimuli are transmitted to the nucleus through intracellular signaling that involves phosphorylation of select target proteins, leading to the translocation of the transcription factors NFkB and NF-IL-6 from the cytoplasm to the nucleus. Alteration of the gene expression and release of the cytokines are mediated by the activation of the transcription factors. Secretion of these cytokines can further switch on a substantial inflammatory signal transduction network by acting in an autocrine and a paracrine fashion through binding the relevant cytokine receptor on nearby macrophages, T-lymphocytes, fibroblasts, and endothelial cells [289,294].

D. Hematopoietic Growth Factors

Of the several members of the family of hematopoietic growth factors (colony stimulating factors), GM-CSF, M-CSF, and IL-3 are known to participate in the regulation of bone cell development and function. The epithet ''colony stimulating factor'' reflects their initial identification through cell culture systems designed for the clonal growth of hematopoietic progenitor cells in semisolid media. Indeed, the colony stimulating factors have proven their capacity in inducing clonal growth (formation of colonies of specific lineages) in cultures of marrow hematopoietic cells. They are produced by the stromal cells, osteoblasts, osteoclast progenitors, macrophages, endothelial cells, and T-lymphocytes of the bone marrow. In this micro-environment, they fulfil vital multifunctional roles during hematopoiesis and osteopoiesis. Of special interest in the context of what occurs at bone–implant interfaces is the key role of GM-CSF and M-CSF in the generation and maturation of the bone-resorbing osteoclasts and osteoclast-like multinucleated giant cells [295].

Horowitz and Jilka have shown that, in vitro cultures of intact bone and bone marrow progenitors, GM-CSF and M-CSF affect osteoclast formation and function at diverse stages of their differentiation. These factors have a profound effect at the earliest stages because they regulate the proliferation of the osteoclast progenitors [296]. Particularly, GM-CSF stimulates the cellular fusion and formation of multinucleated giant cells from mononuclear progenitors in bone marrow cultures [238,297]. The role of GM-CSF as a potential regulator of bone resorption is based, inter alia, on its reinforcing effect on 1,25-dihydroxyvitamin D_3, a calcium-regulating hormone that stimulates the formation of osteoclast and osteoclast-like cells [298]. And lastly GM-CSF assumes cytokine function in contributing to the maintenance of activation of cells of the monocyte–macrophage lineage in chronic inflammatory conditions. Histological analysis of interfacial membranes of failed artificial joints discloses the expression of GM-CSF primarily by certain subsets of macrophages and multinucleated giant cells at the early stages of fusion, at which there is a small number of nuclei in these polykaryons. By the way, some fibroblasts and endothelial cells also produce this cytokine under certain circumstances. A significant proportion of the GM-CSF immunoreactive cells displays a specific distribution pattern in that they are concentrated in the lining and sublining layers of the interfacial membranes. This suggests that the macrophages which are recruited in response to the wear debris exhibit constitutive expression of GM-CSF that is required for their activation and for enhanced phagocytic capacity. The constitutive expression of GM-CSF, which occurs in the in vitro cultures of adherent, metallic debris–containing macrophages isolated from interfacial membranes, manifests a morphological transformation that is indicative of the cell's activation status (Fig. 21). It is highly likely that these early recruited phagocytes are derived from cells of the local microenvironment of the bone and bone marrow. An excessive production of GM-CSF in the periprosthetic tissues plays a central role in the proliferation of macrophage subsets and the early steps of cellular fusion at the site of the implant [233,236,299].

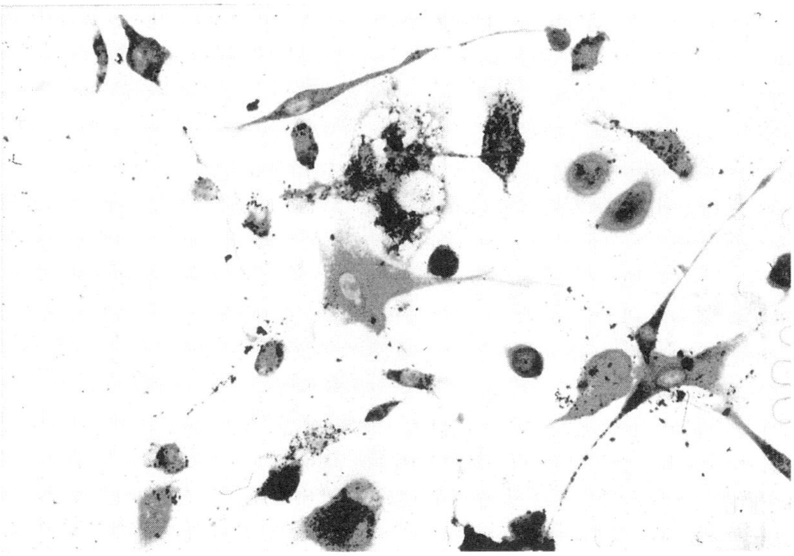

Figure 21 In vitro cultured metallic debris–laden adherent macrophages, isolated from the tissues at the bone–cement interface by enzymatic digestion, manifest the constitutive expression of GM-CSF. Immunohistochemical stain with monoclonal antibody to GM-CSF (R&D Systems, UK), 400×.

Macrophase colony stimulating factor, which is also referred to as CSF-1, is a lineage-specific factor that regulates the differentiation and functional capabilities of macrophages and osteoclasts. In fact, it is a pivotal factor in osteoclastogenesis [266]. In their in vitro culture model, which employs cells derived from arthroplasty specimens as a source of activated macrophage subsets and human trabecular bone–derived cells as a source of precursor cells, Neale et al. have confirmed the capacity of the former to transform into bone-resorbing giant cells in the absence of 1,25-dihydroxyvitamin D_3. The high level of M-CSF release during the cellular fusion and transformation implicates this factor in promoting osteoclastogenesis at the sites of augmented osteolysis associated with aseptic loosening [300]. Using a similar experimental setup, these authors have as well evaluated the differential impact of various cytokines in osteoclastogenesis. By adding neutralizing antibodies to human IL-1 or TNF-α to the coculture system, formation of osteoclasts is not inhibited, whereas adding neutralizing antibodies to M-CSF or IL-6 significantly reduces the number of osteoclasts and reduces their capacity for osteolysis [240]. These findings highlight the particular roles of the cytokines in what happens at the bone–implant interface.

Substantial amounts of M-CSF are produced by particle-laden macrophages residing in all patients' interfacial membranes, regardless of the quantity or type of the wear debris present. Immunohistologically, there is a pronounced expression of the specific receptor for M-CSF (CSF-1 R/c-fms) in the particle-containing macrophages of the periprosthetic tissues. In contrast, multinucleated giant cells holding diverse implant particulates lack the expression of M-CSF and CSF-1 R/c-fms. This down-regulated expression may be due to the internalization of the ligand receptor following phagocytosis and the fusion process. The pattern of expression may also suggest a requirement of M-CSF and its receptor for the genesis and differentiation of osteoclast-like giant cells from the macrophage subsets. However, both M-CSF and CSF-1 R/c-fms are present in detectable levels in osteoclasts and osteoclast-like cells attached to the surfaces of bones removed from patients with aseptically loosened arthroplasties [261]. The requirement of M-CSF and CSF-1 R/c-fms for the bone-resorptive activity of osteoclasts is well established [301]. Bone resorption can only occur when osteoclasts are in close proximity and physically attached to a bone surface. Hence, it appears that the osteoclasts or their counterparts acquire the capacity of expression and synthesis of the cytokine and its receptor after their attachment to the bone. This induction process could be mediated through recognition by distinct cell membrane receptors of the extracellular matrix proteins of the bone.

In addition to its regulatory functions in osteoclastic osteolysis, the M-CSF has further essential roles. By preventing apoptosis it supports the survival of the osteoclasts. Moreover, it stimulates osteoclast chemotaxis, motility, and spreading during bone resorption. On sites undergoing osteolysis, osteoclast activity is regulated as well by exogenous M-CSF secreted by nearby osteoblasts. Hence, the osteoblasts modulate osteoclastic activity by coordinating the balance between resorption and migration in the resorption pit [302].

E. Transforming Growth Factor-α and Epidermal Growth Factor (EGF)

Important aspects of prosthetic failure are a progressive capacity of particle–containing macrophages for fusion and generation of multinucleated giant cells of the osteoclast type. The striking subsistence of these cells indicates that this model of bone erosion predicts the survival and functional activity of the MNGCs. A large number of macrophage-derived cytokines and other mediators have been studied in association with periprosthetic osteolysis. On the other hand, the cytokine profile of particle-induced MNGCs in the periprosthetic tissues has not been adequately addressed in the past. Transforming growth factor-α, epidermal growth factor (EGF), and their common receptor (EGF-R) have been the subject of several studies in recent years [261,303].

The TGF-α and EGF belong to a diverse group of polypeptide growth factors which are involved in the regulation of morphological transformation and modulation of cell division in normal and neoplastic tissues. Their 40% homology is especially pronounced within the receptor-binding domain [304,305]. The high-affinity binding of TGF-α or of EGF to EGF-R leads to phosphorylation of the intrinsic tyrosine kinase located in the cytoplasmic domain of the receptor, triggering a cascade of intracellular events and internalization of the receptor and ligand [306]. The binding of the factors is linked to various cellular responses, which include intercellular adhesion, generation of mitogenic signals, and abnormal cellular transformation in response to chemicals [307,308]. Both cytokines exhibit functional similarities, particularly as to their impact in promoting osteoclastic bone resorption. They induce osteopenia by enhancing the proliferation and fusion of progenitors, leading to an increased number of osteoclasts, or by their inhibitory action on collagen synthesis and osteoblasts' alkaline phosphatase activity [309–311]. The expression of TGF-α and EGF in the periprosthetic tissue hints at the dual role they play in aseptic loosening through promoting osteolytic processes and, at the same time, disrupting the reparative and remodeling processes at the site of implantation.

Studies characterizing expression and tissue distribution have evidenced the constitutive and predominant expression of TGF-α by multinucleated giant cells in the periprosthetic tissues of all cases of aseptic loosening. In fact, they highlight a distinct cytokine profile of the MNGCs. The preponderant expression of cell membrane TGF-α and the lack of expression of the EGF-R on the MNGCs imply that the detected TGF-α is not bound to the receptor but is produced by these cells. Further-more, the prominent and high level of TGF-α expression by particle-induced giant cells at the bone–implant interface, which is accompanied by complete absence of EGF expression on these cells, indicates that TGF-α is the principal and dominating factor in this model of osteolytic processes near orthopedic implants. Its role in periprosthetic bone erosion is corroborated by the powerful correlation in the frequency, antigenic phenotype, and high level of production of this cytokine by the MNGCs, as also by these cells' counterparts in giant cell tumors and malignant bone neoplasms characterized by progressive bone resorption. Ingestion of wear debris, be it polymeric or metallic in nature, by mononucleated and multinucleated phagocytes appears to initiate the production and immobilization of the membrane-anchored TGF-α, which is vital for its autocrine and paracrine functions. Phagocytosis of the particles also contributes to the elevated and constitutive expression of the TGF-α (Fig. 22) and EGF-R (Fig. 23) by the macrophages, while the production of EGF is rather restricted to fibroblasts and types A and B synoviocyte-like cells of the lining layer of the interfacial membrane [261,303].

In view of the key role of TGF-α in osteoclast generation and differentiation [310–312], this cytokine's binding to the EGF-R on the adjacent cells could provide a form of cell–cell adhesion between macrophages and MNGCs, which is pivotal for the genesis of the bone-resorbing giant cells at the implant–bone interface. The binding may also activate the EGF-R, causing juxtacrine stimulation and a prolonged response, relative to that stimulated by the secreted growth factor.

F. Osteoprotegerin and Osteoprotegerin Ligand

Among recently identified cytokines is the osteoprotegerin ligand (OPGL), a polypeptide ligand which binds the receptor osteoprotegerin (OPG). The osteoprotegerin molecule, also known as TNFRSF11B or RANK (receptor activator of NF-kappa B) is a naturally occurring secreted protein with homology to the TNF-receptor family. It is widely known for its inhibitory effect on osteoclastogenesis and associated bone resorption suggesting a pivotal role in neutralizing or inhibiting pathological osteoclastic bone resorption. This effect is accomplished through the binding to the specific ligand osteoprotegerin ligand (OPGL) [313]. The osteoprotegerin ligand

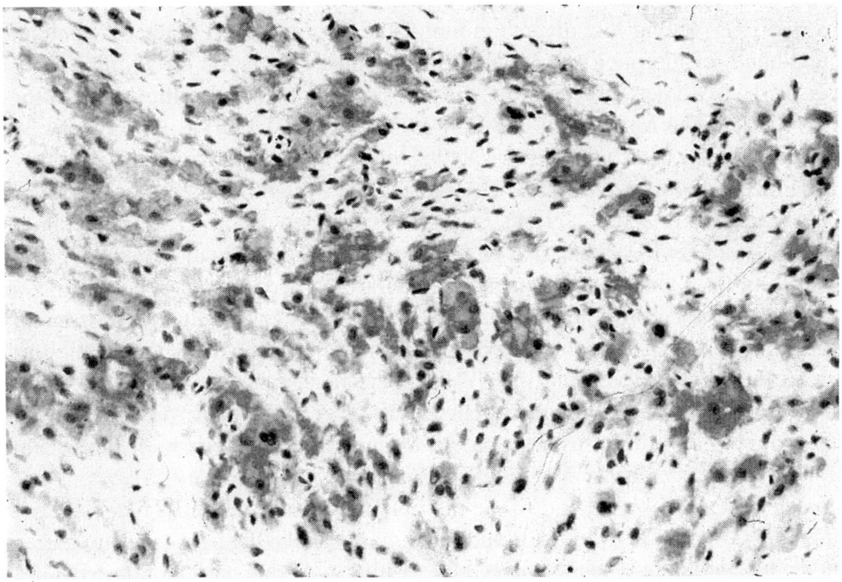

Figure 22 Constitutive expression of TGF-α by monokaryonic and polykaryonic macrophages in an interfacial membrane of a cemented cobalt–chrome alloy–based femoral stem of a failed total hip arthroplasty. Immunohistochemical stain with monoclonal antibody (Ab-2) to TGF-α (Calbiochem-Novabiochem, UK), 300×.

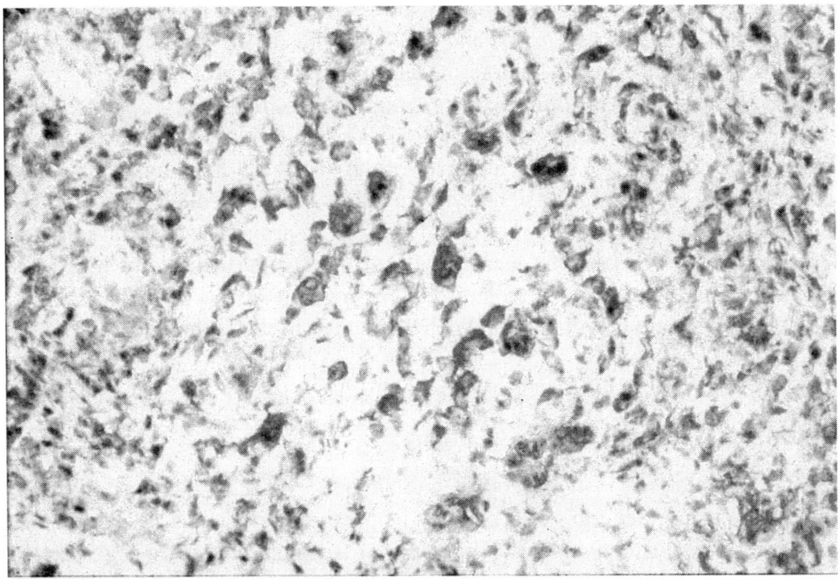

Figure 23 Constitutive expression of EGF-R by all macrophage subsets in the periprosthetic tissue of the femoral component of a failed cemented total knee joint replacement. Immunohistochemical stain with monoclonal antibody (Ab-1) to the EGF-receptor (Calbiochem-Novabiochem, UK), 320×.

is a member of the TNF family of cytokines that exists in transmembrane (cell membrane-bound) as well as soloubl (cleaved) form. By binding to osteoclast lineage–committed hematopoietic progenitor cells, OPGL stimulates the rapid induction of genes that typify osteoclastogenesis in the presence of M-CSF. Both in vitro and in vivo, OPGL activates osteoclasts, leading to enhanced bone resorption. Thus, OPGL exerts a dual function in differentiation and activation of the osteoclasts. Its unique capacity to modulate osteoclastic bone resorption is based on its ability to amply perform in the absence of both vitamin D_3 and osteoblasts or stromal cells, both of which are otherwise essential for the formation and activation of the osteoclasts [314].

It has been repeatedly pointed out that the macrophages accumulating in the interfacial membrane contribute to periprosthetic osteolysis after their activation by the ingested wear particles. The results of a series in vitro culture models have disclosed that peripheral blood monocytes, tissular histiocytes, and wear particle–laden macrophages isolated from periprosthetic tissues of loose prostheses and also their counterparts harvested from metallic or polymeric particle–induced subcutaneous granulomas of experimental animals can transform into bone-resorbing MNGCs. It has again been demonstrated in these systems that coculture with osteoblasts in the presence of vitamin D_3 is imperative for osteoclastogenesis to occur [300,315,316]. In parallel cultures employing cells stemming from interfacial membranes of failed arthroplasties, the addition of exogenous M-CSF is not essential for the differentiation of macrophages to osteoclasts. It is argued that macrophages replete in periprosthetic tissues ab initio produce M-CSF in excessive amounts. However, the presence of an additional potent mediator of osteoclast formation, say, OPGL, may also explain the observation [317].

To delineate macrophage-to-osteoclast kinships, Itonaga et al. have cultured macrophages retrieved from the bone–implant interfaces of loosened implants on glass coverslips or dentine slices in the presence or absence of OPGL. Their findings highlight a selective dependence of the macrophages on this cytokine during the differentiation to osteoclasts. It seems certain by now that OPGL substitutes for osteoblasts and vitamin D3 in this differential process. A significant fraction of the macrophage-like cells of the periprosthetic inflammatory infiltrates constitute OPGL-dependent mononuclear precursors of osteoclasts which, conditionally, acquire the TRAP and vitronectin receptor–positive phenotype as well as the functional capacity for lacunar bone resorption [318]. Analogous relationships existent in rheumatoid arthritis complement our notions, as discussed heretofore: the macrophages of the patient's inflamed joint capsules manifest the same OPGL dependence of osteoclastogenesis, accentuating the regulatory effects of T-lymphocytes, B-lymphocytes, and dendritic cells of the rheumatoid synovium in the induction of OPGL expression and the associated osteolysis [319]. To cap the climax, exposure of macrophages to orthopedic wear particles stimulates the synthesis of mRNA for both OPGL and OPGL-R in in vitro cultures [320].

G. Mast Cells at the Bone-Implant Interface

Whereas the mono- and polykaryonic macrophagic, osteoclastic, osteoblastic, as well as T-celled and B-celled lymphocytic responses to alloplastic breakdown products have been extensively assessed and described, the presence of variable numbers of mast cells in tissues nearby implants has only recently been recognized. In this context, evaluation of the phenotypic characteristics and the functional role of mast cells in the immune reaction to implant's particulates merits consideration in view of the known contributory effects of mast cells in inflammatory, fibrotic, and vascular disorders. Solovieva et al. have applied monoclonal antibodies (specific for mast cell proteases) to quantify the tryptase- and chymase-labeled cells in the interfacial membranes and pseudocapsules obtained at revision operations. Their studies disclose no increased numbers of mast cells in the interfacial membranes when compared to their numbers in the synovial tissues of normal knees. Their distribution is quantitatively similar in the interfacial membranes of loose cemented and noncemented arthroplasties. In contrast, the rate of degranulation of the mast cells is higher in the periprosthetic tissues of loose components than in the control tissues. The localization pattern of the mast cells within well-vascularized, inflamed areas rich in macrophages, fibroblasts, and vascular endothelial cells favors a role of the mast cells in the complex interactive processes involving these cells [321].

Results of recent experimentation imply a pivotal role of the mast cells in the early cellular events that trigger biomaterial particle–mediated inflammatory responses [322,323]. In mice, mast cells abound nearby polyethylene terephthalate (PET) disks implanted intraperitoneally or subcutaneously. With degranulation of the mast cells, the released histamine enhances leukocytic recruitment, explaining, at least partly, the acute inflammatory cell infiltration of the operation sites. This reasoning is amended by demonstrating the activation of mast cells shortly after their contact with biomaterial surfaces and the abruptly ensuing release of histamine and other mediators of mast cell granules. This early acute reaction triggers, in turn, an elevated expression of vascular endothelial adhesion molecules and an enhanced transmigration of phagocytic cells from the blood vessel lumens into the interstitial matrix through the endothelial barrier. A combination of biochemical and cellular mediators appears to participate in the prominent macrophagic infiltration of the tissues, cell–cell attachment, and, finally, fusion to form the foreign body giant cells at the implant's surfaces. Among these mediators are the chemotactic cytokines (the so-called chemokines), that is macrophage inflammatory protein-1α (MIP-1α) and monocyte chemoattractant protein-1 (MCP-1). In addition, the expression of cell membrane integrins is required. Expression of the CD11b/CD18 is essential for macrophage's optimized functional capacities and recognition of implant surfaces. This recognition process depends on the chemical nature of the foreign surfaces, as evidenced, for example, by maximizing adherence of phagocytes to implanted biomaterials through coating their surfaces with fibrinogen [324]. The impact of protein coating of surfaces on the adherence and subsequent cellular reactivities has also been described in relation to the prolonged secretory responses of neutrophils. Adherence of neutrophils to polystyrene surfaces coated with serum, laminin, vitronectin, or fibronectin prompts a massive respiratory burst and the release of reactive oxygen intermediates, such as H_2O_2, in response to TNF-α, when compared to suspended neutrophils in cultures, which are unreactive to TNF-α [325]. The observations suggest that cells that attach to a nonbiological surface respond through specific interactions mediated by cell membrane receptors, leading to assembly of the cytoskeletal microfilaments as well as to the associated cell spreading, intracellular signaling, and initiation of the release of biological mediators.

The bone marrow–derived, multifunctional mast cells participate in the defence mechanisms of the body. They exert their effects by releasing multitudinous cytokines and chemokines, including vasoactive and proinflammatory factors, following their activation with antigens, chemicals, or IgE-dependent factors [326,327]. Mast cells in the interfacial membranes must first come into direct contact with polymeric or metallic particles prior to releasing substantial amounts of phospholipase A2 (as also holds true for a subset of macrophages). Contact with the foreign bodies leads to the mast cell's activation as well as selective and dominating expression of IL-4 and interleukin-5 (IL-5). Phospholipase A_2 is a family of esterases that hydrolyze membrane glycerophospholipids, bringing about the release of free arachidonic acid from membrane phospholipids as well as synthesis of diverse mediators of inflammation, including prostanoids, leukotrienes, and platelet activating factor. By immunolocalization, the expression of phospholipase A_2 is ascribed to a majority of the mast cells in the periprosthetic tissues of failed prostheses. This finding speaks of the capacity of the particles to stimulate the mast cells to produce phospholipase A_2, which is an IgE-independent mode of the inflammatory response. The studies also bear witness to the potential source of IL-4 and IL-5 in the mast cells, a consistent profile existing in relation to metallic debris, cement particles, and fragments of bone (Fig. 24). In its capacity as a regulatory factor, IL-4 controls the growth and differentiation of several hematopoietic cells, including helper T-lymphocytes, B-lymphocytes, and mast cells. Additionally, it has recently been implicated as an inducer of macrophage fusion in that it directs foreign body giant cell formation [244]. Interestingly, the numbers of IL-4–positive mast cells are markedly increased in interfacial membranes in which the giant cells dominate the macrophagic infiltrates, predicting a role of IL-4 in promoting fusion of the monokaryonic to the polykaryonic macrophages [323]. Notable are the extracellular sites of immunoreactive tryptase and chymase adjacent to deposits of wear debris. In association with the many IL-4–positive mast cells present at these locales, it suggests that exocytosis of the mast cells (degranulation) is taking place within the periprosthetic tissues (Fig. 25). An increase of IL-4–positive mast cells in the absence or paucity of giant cells in the interfacial membranes removed at revision operations implies that bone erosion may antedate an emerging foreign body granulomatous reaction. In that they participate in the inflammatory reactions to polymeric (be it of polyethylene or PMMA provenance) and metallic detritus, the activated mast cells constitute the potential candidates in the mediation, augmentation, or both of the bone erosion leading to aseptic loosening of the artificial joints. In addition, interleukin-5 is as well expressed by many mast cells within the inflammatory infiltrate at the surface of the cement mantle (Fig. 26). The role of the mast cells in the processes involved in early osteolysis at implantation sites finds some support in the link between the markedly increased number of mast cells and pathological bone resorption in disease states such as osteopenia in patients with mastocytosis [328], rheumatoid arthritis [329], and postmenopausal osteoporosis [330].

Having reviewed the immunohistological modes of the tissular distribution of the diverse constituents in the periprosthetic tissues, it is of interest to relate these patterns with the widely spread monokaryonic and polykaryonic macrophages that are immunoreactive for interleukin-

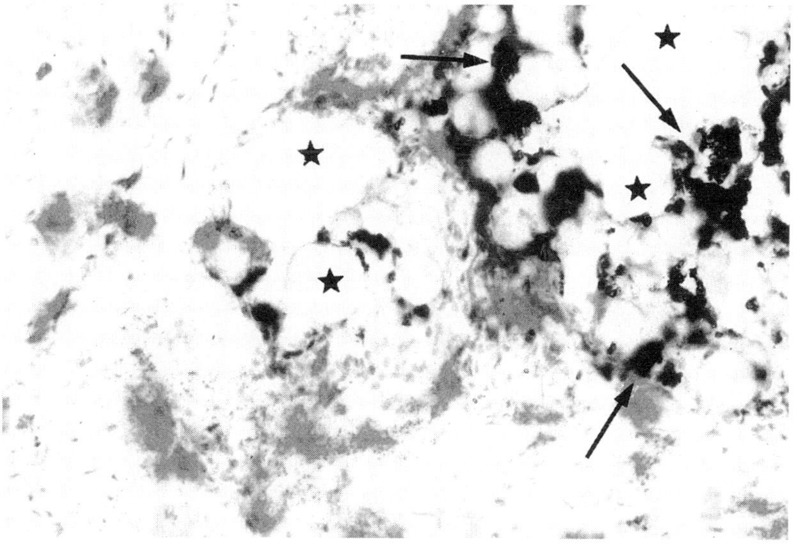

Figure 24 Mast cells constituting a relatively large proportion of the inflammatory cell population at a locale at which polymethylmethacrylate (stars) and metallic particles (arrows) have been deposited in the interfacial membrane of a failed total hip arthroplasty. Cryostat section of operation specimen immunostained with monoclonal antibody to chymase (Chemicon International), 260×.

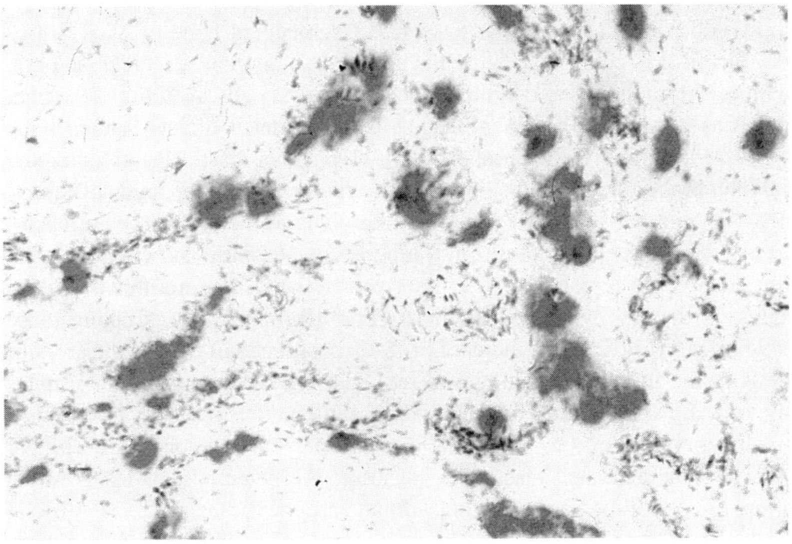

Figure 25 Numerous mast cells make up part of the inflammatory infiltrate, in the absence of polykaryonic macrophages or other giant cell types, at this segment of the periprosthetic tissue at the bone–implant interface of a failed artificial hip joint. Cryostat section of operation specimen immunostained with monoclonal antibody to chymase (Chemicon International), 200×.

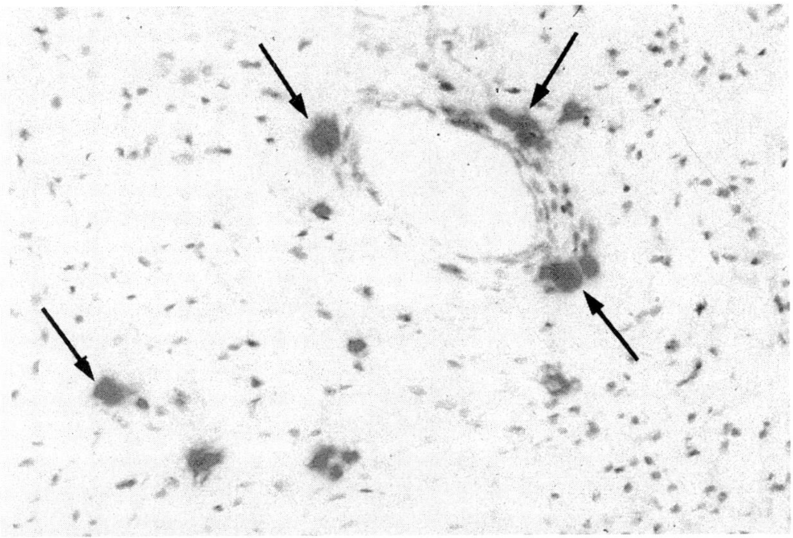

Figure 26 Mast cells expressing interleukin-5 (arrows), among the other types of inflammatory cells in the interfacial membrane, facing the cement of an aseptically loosened acetabular socket of a total hip arthroplasty. Immunohistochemical stain with monoclonal antibody to IL-5 (R&D Systems, UK), 300×.

1. Furthermore, while immunostained cells often reveal cytoplasmic positivity, the cells of the monocyte–macrophage lineage display an exquisite membranous coloring (Fig. 27), expressing a localization of the IL-1 in the cell membranes.

IV. CONCLUDING REMARKS

Replacement of a painful and disabling hip or knee joint with an up-to-date model of cemented artificial articulation deserves its reputation of alleviating most patients1 complaints for many years. The sometime noted pessimism with respect to cemented arthroplasties in young patients is out of place. In a cohort of patients who were 13 to 50 years old at the time of cemented total hip arthroplasties, 5.7 and 2.5% of the acetabular sockets and femoral stems, respectively, were found to be loose at an average of 10 years after surgery [331]. At the 25th postoperative year, ~85% of individuals with a Charnley total hip replacement are still gratified with the function of their original prosthesis. Taking age at operation into account, survivorship of Charnley artificial hip joints at the 25-year follow-up drops to ~70%, but climbs to 100% when the joint replacements are carried out in patients younger than 40 years and older than 80 years, respectively [332]. The long-term results of newly fashioned prostheses, biomaterials, or fixation techniques have to more than measure up with these outcomes to justify their clinical introduction. In addition to patient's specifics — age, gender, and underlying disease — and biomechanical and imaging data [333], knowledge of the morphological and immunohistological scene at the bone–implant interfaces is essential to accomplish comparative analyses of cementless versus cement-fixation systems.

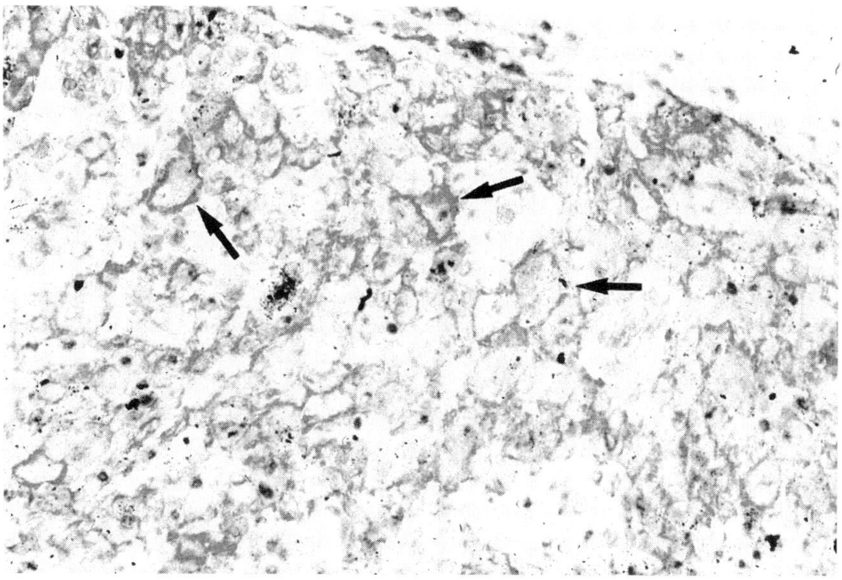

Figure 27 High level of expression of interleukin-1, in a membranous pattern, by all monokaryonic and polykaryonic macrophages (arrows), associated with metallic wear particles, in the interfacial membrane of a loose, cemented femoral stem of a total hip arthroplasty. Immunohistochemical stain with monoclonal antibody to IL-1 (R&D Systems, UK), 400×.

Many problems troubling the pioneers were resolved since the early days of joint replacements, yet wear debris–driven macrophagic activation with its attendant increased osteolysis still plagues patients with artificial joints. It has recently been ascertained that it is not the volume of wear debris by itself that opposes a long-term satisfactory function of artificial joints, but rather the volume of wear particles which are in the 0.2- to 0.8-mm size range, i.e., within the span of biologically active dimensions [334]. It has been claimed that 0.1-mm- to 1.0-mm-sized debris released from the polymer brushes during intramedullary wiping contribute to osteolysis [335]. It is not easy to visualize how small amounts of polymeric particles generated at a one-time event add significantly to the reactions in response to the enormous quantities of prosthetic debris accumulating during many years of service.

A quandary, addressed by many but not answered to the satisfaction of all eminent investigators, is the dilemma of what makes a prosthesis run. Maybe all the characteristics of a prosthesis and its host have to be tiptop in order for implants to perform as anticipated by the planners at the time the components have been on the drawing board. Implants may be ab initio flawed as far as the biomechanical or biochemical properties are concerned. Cases in point are the mismatched size and angle of a femoral head and taper lock [77] and Teflon-based acetabular sockets of total hip arthroplasties. Orthopedic surgeon's skilful execution of the implantation is likewise crucial for a gratifying outcome. Failure at the cement–stem or cement–bone interface may ensue in the wake of excessive relative micromovements. The latter are minimal around titanium alloy–based femoral stems fixated by 3- to 4-mm-thick cement mantles, but the micromovements increase when the cement mantle is thicker than 4 or thinner than 2 mm [336].

In addition to the responses to the wear debris, differences in the osteogenic potential of the bone surrounding the prosthesis could be attributed to various host-related factors as well as material properties. Understanding the responses of the microenvironment to the synthetic materials will assist in modulating the processes involved in periprosthetic osteolysis, which could reduce at least some of the current problems of implant loosening. Both experimental and clinical research invests many resources to develop compounds which delay or eliminate the inflammatory reactions to orthopedic wear debris so as to promote osteogenesis. Practitioners look forward to a time when they will pharmaceutically support bone ongrowth and ingrowth such that the utility of artificial joints will outlive the host's lifespan.

Unravelling the profile of cytokines and receptors produced by cells resident in the tissues nearby artificial joints, whether cemented or otherwise, is essential for a better understanding and characterization of the processes underlying prosthetic failure. There is no doubt that aseptic loosening of the alloplastic components is contingent on activities which are strictly determined by the binding of cytokines to their specific membrane receptors. The synopsized, current knowledge may, first, serve researchers as a baseline with which to compare findings acquired by study of interfacial membranes of cementless arthroplasties and, second, furnish pharmaceutically oriented research on preventing periprosthetic osteolysis by locally or systemically dispensed drugs which intervene in the chain of events kicked off by the synthesis and secretion of signaling proteins upon confrontation of cells of the body's defence mechanisms with foreign materials.

During the past few years tremendous progress has been achieved in the understanding of the role of mediators, including cytokines, growth factors, and their receptors, in the inflammatory granulomatous process at the bone–implant interface of failed prosthetic joints. Their level of expression and cellular sources is awaiting correlation with the diverse cell types of the body's defence mechanism present at the bone–implant interface and with the observed variations in the type, amount, and particle size of the prosthetic wear debris.

REFERENCES

1. Smith-Petersen MN. Evolution of mould arthroplasty of the hip joint. J. Bone Joint Surg 1948; 30-B:59–66.
2. Moore AT. Metal hip joint: a new self-locking Vitallium prosthesis. South. Med. J 1952; 45: 1015–1019.
3. MacIntosh DL. Hemiarthroplasty of the knee using a space occupying prosthesis for painful varus and valgus deformities. J. Bone Joint Surg 1958; 40A:1431–1438.
4. Charnley J. Low Friction Arthroplasty of the Hip: Theory and Practice. Berlin: Springer Verlag, 1979.
5. Gunston FH. Polycentric knee arthroplasty: prosthetic simulation of normal knee movement. J. Bone Joint Surg 1971; 53-B:272–277.
6. Boss JH. Biocompatibility: Review of the concept and its relevance to clinical practice. In: Wise DL, ed. Biomaterials and Bioengineering Handbook. New York: Marcel Dekker, Inc., 2000:1–94.
7. Canale ST. Arthroplasty. In: Campbell's Operative Orthopaedics. 9th ed.. St. Louis: Mosby, 1998: 209–231.
8. Dye SF. The future of anterior cruciate ligament restoration. Clin. Orthop 1996; 325:130–139.
9. Freeman MAR, Tennant R. The scientific basis of cement versus cementless fixation. Clin. Orthop 1992; 276:19–25.
10. Garcia-Cimbrelo E, Cruz-Pardos A, Cordero J, Sanchez-Sotelo J. Low-friction arthroplasty in patients younger than 40 years old: 20- to 25-year results. J. Arthroplasty 2000; 15:825–832.
11. Sochart DH, Hardinge K. Comparison of the Wrightington FC hip with the Charnley low-friction arthroplasty. 10- to 15-year results and survival analysis. J. Bone Joint Surg 1998; 80B:577–584.
12. Gill GS, Joshi AB. Long-term results of cemented, posterior cruciate ligament-retaining total knee arthroplasty in osteoarthritis. Am. J. Knee Surg 2001; 14:209–214.
13. Loon CJ, Wisse MA, de Waal Malefijt MC, Jansen RH, Veth RP. The kinematic total knee arthroplasty. A 10- to 15-year follow-up and survival analysis. Arch. Orthop. Trauma Surg 2000; 120:48–52.
14. Font-Rodriguez DE, Scuderi GR, Insall JN. Survivorship of cemented total knee arthroplasty. Clin. Orthop 1997; 345:79–86.
15. Chockalingam S, Scott G. The outcome of cemented vs. cementless fixation of a femoral component in total knee replacement (TKR) with the identification of radiological signs for the prediction of failure. The Knee 2000; 7:233–238.
16. Reckling PW, Asher MA, Dillon WL. A longitudinal study of the radiolucent line at the bone–cement interface following total joint-replacement procedures. J. Bone Joint Surg 1977; 59A:355–358.
17. Harris WH, McGann WA. Loosening of the femoral component after use of the medullary-plug cementing technique: follow-up note with a minimum five-year follow-up. J. Bone Joint Surg 1986; 68A:1064–1066.
18. Pellicci PM, Salvati EA, Robinson HJ. Mechanical failures in total hip replacement requiring reoperation. J. Bone Joint Surg 1979; 61A:28–36.
19. Jones LC, Hungerford DS. Cement disease. Clin. Orthop 1987; 225:192–206.
20. Müller ME. Lessons of 30 years of total hip arthroplasty. Clin. Orthop 1992; 274:12–23.
21. Akinmade AO, Nicholson JW. Glass-ionomer cement as adhesives. Part I. Fundamental aspects and their clinical relevance. J. Mater. Sci.: Mater. Med 1993; 4:95–101.
22. Markolf KL, Amstutz HC. Penetration and flow of acrylic bone cement. Clin. Orthop 1976; 121: 99–102.
23. Macdonald W, Swarts E, Beaver R. Penetration and shear strength of cement-bone interfaces in vivo. Clin. Orthop 1993; 286:283–288.
24. Johanson NA, Bullough PG, Wilsson PD, Salvati EA, Ranawat CS. The microscopic anatomy of the bone–cement interface in failed total hip arthroplasties. Clin. Orthop 1987; 218:123–135.

25. Mirra JM, Amstutz HC, Matos M, Gold R. The pathology of joint tissues and its clinical relevance in prosthesis failure. Clin. Orthop 1976; 117:221–240.
26. Freeman MAR, Bradley GW, Revell PA. Observations upon the interface between bone and polymethylmethacrylate cement. J. Bone Joint Surg 1982; 64B:489–493.
27. Schmalzried TP, Kwong LM, Jasty M, Sedlacek DC, Haire TC, O'Conner DO, Dragdon CR, Kabo M, Malcolm AJ, Harris WH. The mechanisms of loosening of cemented acetabular components in total hip arthroplasty. Analysis of specimens retrieved at autopsy. Clin. Orthop 1992; 274:60–78.
28. Christel P, Meunier A, Therin M. A method of analysis of cellular distribution in the membrane encapsulating surgically implanted biomaterials. J. Appl. Biomater 1990; 1:205–214.
29. Sennerby L, Thomsen P, Ericson LE. Ultrastructure of the bone–titanium interface in rabbits. J. Mater. Sci.: Mater. Med 1992; 3:262–271.
30. Linder L. Ultrastructure of the bone–cement and the bone–metal interface. Clin. Orthop 1992; 276: 147–156.
31. Aza PN, Luklinska ZB, Anseau MR, Guitian F, De Aza S. Transmission electron microscopy of the interface between bone and pseudo-wollastonite implant. J. Microsc 2001; 201:33–43.
32. Christel P, Meunier A, Therin M. A method of analysis of cellular distribution in the membrane encapsulating surgically implanted biomaterials. J. Appl. Biomater 1990; 1:205–214.
33. Papadimitriou JM. Endocytosis and formation of macrophage polykarya: an ultrastructural study. J. Path 1978; 126:215–219.
34. Linder L, Carlsson A, Marsal L, Bjursten LM, Brånemark PI. Clinical aspects of osseointegration in joint replacement. A histological study of titanium implants. J. Bone Joint Surg 1988; 70B: 550–555.
35. Crowninshield RD, Jennings JD, Laurent ML, Maloney WJ. Cemented femoral component surface finish mechanics. Clin. Orthop 1998; 355:90–102.
36. Goodman SB, Aspenberg P. Effect of amplitude of micromotion on bone ingrowth into titanium chambers implanted in the rabbit tibia. Biomaterials 1992; 13:944–948.
37. Sarmiento A, Ebramzadeh E, Gogan WJ, McKellop HA. Total hip arthroplasty with cement. A long-term radiographic analysis in patients who are older than fifty and younger than fifty years. J. Bone Joint Surg 1990; 72A:1470–1476.
38. Radin EL, Rubin CT, Thrasher EL, Lanyon LE, Crugnola AM, Schiller AS, Paul IL, R, Rose RM. Changes in the bone–cement interface after total hip replacement. An in vivo animal study. J. Bone Joint Surg 1982; 64A:1188–1200.
39. Eid K, Zelicofm S, Perona BP, Sledge CB, Glowacki J. Tissue reactions to particles of bone-substitute materials in intraosseous and heterotopic sites in rats: discrimination of osteoinduction, osteocompatibility, and inflammation. J. Orthop. Res 2001; 19:962–969.
40. Kroon PO, Freeman MAR. Hydroxyapatite coating of hip prostheses. Effect on migration into the femur. J. Bone Joint Surg 1992; 74B:518–522.
41. Bishop NE, Ferguson S, Tepic S. Porosity reduction in bone cement at the cement–stem interface. J. Bone Joint Surg 1996; 78B:349–356.
42. Kwak BM, Lim OK, Kim YY, Rim K. An investigation of the effect of cement thickness on an implant by finite element stress analysis. Int. Orthop 1979; 2:315–319.
43. Massoud SN, Hunter JB, Holdsworth BJ, Wallace WA, Juliusson R. Early femoral loosening in one design of cemented hip replacement. J. Bone Joint Surg 1997; 79B:603–608.
44. Crowninshield RD, Brand RA, Johnston RC, Milroy JC. The effect of femoral stem cross-sectional geometry on cement stresses in total hip reconstruction. Clin. Orthop 1980; 146:71–77.
45. Lu Z, McKellop H. Effects of cement creep on stem subsidence and stresses in the cement mantle of a total hip replacement. J. Biomed. Mater. Res 1997; 34:221–226.
46. Willert H-G, Ludwig J, Semlitsch M. Reaction of bone to methacrylate after hip arthroplasty: a long-term gross, light microscopic, and scanning electron microscopic study. J. Bone Joint Surg 1974; 56A:1368–1382.
47. Willert H-G, Semlitsch M, Buchhorn G, Kriete U. Materialverschleiss und Gewebereaktion bei klinischen Gelenken. Orthopäde 1978; 7:62–83.
48. Willert H-G, Mueller K, Semlitsch M. The morphology of poly-methylmethacrylate (PMMA) bone cement: surface structures and causes of their origin. Arch. Orthop. Trauma Surg 1979; 94:265–292.

49. Semlitsch M, Willert H-. Clinical wear behaviour of ultra-high molecular weight polyethylene cups paired with metal and ceramic ball heads in comparison to metal-on-metal pairings of hip joint replacements. Proc. Inst. Mech. Eng 1997; 211:73–88.
50. Sugano N, Nishii T, Nakata K, Masuhara K, Takaoka K. Poly-ethylene sockets and alumina ceramic heads in cemented total hip arthroplasty. J. Bone Joint Surg 1995; 77B:548–556.
51. Willert H-G, Buchhorn GH, Göbel D, Köster G, Schaffner S, Schenk R, Semlitsch M. Wear behavior and histopathology of classic cemented metal on metal hip endoprostheses. Clin. Orthop– 1996; 329S.
52. McKellop H, Clarke IC, Markolf KL, Amstutz HC. Friction wear and properties of polymer, metal, and ceramic prosthetic joint materials evaluated in a multi-channel screening device. J. Biomed. Mater. Res 1981; 15:619–653.
53. Gonzales JB, Purdon MA, H, Horowitz SM. In vitro studies on the role of titanium in aseptic loosening. Clin. Orthop 1996; 330:244–250.
54. Agins HJ, Alcock NW, Bansal M, Salvati EA, Wilson PD, Pellicci PM, Bullough PG. Metallic wear in failed titanium-alloy total hip replacements: a histological and quantitative analysis. J. Bone Joint Surg 1988; 70A:347–356.
55. Williams DF. Titanium, epitome of biocompatibility or cause for concern. J. Bone Joint Surg 1994; 76B:348–349.
56. Kwon SY, Takei H, Pioletti DP, Lin T, Ma QJ, Akeson WH, Wood DJ, Sung KL. Titanium particles inhibit osteoblast adhesion to fibronectin coated substrates. J. Orthop. Res 2000; 18:203–211.
57. Goldring SR, Schiller AL, Roelke M, Rourke CM, O'Neill DA, Harris WH. The synovial-like membrane at the bone–cement interface in loose total hip replacements and its proposed role in bone lysis. J. Bone Joint Surg 1983; 65A:575–584.
58. Goldring SR, Jasty M, Roelke MS, Rourke CM, Bringhurst FR, Harris WH. Formation of a synovial-like membrane at the bone–cement interface. Its role in bone resorption and implant loosening after total hip replacement. Arthr. Rheum 1986; 29:836–842.
59. Santavirta S, Konttinen VT, Bergroth V, Eskola A, Tallroth K, Lindholm S. Aggressive granulomatous lesions associated with hip arthroplasty. Immunopathological studies. J. Bone Joint Surg 1990; 72A:252–258.
60. Johansson NA, Bullough PG, Wilson PD, Salvati EA, Ranawat CS. The microscopic anatomy of the bone–cement interface in failed total hip arthroplasties. Clin. Orthop 1987; 218:123–135.
61. Korovessis P, Repanti M. Evolution of aggressive granulomatous periprosthetic lesions in cemented hip arthroplasties. Clin. Orthop 1994; 300:155–161.
62. Schmalzried TP, Jasty M, Rosenberg A, Harris WH. Histologic identification of polyethylene debris using oil red O stain. J. Appl. Biomater 1993; 4:119–125.
63. Campbell P, Ma S, Yeom B, McKellop H, Schmalzried TP, Amstutz HC. Isolation of predominantly submicron-sized UHMWPE wear particles from periprosthetic tissues. J. Biomed. Mater. Res 1995; 29:127–131.
64. McKellop HA, Campbell P, Park S-H, Schmalzried TP, Grigoris P, Amstutz HC, SA, Sarmiento A. The origin of submicron polyethylene wear debris in total hip arthroplasty. Clin. Orthop 1995; 311:3–20.
65. Schmalzried TP, Jasty M, Rosenberg A, Harris WH. Poly-ethylene wear debris and tissue reactions in knee as compared to hip replacement prostheses. J. Appl. Biomater 1994; 5:1185–1190.
66. Kossovsky N, Liao K, Gelman A, Campbell P, Amstutz HC, Finerman M, Nasser S, Thomas BJ. Photon correlation spectroscopy analysis of the submicrometre particulate fraction in human synovial tissues recovered at arthroplasty or revision. In: St John KR, ed. Particulate Debris from Medical Implants: Mechanisms of Formation and Biological Consequences, ASTM STP 1144. Philadelphia: American Society for Testing and Materials, 1992:68–74.
67. Malcolm AJ. Pathology of low-friction arthroplasties in autopsy specimens. In: Older J, ed. Implant Bone Interface. London: Springer Verlag, 1990:79–82.
68. Benz EB, Federman M, Godleski JJ, Bierbaum BE, Thornhill TS, Spector M. Transmission electron microscopy of intracellular particles of polyethylene from joint replacement prostheses: size distribution and cellular response. Biomaterials 2001; 22:2835–2842.

69. Boss JH, Shajrawi I, Mendes DG. Histological patterns of the tissue reaction to polymers in orthopaedic surgery. Clin. Mater 1993; 13:11–17.
70. Murray DW, Rushton N. Macrophages stimulate bone resorption when they phagocytose particles. J. Bone Joint Surg 1990; 72B:988–992.
71. Knoch M, Buchhorn G, von Knoch F, Köster G, Willert H-. Intracellular measurement of polyethylene particles. A histomorphometric study. Arch. Orthop. Trauma Surg 2001; 121:399–402.
72. Haddad FS, Cobb AG, Bentley G, Levell NJ, Dowd PM. Hypersensitivity in aseptic loosening of total hip replacements. The role of constituents of bone cement. J. Bone Joint Surg 1996; 78B: 546–549.
73. Ingham E, Green TR, Stone MH, Kowalski R, Watkins N, Fisher J. Production of TNF-alpha and bone resorbing activity by macrophages in response to different types of bone cement particles. Biomaterials 2000; 21:1005–1013.
74. Willert H-G, Bertram H, Buchhorn GH. Osteolysis in allo-arthroplasty of the hip. The role of bone cement fragmentation. Clin. Orthop 1990; 258:108–121.
75. Isaac GH, Wroblewski BM, Atkinson JR, Dowson D. A tribological study of retrieved hip prostheses. Clin. Orthop 1992; 276:115–125.
76. Betts F, Wright T, Salvati EA, Boskey A, Bansal M. Cobalt-alloy metal debris in peri-articular tissues from total hip revision arthroplasties. Metal contents and associated histologic findings. Clin. Orthop 1992; 276:75–82.
77. Buchert PK, Vaughin BK, Mallory TH, Engh CA, Bobyn D. Excessive metal release due to loosening and fretting of sintered particles on porous-coated hip prostheses. J. Bone Joint Surg 1986; 68A: 606–609.
78. Boss JH, Mendes DG, Misselevich I. Particulate metal in late aseptic loosening of cemented total hip arthroplasties. In: Wise DL, Gresser JD, Trantola DJ, Cattaneo MV, Lewandrowski K-U, Yaszemski MJ(), eds. Biomaterials Engineering and Devices: Human Applications. Vol. 2. Orthopedic, Dental, and Bone Graft Applications. Totowa, NJ: Humana Press, 2000:281–289.
79. Milosev L, Antolic V, Minovic A, Cör A, Herman S, Pavlovcic V, Campbell P. Extensive metallosis and necrosis in failed prostheses with cemented titanium-alloy stems and ceramic heads. J Bone Joint Surg 2000; 82B:352–357.
80. Scales JT. Black staining around titanium alloy prostheses an orthopaedic enigma. J. Bone Joint Surg 1991; 73B:534–536.
81. Lalor PA, Revell PA, Gray AB, Wright S, Railton GT, Freeman MAR. Sensitivity to titanium: a cause of implant failure?. J. Bone Joint Surg 1991; 73B:25–28.
82. Lalor PA, Revell PA. T-lymphocytes and titanium aluminium vanadium (TiAlV) alloy: evidence for immunological events associated with debris deposition. Clin Mater 1993; 12:57–62.
83. Toumbis CA, Kronick JL, Wooley PH, Nasser S. Total joint arthroplasty and the immune response. Semin. Arthritis Rheum 1997; 27:44–47.
84. Al-Saffar N, Mah JT, Kadoya Y, Revell PA. Neovascularisation and the induction of cell adhesion molecules in response to degradation products from orthopaedic implants. Ann. Rheum. Dis 1995; 54:201–208.
85. Merritt K, Rodrigo JJ. Immune response to synthetic materials. Sensitization of patients receiving orthopaedic implants. Clin. Orthop 1996; 326:71–79.
86. Rostoker G, Robin J, Binet O, Blamoutier J, Paupe J, Lessana-Leibowitch M, Bedouelle J, Sonneck JM, Garrel JB, M, Millet P. Dermatitis due to orthopaedic implants. A review of the literature and report of three cases. J. Bone Joint Surg 1987; 69A:1408–1402.
87. Boss JH, Shajrawi I, Soudry M, Mendes DG. Histomorphological reaction patterns of the bone to diverse particulate implant materials in man and experimental animals. In: St John KR, ed. Particulate Debris from Medical Implants: Mechanisms of Formation and Biological Consequences, ASTM STP 1144. Philadelphia: American Society for Testing and Materials, 1992:90–108.
88. Koerten HK, Onderwater JJ, Koerten EW, Bernoski FP, Nelissen RG. Observations at the articular surface of hip prostheses: an analytical electron microscopy study on wear and corrosion. J. Biomed. Mater. Res 2001; 15:591–596.
89. Weissman BN, Scott RD, Brick GW, Corson JM. Radiographic detection of metal-induced synovitis as a complication of arthroplasty of the knee. J. Bone Joint Surg 1991; 73A:1002–1007.

90. Santavirta S, Ida H, Ishii M, Takei I, Niissalo S, Ogino T, Konttinen YT. High-turnover periprosthetic bone remodeling and immature bone formation around loose cemented total hip joints. J. Bone Miner. Res 2001; 16:79–88.
91. Kamioka H, Honjo T, Takano-Yamamoto T. A three-dimensional distribution of osteocyte processes revealed by the combination of confocal laser scanning microscopy and differential interference contrast microscopy. Bone 2001; 28:145–149.
92. Jasty M, Bragdon C, Jiranek W, Chandler H, Maloney W, Harris WH. Etiology of osteolysis around porous-coated cementless total hip arthroplasties. Clin. Orthop 1994; 308:111–126.
93. Boss JH, Shajrawi I, Luria M, Mendes DG. Necrotizing granulomas within the periprosthetic tissues of cemented and cementless total arthroplasties. Veter. Comp. Orthop. Traumatology 1995; 8: 107–113.
94. Schmalzried TP, Jasty M, Harris WH. Periprosthetic bone loss in total hip arthroplasty. Polyethylene wear debris and the concept of the effective joint space. J. Bone Joint Surg 1992; 74A:849–863.
95. Schwarz EM, Lu AP, Goater JJ, Benz EB, Kollias G, Rosier RN, Puzas JE, O'Keefe RJ. Tumor necrosis factor-alpha/nuclear transcription factor-kappaB signaling in periprosthetic osteolysis. J. Orthop. Res 2000; 18:472–480.
96. Lassus J, Waris V, Xu JW, Li TF, Hao J, Nietosvaara Y, Santavirta S, Konttinen YT. Increased interleukin-8 (IL-8) expression is related to aseptic loosening of total hip replacement. Arch. Orthop. Trauma Surg 2000; 120:328–332.
97. Stea S, Visentin M, Granchi D, Melchiorri C, Soldati S, Sudanese A, Toni A, Montanaro L, Pizzoferrato A. Wear debris and cytokine production in the interface membrane of loosened prostheses. J. Biomater. Sci. Polym. Ed 1999; 10:247–257.
98. Nakashima Y, Sun DH, Trindade MC, Chun LE, Song Y, Goodman SB, Schurman DJ, Maloney WJ, Smith RL. Induction of macro-phage C–C chemokine expression by titanium alloy and bone cement particles. J. Bone Joint Surg 1999; 81B:155–162.
99. Xu JW, Konttinen YT, Li TF, Waris V, Lassus J, Matucci-Cerinic M, Sorsa T, Santavirta TS. Production of platelet-derived growth factor in aseptic loosening of total hip replacement. Rheumatol. Int 1998; 17:215–221.
100. Kim KJ, Chiba J, Rubash HE. In vivo and in vitro analysis of membranes from hip prostheses inserted without cement. J. Bone Joint Surg 1994; 76A:172–180.
101. Shanbag AS, Jacobs JJ, Black J, Galante JO, Glant TT. Cellular mediators secreted by interfacial membranes obtained at revision total hip arthroplasty. J. Arthroplasty 1995; 10:498–506.
102. Zenni GC, Ellinger J, Lam TM, Greisler HP. Biomaterial-induced macrophage activation and monokine release. J. Invest. Surg 1994; 7:135–141.
103. Peters PC, Engh GA, Dwyerm KA, Vinh TN. Osteolysis after total knee arthroplasty without cement. J. Bone Joint Surg 1992; 74A:864–876.
104. Pollice PF, Hsu J, Hicks DG, Bukata S, Rosier RN, Reynolds PR, Puzas JE, O'Keefe RJ. Interleukin-10 inhibits cytokine synthesis in monocytes stimulated by titanium particles: evidence of an anti-inflammatory regulatory pathway. J. Orthop. Res 1998; 16:697–704.
105. Nivbrant B, Karlsson K, Karrholm J. Cytokine levels in synovial fluid from hips with well-functioning or loose prostheses. J. Bone Joint Surg 1999; 81B:163–166.
106. Jasty MJ, Floyd WE, Schiller AL, Goldring SR, Harris WH. Localized osteolysis in stable non-septic total hip replacement. J. Bone Joint Surg 1986; 68A:912–919.
107. Ohlin A, Johnell O, Lerner UH. The pathogenesis of loosening of total hip arthroplasties. The production of factors by periprosthetic tissues that stimulate in vitro bone resorption. Clin. Orthop 1990; 253:287–296.
108. Shanbag AS, Jacobs JJ, Black J, Galante JO, Glant TT. Human monocyte response to particulate biomaterials generated in vivo and in vitro. J. Orthop. Res 1995; 13:792–801.
109. Pals SD, Wilkins RM. Giant cell tumor of bone treated by curettage, cementation, and bone grafting. Orthopedics 1992; 15:703–708.
110. Puleo DA, Holleran LA, Doremus RH, Bizios R. Osteoblast responses to orthopedic implant materials in vitro. J. Biomed. Mater. Res 1991; 25:711–723.
111. Kwong LM, Jasty M, Mulroy RD, Maloney WJ, Bragdon C, Harris WH. The histology of the radiolucent line. J. Bone Joint Surg 1992; 74B:67–73.

112. Horowitz SM, Gonzales JB. Inflammatory response to implant particulates in a macrophage/osteoblast coculture model. Calcif. Tissue Int 1996; 59:392–396.
113. Sabokbar A, Fujikawa Y, Brett J, Murray DW, Athanasou NA. Increased osteoclastic differentiation by PMMA particle–associated macrophages. Acta Orthop. Scand 2000; 67:593–598.
114. Goodman SB, Kang T, Smith RL. Lysosomal enzyme production at the interface surrounding loose and well-fixed cemented tibial hemiarthroplasty in the rabbit knee. J. Invest. Surg 1993; 6:413–418.
115. Lacey DL, Timms E, Tan HL, Kelley MJ, Dunstan CR, Burgess T, Elliott R, Colombero A, Elliott G, Scully S, Hsu H, Sullivan J, Hawkins N, Davy E, Capparelli C, Eli A, Qian YX, Kaufman S, Sarosi I, Shalhoub V, Senaldi G, Guo J, Delaney J, Boyle WJ. Osteoprotegerin ligand is a cytokine that regulates osteoclast regulation and activation. Cell 1998; 93:165–176.
116. Yasuda H, Shim N, Nakagawa N, Yamaguchi K, Kinosaki M, Mochizuki S, Tomoyasu A, Yano K, Goto M, Murakami A, Tsuda E, Morinaga T, Higashio K, Udagawa N, Takahashi N, Suda T. Osteoclast differentiation factor is a ligand for osteoprotegerin/osteoclastogenesis-inhibitory factor and is identical to TRANCE/RANKL. Proc. Natl. Acad. Sci 1998; 95:3597–3602.
117. Nakagawa N, Kinosaki M, Yamaguchi K, Shima N, Yasuda H, Yano K, Morinaga T, Higashio K. RANK is the essential signaling receptor for osteoclast differentiation factor in osteoclastogenesis. Biochem. Biophys. Res. Commun 1998; 253:395–400.
118. Matsuzuki K, Udagawa N, Takahashi N, Yamaguchi K, Yasuda H, Shima N, Morinaga T, Toyama Y, Yabe Y, Higashio K, Suda T. Osteoclast differentiation factor (ODF) induces osteoclast-like cell formation in human peripheral blood mononuclear cell cultures. Biochem. Biophys. Res. Commun 1998; 246:199–204.
119. Yasuda H, Shima N, Nakagawa N, Mochizuki SI, Yano K, Fujise N, Sato Y, Goto M, Yamaguchi K, Kuriyama M, Kanno T, Murakami A, Tsuda E, Morinaga T, Higashio K. Identity of osteoclastogenesis inhibitory factor (OCIF) and osteoprotegerin (OPG): a mechanism by which OPG/OCIF inhibits osteoclastogenesis in vitro. Endocrinology 1998; 139:1329–1337.
120. Haynes DR, Crotti TN, Potter AE, Loric M, Atkins GJ, Howie DW, Findlay DM. The osteoclastogenic molecules RANKL and RANK are associated with periprosthetic osteolysis. J. Bone Joint Surg 2001; 83B:902–911.
121. Shoji H, Karube S, D'ambrosia RD, Dabezies EJ, Miller DR. Biochemical features of pseudomembrane at the bone–cement interface of loosened total hip prostheses. J. Biomed. Mater. Res 1983; 17:669–678.
122. Boss JH, Shajrawi I, Soudry M, Mendes DG. Histological features of the interface membrane of failed isoelastic cementless prostheses. Intern. Orthop 1990; 14:399–403.
123. Mirra JM, Marder RA, Amstutz HC. The pathology of failed total joint arthroplasty. Clin. Orthop 1982; 170:175–185.
124. Linder L, Lindeberg L, Carlsson Å. Aseptic loosening of hip prostheses. A histologic and enzyme histochemical study. Clin. Orthop 1983; 175:93–104.
125. Williams P, MacQueen DA. A histopathologic study of late aseptic loosening of cemented hip prostheses. Clin. Orthop 1992; 275:174–179.
126. Goodman SB. Polymer particles potentiate loosening of joint arthroplasties. Orthopaedics Intern. Edn 1993; 1:316–321.
127. Linder L, Carlsson ÅS. The bone–cement interface in hip arthroplasty. A histologic and enzyme study of stable components. Acta Orthop. Scand 1986; 57:495–500.
128. Boss JH, Shajrawi I, Dekel S, Mendes DG. The bone–cement interface: Histological observations of the interface of cemented arthroplasties within the immediate and late phases. J. Biomater. Sci. Polymr. Ed 1993; 3:221–230.
129. Feith R. Side-effects of acrylic cement implanted into bone. A histological, (micro)angiographic, fluorescence-microscopic and autoradiographic study in the rabbit femur. Acta Orthop. Scand 1975; 161(Suppl):3–136.
130. Mjöberg B, Pettersson H, Rosenqvist R, Rydholm A. Bone cement, thermal injury and the radiolucent zone. Acta Orthop. Scand 1984; 55:597–600.
131. Swenson LW, Schurman DJ, Piziali R. Finite element temperature analysis of a total hip replacement and measurement of PMMA curing temperatures. J. Biomed. Mater. Res 1981; 15:83–96.

132. Radin EL, Martin RB, Orr RB, Rose RM. Bone changes in the proximal femur following total hip replacement prosthetic insertion. Orthop Trans 1981; 5:285.
133. Boss JH, Shajrawi I, Mendes DG. The nature of the bone–implant interface. The lessons learnt from implant retrieval and analysis in man and experimental animal. Med. Prog. Technol 1994; 20: 119–142.
134. Amir D, Müller-Mai C, Wendland H, Gross U, Sela J. Effect of glass-ceramic implants on primary calcification in rat tibial bone after injury. Biomaterials 1989; 10:585–589.
135. Tibrewal SB, Grant KA, Goodfellow JW. The radiolucent line beneath tibial components of the Oxford meniscal knee. J. Bone Joint Surg 1984; 66B:523–528.
136. Charnley J, Crawford WJ. Histology of bone in contact with self-curing acrylic cement. J. Bone Joint Surg 1968; 50B:228–236.
137. O'Connor J, Goodfellow J, Perry N. Fixation of the tibial components of the Oxford knee. Orthop. Clin. North. Am 1982; 13:65–87.
138. Mjöberg B. Loosening of the cemented hip prosthesis. The importance of heat injury. Acta Orthop. Scand 1986; 221Suppl:1–40.
139. Salvati EA, Betts F, Doty SB. Particulate metallic debris in cemented total hip arthroplasty. Clin. Orthop 1993; 293:160–173.
140. Eschenroeder HC, Krackow KA. Late onset femoral stress fracture associated with extruded cement following hip arthroplasty. Clin. Orthop 1988; 236:210–213.
141. Lorenz K. Behind the Mirror: A Search for a Natural History of Human Knowledge. Bungay. UK: Chaucer Press, 1977.
142. Song Y, Beaupre G, Goodman SB. Osseointegration of total arthroplasties: studies in humans and animals. J. Long-Term Eff. Med. Impl 1999; 9:77–112.
143. Boss JH, Shajrawi I, Mendes DG. The nature of the interface of orthopedic implant devices: a viewpoint of pathologists. Orthopaedics Intern. Edn 1993; 1:325–331.
144. Morberg PH, Johansson CB, Reigstad A, R, Rokkum M. Vital staining in stable, retrieved femoral surface replacement prostheses: a microscopic study of undecalcified ground sections. J. Arthroplasty 2001; 16:1004–1009.
145. Williams DF. Introduction. In: Williams DF, ed. Biocompatibility of Orthopaedic Implants. Boca Raton: CRC Press, 1985:2–3.
146. Amstutz HC, Campbell P, Kossovsky N, Clarke IC. Mechanism and clinical significance of wear debris-induced osteolysis. Clin. Orthop 1992; 276:7–18.
147. Urban RM, Jacobs JJ, Tomlinson MJ, Gavrilovic J, Black J, Peoch M. Dissemination of wear particles to the liver, spleen, and abdominal lymph nodes of patients with hip or knee replacement. J. Bone Joint Surg 2000; 82A:457–476.
148. Gil-Albarova J, Lacleriga A, Barrios C, Canadell J. Lymphocyte response to polymethylmethacrylate in loose total hip prostheses. J. Bone Joint Surg 1992; 74-B:825–830.
149. Salter DM, Krajewski AS, Robertson S. Lymphocytes in pseudo-membranes of late prosthetic joint failure. J. Pathology 1992; 166:271–275.
150. Wayand CM, Geisler A, Brack A, Bolander ME, Goronzy JJ. Oligoclonal T cell proliferation and interferon-gamma production in periprosthetic inflammation. Lab. Invest 1998; 78:677–685.
151. Al-Saffar N. Early clinical failure of total joint replacement in association with follicular proliferation of B-lymphocytes: a report of two cases. J. Bone Joint Surg 2002; 84-A:2270–2273.
152. Wooley PH, Petersen S, Song Z, Nasser S. Cellular immune responses to orthopaedic implant materials following cemented total joint replacement. J. Orthop. Res 1997; 15:874–880.
153. Farber A, Chin R, Song Y, Goodman SB. Chronic antigen-specific immune-system activation may potentially be involved in the loosening of cemented acetabular components. J. Biomed. Mater. Res 2001; 55:433–441.
154. Granchi D, Ciapetti G, Savarino L, Cenni E, Pizzoferrato A, Baldini N, Giunti A. Effects of bone cement extracts on the cell-mediated immune response. Biomaterials 2002; 23:1033–1041.
155. Jasty M, Jiranek W, Harris W. Acrylic fragmentation in total hip replacements. Clin. Orthop 1992; 285:116–128.
156. DeHeer DH, Engels JA, DeVries AS, Knapp RH, Beebe JD. In situ complement activation by polyethylene wear debris. J. Biomed. Mater. Res 2001; 54:12–19.

157. Geesink RG, de Groot K, Klein CP. Chemical implant fixation using hydroxyl-apatite coatings. The development of a human total hip prosthesis for chemical fixation to bone using hydroxyl-apatite coatings on titanium substrates. Clin. Orthop 1987; 225:147–170.
158. Ryd L, Hansson U, Blunn G, Lindstrand A, Toksvig-Larsen S. Failure of partial cementation to achieve implant stability and bone ingrowth: a long-term roentgen stereophotogrammetric study of tibial components. J. Orthop. Res 1999; 17:311–320.
159. Spector M, Shortkroff S, Hsu HP, Lane N, Sledge CB, Thornhill TS. Tissue changes around loose prostheses. A canine model to investigate the effects of an antiinflammatory agent. Clin. Orthop 1990; 261:140–152.
160. Galante JO, Lemons J, Spector M, Wilson PD, Wright TM. The biologic effects of implant materials. J. Orthop. Res 1991; 9:760–775.
161. Maguire JK, Coscia MF, Lynch MH. Foreign body reaction to polymeric debris following total hip arthroplasty. Clin. Orthop 1987; 216:213–223.
162. Dyce J, Tompkins D, Bhushan B, Olmstead ML. Polyethylene wear in retrieved canine acetabular components. A tribological study. Vet. Comp. Orthop. Traumatol 2000; 13:78–86.
163. Nguyen NC, Maloney WJ, Dauskardt RH. Reliability of PMMA bone cement fixation: fracture and fatigue crack-growth behaviour. J. Mater. Sci.: Mater. Med 1997; 8:473–483.
164. Black J, Oppenheimer P, Morris DM, Peduto AM, Clark CCJ. Release of corrosion products by F-75 cobalt base alloy in the rat. III. Effects of a carbon surface coating. J. Biomed. Mater. Res 1987; 21:1213–1230.
165. Rae T. The toxicity of metals used in orthopaedic prostheses. An experimental study using cultured human synovial fibroblasts. J. Bone Joint Surg 1981; 63B:435–440.
166. Jones DA, Lucas HK, O'Driscoll M, Price CH, Wibberley B. Cobalt toxicity after McKee hip arthroplasty. J. Bone Joint Surg 1975; 57B:289–296.
167. Benson MK, Goodwin PG, B, Brostoff J. Metal sensitivity in patients with joint replacement arthroplasties. Br. Med. J 1975; 5993:374–375.
168. Charosky CB, Bullough PG, Wilson PD. Total hip replacement failures. A histological evaluation. J. Bone Joint Surg 1973; 55A:49–58.
169. Haynes DR, Hay SJ, Rogers SD, Ohta S, Howie DW, Graves SE. Regulation of bone cells by particle-activated mononuclear phagocytes. J. Bone Joint Surg 1997; 79B:988–994.
170. Haynes DR, Crotti TN, Haywood MR. Corrosion of and changes in biological effects of cobalt chrome alloy and 316L stainless steel prosthetic particles with age. J. Biomed. Mater. Res 1999; 49:167–175.
171. Winter GD. Tissue reactions to metallic wear and corrosion products in human patients. J. Biomed. Mater. Res 1974; 5:11–26.
172. Stea S, Visentin M, Granchi D, Cenni E, Ciapetti G, Sudanese A, Toni A. Apoptosis in periimplant tissue. Biomaterials 2000; 21:1393–1398.
173. Black J, Sherk H, Bonini J, Rostoker WR, Schajowicz F, Galante JO. Metallosis associated with a stable titanium-alloy femoral component in total hip replacement. J. Bone Joint Surg 1990; 72A: 126–130.
174. Bullough PG. Metallosis. J. Bone Joint Surg 1994; 76B:687–678.
175. Betts F, Wright T, Salvati EA, Boskey A, Bansal M. Cobalt-alloy metal debris in periarticular tissues from total hip revision arthroplasties. Metal contents and associated histologic features. Clin. Orthop 1992; 276:75–82.
176. Maloney WJ, Smith RL, Schmalzried T/, Chiba J, Huene D, Rubash H. Isolation and characterization of wear particles generated in patients who have had failure of a hip arthroplasty without cement. J. Bone Joint Surg 1995; 77A:1301–1310.
177. Kovacik MW, Gradisar IA, Sylvester AM, Haprian JJ, Alexander TS. A comparison of interleukin-1 beta in human synovial fluid of osteoarthritic and revision total knee arthroplasty. Biomed. Sci. Instrum 1997; 33:519–523.
178. Black J. Metal on metal bearings. A practical alternative to metal on polyethylene total joints?. Clin. Orthop 1996; 329 Suppl:S244–S255.
179. Amstutz HC, Grigoris P. Metal on metal bearings in hip arthroplasty. Clin. Orthop 1996; 329 Suppl: S11–S34.

180. Kelley SS, Johnston RC. Debris from cobalt–chrome cable may cause acetabular loosening. Clin. Orthop 1992; 285:140-146.
181. Budde JK, Orosz JF, Bonfiglio TA, Pellegrini VD. Particulate titanium and cobalt–chrome metallic debris in failed total knee arthroplasty. A quantitative histologic analysis. J. Arthroplasty 1994; 93: 291–304.
182. Bobyn JD, Tanzer M, Krygier JJ, Dujovne AR, Brooks CE. Concerns with modularity in total hip arthroplasty. Clin. Orthop 1994; 298:27–36.
183. Christie MJ, DeBoer DK, Trick LW, Brothers JC, Jones RE, Vise GT, Gruen TA. Primary total hip arthroplasty with use of the modular S-ROM prosthesis. Four to seven-year clinical and radiographic results. J. Bone Joint Surg 1999; 81A:1707–1716.
184. Jacobsson S-A, Djerf K, Wahlström O. 20-Year results of McKee-Farrar Versus Charnley Prosthesis. Clin. Orthop 1996; 329S:S60–S68.
185. Schmalzried TP, Peters PC, Maurer BT, Bragdon CR, Harris WH. Long-duration metal-on-metal total hip arthroplasties with low wear of the articulating surfaces. J. Arthroplasty 1996; 11:322–331.
186. Klapperich C, Graham J, Pruitt L, Ries MD. Failure of a metal-on-metal total hip arthroplasty from progressive osteolysis. J Arthroplasty 1999; 14:877–881.
187. Jacobs JJ, Skipor AK, Doorn PF, Campbell P, Schmalzried TP, Black J, Amstutz HC. Cobalt and chromium concentrations in patients with metal on metal total hip replacements. Clin. Orthop 1996; 329 Suppl:S256–S263.
188. McKellop H, Park SH, Chiesa R, Doorn P, Lu B, Normand P, Grigoris P, Amstutz H. In vivo wear of three types of metal on metal hip pros-theses during two decades of use. Clin. Orthop 1996; 329 Suppl:128–140.
189. Dorr LD, Wan Z, Longjohn DB, Dubois B, Murken R. Total hip arthroplasty with use of the Metasul metal-on-metal articulation. Four- to seven-year results. J. Bone Joint Surg 2000; 82A:789–798.
190. Kovacik MW, Gradisar IA, Sylvester AM, Haprian JJ, Alexander TS. A comparison of interleukin-1 beta in human synovial fluid of osteoarthritic and revision total knee arthroplasty. Biomed. Sci. Instrum 1997; 33:519–523.
191. Chen FS, Scher DM, Clancy RM, Vera-Yu A, Di Cesare PE. In vitro and in vivo activation of polymorphonuclear leukocytes in response to particulate debris. J. Biomed. Mater. Res 1999; 48: 904–912.
192. Manlapaz M, Maloney WJ, Smith RL. In vitro activation of human fibroblasts by retrieved titanium alloy wear debris. J. Orthop. Res 1996; 14:465–472.
193. Yao J, Glant TT, Lark MW, Mikecz K, Jacobs JJ, Hutchinson NI, Hoerrner LA, Kuettner KE, Galante JO. The potential role of fibro-blasts in periprosthetic osteolysis: fibroblast response to titanium particles. J. Bone Miner. Res 1995; 10:1417–1427.
194. Lohmann CH, Schwartz Z, Koster G, Jahn U, Buchhorn GH, MacDougall MJ, Casasola D, Liu Y, Sylvia VL, Dean DD, Boyan BD. Phagocytosis of wear debris by osteoblasts affects differentiation and local factor production in a manner dependent on particle composition. Biomaterials 2000; 21: 551–561.
195. Collier JP, Bauer TW, Bloebaum RD, Bobyn JD, Cook SD, Galante JO, Harris WH, Head WC, Jasty MJ, Mayor MB, Sumner DR, White-side LA. Results of implant retrieval from postmortem specimens in patients with well-functioning long-term total hip replacement. Clin. Orthop 1992; 274:97–112.
196. Chiba J, Schwendeman LJ, Booth RE, Crossett LS, Rubash HE. A biochemical, histologic, and immunohistologic analysis of membranes obtained from failed cemented and cementless total knee arthroplasty. Clin. Orthop 1994; 299:114–124.
197. Anwar MM, Sugano N, Masuhara K, Kadowaki T, Takaoka K, Ono K. Total hip arthroplasty in the neglected congenital dislocation of the hip. A five- to 14-year follow-up study. Clin. Orthop 1993; 295:127–134.
198. Zicat B, Engh CA, Gokcen E. Patterns of osteolysis around total hip components inserted with and without cement. J. Bone Joint Surg 1995; 77A:432–439.
199. Callaghan JJ, Forest EE, Olejniczak JP, Goetz DD, Johnston RC. Charnley total hip arthroplasty in patients less than fifty years old. A twenty- to twenty-five year follow-up note. J Bone Joint Surg 1999; 81-A:437.

200. Charnley J, Cupic Z. The nine and ten years results of the low-friction arthroplasty of the hip. Clin. Orthop 1973; 95:9–25.
201. Bell RS, Ha'eri GB, Goodman SB, Fornasiers VL. Osteolysis of the ileum associated with a loose acetabular cup following total hip arthroplasty, secondary to foreign body reaction to polyethylene and methyl methacrylate. Skeletal Radiol 1983; 10:201–204.
202. Maguire JK, Coscia MF, Lynch MH. Foreign body reaction to polymeric debris following total hip arthroplasty. Clin. Orthop 1987; 216:213–223.
203. Pazzaglia U, Byers PD. Fractured femoral shaft through an osteolytic lesion resulting from the reaction to a prosthesis. J. Bone Joint Surg 1984; 66B:337–339.
204. Anthony PP, Gie GA, Howie CR, Ling RS. Localized endosteal bone lysis in relation to the femoral components of cemented total arthroplasties. J. Bone Joint Surg 1990; 72B:971–979.
205. Wroblewski BM, Lynch M, Atkinson JR, Dowson D, Isaac GH. External wear of the polyethylene socket in cemented total hip arthroplasty. J. Bone Joint Surg 1987; 69B:61–63.
206. Hubble MJ, Smith EJ. Revision of failed total hip replacement. Br. J. Hosp. Med 1996; 55:432–436.
207. Maloney WJ, Paprosky W, Engh CA, Rubash H. Surgical treatment of pelvic osteolysis. Clin. Orthop 2001; 393:78–84.
208. Harris WH. Wear and periprosthetic osteolysis: the problem. Clin. Orthop 2001; 393:66–70.
209. Smith SE, Estok DM, Harris WH. 20-year experience with cemented primary and conversion total hip arthroplasty using so-called second generation cementing techniques in patients aged 50 years or younger. J. Arthroplasty 2000; 15:263–273.
210. Bloebaum RD, Beeks D, Dorr LD, Savory CG, DuPont JA, Hofmann AA. Complications with hydroxyapatite particulate separation in total hip arthroplasty. Clin. Orthop 1994; 298:19–26.
211. Boss JH, Behar J, Misselevich I, Mendes DG. The histologic features of the interfacial membrane of intramedullary nails. J. Long-Term Eff. Med. Impl 1995; 5:169–183.
212. Mathiesen EB, Lindgren JU, Blomgren GG, Reinholt FP. Metallographic examination of the prosthetic heads revealed structural imperfections and inhomogeneity of the metal. J. Bone Joint Surg 1991; 73B:569–575.
213. Wright TM, Rimnac CM, Faris PM, Bansal M. Analysis of surface damage in retrieved carbon-reinforced and plain polyethylene tibial components from posterior stabilized total knee replacements. J. Bone Joint Surg 1988; 70A:1312–1319.
214. Mendes DG, Soudry M, Levine E, Silbermann M, Shoshan S, Gross U, Shajrawi I, Boss JH. The collagen composition of the carbon fiber composite ligament. An electron microscopical, biochemical, and immunohistological study. J. Long-Term Eff. Med. Impl 1992; 1:305–319.
215. Tompkins GS, Lachiewicz PF, DeMasi R. A prospective study of a titanium femoral component for cemented total hip arthroplasty. J. Arthroplasty 1994; 9:623–630.
216. Berman AT, Avolio A, DelGallo W. Acetabular osteolysis in total hip arthroplasty: prevention and treatment. Orthopedics 1994; 17:963–965.
217. Herberts P, Malchau H. Long-term registration has improved the quality of hip replacement: A review of the Swedish THR Register comparing 160,000 cases. Acta Orthop. Scand 2000; 71: 111–121.
218. Gruen TA, Markolf KL, Amstutz HC. Effects of laminations and blood entrapment on the strength of acrylic bone cement. Clin. Orthop 1976; 119:250–255.
219. Jaffe WL, H, Hawkins CA. Normalized and proportionalized cemented femoral stem survivorship at 15 years. J. Arthroplasty 1999; 14:708–713.
220. Heck DA, Partridge CM, Reuben JD, Lanzer WL, Lewis CG, Keating EM. Prosthetic component failures in hip arthroplasty surgery. J. Arthroplasty 1995; 10:575–580.
221. Ebramzadeh E, Sarmiento A, McKellop HA, Llinas A, Gogan WJ. The cement mantle in total hip arthroplasty. Analysis of long-term radiographic results. J. Bone Joint Surg 1994; 76A:77–87.
222. D'Lima DD, Oishi CS, Petersilge WJ, Colwell CW, Walker RH. 100 cemented versus 100 noncemented stems with comparison of 25 matched pairs. Clin. Orthop 1998; 348:140–148.
223. Gill GS, Mills DM. Long-term follow-up evaluation of 1000 consecutive cemented total knee arthroplasties. Clin. Orthop 1991; 273:66–76.
224. Knight JL, Atwater RD, Guo J. Early failure of the porous coated anatomic cemented unicompartmental knee arthroplasty. J. Arthroplasty 1997; 12:11–20.

225. Nistor L, Blaha JD, Kjellström U, Selvik G. In vivo measurements of relative motion between an uncemented femoral total hip component and the femur by roentgen stereophotogrammetric analysis. Clin. Orthop 1991; 269:220–227.
226. éMjöberg B, Önnerfält R. Early loosening of femoral components after cemented revision. A roentgen stereophotogrammetric study. J. Bone Joint Surg 1992; 74B:721–724.
227. Kärrholm J, Borssén B, Löwenhielm G, Snorrason F. Does early micromotion of femoral stem prostheses matter? 4–7-year stereoradiographic follow-up of 84 cemented prostheses. J. Bone Joint Surg 1994; 76B:912–917.
228. Bundy KJ, Penn RW. The effect of surface preparation on metal/bone cement interfacial strength. J. Biomater. Mater. Res 1987; 21:773–805.
229. Chen CQL, Scott W, Baker TM. Effect of metal surface topography on mechanical bonding at simulated total hip stem–cement interfaces. J. Biomed. Mater. Res., Appl. Biomater 1999; 48: 440–446.
230. Skripitz R, Aspenberg P. Attachment of PMMA cement to bone: Force measurements in rats. Biomaterials 1999; 20:351–356.
231. Verdonschot N, Tanck E, Huiskes RJ. Effect of prosthesis surface roughness on the failure process of cemented hip implants after stem–cement debonding. J. Biomed. Mater. Res 1998; 42:554–559.
232. Arola DD, Yang DT, Stoffel KA. The apparent volume of interdigitation: A new parameter for evaluating the influence of surface topography on mechanical interlock. J. Biomed. Mater. Res., Appl. Biomater 2001; 58:519–524.
233. Haraguchi K, Sugano N, Nishii T, Sakai T, Yoshikawa H, Ohzono K. Influence of polyethylene and femoral head surface quality on wear: a retrieval study. Int. Orthop 2001; 25:29–34.
234. Goldring SR, Roelke M, Glowacki J. Multinucleated cells elicited in response to implants of devitalized bone particles possess receptors for calcitonin. J. Bone Miner Res 1988; 3:117–120.
235. Kadoya Y, Al-Saffar N, Kobayashi A, Revell PA. The expression of osteoclast markers on foreign body giant cells. Bone Miner 1994; 27:85–96.
236. Al-Saffar N, Khwaja HA, Kadoya Y, Revell PA. Assessment of the role of GM-CSF in the cellular transformation and the development of erosive lesions around orthopaedic implants. Am. J. Clin. Pathol 1996; 105:628–639.
237. Kurihara N, Gluck S, Roodman GD. Sequential expression of phenotype markers for osteoclasts during differentiation of precursors for multi-nucleated cells formed in long term human marrow cultures. Endocrinology 1990; 127:3215–3221.
238. Shinar DM, Sato M, Rodan GA. The effect of haemopoietic growth factors on the generation of osteoclast-like cells in mouse bone marrow culture. Endocrinology 1991; 126:1728–1735.
239. Kodama H, Yamasaki A, Nose M, Niida S, Ohgame Y, Abe M, Kumegawa M, Suda T. Congenital osteoclast deficiency in osteopetrotic (op/op) mice is cured by injections of macrophage colony-stimulating factor. J. Exp. Med 1991; 173:269–272.
240. Neale SD, Sabokbar A, Howie DW, Murray DW, Athanasou NA. Macrophage colony-stimulating factor and interleukin-6 release by periprosthetic cells stimulates osteoclast formation and bone resorption. J. Orthop. Res 1999; 17:686–694.
241. Hassan NF, Kamani N, Meszaros MM, Douglas SD. Induction of multinucleated giant cell formation from human blood-derived monocytes by phorbol myristate acetate in in vitro culture. J. Immunol 1989; 143:2179–2184.
242. Kreipe H, Radzun HJ, Rudolph P, Barth J, Hansmann ML, Heidorn K, Parwaresch MR. Multinucleated giant cells generated in vitro: terminally differentiated macrophages with down regulated c-fms expression. Am. J. Pathol 1988; 130:232–243.
243. McInnes A, Rennick DM. Interleukin-4 induces cultured monocytes/macrophages to form giant multinucleated cells. J Exp Med 1988; 167:598–611.
244. McNally AK, Anderson JM. Interleukin-4 induces foreign body giant cells from human monocytes/macrophages. Differential lymphokine regulation of macrophage fusion lead to morphological variants of multinucleated giant cells. Am. J. Pathol 1995; 147:1487–1499.
245. Most J, Neumayer HP, Dierich MP. Cytokine-induced generation of multinucleated giant cells in vitro requires interferon-a and expression of LFA-1. Eur. J. Immunol 1990; 20:1661–1667.

246. McNally AK, Anderson JM. Complement C3 participation in monocyte adhesion to different surfaces. Proc. Natl. Acad. Sci 1994; 91:10119–10123.
247. Zhao Q, Agger MP, Fitzpatrick M, Anderson JM, Hiltner A, Stokes K, Urbanski P. Cellular interactions with biomaterials. In vivo cracking of pre-stressed pellethane. J. Biomed. Mater. Res 1990; 24:621–637.
248. Hattersley G, Chambers TJ. Calcitonin receptors as markers for osteoclastic differentiation: correlation between generation of bone-resorptive cells and cells that express calcitonin receptors in mouse bone marrow cultures. Endocrinology 1989; 125:1606–1612.
249. Horton MA, Lewis D, McNulty K, Pringle JA, Chambers TJ. Monoclonal antibodies to osteoclastomas (giant cell bone tumors): definition of osteoclast-specific cellular antigens. Cancer Res 1985; 45:5663–5669.
250. Gattei V, Antonio Bernabei P, Pinto A, Bezzini R, Ringressi A, Formigli L, Tanini A, Attadia V, Brandi ML. Phorbol ester induced osteoclast-like differentiation of a novel human leukemic cell line (FLG29.1). J. Cell Biol 1992; 116:437–447.
251. Matayoshi A, Brown C, DiPersio JF, Haug J, Abu-Amer Y, Liapis H, Kuestner R, Pacific R. Human blood–mobilized hematopoietic precursors differentiate into osteoclasts in the absence of stromal cells. Proc. Natl. Acad. Sci 1996; 93:10785–10790.
252. Formigli L, Zecchi Orlandini S, Benvenuti S, Masi L, Pinto A, Gattei V, Bernabei PA, Gehron Robey P, Collin-Osbody P, Brandi ML. In vitro structural and functional relationships between preosteoclastic and bone endothelial cells: a juxtacrine model for migration and adhesion of osteoclast precursors. J. Cellular Physiol 1995; 162:199–212.
253. Skjodt H, Russell G. Bone cell biology and the regulation of bone turnover. In: Gowen M, ed. Cytokines and Bone Metabolism: CRC Press, 1993:1–70.
254. Tezuka K-I, Nemoto K, Tezuka Y, Sato T, Ikeda Y, Kobori M, Kawashima H, Eguchi H, Hakeda Y, Kumegawa M. Identification of matrix metalloproteinase 9 in rabbit osteoclasts. J. Biol. Chem 1994; 269:15006–15009.
255. Hasegawa T, Hirose T, Seki K, Sano T, Hizawa K. Trans-forming growth factor alpha and CD68 immunoreactivity in giant cell tumours of bone: a study on the nature of stromal and giant cells and their interrelations. J. Pathol 1993; 170:305–310.
256. Davies J, Warwick J, Totty N, Philip R, Helfrich M, Horton M. The osteoclast functional antigen, implicated in the regulation of bone resorption, is biochemically related to the vitronectin receptor. J. Cell Biol 1989; 109:1817–1826.
257. King GJ, Holtrop MF. Action-like filaments in bone cells of cultured mouse calvaria as demonstrated by binding to heavy meromyosin. J. Cell Biol 1975; 66:445–451.
258. Marchisio PC, Cirillo D, Naldini L, Primavera MV, Teti A, Zambonin-Zallone A. Cell–substratum interaction of cultured avian osteoclasts is mediated by specific adhesion structures. J. Cell Biol 1984; 99:1696–1705.
259. Blair HC, Kahn AJ, Crouch EC, Jeffrey JJ, Teitelbaum SL. Isolated osteoclasts resorb the organic and inorganic components of bone. J. Cell Biol 1986; 102:1164–1172.
260. Al-Saffar N, Revell PA. Interleukin-1 production by activated macrophages surrounding loosened orthopaedic implants: a potential role in osteolysis. Br. J. Rheumatol 1994; 33:309–316.
261. Al-Saffar N, Revell PA. Differential expression of transforming growth factor-a and macrophage colony stimulating factor/colony-stimulating factor-1R (c-fms) by multinucleated giant cells involved in pathological bone resorption at the site of orthopaedic implants. J. Orthop. Res 2000; 18: 800–807.
262. Al-Saffar N, Burrows MG, Revell PA. Invasion and erosion of periprosthetic bone by matrix metalloproteinases (MMPs) and angiogenic factors. Sixth World Biomaterials Congress. Hawaii, 2000: 15–20.
263. Vermes C, Roebuck KA, Chandrasekaran R, Dobai JG, Jacobs JJ, Glant TT. Particulate wear debris activates protein tyrosine kinases and nuclear factor kappa B, which down-regulates type 1 collagen synthesis in human osteoblasts. J. Bone Miner. Res 2000; 15:1756–1765.
264. Vermes C, Chandrasekaran R, Jacobs JJ, Galante JO, Roebuck KA, Glant TT. The effects of particulate wear debris, cytokines and growth factors on the function of MG-63 osteoblasts. J. Bone Joint Surg 2001; 83A:201–211.

265. Pioletti DP, Leoni L, Genini D, Takei H, Du P, Corbeil J. Gene expression analysis of osteoblastic cells contacted by orthopaedic implant particles. J. Biomed. Mater. Res 2002; 61:408–420.
266. Yoshida H, Hayashi S-I, Kunisada T, Ogawa M, Nishikawa S, Okamura H, Sudo T, Shultz LD, Nishikawa S-I. The murine mutation osteopetrosis is in the coding region of the macrophage colony stimulating factor gene. Nature 1990; 345:442–444.
267. Takahashi N, Akatsu T, Udagawa N, Sasaki T, Yamaguchi A, Moseley JM, Martin TJ, Suda T. Osteoblastic cells are involved in osteoclast formation. Endocrinology 1988; 123:2600–2602.
268. Udagawa N, Takahashi N, Akatsu T, Sasaki T, Yamaguchi A, Kodama H, Martin TJ, Suda T. The bone marrow derived stromal cell line MC3T3-G2/PA6 and ST2 support osteoclast-like cell differentiation in co-cultures with mouse spleen cells. Endocrinology 1989; 125:1805–1818.
269. Gowen M, Wood DD, Ihrie EJ, McGuire MKB, Graham R, Russell G. An interleukin-1 like factor stimulates bone resorption in vitro. Nature 1983; 306:378–380.
270. Dewhirst FE, Stashenko PP, Mole JE, Tsurumach T. Purification and partial sequence of human osteoclast activating factor: identity with interleukin 1a. J Immunol 1985; 135:2562–2567.
271. Sato K, Fuji Y, Kasono K, Ozawa M, Imamura H, Kanaji Y, Kurosawa H, Tsushima T, Shizume K. Parathyroid hormone related protein and interleukin-1a synergistically stimulate bone resorption in vitro and increase the serum calcium concentration in mice in vivo. Endocrinology 1989; 124: 2172.
272. Pfeilschifter J, Chenu C, Biral A. Interleukin-1 and tumour necrosis factor stimulate the formation of human osteoclast like cells in vitro. J. Bone Miner. Res 1989; 4:113–118.
273. Uy HL, Dallas M, Calland JW, Boyce BF, Mundy GR, Roodman GD. Use of an in vivo model to determine the effects of interleukin-1 on cells at different stages in the osteoclast lineage. J. Bone Miner Res 1995; 10:295–301.
274. Brennan FM, Maini RN, Feldmann M. TNF-a—a pivotal role in rheumatoid arthritis. Br. J .Rheumatol 1992; 31:293–298.
275. Cohen M, Cohen S. Cytokine function. A study in biologic diversity. Am. J. Clin. Pathol 1996; 105:589–598.
276. Nathan C, Sporn M. Cytokines in context. J. Cell Biol 1991; 113:981–986.
277. Horowitz SM, Doty SB, Lane JM, Burstein AH. Studies of the mechanism by which the mechanical failure of polymethylmethacrylate leads to bone resorption. J. Bone Joint Surg 1993; 75A:802–814.
278. Jiranek WA, Machado M, Jasty M, Jevsevar D, Wolfe H, Goldring SR, Goldberg MJ, Harris WH. Production of cytokines around loosened cemented acetabular components. J. Bone Joint Surg 1993; 75A:863–879.
279. AL-Saffar N, Revell PA, Khwaja HA, Bonfield W. Assessment of the role of cytokines in bone resorption in patients with total joint replacement. J. Mater. Sci., Mater. Med 1995; 6:762–767.
280. Herman JH, Sowder WG, Anderson D, Appel AM, Hopson CN. Polymethylmethacrylate-induced release of bone-resorbing factors. J. Bone Joint Surg 1989; 71A:1530–1541.
281. Merkel KD, Erdmann JM, McHugh KP, Abu-Amer Y, Patrick Ross F, Teitelbaum SL. Tumor necrosis factor-a mediates orthopaedic implant osteolysis. Am. J. Pathol 1999; 154:203–210.
282. Clohisy JC, Teitelbaum S, Chen S, Erdmann JM, Abu-Amer Y. Tumor necrosis factor-alpha mediates polymethylmethacrylate particle-induced NF-kappaB activation in osteoclast precursor cells. J. Orthop. Res 2002; 20:174–181.
283. Akira S, Taga T, Kishimoto T. Interleukin-6 in biology and medicine. Adv. Immunol 1993; 54: 1–78.
284. Lowik CW, van der Pluijm G, Bloys H, Hoekman K, Bijvoet OL, Aarden LA, Papapoulos SE. Parathyroid hormone (PTH) and PTH-like protein (PLP) stimulate interleukin-6 production by osteogenic cells: a possible role of interleukin-6 in osteoclastogenesis. Biochem. Biophys. Res. Commun 1989; 162:1546–1552.
285. Kurihara N, Bertolini D, Suda T, Akiyama Y, Roodman GD. IL-6 simulates osteoclast-like multinucleated cell formation in long-term human marrow cultures by inducing IL-1 release. J. Immunol 1990; 144:4226–4230.
286. Agematsu K, Takeuchi S, Ichikawa M, Yabuhara A, Uehara Y, Kawai H, Miyagawa Y, Ishii K, Katsuyama T, Nakahata T. Spontaneous production of interleukin-6 by KI-1-position large-cell anaplastic lymphoma with extensive bone destruction. Blood 1991; 77:2299–2301.

287. Blain TA, Rosier RN, Puzas JE, Looney RJ, Reynolds PR, Reynolds SD, O'Keefe RJ. Increased levels of tumor necrosis factor-α and interleukin-6 protein and messenger RNA in human peripheral blood monocytes due to titanium particles. J Bone Joint Surg [Am] 1996; 78-A:1181–1192.
288. Al-Saffar N, Revell PA, Kobayashi A. Modulation of the phenotypic and functional properties of phagocytic macrophages by wear particles from orthopaedic implants. J. Mater. Sci., Mater. Med 1997; 8:641–648.
289. Roebuck KA, Jacobs JJ, Glant TT. New horizons in orthopaedic research: elucidation of cellular signal transduction pathways. J. Bone Joint Surg 1999; 81A:599–602.
290. Sack U, Kinne RW, Marx T, Heppt P, Bender S, Emmrich F. Interleukin-6 in synovial fluid is closely associated with chronic synovitis in rheumatoid arthritis. Rheumatol. Int 1993; 13:45–51.
291. Sabokbar A, Rushton N. Role of inflammatory mediators and adhesion molecules in the pathogenesis of aseptic loosening in total hip arthroplasties. J. Arthroplasty 1995; 10:810–816.
292. Nivbrant B, Karlsson K, Karrholm J. Cytokine levels in synovial fluid from hips with well-functioning or loose prostheses. J. Bone Joint Surg 1999; 81B:163–166.
293. Horowitz SM, Purdon MA. Mechanisms of cellular recruitment in aseptic loosening of prosthetic joint implants. Calcif. Tissue Int 1995; 57:301–305.
294. Nakashima Y, Sun D-H, Trindade MCD, Maloney WJ, Goodman SB, Schurman DJ, Smith RL. Signalling pathways for tumor necrosis factor-a and interleukin-6 expression in human macrophages exposed to titanium-alloy particulate debris in vitro. J. Bone Joint Surg 1999; 81A:603–615.
295. Clark SC, Kemen R. The human hematopoietic colony-stimulating factors. Science 1987; 236: 1229–1237.
296. Horowitz MC, Jilka R. In: Gowen M, ed. Cytokines and Bone Metabolism. Boca Raton, EFL: CRC Press, 1992:185–227.
297. Lorenzo JA, Sousa SL, Fonseca JM, Hock JM, Medlock ES. Colony-stimulating factors regulate the development of multinucleated osteoclasts from recently replicated cells in vitro. J. Clin. Invest 1987; 80:16–64.
298. Roodman GD, Ibbotson KJ, MacDonald BR, Kuehl TJ, Mundy GR. 1,25(OH)2 vitamin D3 causes formation of multinucleated cells with several osteoclast characteristics in cultures of primate marrow. Proc. Natl. Acad. Sci 1985; 82:8213–8217.
299. Xu JW, Konttinen YT, Waris V, Gomez-Barrena E, Sorsa T, Santavirta S. Granulocyte-macrophage colony-stimulating factor in the aseptic loosening of total hip arthroplasty. J. Orthop. Rheumatol 1996; 9:128–134.
300. Neale SD, Fujikawa Y, Sabokbar A, Gundle R, Murray DW, Graves SE, Howie DW, Athanasou NA. Human bone–derived cells support formation of human osteoclasts from arthroplasty-derived cells in vitro. J. Bone Joint Surg 2000; 82B:892–900.
301. Felix R, Cecchini MG, Fleisch H. Macrophage colony-stimulating factor restores in vivo bone resorption in the op/op osteopetrotic mouse. Endocrinology 1990; 127:2592–2594.
302. Fuller K, Owens JM, Jagger CJ, Wilson A, Moss R, Chambers TJ. Macrophage colony-stimulating factor stimulates survival and chemotactic behavior in isolated osteoclasts. J. Exp. Med 1993; 178: 1733–1744.
303. Bainbridge JA, Revell PA, Al-Saffar N. Costimulatory molecule expression following exposure to orthopaedic implants wear debris. J. Biomed. Mater. Res 2001; 54:328–334.
304. Todaro GJ, Fryling C, De Larco JE. Transforming growth factors produced by certain human tumor cells: Polypeptides that interact with epidermal growth factor receptors. Proc. Natl. Acad. Sci 1980; 77:5258–5262.
305. Marquardt H, Hunkapiller MW, Hood IF, Twardzik DR, De Larco JE, Stephenson JR, Todaro GJ. Transforming growth factors produced by retrovirus-transformed rodent fibroblasts and human melanoma cells:amino acid sequence homology with epidermal growth factor. Proc. Natl. Acad. Sci 1983; 80:4684–4688.
306. Carpenter G. Receptors for epidermal growth factor and other polypeptide mitogens. Annu. Rev. Biochem 1987; 56:881–914.
307. Sporn MB, Todaro GJ. Autocrine secretion and malignant transformation of cells. N. Engl. J. Med 1980; 303:878–880.

308. Anklesaria P, Teixido J, Laiho M, Pierce JH, Greenberger JS, Massagué J. Cell–cell adhesion mediated by binding of membrane-anchored transforming growth factor-α to epidermal growth factor receptors promotes cell proliferation. Proc. Natl. Acad. Sci 1990; 87:3289–3293.
309. Raisz LG, Simmons HA, Sandberg AL, Canalis E. Direct stimulation of bone resorption by epidermal growth factor. Endocrinol 1980; 107:270–273.
310. Stern PH, Krieger N, Nissenson RA, Williams RD, Winkler ME, Derynch R, Strewler GJ. Human transforming growth factor-alpha stimulates bone resorption in vitro. J. Clin. Invest 1985; 76: 2016–2019.
311. Ibbotson KJ, Harrod J, Gowen M, D'Souza S, Smith DD, Winkler ME, Derynck R, Mundy GR. Human recombinant transforming growth factor-a stimulates bone resorption and inhibits formation in vitro. Proc. Natl. Acad. Sci 1986; 83:2228–2232.
312. Takahashi N, MacDonald BR, Hon J, Winkler ME, Derynck R, Mundy GR, Roodman GD. Recombinant human transforming growth factor-alpha stimulates the formation of osteoclast-like cells in long-term human marrow cultures. J. Clin. Invest 1986; 78:894–898.
313. Simonet WS, Lacey DL, Dunstan CR, Kelly M, Chang MS, Luthy R, Nguyen HQ, Wooden S, Bennett L. Osteoprotegerin: a novel secreted protein involved in the regulation of bone density. Cell 1997; 89:309–319.
314. Lacey DL, Timms E, Tan H-L, Kelley MJ, Dunstan CR, Burgess T, Elliott R, Colombero A, Elliott G, Scully S. Osteoprotegerin ligand is a cytokine that regulates osteoclast differentiation and activation. Cell 1989; 93:165–176.
315. Quinn JMW, Joiner C, Triffitt JT, Athanasou NA. PMMA induced inflammatory macrophages resorb bone. J. Bone Joint Surg 1992; 74B:652–658.
316. Quinn JMW, Sabokbar A, Athanasou NA. Cells of the mono-nuclear phagocyte series differentiate into osteoclastic lacunar bone resorbing cells. J. Pathol 1996; 179:106–111.
317. Sabokbar A, Fujikawa Y, Neale S, Murray DW, Athanasou NA. Human arthroplasty derived macrophages differentiate into osteoclastic bone resorbing cells. Ann. Rheum. Dis 1997; 56:414–420.
318. Itonaga I, Sabokbar A, Murray DW, Athanasou NA. Effect of osteoprotegerin and osteoprotegerin ligand on osteoclast formation by arthroplasty membrane derived macrophages. Ann. Rheum. Dis 2000; 59:26–31.
319. Itonaga I, Fujikawa Y, Sabokbar A, Murray DW, Athanasou NA. Rheumatoid arthritis synovial macrophage-osteoclast differentiation is osteoprotegerin ligand-dependent. J. Pathol 2000; 192: 97–104.
320. Haynes DR, Potter AR, Atkins GJ, Findlay DM, Crotti T, Howie DW. Metal particles stimulate expression of regulators of osteoclast development including osteoclast differentiation factor (RANKL/TRANCE), osteoprotegerin and M-CSF. 45th Annual Meeting of the Orthopaedic Research Society. Trans. Orthop. Res. Soc 1999; 24.
321. Solovieva SA, Ceponis A, Konttinen YT, Takagi M, Suda A, Eklund K, Sorsa T, Santavirta S. Mast cells in loosening of totally replaced hips. Clin. Orthop 1996; 322:158–165.
322. Tang L, Jennings TA, Eaton JW. Mast cells mediate acute inflammatory responses to implanted biomaterial. Proc. Natl. Acad. Sci 1998; 95:8841–8846.
323. Al-Saffar N, Iwaki H, Revell PA. Direct activation of mast cells by prosthetic biomaterial particles. J. Mater. Sci.: Mater. Med 1998; 9:849–853.
324. Tang L, Ugarova TP, Plow EF, Eaton JW. Molecular determinants of acute inflammatory responses to biomaterials. J. Clin. Invest 1996; 97:1329–1334.
325. Nathan CF. Neutrophil activation on biological surfaces. Massive secretion of hydrogen peroxide in response to products of macrophages and lymphocytes. J. Clin. Invest 1987; 80:1550–1560.
326. Baghestanian M, Hofbauer R, Kiener HP, Bankl HC, Wimazal F, Willheim M, Scheiner O, Fureder W, Muller MR, Bevec D, Lechner K, Valent P. The c-kit ligand stem cell factor and anti-IgE promote expression of monocyte chemoattractant protein-1 in human lung mast cells. Blood 1997; 90:4438–4449.
327. Al-Saffar N, Revell PA. Demonstration of phospholipase A2 (PLA-2) production by mast cell provides insight into mast cell regulation of inflammatory joint disease. J. Pathol 1997; 181(Suppl): S12A.

328. Graves L, Stechschulte DJ, Morris DC, Lukert BP. Inhibition of mediator release in systemic mastocytosis is associated with reversal of bone changes. J. Bone Miner. Res 1990; 5:1113–1119.
329. Crisp AJ, Chapman CM, Kirkham SE, Schiller AL, Krane SM. Articular mastocytosis in rheumatoid arthritis. Arthritis Rheum 1984; 27:845–851.
330. Fallon MD, Whyte MP, Craige RB, Teitelbaum SL. Mast-cell proliferation in postmenopausal osteoporosis. Calcif. Tissue Int 1983; 35:29–31.
331. Wroblewski BM, Siney PD. Charnley low-friction arthroplasty in the young patient. Clin. Orthop 1992; 285:45–47.
332. Berry DJ, Harmsen WS, Cabanela ME, Morrey BF. Twenty-five-year survivorship of two thousand consecutive primary Charnley total hip replacements. J. Bone Joint Surg 2002; 84A:171–177.
333. önsten I, Carlsson ÅS, Ohlin A, Nilsson JÅ. Migration of acetabular components, inserted with and without cement, in one-stage bilateral hip arthroplasty. A controlled randomized study using roentgen stereophotogram-metric analysis. J. Bone Joint Surg 1994; 76A:185–194.
334. Ingham E, Fisher J. Biological reactions to wear debris in total joint replacement. Proc. Inst. Mech. Eng 2000; 214:21–37.
335. Germain MA, Tipper JL, Ingham E, Makarem R, Wroblewski BM, Fisher J. Analysis of debris from brushing the femoral canal with a plastic brush: A potential cause of loosening in total hip replacement. Proc. Inst. Mech. Eng 1999; 213:503–506.
336. Ramaniraka NA, Rakotomanana LL, Leyvraz P-F. The fixation of the cemented femoral component. Effects of stem stiffness, cement thickness and roughness of the cement-bone surface. J. Bone Joint Surg 2000; 82B:297–303.

15

Ceramic Spine Prostheses

Noboru Hosono, Hironobu Sakaura, and Hideki Yoshikawa
Osaka University Graduate School of Medicine
Osaka, Japan

Tetsuo Ohwada
Osaka Koseinenkin Hospital
Osaka, Japan

Kazuo Yonenobu
National Osaka-Minami Hospital
Osaka, Japan

I. INTRODUCTION

Spine surgery has become more aggressive and technically demanding due to the development of methods for procedures such as extensive resection of tumors and reconstruction of disrupted spines. Consequently, various stabilization instruments and biomaterials for replacing spine structures have been developed; a prototype vertebral prosthesis for replacing destroyed vertebral bodies is a typical example of the latter. An increasing number of patients are seeking palliative treatment for severe pain due to skeletal cancer metastases and spine destruction. Since the 1960s, the quest for immediate pain relief and early return to normal daily life has led to the design of vertebral prostheses that can bear heavy weight, bending, and torsional strain while remaining firmly anchored to adjacent unaffected vertebral bodies with bone cement (polymethylmethacrylate). To protect the spinal cord from heat injury during polymerization of bone cement, metal prostheses have been replaced with alumina ceramic prostheses.

Ceramic lamina prostheses were originally designed to replace lamina structures after extensive laminectomy, which is indicated when intraspinal tumors, widespread tumorous conditions, or degenerative spinal lesions expand over multiple segments and compromise the spinal cord. Curiously, degenerative spinal lesions occur more frequently in Japan than in other countries. Ossification of the posterior longitudinal ligament (mostly in the cervical region) or the yellow ligament (mostly in the thoraco-lumbar transition) is an example of degenerative spinal lesions; such disorders have been extensively studied, and therapeutic measures have been developed to treat them. Degenerative conditions of the spine, such as spondylosis, often require surgical intervention when patients have multisegmental lesions combined with a narrow spinal canal. For treatment of compression myelopathy in cervical or thoracic segments, extensive laminectomy or a newly devised method of laminoplasty is commonly performed in Japan. Consequently, various types of ceramic lamina prostheses have been developed there.

II. BONE SUBSTITUTE FOR INTERBODY FUSION OF SPINE

A. Drawbacks of Autograft and Allograft

Interbody fusion is a universal spinal technique performed at cervical, thoracic and lumbar levels. Disc excision with arthrodesis is indicated when disc herniation, spondylosis, or ossification of a posterior longitudinal ligament compromises the spinal cord and/or nerve roots. Although autogenous bone has been widely used as graft material, because it is the material that most readily fuses to host bone, reported morbidity rates from use of autografts are as high as 21% [1]. Complications include graft fracture, extrusion, collapse, and pseudoarthrosis. Reported rates of collapse of cervical spine grafts and subsequent nonunion range from 0 to 26% [2]. Graft bone harvested from the iliac crest is often too weak to bear the axial load of the spine, especially in patients with generalized disease such as osteoporosis or rheumatoid arthritis. Allografts are strong enough to bear axial loading, but they have only limited osteogenic potential. Zdeblick and Ducker found that, in single-level fusions of cervical spine, there was little difference in radiographic or clinical results between autografts and freeze-dried allografts; however, in two-level fusions, radiographic results were markedly better for autografts [3]. Occult infection of an allograft may predispose a patient to pseudoarthrosis. Also, allografts are not readily available in some parts of the world.

B. Complications of Donor Sites

Complications of donor sites are not uncommon. Whitecloud reported a 20% complication rate at the donor site [4]. In 10% of patients who undergo bone harvesting from the anterior iliac crest, injury to the lateral femoral cutaneous nerve occurs due to normal anatomical variation of the nerve. Injury to the lateral femoral cutaneous nerve results in persistent painful dysesthesia along the anterolateral thigh. Complications such as hematoma and infection of the iliac crest have also been reported.

C. What Is the Best Ceramic for Interbody Fusion?

All the above complications of donor sites can be avoided by using hydroxyapatite ceramics instead of autografts. Hydroxyapatite (HA) ceramics are chemically identical to the inorganic component of bone and have been shown to be optimally biocompatible, nontoxic, nonantigenic, and osteotropic. The strength of hydroxyapatite ceramics against heavy loading is related to their degree of porosity: the greater the porosity, the weaker the ceramic. In contrast, osteoconduction increases with increasing porosity. Many researchers have sought the optimal combination of strength and porosity of HA for use as a bone substitute in interbody fusion. Optimal pore size for in vivo osteoconduction ranges from 150 to 500 μm [5,6].

Thalgott et al. reported the outcome of spinal fusion with coralline HA in 26 patients. All patients attained solid fusion, and there was no evidence of nonunion at the final follow-up. The three-dimensional structure of the HA they used was similar to that of cortical bone. In radiographic analysis, they found two distinct patterns of incorporation: "loss of lucency" and "diffuse amalgamation" [7]. However, compared to coralline HA made from sea coral, HA synthesized from limestone is superior in that it has fewer contaminants and therefore can be formed into larger implants with uniform structure. Below, we discuss our newly developed sintered HA ceramics, which have outstanding osteoconductivity.

III. CERAMIC SPACER FOR CERVICAL LAMINOPLASTY

A. Autograft as a Spacer

Laminoplasty is the gold standard for treatment of multisegmental compression of the spinal cord, especially in patients with a developmentally narrow spinal canal. In laminoplasty, the laminae are elevated like a door to expand the spinal canal and decompress the spinal cord (Fig. 1), whereas laminae are completely removed in laminectomy. Unlike laminectomy, laminoplasty can prevent undesired postoperative kyphosis or epidural membrane formation, which could compromise neurological status. Laminoplasty can be roughly classified into two procedures: the unilateral hinge (open-door) method and the bilateral hinge (mid-sagittal splitting) method. In both procedures, elevation of laminae should be maintained in situ to avoid reclosure and concomitant encroachment on the cord. Some surgeons ligate laminae to the capsule of facet joints on the hinged side, and others put spacers between elevated laminae and lateral masses. Autografts from resected spinous processes are generally used for spacers. However, some patients have small spinous processes from C3 to C7, an area routinely exposed in laminoplasty. Also, it is often complicated and time consuming to mold a resected spinous process so that it effectively anchors at the opened space. Graft harvesting from the iliac crest is rather difficult, partly due to the necessity of a second skin incision on the lower back. Furthermore, if the hinged side gutter is improperly constructed, the autograft could be absorbed by pressure from the lamina or spinous process.

B. Clinical Application of Ceramic Spacers

Hase et al. developed a ceramic spacer for bilateral open laminoplasty. Their spacer has two legs on each side which grip a thin lamina between them. Their postoperative CT scans showed no displacement of elevated laminae or spacers [8]. Nakano et al. used a unique, trapezoid-shaped, ceramic spacer for spinous process–splitting laminoplasty in a dog. In histological examinations, they found bone ingrowth into the ceramic spacer, but the ingrowth was insufficient to attain permanent bonding to the recipient site [9]. However, some authors caution against excessive use of HA spacers. Hoshi et al. measured the length of the spinous process after expansive laminoplasty using autogenous bone and a HA spacer. They observed shortening of the spinous process in 80% of their HA group and 9% of their autograft group. They concluded that shortening of the spinous process may affect the function of the nuchal muscles attached to it, and that the nature of the initial contact between the HA spacer and the bone is a factor in shortening of the spinous process [10].

C. Incorporation of Ceramic Spacer

Tsuzuki et al. used CT to examine incorporation of HA spacers implanted during laminoplasty. They found new bone formation in the lamina–spacer gap in about 60% of the spacers, and incorporation progressed faster in patients with ossification of the posterior longitudinal ligament than in patients with cervical spondylosis [11]. In contrast, there have been only a few reports of histological examination of incorporation of ceramic lamina spacers implanted into the spine. Kokubun et al. were able to retrieve an implanted HA spacer a year after spinous process–splitting laminoplasty. They histologically observed direct bonding of bone to the spacer in three out of six bone–HA interfaces. However, bone ingrowth into the HA surface was not very deep, and bone resorption was evident at the surface of spinous processes that failed to bond to a HA spacer. They speculated that successful bonding to bone can be attained by better fixation of the spacer to the bone [12].

(A)

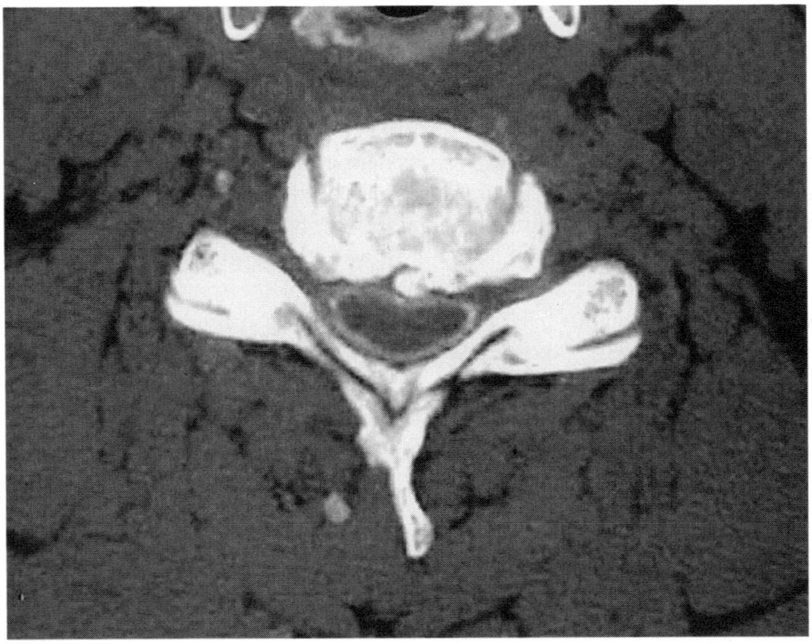

(B)

Figure 1 Ceramic spacer for cervical laminoplasty. (A) The spacer is made of an original ceramic developed by the present authors: interconnected porous calcium hydroxyapatite (IP-CHA). The spacer comes in three lengths: 10, 12, and 14 mm. (B) CT scan with myelography before surgery. The spinal cord is severely constricted by the narrow spinal canal, resulting in tetraparesis of this 49-year-old patient. (C) Lateral radiograph before surgery. (D) CT scan immediately after laminoplasty. A ceramic spacer is inserted between elevated lamina and lateral mass. The spinal canal is fully enlarged to facilitate decompression of the spinal cord. (E) Lateral radiograph after laminoplasty. Anteroposterior canal diameter is increased with ceramic spacers from C3 to C7.

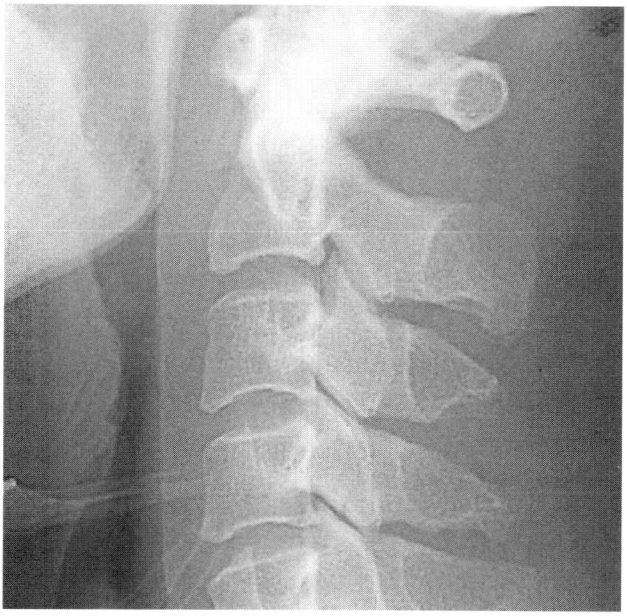

(C)

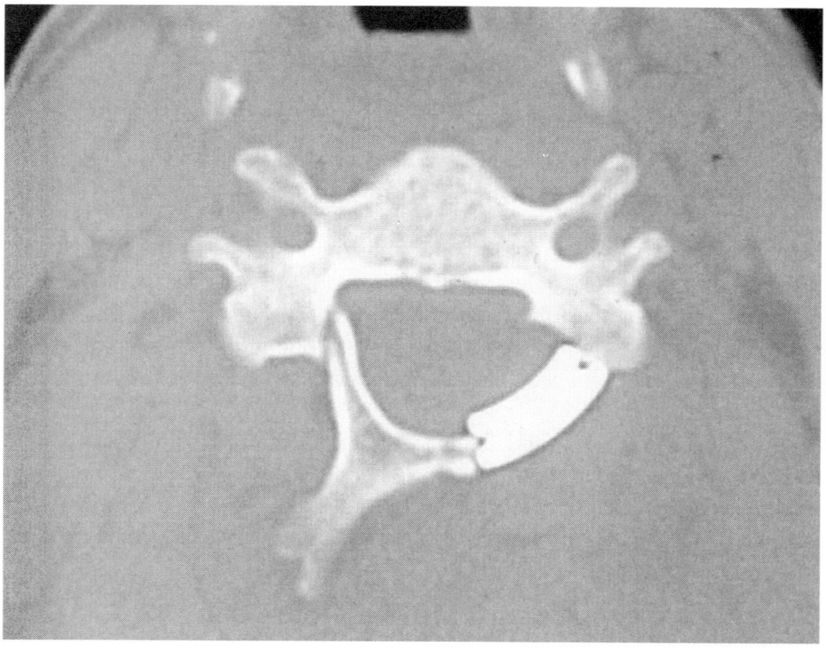

(D)

Figure 1 Continued.

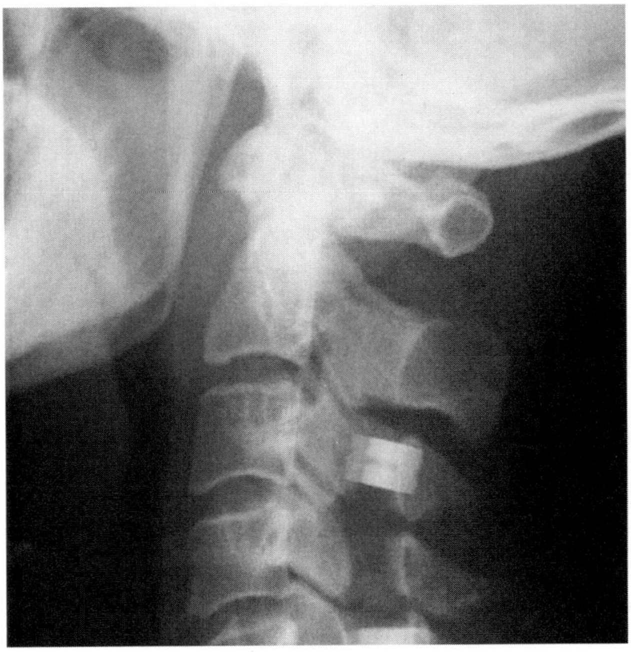

(E)

Figure 1 Continued.

D. Better Design, Better Fixation, and Better Material

Design and fixation technique are key factors in facilitating rigid fixation of the spacer [13]. Takayasu et al. have recently reported a unique method of fixing a HA spacer with titanium screws. In their study, pull-out strength of the HA spacer was significantly greater in the screw fixation group than in the suture fixation group [14]. However, long-term follow-up results of their procedure are required to determine whether there was successful bone ingrowth into the HA spacer.

Wang et al. constructed a unique artificial lamina with a dense interior surface and porous exterior, fabricated from biphasic calcium phosphate ceramic (70% HA, 30% tricalcium phosphate [TCP]). Their animal experiments showed degradation of the TCP phase and bone ingrowth into the porous structure of the implant as early as 1 month after surgery. Within 1 year after surgery, their implant had completely fused with bone [15].

IV. VERTEBRAL PROSTHESES

A. Historical Background

Cancer metastasis to the spine causes spinal cord compression, leading to severe pain and paralysis below the lesion. Until approximately 30 years ago, spinal metastases were considered terminal events of cancer patients, and irradiation or palliative care were the only alternatives for their treatment. Ono, one of our colleagues, found that patients with spinal metastases were not a homogenous group, and that surgical intervention could improve quality of life in some cases. Because most spinal metastases primarily affect vertebral bodies, early replacement surgery was performed using a metal vertebral prosthesis, and later was performed using a ceramic vertebral prosthesis. There have been several reports of the surgical outcome of such cases [16–18]. Replacement of collapsed vertebrae has been attempted using various materials, including autogenous bone grafts. However, incorporation of grafted bone is difficult because ubiquitous use of radiation to prevent local recurrence significantly affects bony union. For patients with a short time to live, until bony fusion is complete, postoperative constraint is imposed for several months. Bone cement is another alternative for reconstruction of collapsed vertebrae: it is bioinert; it can be molded to the desired shape; and its solidity is not affected by irradiation. Use of cement ensures immediate stability. However, it is not a durable material, and fatigue fractures occasionally occur in implanted cement [19].

B. Clinical Application of a Vertebral Prosthesis Developed by the Present Authors

In the late 1960s, we developed a novel ceramic prosthesis (Kyocera Co. Kyoto, Japan) as a solid spacer to replace the affected vertebra. At that time, alumina ceramics were the only available ceramics, and the bond to the adjacent vertebra was generally neither biological nor strong if the ceramic was simply implanted in the cleaned-out space. We use bone cement to anchor this ceramic prosthesis, which has three windows for passage of bone cement: top, bottom, and anterior (Fig. 2).

After resection of the diseased vertebra, a hollow is made at the center of the endplate of both adjacent vertebrae, as an anchor hole for bone cement. A ceramic prosthesis of the appropriate size is then introduced into the intervertebral space, and bone cement is packed with pressure through the anterior window until it fills the prosthesis and then passes out into the anchor holes of the adjacent vertebrae (Figs. 2a, b). When polymerized, the bone cement fixes the prosthesis rigidly between the adjacent vertebrae. The prosthesis protects the spinal cord from heat injury during polymerization of the bone cement. Biomechanical testing has shown that spinal segments implanted with this prosthesis are sufficiently stable. From 1972 to 1993, 90 of these ceramic prostheses were used in 84 patients with spinal metastasis, and the average follow-up period was 26.2 months. The primary tumor was thyroid cancer in 13 patients, breast cancer in 12, multiple myeloma in 8, renal cell cancer in 8, gastrointestinal cancer in 8, and lung cancer in 8. Pain relief was achieved in 94% of patients, motor function improved in 81%, and ambulation recovered in 64%. There were no serious complications associated with the procedure, and the operative benefit lasted until the terminal stage in the vast majority of patients. Because the pores in our ceramic prosthesis do not permit bone ingrowth, there was frequent radiographic loosening of the prosthesis with or without local recurrence of the tumor. However, loosening of the prosthesis did not directly lead to clinical symptoms during the remaining life of the patients, and only one out of the 84 patients required additional posterior surgery to restabilize the prosthesis [18].

C. Other Vertebral Ceramic Prostheses

Matsui et al. developed a modular type of vertebral prosthesis reinforced by a metal plate and screws. Like our prosthesis, their prosthesis is made of alumina ceramic (Kyocera Co. Kyoto, Japan), but theirs has a larger trough on the anterior aspect to facilitate bone cement packing. They reported satisfactory results with this prosthesis, including firm contact between the prosthesis and surrounding tissue on postmortem autopsy. However, metal artifacts caused by the

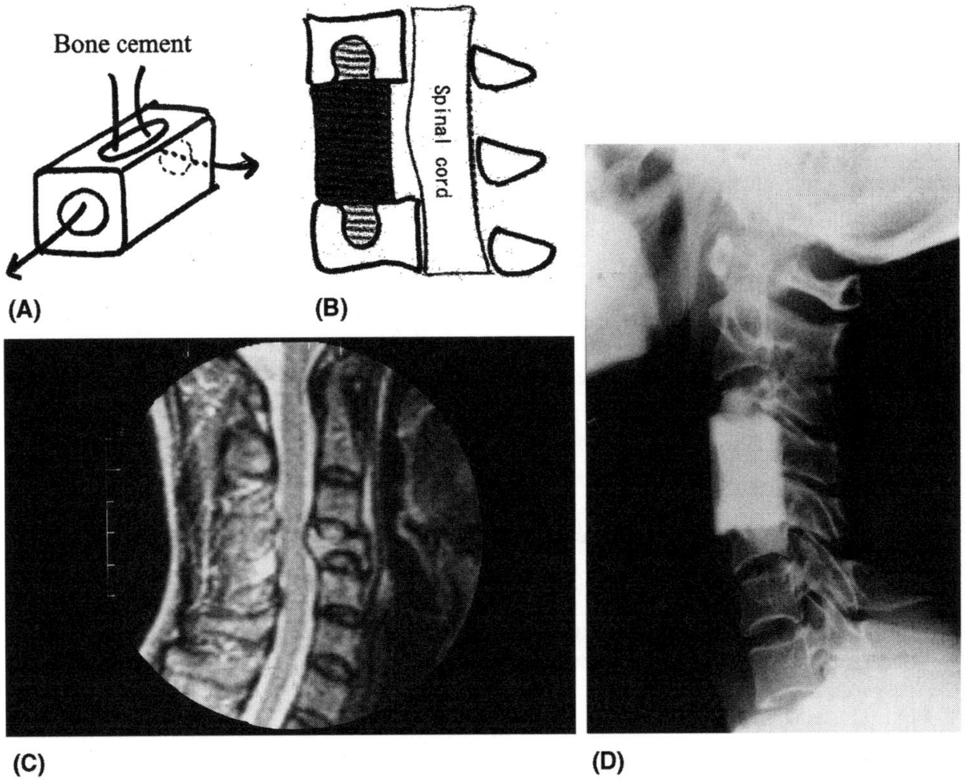

Figure 2 Ceramic vertebral prosthesis developed by the present authors. (A) The prosthesis (Kyocera Co. Kyoto, Japan) is made of alumina ceramics. It has three windows for bone cement passage. (B) Bone cement is packed with pressure through the anterior window until it fills the prosthesis and then passes out into the anchor holes of the adjacent vertebrae. (C) MRI indicates collapse of the fourth cervical vertebra due to metastasis from gastric cancer. The spinal cord is markedly shifted to the posterior. (D) The affected vertebra was replaced with our ceramic prosthesis, and the patient's symptoms were completely relieved.

plate and screws may interfere with detection of local recurrence on MRI or CT scans after surgery. Although inflammatory tissue reaction was not observed histologically, a fibrous membrane formed at the bone–prosthesis interface [20].

Apatite- and wollastonite-containing glass-ceramic (A-WGC) was developed in 1982 at Kyoto University. It is an appropriate replacement material for the vertebral body; its strength is greater than that of HA or human cortical bones. Yamamuro et al. reported the results of implantation of A-WGC prostheses in sheep lumbar spine. Bonding between prosthesis and bone was observed in half the cases, and all surfaces of unbonded prostheses were covered by an apatite layer about 60 μm thick. They speculated that combining this prosthesis with rigid internal fixation would improve its bonding rate [21]. Kaneda et al. implanted A-WGC vertebral prostheses in 22 patients with osteoporotic vertebral collapse, and reported solid fusion in 20 of them; however, histological data were not available. They used their original instrumentation to stabilize the implanted segment, and this may be partly responsible for their high success rate under such unfavorable conditions [22]. Although the high osteoconductivity and high strength of A-WGC make it an ideal bone replacement material, it is no longer commercially available, because its deficit-ridden manufacturer has ceased producing bone substitutes.

V. ARTIFICIAL LAMINA FOR THORACIC SPINE

A. Complications of Laminectomy in Thoracic Spine

Posterior decompression of the spinal cord, or cauda equine, is one of the most common procedures in spine surgery. Removing laminae and spinous processes allows surgeons to enlarge a stenotic spinal canal or remove pathologies from the spinal canal. On the other hand, removal of laminae results in loss of their protective function in the spine and may cause neurological deterioration secondary to hematoma during late postoperative periods, especially when laminae are aggressively resected. Postoperative neurological deterioration is one of the most devastating complications of laminectomy. The thoracic spine has inherent kyphosis, which renders the cord susceptible to direct pressure from paraspinal muscles in the supine position. This type of morbidity is rather rare at the cervical or lumbar spinal level.

B. Artificial Lamina Designed by the Present Authors

To protect the exposed dura, we devised an artificial lamina made of ceramics (Fig. 3). Our original artificial lamina was made of A-WGC, and the current version is made of hydroxyapatite with 30% porosity (Apaceram, Pentax Asahi Optical Co. Ltd., Tokyo). The aspect facing the dura has a slight concave curvature, and small projections on each end help to anchor the ceramic lamina between the edges of the laminectomy. The artificial lamina is tightly sutured with thread at both ends.

C. Clinical Application of Artificial Lamina

Thirty-seven patients have undergone lamina reconstruction using our artificial lamina, which is placed every two segments. CT scans confirmed consolidation of artificial laminae with bone, and no neurological deterioration occurred during the follow-up periods. No surgical complications, including dislodgement, fracture, and sinking of artificial laminae, were encountered. There were no neurological complications associated with hematoma. This implant allows us to easily reconstruct the protective function of the lamina.

VI. THE FUTURE OF CERAMIC SPINE PROSTHESES: THE QUEST FOR SUPERIOR OSTEOCONDUCTION

Hydroxyapatite ceramics have a long history as bone substitutes due to their chemical identity with inorganic bone components. However, increasing the porosity of HA to promote bone ingrowth affects the strength of this graft material. Biomaterial specialists have not yet succeeded in finding the ideal form of HA, with the perfect balance between porosity and strength. Optimal pore size for in vivo osteoconduction is thought to range from 150 to 500 μm. However, for promotion of bone ingrowth, interconnection of pores is a more important factor than pore size itself. We have developed a novel sintered HA ceramic that we call interconnected porous calcium hydroxyapatite (IP-CHA), which is fabricated by mixing two crosslinking agents using a foam-gel technique during the production process (Fig. 1a). The foam-gel technique solidifies the foamed slurry almost immediately, and creates a uniform spongy structure throughout the material. We found 3-mm-deep bone ingrowth into the surface of this new HA ceramic after implanting it into rabbit femurs, whereas Ayers et al. found a bone ingrowth depth of only 300 μm after implantation of the same material in humans [23,24,25]. We believe that the HA ceramics we have developed provide the best available osteoconduction in vivo.

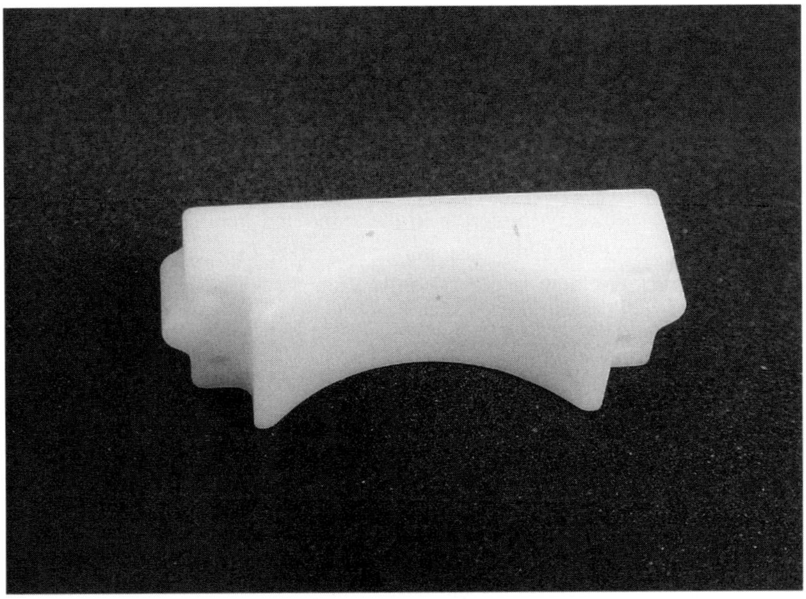

(A)

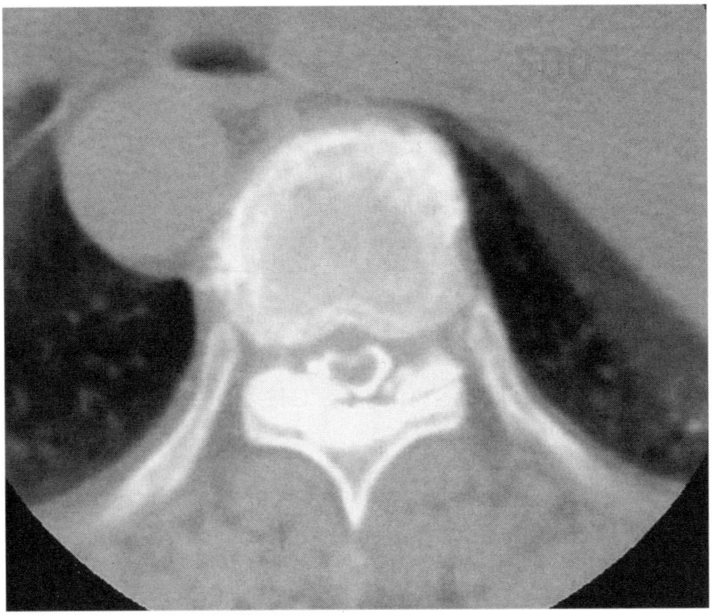

(B)

Figure 3 Artificial lamina. (A) This is an older type of artificial lamina made from apatite- and wollastonite-cotaining glass-ceramic (A-WGC). It comes in four lengths: 24, 26, 28, and 30 mm. It is mainly applied at thoracic or lumbar spine levels after laminectomy. Because A-WGC is not currently available, we used an artificial lamina (of the same shape and sizes) that we constructed from hydroxyapatite (Apaceram, Pentax Asahi Optical Co. Ltd., Tokyo, Japan). (B) CT scan with myelography before surgery. Ossification of yellow ligament compresses the spinal cord in this 72-year-old female. (C) Anteroposterior radiograph after laminectomy. (D) MRI after surgery. On MRI, ceramic spacers are generally delineated as nonsignal areas without halation artifacts, facilitating close examination of the spinal cord after surgery. (E) CT scan 4 years after implantation of an artificial lamina. Incorporation of the ceramic is complete.

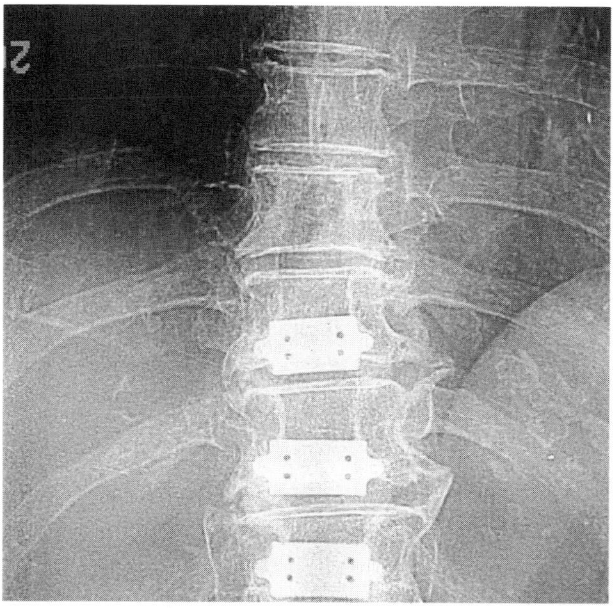

(C)

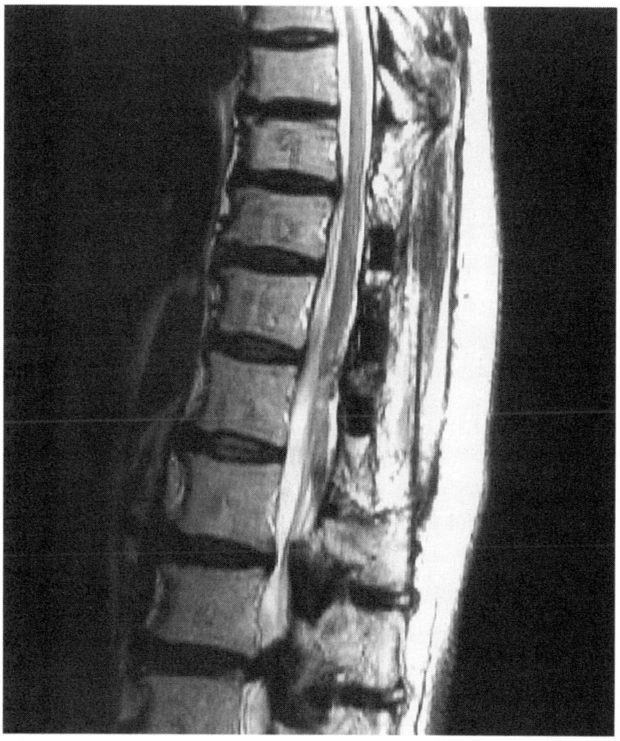

(D)

Figure 3 Continued.

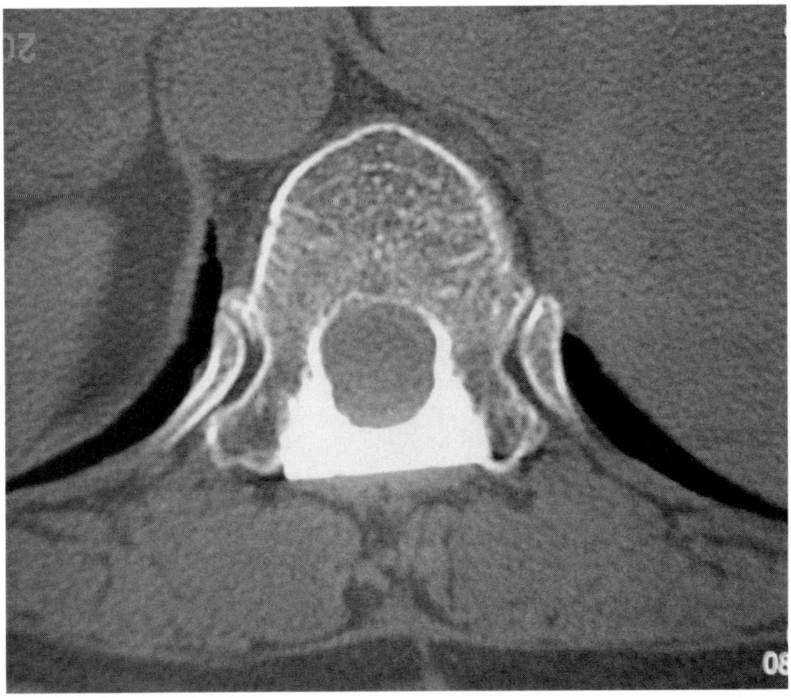

(E)

Figure 3 Continued.

REFERENCES

1. DePalma AF, Rothman LH, Lewinnek GE, Kanale TA. Anterior interbody fusion for severe cervical disc degeneration. Surg Gynecol Obstet 1972; 134:755–758.
2. Aronson NI. The management of soft disc protrusion using the Smith-Robinson approach. Clin Neurosurg 1973; 20:253–258.
3. Zdeblick TA, Ducker TB. The use of freeze-dried allograft bone for anterior cervical fusions. Spine 1991; 16:726–729.
4. Whitecloud TS. Complications of anterior cervical fusion. Inst Course Lect 1978; 27:223–227.
5. Hulbert SF, Morrison SJ, Klawitter JJ. Tissue reaction to three ceramics of porous and non-porous structures. J Biomed Mater Res 1972; 6:347–374.
6. Flatley TJ, Lynch KL, Benson M. Tissue response to implants of calcium phosphate ceramic in the rabbit spine. Clin Orthop 1983; 179:246–252.
7. Thalgott JS, Fritts K, Giuffre JM, Timlin M. Anterior interbody fusion of the cervical spine with coralline hydroxyapatite. Spine 1999; 24:1295–1299.
8. Hase H, Watanabe T, Hirasawa Y, Hashimoto H, Miyamoto T, Chatani K, Kageyama N, Mikami Y. Bilateral open laminoplasty using ceramic laminas for cervical myelopathy. Spine 1991; 16: 1269–1276.
9. Nakano K, Harata S, Suetsuna F, Araki T, Itoh J. Spinous process–splitting laminoplasty using hydroxyapatite spinous process spacer. Spine 1992; 17:S41–S43.
10. Hoshi K, Kurokawa T, Nakamura K, Hoshino Y, Saita K, Miyoshi K. Expansive cervical laminoplasties—Observations on comparative changes in spinous process lengths following longitudinal laminal divisions using autogenous bone or hydroxyapatite spacers. Spinal Cord 1996; 34:725–728.

11. Tsuzuki N, Abe R, Saiki K, Iizuka T. Tension-band laminoplasty of the cervical spine. International Orthopaedics (SICOT) 1996; 20:275–284.
12. Kokubun S, Kashimoto O, Tanaka Y. Histological verification of bone bonding and ingrowth into porous hydroxyapatite spinous process spacer for cervical laminoplasty. Tohoku J Exp Med 1994; 173:337–344.
13. Hirabayashi S, Kumano K. Contact of hydroxyapatite spacers with split spinous process in double-door laminoplasty for cervical myelopathy. J Orthop Sci 1999; 4:264–268.
14. Takayasu M, Takagi T, Nishizawa T, Osuka K, Nakajima T, Yoshida J. Bilateral open-door cervical expansive laminoplasty with hydroxyapatite spacers and titanium screws. J Neurosurg (Spine I) 2002; 96:22–28.
15. Wang J, Chen W, Li Y, Fan S, Weng J, Zhang X. Biological evaluation of biphasic calcium phosphate ceramic vertebral laminae. Biomaterials 1998; 19:1387–1392.
16. Ono K, Tada K. Metal prosthesis of the cervical vertebra. J Neurosurg 1975; 42:562–566.
17. Ono K, Yonenobu K, Ebara S. Prosthetic replacement surgery for cervical spine metastasis. Spine 1988; 13:817–822.
18. Hosono N, Yonenobu K, Fuji T, Ebara S, Yamashita K, Ono K. Vertebral body replacement with a ceramic prosthesis for metastatic spinal tumors. Spine 1995; 20:2454–2462.
19. McAfee PC. Failure of methylmethacrylate stabilization of the spine. In The Cervical Spine. 2nd ed. (Sherk HH, ed, 1989:838–850.
20. Matsui H, Tatezaki S, Tsuji H. Ceramic vertebral body replacement for metastatic spine tumors. J Spinal Disord 1994; 7:248–254.
21. Yamamuro T, Shikata J, Okumura H, Kitsugi T, Kakutani Y, Matsui T, Kokubo T. Replacement of the lumbar vertebrae of sheep with ceramic prostheses. J Bone Joint Surg[Br] 1990; 72-B:889–893.
22. Kaneda K, Asano S, Hashimoto T, Satoh S, Fujiya M. The treatment of osteoporotic-posttraumatic vertebral collapse using the Kaneda device and a bioactive ceramic vertebral prosthesis. Spine 1992; 17:S295–S303.
23. Tamai N, Myoui A, Tomita T, Nakase T, Tanaka J, Ochi T, Yoshikawa K. Novel hydroxyapatite ceramics with an interconnective porous structure exhibit superior osteoconduction in vivo. J Biomed Mater Res 2001: In press.
24. Myoui A, Furuno M, Araki N, Fujii S, Tomita T, Tamai N, Ochi T, Yoshikawa H. A clinical experience with novel interconnected porous hydroxyapatite ceramics for the treatment of bone defects. Rinshoseikeigeka 2001; 36:1381–1388.
24. Ayers RA, Simske SJ, Nunes CR, Wolford LM. Long-term bone ingrowth and residual microhardness of porous block hydroxyapatite implants in humans. J Oral Maxillofac Surg 1998; 56:1297–1301.

16
Safety Aspects of Alumina and Zirconia Ceramics in Hip Surgery

Günther Heimke*
Clemson University
Clemson, South Carolina, U.S.A.

I. INTRODUCTION

In 1960 Charnley [1] opened the age of low friction arthroplasty with the introduction of polyethelene sockets in articulation with metal balls; in 1970 Boutin [2] commenced the period of low wear arthroplasty by the application of alumina ceramic to hip surgery. This ceramic had been recommended for reconstructive surgery after preliminary studies by Hulbert et al. [3], Kawahara [4], and Sandhaus [5] in the late 1960s. Boutin [6] was the first to replace polyethylene sockets and metal balls by Al_2O_3 ceramic components for total hip joint replacements, soon followed by Langer [7] with partial tibia plateau inserts. Independently, Griss et al. [8] and Mittelmeier et al. [9] initiated extensive compatibility studies which confirmed the bioinertness of this material and indicated the possibility of a marked reduction of all wear-related phenomena and a direct, cement-free anchorage of the acetabular component if the shape of the implant accounts for the remodeling ability under the influence of the stress and strain field created by the implant in the adjacent bony tissue [10]. More than three million alumina ceramic components of different designs have been inserted since, mostly as joint replacements and dental implants. The expected wear reduction in total joint replacements [11,12] as well as the direct anchorage of alumina ceramic dental implants [13] were widely confirmed. The cement-free anchorage of these monolithic ceramic hip sockets, however, was abandoned after about 10 years because of a too high number of late cup migrations, and was replaced by composite sockets consisting of a metal backing providing a stable anchorage in the pelvis and a ceramic inlay for minimizing all wear-related problems (for a review see Ref. 14).

During the 1970s the mechanical strength of zirconia ceramics was improved considerably, reaching a level of more than twice of that of alumina ceramics [15]. It was subsequently considered for implant purposes [16–19], originally in particular for smaller-sized devices as needed, e.g., for Japanese patients. Following the old rule in surgery to sacrifice not more healthy tissue than absolutely necessary, this material was also introduced elsewhere [20–22]. During the last decade the number of implanted zirconia ceramic devices also reached high numbers,

---

* Retired.

on the order of magnitude of half a million. However, a recent set of recalls of zirconia ceramic balls [23] demands additional consideration in order to avoid further damage to all concerned.

II. MATERIALS

Alumina as well as zirconia ceramic belong to the group called oxide ceramics, consisting essentially of pure metal oxides, in contrast to all kinds of silica-based pottery like stoneware, china, and porcelain, the so-called classical ceramics [24]. The metals concerned belong to the least noble ones, thus they have a high binding energy on which the chemical stability and biological compatibility of their oxides is based. Following the in vitro and in vivo studies mentioned above and the extended clinical experience, alumina ceramic has even become the prototype of a biologically inert material.

As the raw materials for these oxide ceramics are fine grained powders, most of the shaping processes well established for the clay-containing classical ceramics cannot be used. Rather, the dry powders of the metal oxides are rendered into a ''green'' body by compression, parts for medical purposes mostly by pressing isostatically. The actual components are worked from the green bodies by turning or grinding, accounting for the shrinking during the transformation of the shaped powder agglomerate into a polycrystalline solid by firing. This kind of densification by heating to between 1200 to 1500°C is also called sintering. During this operation the tiny powder particles interact with each other by several diffusion processes resulting in the increase in density and some grain growth. Since about 1980 the sintering temperatures have been chosen somewhat below the values necessary for reaching the final density in order to confine the grain growth, essentially to values below 2 μm. The final density can be reached by hot isostatic pressing (HIPing) nearly without any additional grain growth. The parts thus made are finished by grinding and polishing where necessary.

A. Crystallographic Phases and Their Transformation

For alumina, the only thermodynamically stable phase below its melting temperature is the alpha phase with the corundum crystal structure. Thus, phase changes are not an issue at all for alumina ceramics. Rather, the grain size and grain size distribution, its homogeneity, and the density are their main structural features [25] on which the mechanical properties rest.

In contrast to alumina ceramic, the very existence of zirconia ceramics as well as their favorable mechanical properties completely rely on their polymorphism and even the metastability of their crystal phases. In order to understand the problems concerned, the concept of crystallographic phases and their transformations will be explained.

Different regular arrangements of identical atoms or ions (crystal lattices, phases) exist because of differences in their energy of formation, their free energy, G. Under given thermodynamic conditions of composition, temperature, and pressure the arrangement with the lowest free energy has the highest chance to survive. If one or even several of the conditions change, another arrangement of the same atoms may have a lower free energy and, consequently, a phase transition will be initiated. In other words, the driving forces for phase changes are the differences of the free energies, ΔG, of the phases concerned. If these differences are large, phase transformations can be expected to occur instantaneously and completely; if they are extremely small, they may be delayed depending on local variations of one or the other or some of the three conditions mentioned above. The terms ''large'' and ''small'' must be seen in relation to the thermal energy, kT (k is Boltzmann constant-1.3807 10^{-23} J/K; T is temperature in K), available at the transition temperature. These differences between the phases allow stable

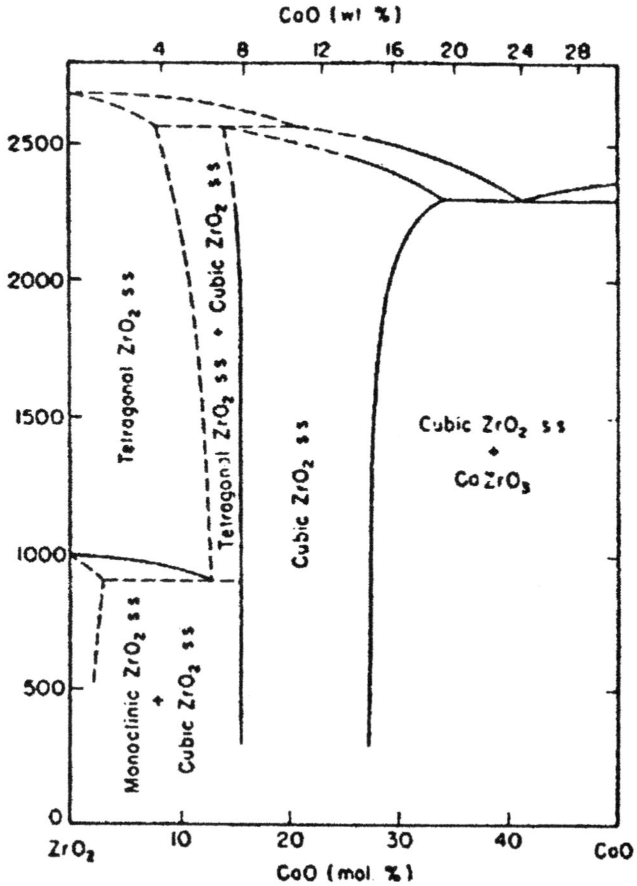

Figure 1 The zirconia/calcia (lime) phase diagram (ss, solid solution). (Adapted from Ref. 24.)

conditions, to be established as represented in the usual phase diagrams as functions of composition and temperature (shown in Fig. 1 for the zirconia/calcia system in a simplified version). The presence of two phases within some regions indicates equal or nearly equal free energies for these phases. In changing the thermodynamic conditions, the volume ratios of these phases may change also. This simple picture is based on the assumption that the state of lowest free energy could always be reached completely. Thus, these are equilibrium diagrams.

However, besides the actual differences of the free energies themselves ΔG, some kind of activation energies, ΔG^*, must be considered. This is demonstrated in Fig. 2 schematically: if one of the thermodynamic conditions is changed, the free energy of the phases present will change also. But this change need not be monotonically raising or falling, rather there might be an intermediate maximum separating two adjacent phases. Thus, if the thermodynamic condition is assumed to be the temperature decreasing along the horizontal axis, phase I will be transformed into phase II only if the thermal energy is sufficiently high, $kT_1 >> \Delta G_1^*$, which usually will be true at high temperatures. In reducing the temperature further, the transformation from phase II into phase III can be delayed or even prevented because here the thermal energy may be relatively small, $kT_2 < \Delta G_2^*$. Thus, phase II will be present below its equilibrium temperature; it will be metastable. Of course, any kind of additional energy, e.g., from compositional fluctuations or pressure applications, can trigger this phase transformation and thus change the structure of the material. It is, therefore, nearly impossible to establish an equilibrium phase diagram for such a material.

This is exactly the situation in the systems of the stabilized zirconias regarding the tetragonal to monoclinic phase transitions and their relation to the cubic solid solution. Figure 3 [26] summarizes different versions of the zirconia/yttria system; Fig. 4 [26,27] contains seven versions of phase distributions for the zirconia-rich portion of this phase diagram. This clearly indicates that it is nearly impossible to come close to any real equilibrium and well established phase boundaries reliably. In contrast to the calcia and magnesia-stabilized systems, the transformation temperature from the tetragonal to the monoclinic phase decreases relatively steeply with increasing yttria content in the zirconia/yttria system, down to about 500°C, thus increasing the sensitivity of this material to any compositional variation as, e.g., can be caused by surface reaction with steam or water. Even a change in the humidity between the heat treatments of two batches can result in slightly different precipitation. This high sensitivity to only slight variations in the production processes are of great importance in the medical applications as we shall see below.

There are differences regarding the details of the phase distributions for the differently stabilized materials. While the calcia-containing ceramics are excluded from medical applications because of the too high reactivity of lime, the MgO-stabilized versions were tested extensively in the 1980s, but interest was lost thereafter. This might be due to the higher firing temperatures necessary in this system and the problems in the continuous reproduction of the required specifications. The larger grain sizes and some unavoidable porosity may also have contributed because of their influence on the wear and tear problems.

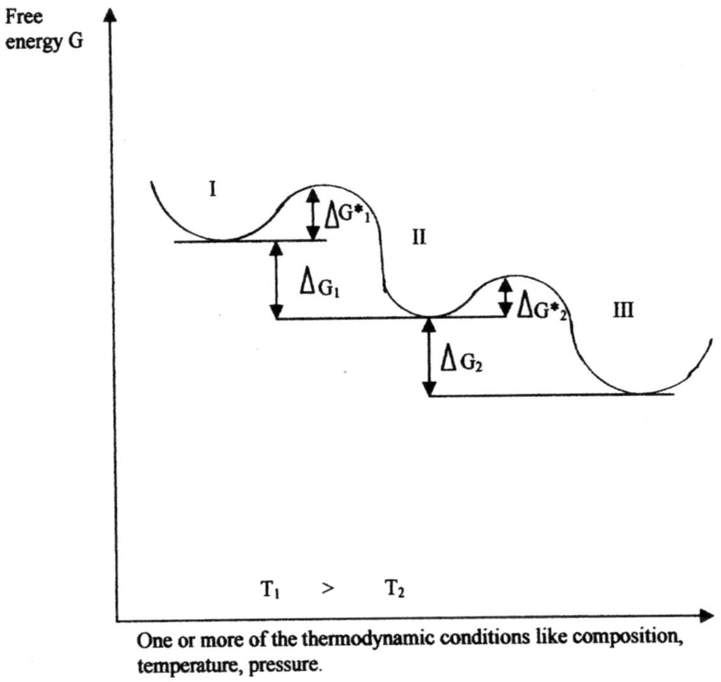

Figure 2 Schematic of the free energy of crystallographic phases as a function of thermodynamically relevant conditions indicating free energy differences of different phases (ΔG_1 and ΔG_2 regarding the phase transitions I to II and II to III) and the activation energies ΔG_1* and ΔG_2*.

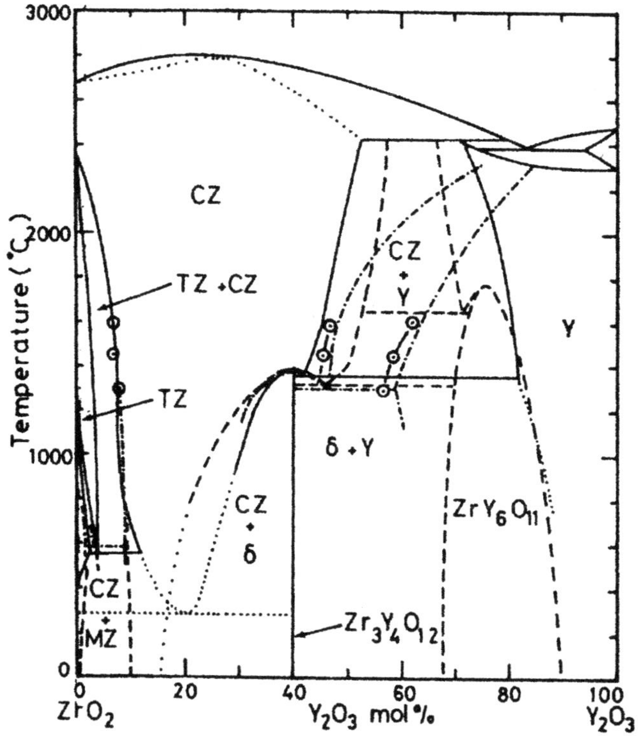

Figure 3 Composition of different phase diagrams of the zirconia–yttria system indicating the problems caused by metastable phases. (Adapted from Ref. 26.)

B. Alumina Ceramics

Alumina—or Al_2O_3—ceramics, as mentioned above, solely consist of densely packed tiny corundum crystals. This crystallographic phase is also called alpha-alumina. Corundum is crystallographically identical with ruby and sapphire, but does not contain the traces of metal ions which render these precious stones so nicely colourful. The crystal lattice is a hexagonally close packed arrangement of the negatively charged oxygen ions within which the Al^{3+} ions occupy two-thirds of the octahedral interstitials. The melting temperature of alumina is 2050°C. On cooling the melt corundum directly solidifies and maintains its structure down to room temperature. Thus, alumina ceramic is a strictly monophased material. Only if molten alumina is cooled very fast, as happens with the droplets in flame or plasma spraying the other, can the less stable beta-phase precipitate.

Corundum is one of the hardest materials, surpassed in nature only by diamond. Single crystals of corundum are used, e.g., as scratch-proof glasses of wristwatches and were used in ophthalmology as the optical cylinders in a particular type of keratoprostheses [28]. In Japan, bone screws (for an early survey see Ref. 29), and dental implants [30] made of corundum single crystals were used in the 1980s.

The possibility for sintering alumina ceramics was discovered in the mid-1920s. Its first large-scale application was in spark plugs, where nearly the whole combination of properties was used: besides the electrical resistivity being nearly equivalent to amber, good thermal con-

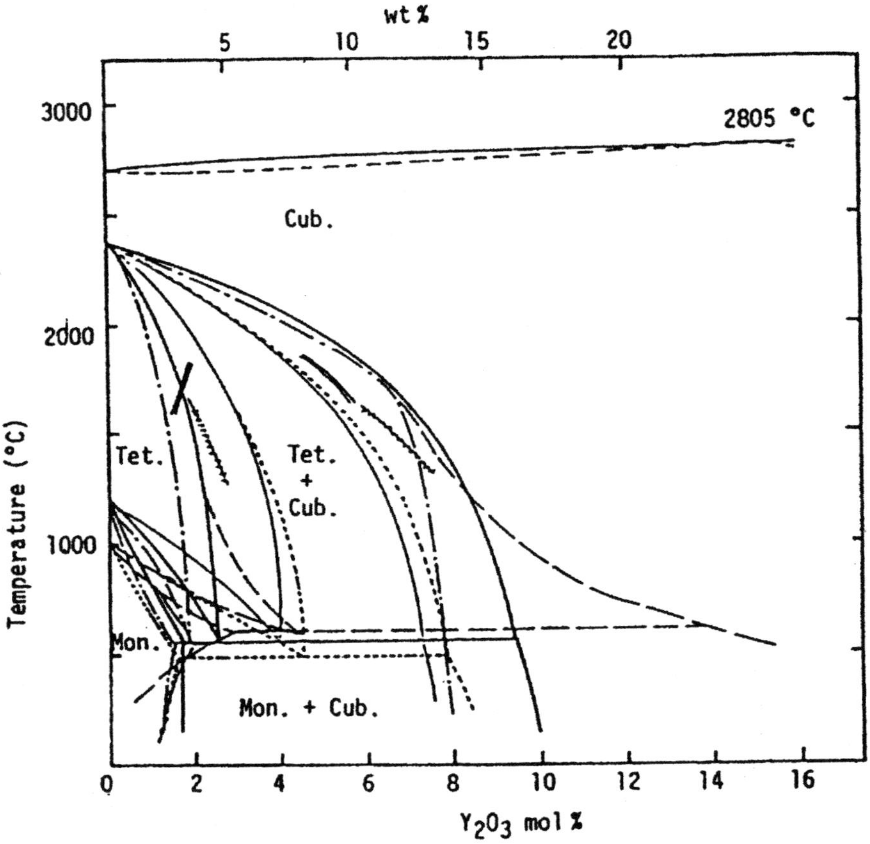

Figure 4 The zirconia-rich portion of the zirconia–yttria phase diagram summarizing seven different versions as an indication of the difficulties of establishing correct phase relations in this system. (Adapted from Ref. 26.)

ductivity and the high mechanical strength. But soon the extremely favorable gliding and tribological properties were also employed in seals of rotating axis through walls of vessels in chemical engineering and similar applications [31]. In sintering these alumina ceramics for engineering applications, the grain growth and the density increase were mostly controlled by small additions of so-called grain growth inhibitors like silica or magnesia. These materials react with the surfaces of the corundum crystals forming thin layers of spinel preventing exaggerated grain growth. The sintering temperatures are on the order of 1300 to 1600°C, depending on the grain sizes and activities of the raw material used and the desired properties. Because of the correlation between grain size, d, and mechanical strength, Φ—$\sigma = c/\sqrt{d}$ (with a constant c)—the grain size and grain size distribution must be kept as small as possible for the medical grade alumina ceramics. This necessitates a closely controlled temperature versus time cycle during the sintering process in order to achieve the required high density without too much grain growth, in particular avoiding the so-called exaggerated grain growth. The control of all processes of grain growth and densification could be improved by the introduction of the additional production step of hot isostatic pressing, the so-called HIPing. It allows the choice of a sintering temperature somewhat lower, thus keeping the grain size reliably smaller at the expense of reaching the final density. The ultimate densification is achieved in the HIPing process without additional grain growth. However, there is one point of view the importance of which will become clear in the discussion of the processes necessary to achieve the required mechanical properties in zirconia ceramics: the production step essential for achieving the required mechanical strength of alumina ceramics is performed at temperatures well above 1300°C. In this temperature range the heat transfer in the furnaces or kilns is essentially by radiation; the contribution of thermal conduction and gas flows is reduced. Thus, homogenous temperature distributions can be relatively easily and reliably established in the furnaces concerned.

The property combinations of the different versions of alumina ceramics used in these engineering applications were optimized for their specific purposes. For medical applications, those developed for mechanical purposes were considered most likely suited for bone replacements and were tested in vitro and in vivo as solid pieces [8] and as powders [32]. Even their carcinogenicity was studied [33]. The material thus selected was used for the first generations of hip and dental implants. In spite of the fact that the first mechanical failures [34] could be traced back nearly completely to handling deviations from the well-documented clinical standard procedures, the design of the components and the quality of the ceramics were reevaluated using newly available information from materials science [35]. A survey on the properties of the medical grade alumina ceramics is given in Table 1.

C. Zirconia Ceramics

As a result of the complicated phase relations of zirconia described above, pure zirconia (ZrO_2) powders cannot be transformed into polycrystalline solids by sintering because of several phase changes occurring on cooling from the molten state to room temperature [36]. In passing down through the melting temperature at 2706°C a cubic phase is formed which changes into a tetragonal phase at 2370°C. On further cooling a monoclinic structure appears between 1170 and about

Table 1 Properties and Standards of the Different Generations of Medical Grade Alumina Ceramics

Property	Unit	First generation(about 1974 to 1980)	Presently used material (e.g., Biolox® forte)	Standard ISO 6474, issue 1994[a]
Density	(g/cm^3)	>3.94	>3.98	3.94
Alumina content	(%)	>99.5	>99.7	≥99.5
Impurities (sum of SiO_2+CaO+Na_2O)	(%)	<0.1	<0.05	≤0.1
Additives	(%)	<0.3	<0.25	≤0.3
Porosity	(%)	[b]	0	
Microstructure (mean grain size)	(μm)	<4.5	<1.8	4.5
Hardness	(HV)	[b]	2300	
Flexural strength	(MPa)	>250[c]	580[d]	≥250[c]
Elastic modulus	(GPa)		>380	
Fracture toughness	(MPa $m^{1/2}$)		4	

[a]Type A.
[b]No values mentioned.
[c]Biaxial test.
[d]Four point bending test.

950°C. This diffuse phase transition is associated with a volume change of up to 9%. Some data on the phases concerned are summarized in Table 2.

If, however, some other oxides like calcia (CaO), magnesia (MgO), ceria (CeO_2), or yttria (Y_2O_3) are added in amounts of about 1 to 15 mol%, a cubic solid solution is formed at high temperatures which can be partially maintained or ''stabilized'' down to room temperature. If the amount of the stabilizing oxides is chosen correctly partially stabilized zirconia (PSZ) ceramics are possible. They consist of this cubic solid solution phase into which tetragonal and/or monoclinic precipitates are embedded, either at grain boundaries or inside the cubic phase grains. This was first realized for calcia additions in the late 1920s [38] and used in refractory materials and carried on essentially by Garvie et al. [39] for mechanical engineering applications in the early 1970s.

These PSZ ceramics can also be obtained by stabilization with yttria and magnesia. In the yttria-stabilized system ceramics can be made with a high concentration of tiny (less than 1 μm) tetragonal zirconia particles (TZP), which are metastable at room temperature [40]. The percentage of particles of this phase, however, strongly depends on their grain sizes, the composition, and on the stresses present. Ceramics of this kind containing 3% yttria can be sintered to full density below 1400°C while maintaining the required fine grain size distribution [37].

Because of its surprisingly high oxygen ion conduction at moderately elevated temperatures (starting at about 350°C) and, thus, acting as a solid electrolyte, this material has also been considered for fuel cells and oxygen partial pressure sensors [41]. A large-scale application is the ''lamda'' oxygen sensor controlling the exhaust gases of our cars.

In mechanical engineering, the considerably improved mechanical strength and fracture toughness of the yttria- and magnesia-stabilized materials beyond that of alumina ceramic raised the interest in these materials for biomedical applications. This strength increase is caused by a similar phenomenon that prevents sintering of pure zirconia into a ceramic. The precipitation process of the tetragonal phase inside the cubic solid solution grains can be controlled to result in finely and evenly dispersed metastable tetragonal particles under some compression, thus preventing the transformation into the thermodynamically more stable and larger monoclinic phase. This prestressing of the cubic matrix grains contributes to the higher flexural strength of these materials. If, however, these same metastable precipitates are touched by the tip of a propagating crack, they are freed from their matrix constrains and can expand and transform into the monoclinic phase, thus absorbing much of the excessive stress and strain energy at the tip of the crack and thereby stop it. This crack-stopping ability of the metastable precipitates is

Table 2 Some Crystallographic Details on the Phases of Pure Zirconia

Phase	Temperature range of stability (°C)	Dimensions of unit cell (Å)	Density (g/cm^3)
Tetragonal	2370–1170	a = 5.094 c = 5.177	6.10
Monoclinic	<1070	a = 5.156 b = 5.191 c = 5.304 β = 98.9°	5.83
Cubic		a = 5.124	6.09

Source: Ref. 37.

a major contribution to the higher fracture toughness of these materials as compared to alumina ceramics. As the well-balanced distribution of the metastable phases is the essential requirement for the stability of this ceramic generally, its mechanical properties, and its fatigue behavior, these materials are extremely sensitive to any variation in their raw materials, in their handling during processing, and in their application.

Because of its attractive mechanical properties and biological compatibility, the zirconia/yttria ceramics and in particular the TZP versions mentioned above have found widespread attention in reconstructive surgery. Properties of this material relevant for medical applications are given in Table 3. Of course, here too the problems of the transformation of the metastable phases are well observed. In these applications, the stability of the material is additionally endangered by many external influences. As the different phases present in this system do have markedly different chemical reactivity with different temperature dependencies, in particular in humid environments during the firing and machining process, but also during their application, not only must all production steps be controlled extremely closely, but also their storage and handling on the way to the consumer and in the final application. Therefore, a large number of studies have focused on the aging behavior of these ceramics. In a recent review [42], 13 publications on the degradation of the modulus of rupture (or equivalent) under different conditions as a function of time have been summarized . In nearly all cases, with the presence of air, water, steam, or Ringer's or saline solutions, mostly at 37°C, one at 100°C, and another at 140°C, a reduction was found after some days to up to 30 months with the exception of four cases in Ringer's solution at 37°C for 90 to 783 days. One group of authors [19] reports about strength increases in Ringer's solution as well as in bone marrow and subcutaneous after 6 to 12 months for a magnesia-stabilized ceramic. This large scattering of results must be regarded as an additional indication of the importance of the phase distributions in these materials.

In the manufacturing of zirconia ceramics, the sintering and actual densification process is separated from the precipitation processes controlling the mechanical properties by a wide temperature gap. While the sintering temperatures are mostly well above 1350°C, the essential phase transformations are activated at much lower temperatures. For the presently most-used ceramic in bioengineering, the yttria/zirconia TZP system, the precipitation processes reach down to the temperature range below 500°C for the compositions mentioned in the standards (Table 4), according to the phase diagram presented in Fig. 4. At this temperature level, heat

Table 3 Comparison of Properties of Alumina, PSZ Ceramics, and the New AMC Ceramic

Property	Alumina ceramic	Mg-stabilized zirconia ceramic	Yttria-stabilized zirconia ceramic	AMC ceramic[a]
Composition (%)	> 99.97 alumina	91 zirconia, 9 magnesia	97 zirconia, 3 yttria	75 alumina, rest Z + Sr, Y,Cr oxides
Density (g/cm^3)	>3.98	5.75	>6	4.37
Bend strength (MPa)	580[b]	450–700	900–1200	1200
Young's molulus (GPa)	380	200	210	350
Fracture toughness ($MPa\ m^{-1/2}$)	4	7–15	7–10	7
Thermal conductivity (WmK^{-1})	30	2	2	
Hardness (HV0.1)	2300	1200	1200	1800

[a]Alumina Matrix Composite ceramic. Also called zirconia and platelet toughened alumina (ZPTA) ceramic [[82]].
[b]In four point bending.
Source: Adapted from Ref. 42.

Table 4 Properties of Y-TZP Ceramics for Medical Applications According to ISO 13356

Property	Value according to standard	Comment
Density (g/cm^3)	≥ 6.00	Influences Young's modulus, all mechanically relevant properties, corrosion resistance, wear, and tear
$ZrO_2 + HfO_2 + Y_2O_3$ (wt %)	$> 99{,}0$	
Hafnia content (wt %)	≤ 5	Influences density
Yttria content (wt %)	4.5–5.4	Influences phase stability, density, and mechanical properties
Contaminations (wt %)	<0.5	Reduces mechanical strength
Alumina (wt %)	<0.5	Influences corrosion resistance
Traces of radioactivity (Bq/kg)	<200	Radiation damage
Average grain size (μm)	≤ 0.6	Influences mechanical properties, phase stability, wear resistance
Four-point bend strength (MPa)	≥ 900	Mechanical strength, safety aspects
Double ring bend strength (MPa)	≥ 599	As above

transfer via radiation plays a minor role only; rather thermal homogeneity is controlled by the flow of gases and by thermal conduction. Both these processes are relatively slow, thus thermal equilibrium will be a much more critical issue in all these annealing treatments. This applies to chamber furnaces as well as to tunnel kilns and to electrically heated as well as to gas-fired units.

III. THE PERFORMANCE OF ALUMINA AND ZIRCONIA CERAMICS IN HARD TISSUE REPLACEMENT

Both these engineering ceramics, of course, share the basic difference of all ceramics as compared to metals—the lack of ductility, or in other words both are brittle materials. Thus, both are relatively sensitive to highly localized mechanical stress application like shocks, e.g., working on them with iron hammers must be strictly avoided. Both materials have also been shown to allow for a considerable reduction of all wear phenomena as compared to the metal against polyethylene combinations [43]; however, this advantage can be used beneficially only if all other possible causes for problems related to the specific properties of the ceramics are closely controlled and thus kept much smaller than all other causes for reoperations.

A. Summary of Severe Clinical Problem Cases with Alumina Ceramics

For the alumina ceramic, having been introduced into surgery in the early 1970s, the introduction or learning phase was followed in the early 1980s by the application of the second generation materials, characterized, e.g., by the addition of the HIPing procedure mentioned above.

The standardization process for the medical grade alumina ceramic had been initiated as early as 1976. The resulting national and international standards have been adjusted since several times, following the progress in materials science and engineering [44]. The presently valid specifications are mentioned in the last column of Table 1. This table also indicates that the most reliable presently offered materials surpass the standard values in all relevant data [45].

In addition to the first alumina ceramic implants in hip surgery consisting of monolithic acetabular cups and femoral balls fixed to metal stems, the combination of alumina ceramic balls with polyethylene sockets [46] contributed to a considerable reduction of all wear-related problems. This latter step with its improvement as compared to the polyethylene against metal combination by about a factor of at least two was a major breakthrough for the large-scale application of ceramics in surgery. However, in order to further take advantage of the much lower wear rates of the alumina against alumina combination already observed clinically with the monolithic sockets [47,48] as compared to the articulation against polyethylene, modular socket systems were introduced during the 1980s consisting of some kind of metal caging into which a ceramic inlay is positioned as the partner for articulation with the ceramic ball [49,14]. The more than 20 years of experience with these many different designs of the two material combinations for the articulating surfaces have fully confirmed the original expectations. Alumina ball heads in combination with polyethylene sockets do not show any signs of wear if explanted after 10 years [50]. Now the wear rate of the polyethylene is down by a factor of 2 to 4 as compared with the articulation against metals [51]. The by far lowest wear rates can be achieved in the sliding combination of alumina ceramic against itself [52,53].

The rates of catastrophic failures of the alumina ceramics during the early period in the 1970s have been documented early: for the group who first started using alumina ceramics in hip arthroplasty in France, a fracture rate of 8% is reported for the period from 1970 to 1976; for the years 1977 to 1989 the rate was 6.5%; for 1990 to 1997 it was 2.5%, and thereafter 0% [54]. The Mannheim team (Frialit alumina ceramic) reported on a series of ceramic femoral head fractures amounting to 6% of the 130 prostheses implanted during the first 6 years [34]. These fractures were concentrated toward the end of this period and confined to those most recently implanted. The Mittelmeier team mentions 25 fractures out of 6862 ceramic components, or 0.36% implanted during the period from October 1974 to December 1993 [55]. The overall failure rate of the Biolox® alumina balls in combination with ceramic socket components as well as with polyethylene cups is 0.02% for up to 20 years, thus including the initial learning phase [56]. In some more detailed data for this material having been given more recently for the generation manufactured until 1992 the overall fracture rate was 0.026%, for the next generation from 1992 to 1995 it was 0.014%, and for the presently manufactured generation it is 0.004% [57,58]. Regarding the causes for the remaining later or more recent failures, a large—although not well defined—number of them must be attributed to influences outside ceramic technology, e.g., to metallic cones that had been machined by unauthorized shops [59].

The failure rates mentioned above are summarized in Fig. 5. These numbers and their distribution in time clearly indicate that the teachings of the initial or learning period of the 1970s had been well understood by all dealing with medical application of alumina ceramics. This, of course, is not only the result of the considerable improvements made possible by progress in materials science and in the more and more detailed understanding of the engineering problems posed by the conical fixation of the ceramic balls on the metal stems, but also it was achieved to a large degree by careful adjustments of the clinical procedure to the necessities of this material. The most important parts of this wealth of experience had already been available when the zirconia ceramics were first considered for total joint replacements.

B. Severe Clinical Problems with Zirconia Ceramics

While zirconia ceramic components had already been considered for reconstructive surgery in the early 1980s [17,18,19] their large-scale application did not commence until toward the end of that decade [20] with several improvements of the materials since. The introduction of the

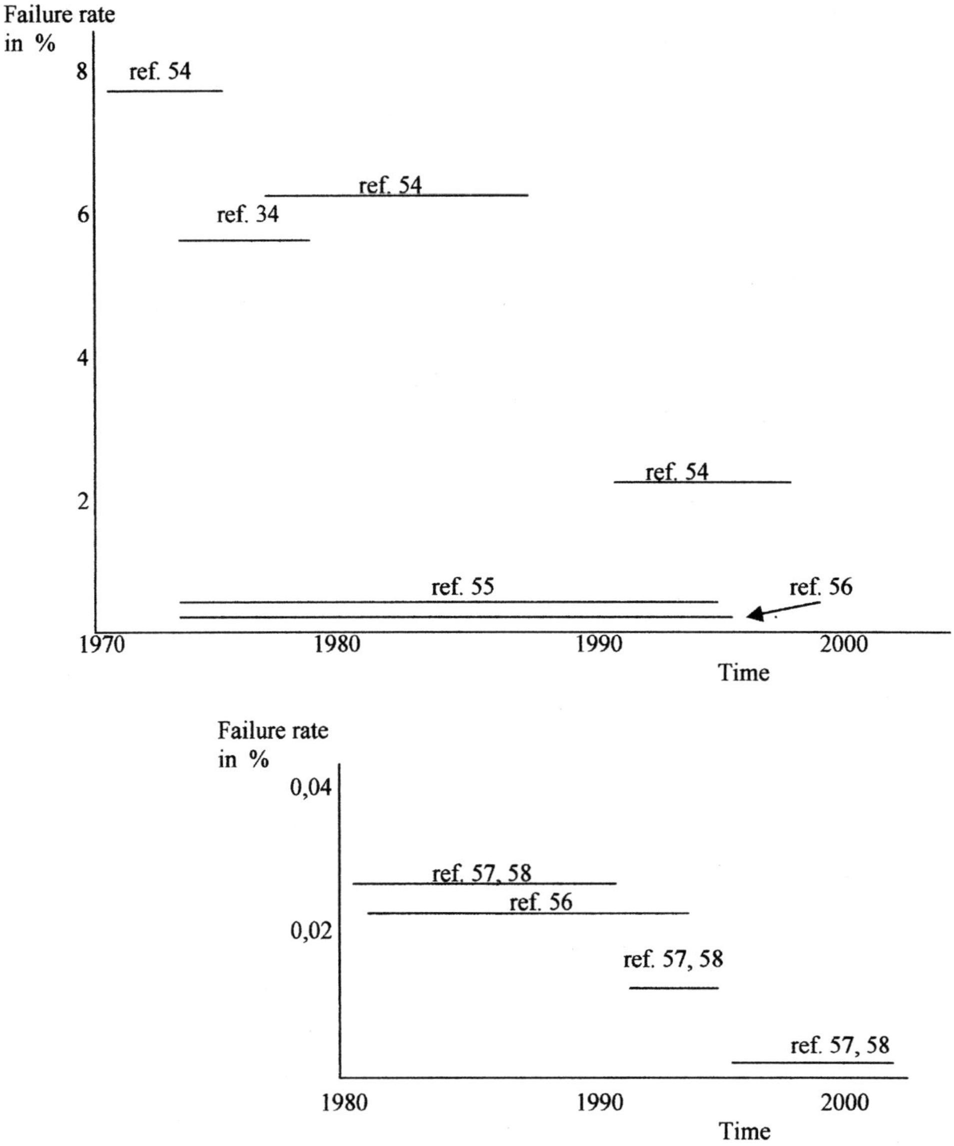

Figure 5 Approximate presentation of the failure rates due to fracture of alumina ceramic components as reported in literature (see text) over time. Values essentially below 0.1% are summarized in the lower portion.

HIPing treatment also contributed to the stability of the material, and a further step was expected from additions of 0.15 to 0.25 wt.% alumina [60].

Of the many stabilised zirconia ceramics the Y-TZP version has recently been covered by an ISO standard (ISO 13356). The essential specifications are listed in Table 4 together with some comments, following a critical evaluation of this standard [61].

When the large-scale application of zirconia ceramics in orthopedic surgery was initiated in about 1990, it could be based on many experiences already available from more than 15 years of use of alumina ceramics in this field. Many clinics were already familiar with the handling of ceramic materials.

By far most of the presently used zirconia ceramics are yttria-stabilized materials (Y-TZP), essentially consisting of tiny (less than 1 μm) grains with the metastable tetragonal crystal structure. As mentioned above, it is this metastability that accounts for the favorable mechanical properties as well as for the sensitivity against slight variations of the heat treatments during production and all outside influences like changes of pressure and humidity. Therefore, the aging behavior of these ceramics has been studied extensively (reviewed in Ref. 42). Their sensitivity to any kind of humidity in the temperature range of about 80°C and above can lead to hydroxide formations, destabilizing the tetragonal crystals and transforming them to the monoclinic modification resulting in a severe strength reduction and, as far as sliding surfaces are concerned, to roughening [62]. This sensitivity also accounts for the recommendation to avoid steam sterilization of zirconia ceramic components [63].

Regarding the wear-related aspects, another property of zirconia ceramic must also be taken into account: its relatively low thermal conductivity as compared to alumina ceramic (2 W/mK vs. 25 W/mK) can raise the maximal temperature in the articulating surfaces. For alumina components temperatures of up to slightly above 42°C have been measured in vivo [64]. For zirconia ceramics with its lower heat conductivity these contact temperatures have been calculated coming close to the hydroxide formation temperature [65]. A still higher temperature was measured inside polyethylene cups articulating against zirconia ceramic balls in simulator experiments (99°C), while with CoCr alloy balls 60°C and with alumina ceramic balls 45°C were found [66]. Of course, all aspects of the wear behavior of zirconia ceramics have been considered carefully for all possible articulating material combinations like against itself, versus alumina ceramic and versus polyethylene [67] (for a summary, also see Ref. 42). As a result, it has been suggested to confine the use of zirconia ceramic in total hip surgery to the combination with polyethylene sockets [68]. However, even here some caution may be justified because of the problems encountered with steam sterilization [69]. Quite some clinical observations justify this caution [70].

But in spite of these considerations, Y-TZP ceramics have found wider applications in hip surgery because of their considerably higher mechanical strength and toughness, both being about two times higher than for alumina ceramics (see Tables 1 and 3), in particular as all burst strength experiments of zirconia ceramic balls on metal tapers [22] point in the same direction.

Regarding in vivo fracture cases of zirconia ceramic components, a recent survey mentions 28 events of this kind out of 280,000 implanted for balls manufactured between 1987 and 1993 [71]. This gives a failure rate of exactly 0.01%. None of these balls were HIPed, most of them not proof tested. There was no correlation of the frequency of occurrence of these fractures with the size and type of ball design. All these balls fractured within 3 years after implantation [72]. In this paper two other reports are mentioned in which two fractures out of 189 [73] or 1.05% in 1995 and ten others (without giving the total number implanted) [74] are reported in 1997. In a recent note the largest European manufacturer mentions 247 fractures out of 2433 balls produced since 1998 [75], amounting to 10.15%. In a recent warning issued by the French Orthopaedic Society [76] a fracture rate of more than 10% is mentioned also for some newly manufactured lots. The total number of fractures mentioned above sums up to 287 cases. As the main manufacturer claims to have sold 480,000 balls since 1987 [72], the overall fracture rate is 0.06%. These failure rates for zirconia ceramic heads are plotted over time in Fig. 6

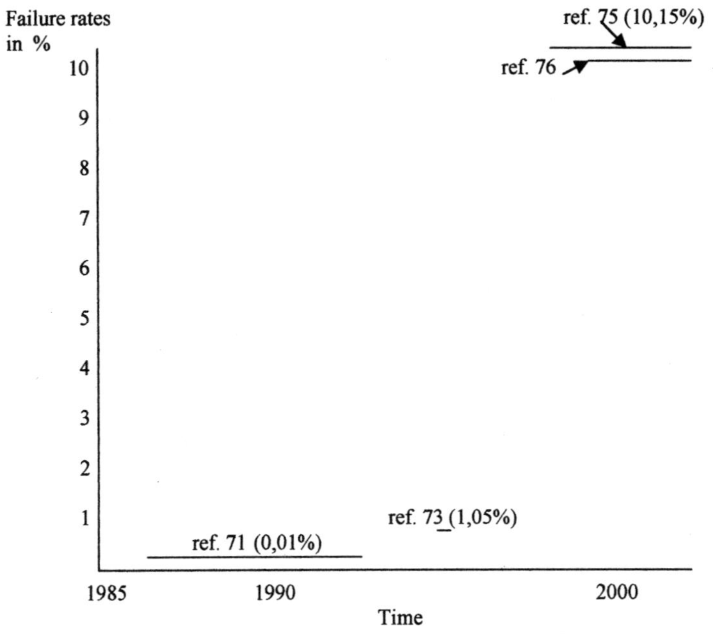

Figure 6 Approximate presentation of the failure rates due to fracture of zirconia ceramic heads reported in literature (see text) over time. In view of the values of Refs. 75 and 76, data below 0.1% are not shown separately.

similarly to the presentation in Fig. 5. The different frequency of fractures of zirconia and alumina ceramic and their time dependence will be discussed below.

IV. EXPERIENCE-BASED COMPARISON OF THE SAFETY OF ALUMINA AND ZIRCONIA CERAMICS IN HIP SURGERY

The distribution of catastrophic failures of zirconia ceramic components over time as evident from Fig. 6 clearly lacks the higher values of the initial or learning period for the early times of application which can be clearly seen in Fig. 5 for alumina ceramic balls. The figures for zirconia ceramic rather increase with time. This difference becomes more striking still in view of the fact that the original alumina ceramic implant systems covered by Refs. 34, 54, and 55 had been taken off the market in the late 1980s. Therefore, the average fracture rates mentioned above for the last decade and valid presently of 0.004% for alumina and 0.06% (not to mention the new 10% figure of refs. 75 and 76) for zirconia ceramic must be taken as a serious estimate of the safety of the materials. Thus, the two materials compare by a factor of more than 10 in favor of the medical grade alumina ceramic.

As mentioned above, when zirconia ceramic balls were introduced in the late 1980s, the essential initial experiences had already been available from alumina ceramic components. As one can assume that reasons for failures not directly associated with the ceramics (e.g., wrongly machined metal cones) are evenly distributed between both these types of ceramics, the cause for those differences must be accounted for by the differences of the two materials themselves. This is the more important as all reports about the mechanical strength and fracture toughness have been in favor of the zirconia ceramics [77]. This applies not only to measurements of original test pieces, but also to extended fatigue studies including those under humid conditions at elevated temperatures [78].

All the recently fractured zirconia ceramic balls are made of the Y-TZP structure consisting, as mentioned above, to more than 90% of small, essentially below 1 μm, metastable tetrahedral crystals. These crystals have been shown [77] to change into the monoclinic phase in the presence of water at slightly elevated temperatures already. This kind of reaction is thought to be the main reason for the surface roughening observed experimentally [62] and shown schematically in Fig. 7A. Here, the regions of transformed crystals are assumed to have a somewhat shallow contour creating a relatively smooth surface undulation. However, obviously the clusters of transformed crystals can as well assume a shape as indicated in Fig. 7B. Such a wedge-shaped cluster of transformed crystals must be regarded as a kind of a pseudocrack, even as some kind of active crack, because of the volume expansion associated with the tetrahedral to monoclinic phase transition in zirconia ceramics. The resulting forces and a possibly applied outside force are indicated by arrows in Fig. 7B. It is obvious that the formation of such pseudocracks must be regarded as a contribution to the probability of failures.

But things are worse still: in front of the tip of a crack, stresses are concentrated. This concentration of mechanical energy is, of course, available for triggering the transformation of any metastable crystal into its thermodynamically stable phase. Thus, if the first grain of tetrahedral crystals in front of the tip of such a pseudocrack transforms, this crack will grow, triggering the next grain into transformation and, consequently, an overcritical crack propagation will be

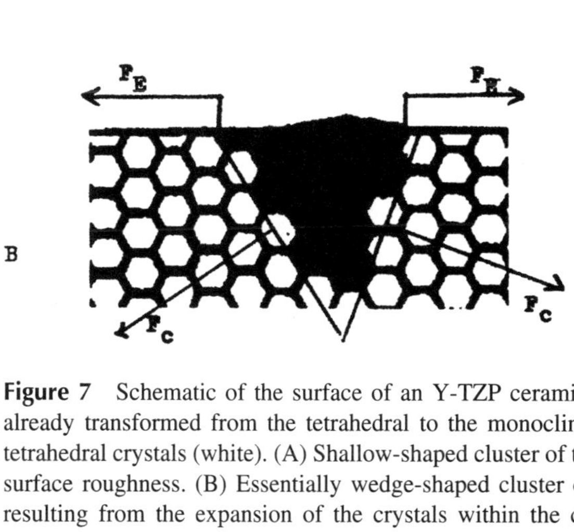

Figure 7 Schematic of the surface of an Y-TZP ceramic and the cross-sections of clusters of crystals already transformed from the tetrahedral to the monoclinic phase (black) inside a matrix consisting of tetrahedral crystals (white). (A) Shallow-shaped cluster of transformed crystals resulting in some increased surface roughness. (B) Essentially wedge-shaped cluster of transformed crystals creating the forces, F_c, resulting from the expansion of the crystals within the cluster on their transformation and having the tendency to widen the pseudocrack thus formed. This can be reinforced by the external tensions, F_E, acting along the surface.

initiated. This cause of catastrophic failure has an additional severe consequence: it renders proof testing worthless. In proof testing, the pieces are taken up to a stress and strain level in the most critically loaded area in their later application, which is just sufficient to initiate growth for all cracks of more than the predetermined critical length. In Y-TZP ceramics, however, cracks of the kind described above do not necessarily exist during the testing procedure. Rather, these pseudocracks can be the result of the actual service conditions. Therefore, proof testing of this kind of ceramic is rather useless.

V. CONCLUSIONS

The evaluation of the occurrence of fractures of alumina and zirconia ceramic components in hip surgery clearly indicates an obvious difference in their time dependence. While the relatively high fracture rates of the initial learning period of alumina ceramics could be avoided for zirconia ceramic components, the fracture rates for the later periods of application of zirconia ceramic remain higher by about one order of magnitude and do not show any decrease. This observation is in contrast to the much higher mechanical strength and toughness of the zirconia ceramics. The discussion of this discrepancy considering the essential structural differences of the two materials leads to the conclusion that the problem of the Y-TZP ceramic is inherent with this material [79]. As, at present at least, no chance for a remedy of this situation nor improvement can be seen, the use of this material in reconstructive surgery should be reconsidered, as has been suggested previously [80,81].

But such a reconsideration does not necessarily indicate that all hopes for improvements of the safety of ceramic components in reconstructive surgery beyond those presently achieved with alumina ceramic must be abandoned. The recently developed alumina matrix precipitation hardened ceramics have reached levels of mechanical strength and toughness on the order of magnitude of the best zirconia ceramics (see last column in Table 3 [82]) and, because of their stable alumina matrix, have the potential of avoiding the corrosion problems of the zirconia ceramics.

REFERENCES

1. Charnley J. Arthroplasty of the hip, a new operation. Lancet 1961:1129–1132.
2. Boutin P. L'alumine et son utilisation en chirugie de la hanche (étude experimentale). Presse Med 1971; 79:14–19.
3. Hulbert SF, Young FA, Mathews RS, Klawitter JJ, Talbert CD, Stelling FH. Potential of ceramic material as permanently implantable skeletal prostheses. J. Biomed. Mater. Res 1970; 4:433–439.
4. Kawahara H. Biological consideration of ceramics. Dent. Eng. (Japan) 1969; 9:28–31.
5. Sandhaus S. Bone implants and drills and taps for bone surgery. Brit Patent No. 1080769, 1967.
6. Boutin P. Arthroplasty totale de la hanch par des prothèse en alumine frittée. Étude expérimentale et première applications cliniques. Rev. Chir. Orthop 1973; 58:229–246.
7. Langer G. Alloplastik am Kniegelenk mit Keramikprothesen. Paper presented at Fall Meeting of Medical and Scientific Society of Friedrich Schiller University Jena. Erfurt, 8th Dec. 1972.
8. Griss P, Andrian-Werburg Hv, Krempien B, Heimke G. Biological activity and histocompatibility of dense Al_2O_3/MgO ceramic implants in rats. J. Biomed. Mater. Res. Symp 1973; 4:453–462.
9. Mittelmeier H. Zementlose Verankerung von Endprothesen nach dem Tragrippenprinzip. Z. Orthop 1974; 112:27–33.

10. Griss P, Silber R, Merkle B, Haehner K, Heimke G, Krempien B. Biomechanically induced tissue reactions after Al_2O_3-ceramic hip joint replacements. Experimental and early clinical results. J. Biomed. Mater. Res. Symp 1976; 7:519–528.
11. Griss P, Heimke G, Werner E, Bleicher J, Jentaschura G. Was bedeutet die Resorption des Calcar femoris nach der Totalendoprothesenoperation der Hüfte? Eine vergleichende Studie an Charnley-Müller und Oxidkeramikendoporthesen (Typ Lindenhof). Arch. Orthop. Traumat. Surg 1978; 92: 225–232.
12. Harms J, Mäusle E. Tissue reaction to ceramic implant material. J. Biomed. Mater. Res 1979; 13: 67–87.
13. Schulte W. Das enossale Tübinger Implantat aus Al_2O_3 (Frialit): Der Entwicklungsstand nach 6 Jahren. Vol. 71: Zahnärztl. Mitt, 1981.
14. Heimke G, Willmann G. Follow-up-study–based wear debris reduction with ceramic-metal–modular hip replacements. In: Wise DL, Trantolo DJ, Lewandrowski K-U, Gresser JD, Cattaneo MV, Yaszemski MJ, eds. Biomaterials Engineering and Devices: Human Applications. Vol. 2. Totowa. NJ: Humana Press, 2000:223–251.
15. Claussen N. Umwandlungsverstärkte keramische Werkstoffe. Z. Werstofftech 1982; 13:138–145.
16. Fujiki H, Shiraishi K, Ikeda S, Iwasaki K, Suzuki R, Kurita S. Characteristics of the zirconia ceramics. In: Oonishi H, Ooi Y(), eds. Othopaedic Ceramic Implants. Vol. 3. Tokyo: Japanese Soc. Orthop. Ceram. Implants, 1983:77–80.
17. Shiraishi K, Fujiki H, Ikeda S, Iwasaki K, Suzuki R, Kurita S. The affinity of zirconia for bone idem, 1983:81–85.
18. Oonishi H, Tsuji E, Nabeshima T, Koike T, Matsuda T, Masaoka S. Studies on an application of zirconia to orthopaedic implants—biological reactions. idem, 1983:87–94.
19. Garvie RC, Urbani C, Kennedy DR, McNeuer JC. Biocompatibility of magnesia partially stabilized (Mg-PSZ) ceramics. J. Mater. Sci 1984; 19:3224–3229.
20. Christel PS. Zirconia: the second generation of ceramics for total hip replacement. Bull. Hosp. Joint Dis. Orthop. Inst 1989; 49:170–177.
21. Willmann G. Zirconia—a medical grade material? In: Ducheyne P, Christiansen D, eds. Bioceramics. Vol. 6. Oxford: Butterworth-Heinemann, 1993:271–276.
22. Cales B, Stefani Y. Yttria-stabilized zirconia for improved orthopedic prostheses. In: Wise DL, Trantolo DJ, Altobelli DE, Yaszemski MJ, Gresser JD, Schwartz AR, eds. Encyclopedic Handbook of Biomaterials and Bioengineering, Part B. Vol. 1. New York: Marcel Dekker, 1995:415–452.
23. Recall of Zirconia Ceramic Femoral Heads for Hip Implants. FDA, Sept. 1, 2001. Recall of Specified Batches of Zirconia Ceramic Femoral Heads for Use in Hip Replacements. MDA DA2001 (05). See also Agence Francaise de Securite Sanitaire des Produits de Sante, dated 16th August 2001.
24. Heimke G. Structural characteristics of metals and ceramics. In: Ducheyne P, Hastings GW, eds. Metal and Ceramic Biomaterials. Vol. I. Boca Raton, FL: CRC Press, 1984:7–61.
25. Willmann G. Ceramic components for total hip arthroplasty. Orthopaedics International Edition 1997; 5:269–277.
26. Yoshimura M. Phase stability of zirconia. Ceramic Bulletin 1988; 67:1950–1955.
27. Stevens R. Zirconia and zirconia ceramics. Twickenham. UK: Magnesium Electron Ltd., 1986.
28. Polack F, Heimke G. Ceramic keratoprostheses. Ophthalmology 1980; 87:693–698.
29. Clarke I, Ebramzadeh E, Oonishi H, Oette PH, Kitamura Y. Clinical performance criteria for sapphire bone screws. In: Oonishi H, Ooi Y, eds. Orthopaedic Ceramic Implants Vol. 3, Japanese Soc. Orthop: Ceramic Implants, 1983:297–307.
30. Kawahara H, Imai K, Kawahara D. Porous polycrystal alumina implant reinforced with single crystal alumina. In: Vincenzini P, ed. Ceramics in Clinical Applications. Amsterdam: Elsevier, 1987:53–264.
31. Ryskewitch E. Oxide Ceramics. New York: Academic Press, 1960.
32. Griss P, Krempien B, Andrian-Werburg Hoon, Heimke G, Fleiner R. Experimentelle Untersuchung zur Gewebsverträglichkeit oxidkeramischer (Al_2O_3) Abriebteilchen. Arch orthop. Unfall-Chir 1973; 76:270–279.
33. Griss P, Werner E, Buchiner R, Büsing C-M, Heimke G. Zur Frage der unspezifischen Sarkomentstehung um Al_2O_3-keramische Implantate. Arch. Orthop. Unfall-Chir 1977; 90:29–40.

34. Griss P, Heimke G. Five years experience with ceramic-metal-composite hip endoprostheses. I. Clinical evaluation. Arch. Orthop. Traumat. Surg 1981; 98:157–164.
35. Heimke G, Griss P. Five years experience with ceramic-metal-composite hip endoprostheses. II. Mechanical evaluations and improvements. Arch. Orthop. Traumat. Surg 1981; 98:165–171.
36. Javaratner M, Yoshimura M, Sämiya S. Subsolidus phase relations in the pseudoternary system ZrO_2-$YO_{1.5}$-$CrO_{1.5}$ in air. J. Am. Ceram. Soc 1984; 67:[11, C240–C242.].
37. Yoshimura M. Phase stability of zirconia. Ceramic Bulletin 1988; 67:1950–1955.
38. Ruff O, Ebert F, Stephan E. Contributions to the ceramics of high refractory materials: II. System zirconia-lime. Z Anorg. Allg. Chem 1929; 180:215–224.
39. Garvie RC, Hanning RH, Pascoe RT. Ceramic steel?. Nature 1975; 258:703–704.
40. Reed PH, Reed JS, Naumann AW. Fabrication and flexural strength of ultra-fine grained yttria-stabilized zirconia. Bull. Am. Ceram. Soc 1976; 55:717.
41. Petanides K, Marincek B, Heimke G. Messen des freien Sauerstoffs. CAV (1973):167–169 and 198–199.
42. Piconi C, Maccauro G. Review: zirconia as a biomaterial. Biomaterials 1999; 20:1–25.
43. Sauer WL, Anthony ME. Predicting the clinical wear performance of orthopedic bearing surfaces. In: Jacobs JJ, Craig TL, eds. Alternative Bearing Surfaces in Total Joint Replacement, ASTM STP 1346: Am. Soc. for Testing and Materials, 1998:1–29.
44. DIN 58 385. Chirurgische Implanate—Keramische Werkstoffe—Aluminiumoxidkeramik. 1979, Beuth Verlag, Berlin, Käln. ASTM F603–83. Specifications for high-purity dense aluminum oxide for surgical implant application. ASTM, PA 1983. ISO 6474. Implants for surgery—Ceramic materials based on alumina. Philadelphia International Organization for Standardization, Geneva, 1981.
45. Willmann G. Production of medical grade alumina. Brt. Ceramic Trans 1995; 94:38–41.
46. Semlitsch M, Lehmann M, Weber H, Dörre E, Willert HG. New prospects for a prolonged functional lifespan of artificial hip joints using the material combination polyethylene/aluminum oxide ceramic/metal. J. Biomed. Mater. Res 1977; 11:537–542.
47. Mahoney OM, Dimon JH. Unsatisfactory results with a ceramic total hip prosthesis. J. Bone Joint Surg 1990; 72-A:663–671.
48. Heimke G, Berg E, Chillag K. Correspondence to the editor. J. Bone Joint Surg 1991; 73-A: 1112–1113.
49. Puhl W, ed. Die Keramikpaarung BIOLOX in der Hüftendoprothetik. Stuttagrt: Enke Verlag, 1996.
50. Willmann G, Kenner K, Zweymüller K. Investigation of retrieved Biolox heads. In: Andersson ÖH, Yli Urpo A, eds. Bioceramics. Vol. 7. Oxford: Butterworth-Heinemann, 1994:377–381.
51. Willmann G, Pfaff HG, Richter HG. Improvement of the safety of ceramic femoral heads for total hip replacement. Biomed. Technol 1995; 40:342–346.
52. Rueda D. Long-term response (10–15 years) to threaded ceramic acetabular cups. In: Puhl W, ed. Die Keramikpaarung BIOLOX in der Hüftendoprothetik. Stuttgart: Enke Verlag, 1996:59–63.
53. Toni A, Sudanese A, Terzi S, Tabarroni M, Calista F, Giunti A. Ceramics in total hip arthroplasty. In: Wise DL, Trantolo DJ, Atobelli DE, Yaszemski MJ, Gresser JD, Schwartz ER, eds. Encyclopedic Handbook of Biomaterials and Bioengineering, Part B: Applications. Vol. 2. New York: Marcel Dekker, 1995:1501–1544.
54. Blanquaert D. Fractures et reprieses. Qu'en est-il des fractures de tetes et inserts en alumine et de leurs reprise? Symposium Ceraver, Résumé des Conférences. Paris, Sept. 2000:65–67.
55. Mittelmeier H, Frisch E, Heisel J, Siebel T, Deimel D. Entwicklung und 20 Jahre klinische Erfahrung mit Keramik-Hüftendoprothesen (bei über 4000 Fällen). In: Puhl W, ed. Die Keramikpaarung BIOLOX in der Hüftendoprothetik. Stuttgart: Enke Verlag, 1996:4–10.
56. Willmann G. Wie sicher sind keramische Kugelkäpfe für Hüftendoprothesen?. Mat.-Wiss. U. Werkstofftech 1996; 27:280–286.
57. Willmann G. Ceramic ball head retrieval data. In: Sedel L, Willmann G, eds. Reliability and Long-term Results of Ceramic in Orthopaedics. Stuttgart: Thieme Verlag, 1999:62–63.
58. Willmann G, Chamier W. The improvements of the material properties of BIOLOX offer benefits for THR. In: Puhl W, ed. Bioceramics in Orthopaedics—New Applications. Stuttgart: Enke Verlag, 1998:19–24.

59. Heimke G. Safety aspects of the fixation of ceramic balls on metal stems. In: Ducheyne P, Christiansen D, eds. Bioceramics. Vol. 6. Oxford: Butterworth-Heinemann, 1993:283–288.
60. Blaise L, Villermaux F, Calès B. Ageing of zirconia: everything you always wanted to know. In: Moroni A, ed. Bioceramics. Vol. 13. Switzerland: Trans Tech Publications, 2001:553–556.
61. Willmann G. Normung von Zirkonoxidkeramik für Hüftendoprothesen. Biomed. Technik 1997; 42: 342–346.
62. Grant KL, Rawlings RD, Sweeney R. Effect of HIPing, stress and surface finish on the environmental degradation of Y-TZP ceramics. J. Mater. Sci: Mater. Med 2001; 12:557–564.
63. Willmann G. Sterilization of bioceramics. Bioceramics-14. 14th International symposium on ceramics in medicine. CA: Palm Springs, Nov 26–30, 2001.
64. Bergmann G, Graichen F, Rohlmann A. Friction induced temperature increase in hip implants. In: Sedel L, Willmann G, eds. Reliability and Long-Term Results of Ceramics in Orthopaedics. Stuttgart: Thieme Verlag, 1999:91–95.
65. Klo H, Willmann G, Woydt M. Temperaturberechnung an den Artikulationsflächen beim künstlichen Hüftgelenk. Mat.-Wiss. U. Werkstofftech 2001; 32:200–210.
66. Lu Z, McKellop H. Frictional heating of bearing materials tested in a hip joint simulator. Proc. Inst Mech Engrs 1997; 211 H:101–108.
67. Willmann G, Früh HJ, Pfaff HG. Wear characteristics of sliding pairs of zirconia (Y-TZP) for hip endoprostheses. Biomaterials 1996; 17:2157–2162.
68. Willmann G. Improving bearing surfaces of artificial joints. Adv. Eng. Mater 2001; 3:135–141.
69. Medical Device Agency, Adverce Incidence Center. Zirconia ceramic heads for modular total hip femoral components: advice to users on sterilization, 1996. London: Notice Medical Device Agency SN 97617, June 1997.
70. Allain J, Le Mouel S, Goutallier D, Voisin MC. Poor eight-year survival of cemented zirconia-polyethylene total hip replacements. J. Bone Joint Surg 1999; 81-B:835–842.
71. Cales B. Fractures des tetes de prothèse de hanch en zircone, Le Journal Francaise de l'Orthopédie. http://www.maitrise-orthop.com/corpusmaitri/orthopaedic/96/cales/cales.shtml. 2001.
72. Villermaux F. Reliability of zirconia hip joint heads. Workshop 7, 6th World Biomaterials Congress. Hawaii: Society for Biomaterials, 2000.
73. Hummer CD, Rothmann RH, Hozack WJ. Catastrophic failure of modular zirconia-ceramic femoral head components after total hip arthroplasty. J. Arthroplasty 1995; 10:848–850.
74. Arnaud JM, Charissoux JL, Setton D, Pecout C, Mechin JF, Dunoyer J. Utilisation de la zircone pour des tetes 22,2 de PTH. A propos de 10 fractures de tete. Presented at the 30th meeting of Société Orthopédique de l'Ouest, Pont-l' Abbé, France, June 1997.
75. SGCA Desmarquest. 2002. Information on recent breakages reported on Prozyr® zircor. June 26th, 2002. http://www.prozyr.com/Pages_UK/Biomedical/positionE.htm.
76. Barthas J. Arbre decisionnel en presence d'un malade porteur d'une tete zircone faisant partie des lots a risque de ruptures. Société Francaise de Chirurgie Orthopédique et Traumatologique, Paris, 25.03.2002.
77. Frangini S, Piconi C, Bartolomeo A, Magnani G. Transformation kinetics of Y-ZTP zirconia ceramics in simulated physiological solution. In: Gianni S, Moroni A, eds. Bioceramics. Vol. 13. Switzerland: Trans Tech Publications, 2001:557–560.
78. Piconi C, Burger W, Richter HG, Cittadini A, Maccauro G, Covacci V, Bruzzese N, Ricci GA, Marmo E. Y-TZP ceramics for artificial joint replacements. Bioceramics 1998; 19:1489–1494.
79. Richter HG, Burger W, Osthues F. Zirconia for medical implants. In: Andersson ÖH, Yli-Urpo A, eds. Bioceramics . Vol. 7: Butterworth-Heinemann, 1994:401–406.
80. Thomson I, Rawlings RD. Mechanical behaviour of zirconia and zirconia-toughened alumina in a simulated body environment. Biomaterials 1990; 11:505–508.
81. Pfaff H-, Willmann G. Stability of Y-TZP zirconia. In Puhl W, ed. Bioceramics in Orthopaedics—New Applications. Stuttgart: Enke Verlag, 1998:29–31.
82. Burger W, Richter HG. High strength and toughness alumina matrix composites by transformation toughening and 'in situ' platelet reinforcement (ZPTA)—the new generation of bioceramics. In: Giannini S, Moroni A, eds. Bioceramics. Vol. 13. Switzerland: Trans Tech Publications, 2001: 545–548.

17

Plasma-Sprayed Hydroxyapatite-Coated and Plasma-Sprayed Titanium-Coated Implants

Y. Yang and J. L. Ong
University of Texas Health Science Center at San Antonio
San Antonio, Texas, U.S.A.

K. Bessho
Kyoto University
Kyoto, Japan

I. INTRODUCTION

Due to its excellent biocompatibility, titanium (Ti) permits good tissue integration [1–3]. However, the success rates and the amount of bone contact to smooth implant surfaces were reported to be lower as compared to rough implant surfaces in low density bone [4]. Since the beginning of 1980s, many researchers have worked on the development of new surface textures in order to improve initial implant stability and bone healing. Popular implant surface modifications have included plasma-spraying of hydroxyapatite (HA) and Ti (TPS).

Many dental and orthopedic implant systems utilized osteoconductive surfaces such as HA coatings to promote implant fixation to bone by direct bonding or osteointegration. In addition, the organic matrix of bone is permeated by HA, which makes up 67% of bone. Increased roughness on TPS implants has been another reason to promote implant fixation through bone ingrowth into the porous structures and provide an alternative to osteconductive surfaces [5]. However, in the past decade, despite many excellent basic science studies and good long-term clinical results, some controversies persist regarding proper application of plasma-sprayed HA coatings and the TPS coatings in dental and orthopedic implants [6–14]. This chapter provides the current research of plasma-sprayed HA and TPS implants used for dental and orthopedic applications, including plasma spraying techniques, in vivo study, application and indication in dental and orthopedic clinics, retrieval studies, and future developments.

II. PLASMA SPRAYING

Of the many techniques used for implant surface modification, plasma spraying is one of the commercially accepted methods for producing HA and TPS coatings. The principle of plasma spraying is shown schematically in Fig. 1 [15]. A DC electric arc is struck between two electrodes, while a stream of gases pass through this arc. This results in an ionized high temperature gas (up to 30,000°C). A large gaseous expansion occurs as a result of the increase in gaseous

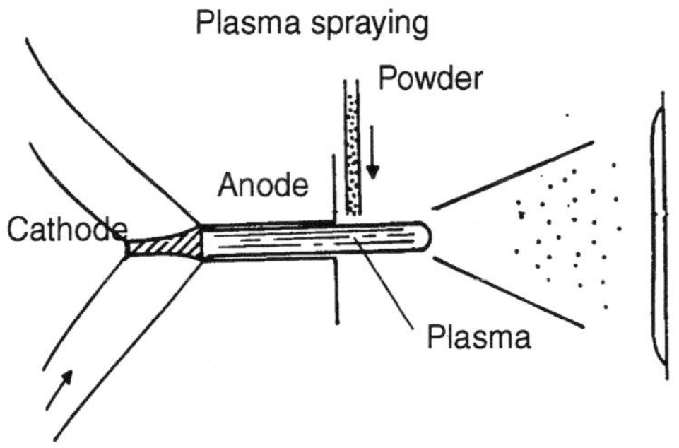

Figure 1 Schematic of the plasma-spray principle. Between two electrodes, a DC power (60–100 V) exists, such that currents of up to several hundred amperes are generated between cathode and anode. The energy released causes a high-speed plasma to be formed, the so-called plasma flame. Ceramic powders injected into this flame can be deposited onto a (metallic) substrate, thus forming a ceramic coating. (From Ref 15.)

temperature, thereby causing the carrier gas stream to pass through the arc at a speed approaching the speed of sound. Coating powder is suspended in the carrier gas stream, which is fed into the plasma flame. The heat content of the plasma flame, and thus the ability to increase the temperature of a particle, depends strongly on the gas used. In addition, the longer a particle resides in the plasma flame, the higher will be the plasma temperature. Ideally, only a thin outer layer of each powder particle gets into the molten plastic state, which unavoidably undergoes some phase transitions. This plastic state, however, is necessary to ensure dense and adhesive coatings. As such, properties of plasma-sprayed coatings depend on the optimal relation between the particle size, the gas used, the speed of the plasma, the distance between plasma nozzle and substrates, and the cooling process for the desired composition and crystallinity.

Operation of DC plasma guns used in plasma spraying is characterized by arc instabilities which affect the plasma jet outside the nozzle. These instabilities are governed by the electrode design, the gas injection mode, the spraying parameters, and the condition of the anode wall [16]. Among the above factors, the electrode design, the gas injection mode, and the condition of the anode wall have been fixed as soon as the plasma-spraying equipment is chosen. In the plasma-spraying process, some basic parameters include power, current, distance between nozzle and substrates, plasma work gas rate, carrier gas rate, and powder feed rate. The properties of coatings are controlled by regulating the above plasma-spraying parameters, whereas spraying time was used to regulate the coating thickness. Table 1 is an example of parameters for plasma-sprayed HA and TPS coatings. Among the above parameters, plasma work gas composition is an important parameter that can be regulated to obtain coatings with the desired properties. First, plasma work gas composition influences the crystallinity of the coating. Argon mixed with hydrogen gives a higher degree of crystallinity. However, without hydrogen, the powder particles cannot enter the gas. This is because the high velocity and viscosity of the argon gas cause the particles to bounce back from the flame instead of entering it. Thus, it is obvious that the plasma-spraying technology allows for the production of either amorphous or (partially) crystalline coatings of HA or TPS on implant surfaces. Second, plasma work gas composition also influences the thickness. For example, the use of nitrogen as the carrier gas resulted in a thicker coating layer as compared to the use of argon.

It is well known that both Ti and apatite are thermodynamically unstable at plasma-spray temperatures. The TPS-coated dental implants are designed for the mandibular symphysis

Table 1 Plasma Spraying Parameters Used for Depositing Different Coatings

Coatings	Power (Kw)	Current (A)	Distance (mm)	Work gas rate (L/min)	Carrier gas rate (L/min)	Powder feed rate (g/min)
HA	~30	500	80~90	Ar: 50 H_2: 3	Ar: 3	~20
TPS	~35	550	120~150	Ar: 50 H_2: 5	Ar: 3	~50
Zirconia	~40	630	70~80	Ar: 50 H_2: 6	Ar: 3	~40

Source: Ref. 26.

anterior to the mental foramen. For most dental implants, the plasma-sprayed layer begins at the implant shoulder. It is from this point that the implant will be completely embedded in jaw bone. During deposition, TPS coatings are usually performed in a vacuum atmosphere in order to avoid the oxidation of metal particles. Babbush et al. provides a sample coating parameter for TPS coatings [17]. The coating was produced using an argon gas flame spraying powdered titanium at 15,000°C onto a core not heated above 220°C. This process gives a porous coating, 0.04 to 0.05 mm thick, on the threaded area of the implant with bond strength of 0.5 kp/mm^2 and no impairment of fatigue resistance. Overgaard et al. demonstrated that porosity of plasma-coated surface led to more rapid healing and substantially less foreign body reaction [18]. Scanning electron microscopy shows that bone is able to grow into these pores without a connective tissue membrane. Additionally, plasma coating results in a sixfold surface enlargement of the anchored implant portion, thereby substantially improving the microanchoring characteristic of the bone and lowering the specific pressure per unit area. However, a very thin TPS coating is not desirable because it cannot provide enough rough surfaces for bone anchoring.

Calcium oxide is usually formed when HA deposition is performed in air. The integrity of the coating produced in air will depend on the amount of calcium oxide formed. The reaction of high concentration of calcium oxide with water usually resulted in poor coating integrity. An increase in calcium oxide formation also occurs when a higher concentration of hydrogen is used as a carrier gas as compared to the use of pure nitrogen as a carrier gas. The use of hydrogen resulted in more flame enthalpy and a lower flame velocity. As such, the particle undergoes more melting and decomposing when a higher concentration of hydrogen is used as a carrier gas.

The deposition of HA coatings on medical devices are between 50 to 100 μm thick. Coatings with thickness greater than 80 μm become brittle, whereas very thin HA coatings may resorb too fast. Bond strength of the HA coating to the metal is difficult to measure, but generally is higher than that of porous-bead coatings. Significant porosity of the HA material lowers the mechanical strength and increases the resorption rate. Good coatings are completely dense and consist of pure HA. The Ca/P ratio in the coating should be as close to that of natural HA ($10/6 = 1.67$). Combinations of HA with tricalcium phosphate have been studied but clinical application of these coatings is rare. This is because tricalcium phosphate has been known to have a much faster degradation rate than hydroxyapatite, and may lead to uncontrolled degradation of the entire coating. Other calcium phosphates such as fluorapatite [19,20], whitlockites [21], or

Table 2 Surface Roughness of Ti Substrate and Coatings

Surface	Average roughness, R_a (μm)	Starting powder size (μm)
HA coating on Ti substrate	4.96 ± 0.43	45~160
TPS coating on Ti substrate	5.65 ± 0.35	100~150
ZrO_2 coating on Ti substrate	3.61 ± 0.39	40~100

Source: Ref. 26.

brushite [22] have been studied as coating material, but no advantages were established to justify their clinical applications [23–25].

In our previous study, the effects of TPS- and zirconia (ZrO_2)-coated Ti substrates on the adhesive, compositional, and structural properties of plasma-sprayed HA coatings were evaluated [26,27]. The coatings evaluated were (1) HA coating on Ti substrate; (2) TPS coating on Ti substrate; (3) ZrO_2 coating on Ti substrate; (4) HA coating on ZrO_2-coated Ti substrate; and (5) HA coating on TPS-coated Ti substrate. Using x-ray diffraction analyses and Fourier transform infrared spectroscopy, apatite-type and α-tricalcium phosphate phases were observed for all HA coatings. Scanning electron analyses indicated that coating surfaces appeared rough and melted. As shown in Table 2, the surface roughness of different coatings was reported to correlate to the size of the starting powder [26]. No significant difference in the Ca/P ratio of HA on Ti substrates and the HA on TPS-coated Ti substrate was observed. However, the Ca/P ratio of HA on ZrO_2-coated Ti substrate was significantly increased. Interfaces between all coatings and substrates were observed to be dense and tightly bound, except for HA coatings on TPS-coated Ti substrate interface. In addition, the adhesive strength of TPS coating to Ti substrate was significantly greater compared to the other groups tested (Table 3). No statistical difference in the adhesive strength of HA coating to Ti substrate and ZrO_2 coating to Ti substrate was observed. The adhesive strengths of HA coating on ZrO_2-coated Ti substrate and HA on TPS-coated Ti substrate were the lowest and less than 30 MPa. At the 0.05 level, significant difference in the adhesive strength between HA coating on Ti substrate and HA coating on ZrO_2-coated Ti substrate was observed ($p=0.02$). Similarly, significant difference in the adhesive strength between the HA coating on Ti substrate and HA coating on TPS-coated Ti substrate was observed ($p=0.005$).

III. IN VIVO STUDY

Numerous initial studies demonstrated direct bone contact with cylindrical screw–shaped Ti implants [28,29], Ti alloy implants [30,31], porous Ti implants [32,33], TPS implants [34,35],

Table 3 Adhesive Strength of Different Coatings to Ti Substrates

Surface	Adhesive strength (MPa)
HA coating on Ti substrate	32.50 ± 3.56
Ti coating on Ti substrate	54.10 ± 1.33
ZrO_2 coating on Ti substrate	32.30 ± 1.80
HA coating on TPS-coated substrate	21.40 ± 1.02
HA coating on ZrO_2-coated Ti substrate	25.10 ± 1.00

Source: Ref. 26.

and HA-coated Ti implants [36,37]. However, these studies primarily involve simple descriptive qualitative studies and lack of comparison investigations. Some of the following studies quantified and compared the actual percentage of bone contact with implants.

A 1-year clinical evaluation of data on 120 alumina-coated Ti cylindrical and Ti blade-type implants placed in the mandible of 30 adult dogs demonstrated healthy bone and gingival tissue responses to 112 of the 120 implants (success rate 93%). However, both the one-stage and two-stage alumina-coated implants exhibited problems for five of the 20 implants placed (success rate 75%) [38]. Using 22 stems, Hacking et al. compared the osseous response to implant surfaces with different roughness in a canine hip model for 6 months [39]. The entire surface of the femoral implant had a 2.9-, 4.2-, or 6.7-μm average surface roughness (R_a) from blasting with 60-, 24-, and 16-grit corundum particles, respectively. No significant difference in bone apposition between the 60-grit stem (31.7%), 24-grit stem (32%), and 16-grit stem (27.9%) was observed [39].

In comparing the percentage of direct bone–implant contact for sandblasted, HA, TPS, and acid-etched implant surfaces, Buser et al. used a pig model [40]. Sandblasted and TPS-coated implants demonstrated 30-40% mean bone contact. The highest percentage of bone–implant contact was observed in HA-coated implants (60–70%), then sandblasted followed by acid etched (HCl/H_2SO_4) implants (50–60%). However, HA-coated implants consistently revealed signs of resorption [40]. In a canine model, Novaes et al. compared the bone–implant contact with smooth (machined) implants, TPS implants, HA implants, and implant surfaces sandblasted with soluble particles (SBM) after 90 days in unloaded condition [41]. Although bone–implant contact for TPS (48.9%) and HA (57.9%) were numerically superior to that for machined implants (41.7%), significant difference was observed only between SBM implants (68.5%) and machined implants ($p<0.05$) [41]. In a rabbit tibia model, no advantage in bone contact percentage was reported when comparing HA surface to Ti [42]. The bone contact to rough HA surface scored similarly to that for TPS surface of similar roughness [42]. In a 2- and 12-week rabbit model, calcium phosphate (CaP) sputtered implants showed a higher amount of bone contact compared to noncoated implants ($p<0.05$) [43]. In another study, total bone apposition ($p=0.01$) and bone ingrowth ($p=0.01$) in the implants coated with HA was significantly greater compared to implants coated with TPS after 4-month implantation in a canine hip model. However, no statistical difference in bone ingrowth or bone apposition was observed between the 50- and 100-μm thickness HA coatings [44].

The above contact length observations were in agreement with our canine study using the mandibles. In addition to bone contact length, we observed no significant difference in mean ultimate interfacial strength (Fig. 2) between HA (2.7 ± 0.3 MPa), TPS (3.1 ± 0.5 MPa), and noncoated Ti implants (2.15 ± 0.2 MPa) at 12 weeks after implant placement (prior to loading). Similarly, after loading for 1 year, no statistical difference in ultimate interfacial strength between the HA (3.2 ± 0.4 MPa), TPS (3.1 ± 0.9 MPa), and Ti implants (3.4 ± 0.1 MPa) was observed. Scanning electron micrographs of the mechanically tested implants at 12 weeks after implantation and at 1 year after loading revealed that fracture between the bone and HA or TPS implants occur at the coating–substrate interface and/or within the coating (Fig. 3). An interesting observation is that there is no clear correlation between bone contact percentage and interfacial strength. Some other investigations also reported that the implants with lower bone contact percentage had higher interface strength [45,46]. These observations suggested that the bonding mechanisms for HA-coated and TPS implants are different, with the HA being more of a chemical bonding and the TPS being more of a mechanical interlocking.

In a 4-week rat model, the push-out strength of the HA-coated implants was higher than that of noncoated implants in 15 ovariectomized and 15 normal control rats ($p<0.0001$). However, the push-out strength of HA-coated implants in the ovariectomised group had a reduction of 40.3%

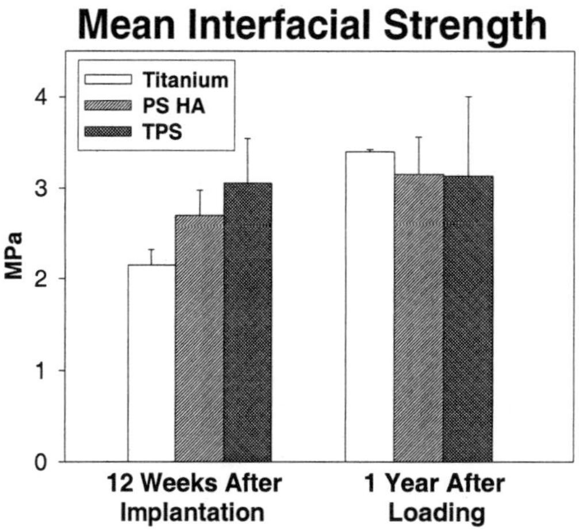

Figure 2 Pull-out strength of different implant surfaces after 12 weeks implantation and after 1 year loading (64 weeks after implantation). Error bar indicates 1 SD.

compared with the control normal rat group ($p<0.0001$) [47]. In another study using TiO_2-gritblasted and CaP-sputtered implants, the sputtered CaP coatings showed improved initial fixation and healing response when implanted into the trabecular bone of the goat [48,49]. In a 14-day goat model, a lipid-stabilized HA/chlorhexidine coating was also reported to be successful in decreasing infection ($p<0.01$) and improving fixation ($p=0.04$) of external fixator pins as compared to noncoated stainless steel and Ti pins [50]. Moroni et al. found better fixation and greater bone–implant interface strength of HA-coated pins in a 6-week sheep model compared to

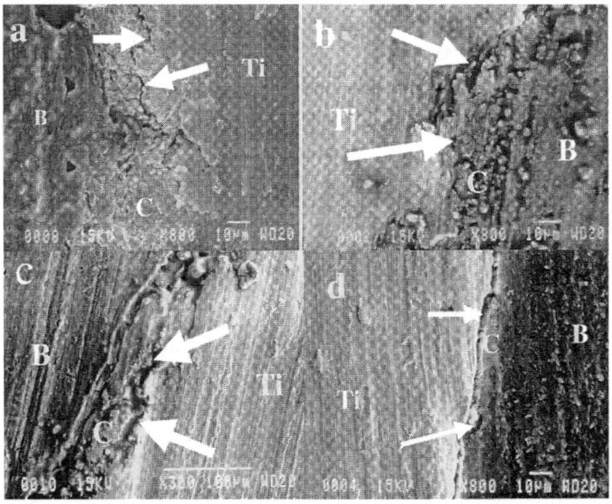

Figure 3 SEM photomicrographs of fracture interface for (a) TPS after 12 weeks implantation, (b) HA after 12 weeks implantation, (c) TPS after 1 year loading, and (d) HA after 1 year loading. C = coatings; B = bone; Ti = titanium substrate. Arrows indicate cracks at the coating-metal interface.

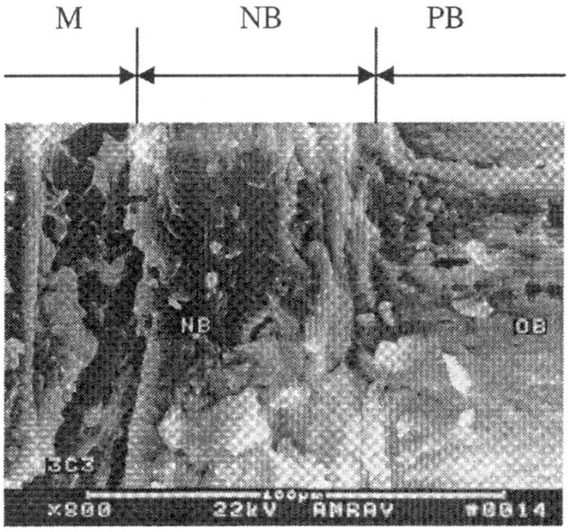

Figure 4 A representative scanning electron micrograph of fractured half of the bone–implant block, with the implant removed. M = implant removed; NB = newly formed bone; PB = pre-existing bone. (From Ref. 60.)

TPS-coated and uncoated pins [51]. Factually, many recent clinical reports have confirmed the remarkable improvements in the clinical results of external fixation [52–59].

In current coating research using the plasma-spraying technology, a hybrid plasma-spraying processing including Ti plasma-spraying and ion implantation of amino (NH_2^+) groups was also used as a means of modifying Ti surfaces [60,61]. It was reported that ionimplanted porous graded titanium coatings had a thick surface oxide layer containing small amount of nitride. *In vivo* study indicated direct bone contact between surface-modified Ti implants and osseous tissues (Figs. 4 and 5). In addition, osseous tissues were observed to grow into the pores inside the coatings (Fig. 6), thereby allowing the formation of a gradual calcium phosphate interface layer. It was concluded from the study that ionimplantation of Ti surfaces with amino groups induced higher concentration of calcium and phosphorus precipitation and more mineralization as compared to non-ion-implanted Ti surfaces [60,61.]

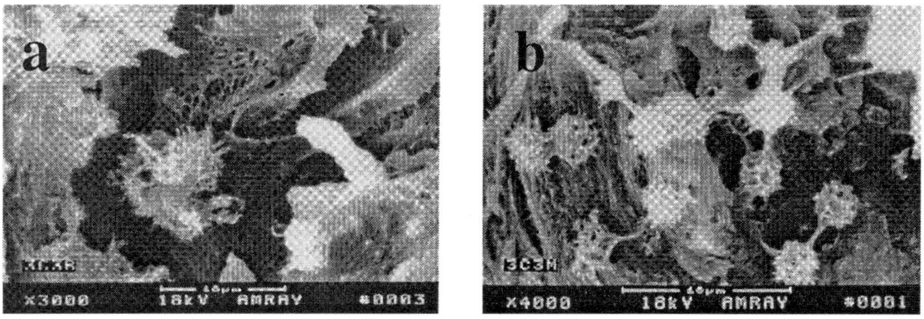

Figure 5 Representative scanning electron micrographs of the two halves of the fractured bone-implant block after 8 weeks implantation. (a) Block containing bone. (b) Block containing an implant embedded in the bone. Osteoblasts and fractured bone tips were observed on both fractured halves. (From Ref. 60.)

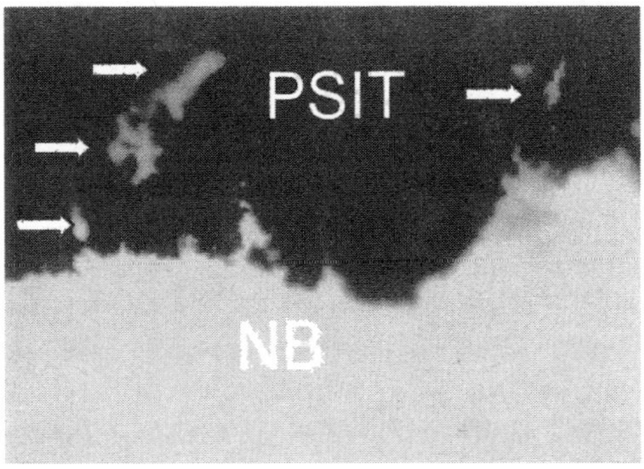

Figure 6 Representative fluorescence micrograph of the PSIT implant–bone interface at 8 weeks after implantation (arrows = NB). Original magnification 100 ×. (From Ref. 60.)

In addition, current research has included the investigation of HA crystallinity on bone ingrowth and fixation [23,62,63]. After 16 weeks postimplantation, HA implants with 50% crystallinity achieved a higher mechanical fixation compared to HA implants with 75% crystallinity. After 32 weeks postimplantation, no difference between HA implants with 50% crystallinity and 75% crystallinity was observed. It was observed that fixation of HA implants with 75% crystallinity increased from 16 to 32 weeks, whereas fixation of HA implants with 50% crystallinity remained unchanged [64]. No significant difference in extraction torque values was observed between screw-type HA implants with 70% crystallinity and 40% crystallinity at 1, 3 and 12 months [65]. However, the extraction torque values of HA-coated screws at 3 and 12 months were significantly increased compared to the extraction values at 1 month [65]. In a comparative investigation using a canine model, involving 36 endosteal dental implants coated with 20-, 40- and 50-μm-thick-HA or left uncoated, bone adaptation on uncoated implants (53.9%) and 20-μm-thick HA-coated implants (54.2%) was significantly higher compared to bone adaptation on 50-μm thick HA-coated implants (47.5%) at 1 month. However, no difference in bone adaptation was observed between uncoated implants, 20-μm-thick HA-coated implants, and 50-μm-thick HA-coated implants after 3 months implantation [66].

Some implant surface modifications of HA and TPS coatings are carried out by combining bioactive factors. In a New Zealand white rabbits model using cranium, the combination of Timesh (1.37 mm^2 and 36%) with recombinant human transforming growth factor-β 1 (rhTGF-β1) was reported to induce orthotopic bone formation as compared to nonloaded implants (0.57 mm^2 and 26%) [67]. In addition, a thin CaP coating (57% bone ingrowth) did not further enhance the bone-conducting properties of a Ti scaffold material (54% bone ingrowth). In a rat model, significantly more bone formation was observed for TPS implants treated with recombinant human bone morphogenetic protein 2 (rhBMP2) as compared to control implants [68]. It was also reported that rhBMP2 remains biologically active after application to TPS implants and may be used to enhance apposition bone formation by direct application to the implant surface.

An interesting observation with the use of HA-coated and TPS-coated implants is the degradation of the coatings. Although various studies showed different degradation behavior of HA coatings, some initial dissolution or degradation of HA coating is essential to obtain rapid biological fixation. Natural remodeling of bone around the implants by the physiological osteoclast and osteoblast cell cycle promotes degradation of coatings, especially the amorphous phase of the coating. In a canine model, the thickness of HA coating was observed to decrease significantly after 4 and 24 weeks implantation time ($p<0.01$) [69]. Resoption of HA coating was reported as early as 16 weeks after placement of the implant [70]. von Vlitterwjk reported resoption of 65% of the 50-μm-thick HA coating during unstable mechanical conditions [71]. However, retrieval from a 10.6-year experimental canine showed significant remnants of the HA coating [13,72]. However, other studies indicated the existence of strong bonding despite the degradation of HA coatings within 12 months of implantation [73].

Although an increased surface area is observed with TPS-coated implants, the release of Ti ions was not observed 8 weeks after implantation [60]. However, 14 months after implantation, tiny Ti granules ranging from 5 to 50 nm in diameter have been found in the adjacent bone when TPS-coated implants were used [74]. In addition, CaP needlelike crystals within the porous Ti were observed, indicating bone ingrowth and a direct bone-binding mechanism to the outer surface of the TPS coating.

IV. APPLICATION OF HA AND TPS COATING IN CLINICS

Plasma-sprayed HA-coated and TPS-coated implants are used as dental implants, pins and total hip arthroplasty [75–78], total knee arthroplasty, total elbow arthroplasty [79], and total shoulder arthroplasty [80].

A. Dental Implants

Since the initial development of TPS- and HA-coated dental implants, numerous studies have demonstrated favorable or superior results for TPS- or HA-coated implants as compared to noncoated Ti implants [81,82]. O'Roark did a large anecdotal case report of survival rate of dental implants in excess of 3000 implants of one practitioner placed over a span of 25 years [83]. The survival rate of dental implants reached 90.5% over 12 years. With over 6 years in a dental school clinical training center, 80 different operators with a wide range of clinical experience inserted 1263 implants in a diverse patient pool of 380 individuals [84]. Analysis of the outcomes showed a cumulative survival rate of 91.3%. The TPS screw implant system for reconstruction of the edentulous mandible has been used successfully for over 25 years, and the technique has demonstrated over 8 years 94% average success [17]. The ITI® dental implant system, with its advantage attaining osteointegration microscopically due to TPS coatings, has been tested in the laboratory and clinically used for almost 20 years [85]. A 6-year clinical follow-up study of Osseotite® implants reported a cumulative implant success rate of 96.6% [86]. The surface complexity of Osseotite is created by thermal acid-etching a commercially pure Ti implant with hydrochloric and sulfuric acids.

The long-term clinical success of HA-coated endosseous dental implants also has been excellent. Results of thousands of implants at greater than 5 to 8 years' follow-up have consistently shown survival rates in excess of 90% [87–91]. Minimal problems related to the HA coatings have been reported. Buchs et al. published an interim clinical study report to evaluate a 5-year postrestoration safety and efficacy of threaded HA-coated implants conducted at three study sites [92]. The 2062 implants placed and restored among 720 patients over the 5-year period included over 600 implants and 200 patients at each study site. Their data included accountability and rigorously applied implant success/failure criteria. The surgical success rate at exposure was 98.3%. The 5-year post-restoration follow-up success rate, determined by means of life table methodology, was 96.0%.

In an 8-year compilation of clinical results obtained with Steri-Oss endosseous implants, a total of 1499 implants placed in 389 women and 216 men had an average success rate of 96.1% [93]. The success rates of 262 Ti screws, 209 HA-coated screws, 810 HA-coated cylinders, and 132 TPS cylinders were 88.5, 97.6, 98 and 99.2%, respectively. A higher success rate for HA and TPS implants was observed.

Several short-term clinical trials of HA-coated implants and Ti implants have been compared to information collected using animal models [94]. Most of these studies seem to agree that HA-coated implants do better than Ti implants, presumably because of a biochemical bond at the bone–implant interface. In a 3-month nonloading clinical evaluation of HA-coated and TPS-coated cylinder dental implants, TPS-coated implants showed a higher but not significant failure rate compared with HA-coated implants ($p = 0.06$, $n = 352$) [94]. The TPS surfaces have been shown to increase the surface area available for osseointegration and to enhance the rate of bone formation. In a 3-year follow-up study of three different endosseous implants in edentulous mandibles, peritest values for HA-coated implants were significantly better than test values for the TPS-coated implants and standard threaded Ti implants after 2 years implantation [95]. However, significant difference was observed between HA-coated implant and standard implant only ($p<0.05$) after 3 years implantion. In addition, when marginal bone resorption was considered, standard implants and HA-coated implants had significantly better scores than TPS-coated implants.

A 7-year clinical evaluation of soft-tissue effects of HA-coated and noncoated subperiosteal Ti implants showed HA-coated implants to be associated with a kinder soft-tissue response, typified by a tighter, healthier soft-tissue seal around the perigingival abutment posts and with no postimplantation strut dehiscence complications [96]. Fish and Misch also reported that a HA-coated subperiosteal implant induced a mandibular bone growth in a 13-year-old patient [97].

B. Pins and Screws

Hydroxyapatite coating on pins was reported to increase stability, thereby reducing the risks for pin tract infection and mechanical failure of fracture fixation [52–57]. In a prospective, randomized clinical study of wrist fractures of 20 female patients, HA-coated pins provided improved fixation in the treatment of wrist fractures in patients with osteoporosis ($p<0.0001$) [53]. The use of HA-coated pedicle screws was reported to improve fixation compared to noncoated screws ($p<0.001$), with reduced risk of loosening of the screws [54].

C. Knee Arthroplasty

In a 1-year follow-up study of HA-coated (16 patients), TPS-coated (15 patients) and noncoated (15 patients), total knee prostheses, HA coating had a strong positive effect on the tibial components' fixation [98]. No prosthesis in the HA groups showed continuous migration. In a 2-year follow-up study of 31 total knee prostheses, the reported micromotions on HA-coated tibial components fixed without cement was similar to that of noncoated tibial components fixed with cement [99]. In addition, the micromotions on noncoated tibial components fixed with cement was significantly lower than noncoated components fixed without cement ($p<0.001$). Therefore, HA may be necessary for the adequate fixation of tibial components when cement is not used. In a 2-year follow-up study of 116 total knee prostheses, TPS-coated implants migrated at a statistically significantly higher rate than HA-augmented or cemented implants [100]. There was no significant difference between HA-coated and cemented implants. HA-augmentation may offer a clinically relevant advantage over a simple porous coating for tibial component fixation, but is no better than cemented fixation.

D. Femoral Stems

Hydroxxapatite and TPS coatings on femoral stems have been widely and successfully used in clinical settings. Tables 4 and 5 list the surface properties of some products and the comparison of long-term follow-up study in the recent literature [101–122].

After a mean follow-up duration of 7.5 years, the prevalence of osteolysis around femoral stem-coated with noncircumferential TPS coating (40%; 50 of 126 hips) was significantly higher than that coated with circumferential TPS coating (10%; 9 of 90 hips) ($p<0.001$) [123]. Osteolysis remote from the joint space (acetabular cup) was found in 11% 14 of the hips with noncircumferential TPS coating. However, no osteolysis remote from the joint space was observed with circumferential TPS coating ($p = 0.0004$). The average total area of osteolysis around the hips with noncircumferential TPS coating (5.0 cm^2) was significantly larger that with circumferential TPS coating (2.9 cm^2 ($p<0.05$). It was suggested that a circumferential TPS coating on the femoral component of a total hipreplacement prosthesis inserted without cement provides an effective barrier preventing wear debris from gaining access to the implant–bone interface.

So far all the 10-year follow-up studies of femoral components with cobalt–chromium (CoCr) porous coating and TPS or HA coating have suggested that femoral components with HA coatings have higher survival rates. In addition, the survival rate of femoral components with TPS coating was reported to be better than CoCr porous coating. The success of femoral stems with HA and TPS coating probably resulted from two aspects: [1] the improvement of initial fixation due to biochemical bonds of HA coating and increased surface area of TPS coating, and [2] the effective sealing of HA and TPS coating to the debris out from the wear.

From a clinical point of view, HA and TPS coatings have excellent success when used as dental implants and pin, femoral stem, knee, and shoulder implants. However, some authors have also reported peri-implant bone loss, rapid bone resorption [124], and a higher failure rate for HA-coated implants [125–128]. Erosion of the surface on HA coatings has been reported [129]. After a minimum follow-up duration of 5 years (mean 7.9 years, range, 5.3 to 9.1 years), one (1%) out of the 131 HA-coated threaded cups, two (2%) out of the 109 porous-coated press-fit cups, and 21 (11%) out of the 188 HA-coated press-fit cups had been revised because of aseptic loosening [130].

In 1982, Adell and coworkers proposed the prerequisites for achieving osseointegration: (1) biocompatible materials; (2) primary stability of the implant; (3) atraumatic surgical procedure; and (4) surgical protocol with submerged implants, which are unloaded during the healing phase of 3 to 6 months [131]. Furlong, who first began clinical trials using the HA-coated implants in 1985, stated emphatically that for a HA-coated implant to become integrated, the conditions of contact, stillness, and stressing must be fulfilled [132]. The HA coating itself does not give initial stability. It is the design of implants and suitable surgical procedures which must provide initial total immobility (primary fixation) so that bone ingrowth (secondary fixation) can occur. On the basis of this concept, a success rate of 99% of HA-coated implants was reported in a mean follow-up study of 10 years. Baumgarten and Chiche also discussed the complications and failures associated with osseointegrated implants [133]. It was reported that failure cases are mostly related to lapses in quality control by manufacturers and the possibilities of mishandling HA-coated implants by surgeons and restorative dentists, rather than the fault of the HA coating [134].

V. RETRIEVAL STUDIES

Steflik et al. carried out a continuous investigation using light and scanning electron microscopy analyses of 51 implant cases in 1994 [135], 135 implant cases in 1999 [136], and 200 implant

Table 4 Surface Characterization of Coating on Different Femoral Components

Femoral name	Femoral component	Company	Surface characterization of coating	Ref.
AML	Anatomic medullary locking	Depuy, Warsaw, IN	The porous surface is made of cobalt–chromium (CoCr) substrate with sintered CoCr beads and a mean pore size of 200 μm.	101
Trilock		Depuy, Warsaw, IN	The porous surface is coated with sintered CoCr beads. The average diameter of the beads is 150 μm (100–250 μm) and forms an irregular porous surface.	106
Taperloc		Biomet, Warsaw, IN	The proximal 40% of the implant is coated with the identical Ti alloy (Ti-6Al-4V) applied with a pressure plasma spray technique. The plasma spray porous coating is between 635 to 889 μm thick. Pore diameter varies throughout the coating from 92.3 μm at the surface to 480.3 μm at the outermost points of the coating surface.	106
PCA	Porous coated, anatomic	Howmedia, Rutherford, NJ	The femoral and acetabular components are composed of chromium–cobalt alloy (vitallium), with a double layer of sintered vitallium beads on the back of the acetabular components and circumferentially on the proximal one-third of the femoral stem. The mean size of the pores of the coatings is 425 μm. The coating has a porosity of 35%, a thickness of 1.5 mm, and a reported shear strength of 7000 lb/in^2.	102–104
HG	Harris-Galante	Zimmer, Warsaw, IN	The femoral component is made from forged Ti alloy with commercial Ti fiber mesh porous ingrowth pads. The mesh has a pore volume of 50% with an average pore size of 300 μm.	119,120
MH	Malloy head	Biomet, Warsaw, IN	The proximal 1/3 of the femoral component is covered circumferentially with a plasma-sprayed coating, whereas the middle 1/3 is grit blasted. The distal surface featured a matte finish.	107,108

(*Continued*)

Table 4 Continued

Femoral name	Femoral component	Company	Surface characterization of coating	Ref.
RPF		Plasma Biotal, Tidewell, UK	The HA coating is applied by plasma spraying (first on Ti alloy implants and latter on CoCr implants). The coatings have a relative crystallinity of 70–90%, a thickness of 55–90 μm, minimum shear bond strength of 25 MPa, purity of 97–98%, and roughness of 6–16 μm Ra.	110,111
Landanger		Landanger, Chaumont, France	The surface of Ti alloy substrate is entirely plasma sprayed with a HA layer (thickness of 155 ± 35 μm).	121,122
Omnifit		Osteonics, Allendale, NJ	The porous coating consists of a double layer of 425- to 500-μm-thick CoCr spherical particles sintered to the substrate stem.	105
Omnifit-HA		Osteonics, Allendale, NJ	Stem surface roughness is 3 to 4.3 μm before the application of HA coatings. After HA coating, stem roughness ranges between 4.3 to 8.1 μm. The HA coating of 50-μm thickness has a porosity below 3%, 97% HA purity, and 65% crystallinity.	112–114
JRI Furlong-HA	JRI Furlong, HA-coated	JRI Instrumentation, London	Furlong Ti-6Al-4V prosthesis is coated with Osprorit HA. The 200-μm-thick HA coating has a bond strength of 40 MPa in both tension and shear tests.	146

cases in 2001 [137,138]. These implants were retrieved from humans and submitted to the American Academy of Implant Dentistry Research Foundation, Medical College of Georgia Implant Retrieval Center. Cases included both orthopedic and dental implants as well as entire mandibles and portions of maxillae obtained at autopsy. These implants failed because of implant fracture. As observed with animal studies, healthy bone supported these implants, with the bone containing an interdigitating canaliculi us network that provided communication between interfacial osteocytes and osteocytes deeper within the remodeled osteonal and trabecular bone. Early dental implants containing a coating of beads showed a connective tissue interface which corresponded to the bead surface of specific orthopedic implants that underwent some degree of micromovement. This was in contrast with the excellent response reported for successful contemporary beaded implants. Significant numbers of osseointegrated fractured HA-coated dental implants demonstrated adequate serviceability of these implants before biomaterial fracture. In contrast, the HA coating was dissociated from retrieved orthopedic implants, leading to extensive cup loosening and case failure.

Rosenlicht and Tarnow did the first human histological case report of implants that were simultaneously placed with a sinus lift and loaded [139]. These implants were retrieved 2.5

Table 5 Comparison of 10-Year Follow-Up Postimplantation of Femoral Components

Author	Femoral name	Number of hips	Average age of patients (range) (years)	Average follow-up (range) (years)	Mechanical failure rate (%)	Thigh pain (%)	Ref.
Engh (1997)	AML (CoCr)	174	54 (16–79)	11 (10–13)	3	3–10	101
Xenos (1999)	PCA (CoCr)	100	58 (22–81)	10–13	5	0	102, 103
Kim (1999)	PCA (CoCr)	116	48.4 (16–69)	10 (10–12)	22	14	104
Hellman (1999)	Omnifit (CoCr)	76	45.4 (19–67)	9.9 (5.1–12.5)	2.6	4	105
Purtill (2001)	Trilock (CoCr)	77	50.4	15 (14.5–16.9)	12	2	106
Purtill (2001)	Taperloc (TPS)	180	60	11 (8–14)	0.5	4	106
Bourne (2001)	MH (TPS)	307	64 ± 10	10–13	1	3	107, 108
McLaughlin (2000)	Taperloc (TPS)	114	37 (20–50)	10.2 (8–13)	4	2	109
Donnelly (1997)	RPF (HA)	115	52.0±11.9	5–10	0		110, 111
D' Antonio (2001)	Omnifit (HA)	314	51	10–13	0.5	1.6	112–114
MaNally (2000)	JRI (HA)	100	45–94	10 (9–12)	0	1.2	115,116
Geesink (2002)	Omnifit (HA)	118	53 (21–73)	12 (11–13)	0	<3	117, 118

years after placement and 2 years after loading. The case report histologically verified that osseous integration can occur when implants are placed simultaneously at the time of antral augmentation.

Coathup et al. investigated implant–bone interfaces around one design of femoral stem, proximally coated with either a plasma-sprayed coating (plain porous), a HA porous coating (porous HA), or which had been grit-blasted (Interlock) using retrieval hip replacements [140]. They treated 165 patients with a mean age of 84.8 years (79 to 92) for fracture of the femoral neck using a Bimetric hip hemiarthroplasty (Biomet, Bridgend, UK). Each patient randomly received a femoral component with either a plain porous, a porous HA, or an interlock surface on the proximal region of the femoral stem. Of 165 patients implanted, 58 specimens were retrieved at postmortem. Bone ingrowth and attachment to the surface of implants were estimated in eight plain porous, seven porous HA, and six interlock. There was significantly more ingrowth ($p = 0.012$) and attachment of bone ($p < 0.05$) to the porous HA surfaces (mean bone ingrowth $29.093 \pm 2.019\%$; mean bone attachment $37.287 \pm 2.489\%$) as compared to implants with plain porous surface (mean bone ingrowth $21.762 \pm 2.068\%$; mean bone attachment 18.941 ± 1.971). However, no significant difference in bone attachment between implants with plain porous surface and interlock surface was observed. In addition, no significant differences in the volume of HA were observed with the passage of time.

Tonino et al. examined six clinically successful HA-coated cementless acetabular components that were retrieved at autopsy between 3.3 and 6.6 years after implantation [141,142]. All of the cups showed bone growth onto the implant, with a mean bone-implant contact of 36.5 ± 13.5%. The extent and thickness of the HA coating were much reduced in the specimens from old patients and in those associated with a longer duration of implantation. Degradation of the HA coating by osteoclasts was observed. Loose HA granules were not observed to be far from the coating, and no adverse tissue reaction to these granules was reported.

Bauer et al. described the histological features of 5-year HA-coated implants, retrieved at autopsy from three patients who died from non-hip-related causes [13,143,144]. Clinical performance of all involved implants was good, and the implantation time varied between 4 and 25 months. HA was present on all explants with an average coating thickness of 40 μm. No delamination of coating material and no correlation between coating thickness and implantation time were reported. There was a uniform layer of bone present over the HA coating with a minimum of 32% and a maximum of 78% circumference. There was a tendency toward more bone fixation in the distal parts of the coating area. The average area of osseointegration was greater than 60% of available HA surface. In the 25-month sample, focal areas of osteoclastic resorption on the HA coating were observed (quantitative less than 10% surface). In these areas, direct osseointegration between bone and Ti was observed. All HA-coated cups had bony fixation.

There was a uniform layer of HA coating present until at least 2 years after surgery [13,143,144]. The lack of correlation between coating thickness and implantation time suggests minimal early loss of coating to extracellular fluid. It was suggested that once bony coverage of the coating was obtained, the process of physicochemical dissolution was arrested. However, a cellular process of coating degradation caused by osteoclastic activity was observed. Some osteoclastic resportion of HA was seen at 2 years that was replaced by bone. This suggested that with longer implantation period, the HA coating will be replaced by autologous bone, while preserving the quality of osseointegration. There was no fibrous tissue formation in these HA-resorption areas. Other retrieval studies reported good osseointegration of HA-coated hip prostheses when implanted in osteoporotic or rheumatoid patients [13,145,146]. Degradation of the HA coating usually was more pronounced in the proximal femur as compared with the more cortical middle and distal stem area. There seems to be a relationship between coating degradation and rate of bone remodeling. The amount and quality of fixation in all retrievals was not altered by any degradation of the HA coatings.

VI. FUTURE DEVELOPMENTS

At present, there is no standard guideline for plasma spraying coatings on medical devices. It is known that the advantages of plasma spraying include a rapid deposition rate and sufficiently low cost [147]. Nevertheless, despite the advantages and the clinical success reported, there are problems associated with plasma-sprayed coatings, especially HA. These problems include poor adhesion and variation in bond strength between the coatings and metallic substrates as well as alterations in HA or CaP structure resulting from the coating process [148,149]. The nature of the substrate plays an important role in the adhesion between plasma-sprayed coatings and metallic substrates [26]. The bonding of plasma-sprayed coatings to the metallic substrate appears to be entirely mechanical in nature, and reports indicate that a highly roughened substrate surface exhibits higher bond strength than a smooth substrate surface [150].

Several experimental deposition processes are available for producing HA and CaP coatings, including electrophoretic codeposition, sputter deposition, and high velocity oxy-fuel combustion spray deposition [151–154]. These experimental processes have been developed in an attempt to improve the adhesive, compositional, and structural properties of the coatings. Cooley et al., Lacefield et al., and de Groot et al. concluded that sputtering may be the method of choice for controlling the physical and chemical properties of HA or CaP coatings on dental and orthopedic implants [15,155,156]. The advantages of RF magnetron sputter coating over other coating processes include the high adhesion of films, the excellent uniform layers, and the ability to coat implants with difficult geometries. Studies indicate a significantly greater coating-metal interfacial strength for sputtered CaP coatings compared to commercially available plasma sprayed HA coatings (40 MPa vs. 9 MPa) [157]. Another important advantage, especially for dental implants, is that no grit blasting of the substrate surface is needed, contrary to plasma spraying [15], In a comparative investigation between sputtered and plasma-sprayed HA coatings in a canine model, the 1-μm-thick HA coating was reported to exhibit significantly more bone adaptation as compared to 40- or 50-μm-thick plasma-sprayed HA implants after 1 month implantation [66].

Another promising technique is biomineraliztion. Recently, biomemetic HA coating, a thin calcium phosphate layer of 10 to 30 μm thick, has been produced in a supersaturated CaP solution or stimulated body fluid using precipitation techniques [158,159]. Using a canine model, a higher bone apposition rate on the biomimetic coating was observed as compared to the plasma-sprayed HA coatings after 4 and 8 weeks implantation [13]. In conclusion, new promising techniques for coating medical devices are continuously being investigated. At present, investigators are exploring nanotechnology and biomimetics in the hope to optimize coating properties for rapid osseointegration. These evaluations have included better materials characterization, furthering, understanding bone responses to implant surfaces of different properties, and the use of various biological factors. Although it has been generally accepted that HA and TPS coatings improve bone strength and initial osseointegration rate, optimal coating properties required to achieve maximal bone response are yet to be reported. As such, the use of well-characterized implant surfaces in cell culture and animal and clinical studies should be well documented to avoid controversial results.

ACKNOWLEDGMENTS

The studies presented were funded by NIH/NIDCR (DE12276), NIH/NIAMS (AR46581), and collaboration was also supported in part by a Grant-in-Aid for Scientific Research B (No. 12470437) of the Japanese Ministry of Education, Science, Sports and Culture. The authors also thank Friatec AG, Germany for some of the implants mentioned in the chapter.

REFERENCES

1. Branemark PI, Brieine V, Adell R, Hansson BO, Lindestrom J, Ohlsson A. Intra-osseous anchorage of dental prosthesis. I. Experimental studies. Scand J Plas Resconstr Surg 1969; 3:81–100.
2. Albrektsson T, Jacobsson M. Bone–metal interface in osseointegration. J Prosth Dent 1987; 57: 597–607.
3. Williams DF. Titanium and titanium alloys. In: Williams DF, ed. Biocompatibility of Clinical Implant Materials, Volume 1: CRC Series in Biocompatibility. Boca Raton, FL: CRC press, 1981: 9–44.

4. Martinez H, P Davarpanah Missika, Celletti R, Lazzara R. Optimal implant stabilization in low density bone. Clin Oral Impl Res 2001; 12:423–432.
5. Cook SD, Rust-Dawicki AM. In vivo evaluation of a CSTi dental implant: a healing time course study. J Oral Impl 1995; XXI:182–190.
6. de Groot K, Geesink R, Klein CP, Serekian P. Plasma sprayed coatings and hydroxyapatite. J Biomed Mater Res 1987; 21:1375–1391.
7. Cook SD, Thomas KA, Kay JF, Jarcho M. Hydroxyapatite-coated titanium for orthopedic implants applications. Clin Orthop 1988; 232:225–243.
8. Kay JF. Calcium phospahte coatings for dental implants: current state and future potential. Dent Clin North Am 1992; 36:1–18.
9. Ducheyne P, Cuckler JM. Bioactive ceramic prosthetic coatings. Clin Orthop 1992; 276:102–114.
10. Cooke FW. Ceramics in orthopedic surgery. Clin Orthop 1992; 276:135–146.
11. Steflik DE. Hydroxyapatite coatings on dental implants: benefits and risks. J Oral Impl 1994; XX: 198–200.
12. Jaffe WL, Scott DF. Current concepts review: total hip arthroplasty with hydroxyapatite-coated prostheses. J Bone Joint Surg 1996; 78A:1918–1934.
13. Geesink RGT. Osteoconductive coatings for total joint arthroplasty. Clin Orthop 2002; 395:53–65.
14. LeGeros RZ. Properties of osteoconductive biomaterials: calcium phosphates. Clin Orthop 2002; 395:81–98.
15. de Groot K, Wolke JGC, Jansen JGC. State of the art: hydroxylapatite coatings for dental implants. J Oral Impl 1994; XX:232–234.
16. Janisson S, Vardelle A, Coudert JF, Fauchais P, Meillot E. Analysis of the stability of DC plasma gun operating with Ar-He-H_2 gas mixtures. Heat and mass transfer under plasma conditions. Ann NY Acad Sci 1999; 891:407–416.
17. Babbush C, Kent J, Misiek D. Titanium plasma-sprayed screw implants for the reconstruction of edentulous mandible. J Oral Maxillofac Surg 1986; 44:274–282.
18. Overgaard S, Lind M, Rahbek O, Bunger C, Soballe K. Improved fixation of porous-coated versus grit-blasted surface texture of hydroxyapatite-coated implants in dogs. Acta Orthop Scand 1997; 68:337–343.
19. Clemens JA, Klein CP, Vriesde RC, Rozing PM, de Groot K. Healing of large (2 mm) gaps around calcium phosphate coated implants. J Biomed Mater Res 1998; 40:341–349.
20. Gineste L, Gineste M, Ranz X, Ellefterion A, Guilhem A, Rouquet N, Frayssinet P. Degradation of hydroxylapatite, fluorapatite and fluorhydroxyapatite coatings of dental implants in dogs. J Biomed Mater Res 1999; 48:224–234.
21. Dhert WJ, Klein CP, Jansen JA, van der Velde E, Vriesde RC, Rozing PM, de Groot K. A histological and histomorphometric investigation of fluorapatite, magnesium whitlockite and hydroxyapatite plasma-sprayed coating in goats. J Biomed Mater Res 1993; 27:127–138.
22. Kumar M, Dasarathy H, Riley C. Electrodeposition of brushite coatings and their transformation to hydroxyapatite in aqueous solution. J Biomed Mater Res 1999; 45:302–310.
23. Overgaard S, Lind M, Glerup H, Grundvig S, Bunger C, Soballe K. Hydroxyapatite and fluorapatite coatings for fixation of weight-loaded implants. Clin Orthop 1997; 336:286–296.
24. Overgaard S, Lind M, Josephsen K, Maunsbach AB, Bunger C, Soballe K. Resorption of hydroxyapatite and fluorapatite ceramic coatings on weight-loaded implants. J Biomed Mater Res 1998; 39: 141–152.
25. Overgaard S, Soballe K, Lind M, Bunger C. Resorption of Hydroxyapatite and fluorapatite coatings in man. J Bone Joint Surg 1997; 79B:654–659.
26. Yang Y, Ong JL. Bond Strength, Compositional and structural properties of HA coating on Ti substrate, ZrO_2-coated Ti substrate, and TPS-coated Ti substrate. J Biomed Mater Res 2003; 64A: 509–516.
27. Yang Y, Tian J, Tian J, Chen Z, Deng X, Zhang D. Preparation of graded porous titanium coatings on titanium implants materials by plasma spraying. J Biomed Mater Res 2000; 52:333–337.
28. Hansson HA, Albrektssson T, Branemark PI. Structural aspects of the interface between tissue and titanium implants. J Prosthet Dent 1983; 50:109–113.
29. Sisk AL, Steflik DE, Parr GR, Hanes PJ, McKinney RV. A light and electron microscopic comparison of osseointegration of six implant types. J Oral Maxillofac Surg 1992; 50:709–716.

30. Lum LB, Beirne OR. Viability of the retained bone core in the Core-Vent dental implant. J Oral Maxillofac Surg 1986; 44:341–345.
31. Lum LB, Beirne OR, Dillings M, Curtis TA. Osseointegration of two types of implants in nonhuman primates. J Prosthet Dent 1988; 60:700–705.
32. Deporter DR, Watson PA, Pilliar RM, Melcher AH, Winslow J, Howley TP. A hisotological assessment of the initial healing response adjacent to porous-surfaced, titanium alloy dental implants in dogs. J Dent Res 1986; 65:1064–1070.
33. Keller JC, Young FA, Natiella JR. Quantitative bone remodeling resulting from the use of porous dental implants. J Biomed Mater Res 1987; 21:305–319.
34. Listharten MA, Buser D, Steinmann SG, Donath K, Lang NP, Weber HP. Light and transmission electron microescopy of the intact interface between non-submerged titanium-coated epoxy resin implants and bone or gingiva. J Dent Res 1992; 71:364–371.
35. Schroederer A, van der Zypen E, Stich H, Sutter F. The reaction of bone, connective tissue and epithelium to endosteal implants with titanium-sprayed surfaces. J Oral Maxillofac Surg 1981; 9: 15–25.
36. Cook SD, Kay JF, Thomas HA, Jarcho M. Interface mechanism and histology of titanium and hydroxyapatite-coated titanium for dental implant applications. J Oral Maxillofac Implants 1987; 2:15–22.
37. De Lange GL, Donath K. Interface between bone tissue and implants of solid hydroxyapatite or hydroxyapatite coated titanium implants. Biomaterials 1989; 10:121–125.
38. Steflik DE, White SL, Parr GR, Sisk AL, Schoen SP, Lake FT, Hanes PJ. Clinical evaluation data from a comparative dental implant investigation in dogs. J Oral Impl 1993; XIX:199–208.
39. Hacking SA, Bobyn JD, Tanzer M, Krygier JJ. The osseous response to corundum blasted implant surafces in a canine hip model. Clin Orthop 1999; 364:240–253.
40. Buser D, Schenk RK, Steinmann Fiorellini JP, Fox CH, Stich CH. Influence of surface characteristics on bone integration of titanium implants. A histomorphometric study in miniature pigs. J Biomed Mater Res 1991; 25:889–902.
41. Novaes AB, Souza SLS, de Oliveira RT, Souza AMMS. Histomorphometric analysis of the bone–implant contact obtained with 4 different implants surafce treatment placed side by side in the dog mandible. Int J Oral Maxillofac Impl 2002; 17:377–382.
42. London RM, Roberts FA, baker DA, Rohrer MD, O'Neal RBO. Histologic comparison of a thermal dual-etched implant surface to machined, TPS, and HA surfaces bone contact in vivo in rabbits. Int J Oral Maxillofac Impl 2002; 17:369–376.
43. Hayakawa T, Yoshinari M, Nemoto K, Wolke JGC, Jansen JA. Effect of surface roughness and calcium phosphate coating on the implant/bone response. Clin Oral Impl Res 2000; 11:296–304.
44. Karabatsos B, Myerthall SL, Fornasier VL, Binnington A, Maistrelli GL. Osseointegration of hydroxyapatite porous-coated femoral implants in a canine model. Clin Orthop 2001; 392:442–449.
45. Cook SD, Rust-Dawicki AM. Preliminary evaluation of titanium–coated PEEK dental implants. J Oral Impl 1995; XXI:176–181.
46. Darimont GL, Cloots R, Heinen E, Seidel L, Legrand R. In vivo behavior of hydroxyapatite coatings on titanium implants: a quantitative study in the rabbit. Biomaterials 2002; 23:2569–2575.
47. Hara T, Hayashi K, Nakashima Y, Kanemaru T, Iwamoto Y. The effect of hydroxyapatite coating on the bonding of bone to titanium implants in the femora of ovariectimised rats. J Bone Joint Surg 1999; 81B:705–709.
48. Vercaigne S, Wolke JGC, Naert I, Jansen JA. A mechanical evaluation of TiO_2-grit blasted and Ca-P magnetron sputter coated implants placed into the trabecular bone of the goat: part 1. Clin Oral Impl Res 2000; 11:305–313.
49. Vercaigne S, Wolke JGC, Naert I, Jansen JA. A histological evaluation of TiO_2-grit blasted and Ca-P magnetron sputter coated implants placed into the trabecular bone of the goat: part 2. Clin Oral Impl Res 2000; 11:314–324.
50. DeJong MES, DeBerardino MTM, Brooks DE, Nelson MBJ, Campbell AA, Bottoni MCR, Pusateri AE, Walton MRS, Guymon CH, McManus AT. Antimicrobial efficacy of external fixator pins coated with a lipid stabilized hydroxyapatite/chlorhexidine complex to prevent pin tract infection in a goat model. J Trauma Inj Infect Crit Care 2001; 50:1008–1014.

51. Moroni A, Toksvig-larsen S, maltarello MC, Orienti L, Stea S, Giannini S. A comparison of hydroxy-apatite-coated, titanium-coated and uncoated tapered external-fixation pins. J Bone Joint Surg 1998; 80A:547–554.
52. Moroni A, Aspernberg P, Toksvig-Larsen A, Falzarano G, Giannini S. Enhanced fixation with hydroxyapatite coated pins. Clin Orthop 1998; 346:171–177.
53. Moronio A, Faldini C, Marchetti S, Manca M, Consoli V, Giannini S. Improvement of the bone–pin interface strength in osteoporotic bone with use of hydroxapatite-coated tapered external-fixation pins. J Bone Joint Surg 2001; 83A:717–721.
54. Sanden B, Olerud C, Petren-Mallmin M, Larsson S. Hydroxyapatite coating improves fixation of pedicle screws. A clinic study. J Bone Joint Surg 2002; 84B:387–391.
55. Moroni A, Heikkila J, Magyer G, Toksvig-Larsen A, Giannini S. Fixation strength and pin tract infection of hydroxyapatite-coated tapered pins. Clin Orthop 2001; 388:209–217.
56. Magyar G, Toksvig-Larsen A, Moroni A. Hydroxyapatite coating of threaded screws enhances fixation. J Bone Joint Surg 1997; 79B:487–489.
57. Pommer A, Muhr G, David A. Hydroxyapatite-coated Schanz pins in external fixators used for distraction osteogenesis. J Bone Joint Surg 2002; 84A:1162–1166.
58. Rahbek O, Overgaard S, Lind M, Bdendix M, Bunger K, Soballe K. Sealing effect of hydroxyapatite coating on peri-implant migration of particles. J Bone Joint Surg 2001; 83B:441–447.
59. Moroni A, Vannini F, Mosca M, Giannini S. Techniques to avoid pin loosening and infection in external fixation. J Orthop Trauma 2002; 16:189–195.
60. Yang Y, Ong JL, Tian J. *In vivo* evaluation of modified titanium implant surfaces produced using a hybrid plasma spraying process. Mater Sci Eng C 2002; 20:117–124.
61. Yang Y, Tian J, Deng L, Ong JL. Morphological behavior of osteoblast-like cells on surface modified titanium in vitro. Biomaterials 2002; 23:1383–1389.
62. Overgaard S, Bromose U, Lind M, Bunger C, Soballe K. The influence of crystallinity of the hydroxyapatite coating on the fixation of implants. J Bone Joint Surg 1999; 81B:725–731.
63. Moroni A, Faldini C, Rocca M, Stea S, Giannini S. Improvement of the bone–screw interface strength with hydroxyapaptite-coated and titanium-coated AO/ASIF cortical screws. J Orthop Trauma 2000; 16:257–263.
64. Overgaard S, Bromose U, Lind M, Bunger C, Soballe K. The influence of crystallinity of the hydroxyapatite coating on the fixation of implants. J Bone Joint Surg 1999; 81B:725–731.
65. Moroni A, Faldini C, Rocca M, Stea S, Giannini S. Improvement of the bone–screw interface strength with hydroxyapaptite-coated and titanium-coated AO/ASIF cortical screws. J Orthop Trauma 2002; 16:257–263.
66. Steflik DE, Lacefield WR, Sisk AL, Parr GR, Lake FT, Patterson JW. Hydroxylapatite-coated dental implants: descriptive histology and quantitative histomorphometry. J Oral Impl 1994; XX:201–213.
67. Vehof JWM, Haus MTU, de Ruijter AE, Spauwen PHM, Jansen JA. Bone formation in transforming growth factor beta-1–loaded titanium fiber mesh implants. Clin Oral Impl Res 2002; 13:94–102.
68. Cole BJ, Bostrom MPG, Pritchard TL, Sumner DR, Tomin E, Lane JM, Weiland AJ. Use of bone morphogenetic protein 2 on ectopic porous coated implants in the rat. Clin Orthop 1997; 345: 219–228.
69. Mochoida Y, Bauer TW, Akisue T, Brown PR. Alendronate does not inhibit early bone apposition to hydroxyapatite-coated total joint implants. J Bone Joint Surg 2002; 84A:226–235.
70. Soballe K, Hansen ES, Brockstedt-Rasmussen H, Burgess A. Torsional stability of HA-coated and grit blasted Ti dental implants. J Bone Joint Surg 1993; 75B:270–8.
71. von Blitterwijk CA, Croke JJ, Kuypers W, Blok-van Hoek CJ, Daems WT. Bioreactions at the tisssue/hydroxyapatite interface. Biomaterials 1985; 6:243–251.
72. Geesink RGT, de Groot K, Klein CPAT. Chemical implant fixation using hydroxylapatite coatings. Clin Orthop 1987; 225:147–170.
73. Wolke JGC, de Blick-Hogervorst JMA, Dhert WJA, Klein CPAT, de Groot K. Studies on the thermal spraying of apatite bioceramics. J Thermal Spray Technol 1992; 1:75–82.
74. Leize EM, Hemmerle J, Leize M. Characterization, at the bone crystal level, of the titanium-coating/bone interfacial zone. Clin Oral Impl Res 2000; 11:279–288.

75. Walker PS. Innovation in total hip replacement – when is new better?. Clin Orthop 2000; 381: 9–25.
76. Walker PS, Culligan SG, Hua J, Muirhead—Allwood SK, Bentley G. Stability and bone preservation in custom designed revision hip stems. Clin Orthop 2000; 373:164–173.
77. Blunn GW, Briggs TWR, Cannon SR, Walker PS, Unwin PS, Culligan S, Cobb JP. Cementles fixation for primary segmental bone tumor endoprostheses. Clin Orthop 2000; 372:223–230.
78. Rokkum M, Reigstad A. Total hip replacement with an entirely hydroxyapatite-coated prosthesis. 5 years' follow-up of 94 consecutive hips. J Arthroplasty 1999; 14:689–700.
79. King GJW, Adams RA, Morrey BF. Total elbow arthroplasty: revison with use of a non-custom semiconstrained prosthesis. J Bone Joint Surg 1997; 79A:394–400.
80. Levy O, Copeland SA. Cementless surface replacement arthroplasty of the shoulder. J Bone Joint Surg 2001; 83B:213–221.
81. Bubbush CA, Kent NJ, Salon JM. A solution for the problematic atropic mandible: the TPS screw implant system. Gerodontics 1986; 2:16–23.
82. Pilliar RM, Deporter DA, Watson PA, Pharoah M, Chipman M, Valiquette N, Carter S, de Groot K. The effect of partial coating with hydroxyapatite on bone remodeling in relation to porous-coated titanium-alloy dental implants in dog. J Dent Res 1991; 70:1338–1345.
83. O'Roark WL. Survival rate of dental implants: an individual practitioner's anecdotal review of 25 years of experience. J Oral Impl 1997; XXIII:90–103.
84. Minsk L, Polson AM, Weisgold A, Rose LF, Sanavi F, Baumgarten H, Listgarten MA. Outcome failure of endosseous implants from a clinial training center. Compend Cont Educ Dent 1996; 17: 848–859.
85. Levine RA, Wilson TG, Weber HP. The ITI® dental implant system. Compend Contin Educ Dent 1994; XV:526–535.
86. Sullivan DY, Sherwood RL, Porter SS. Long-term performance of osseotite implants: A 6-year clinical follow-up. Compend Cont Educ Dent 2001; 22:326–334.
87. Kent JN, Block MS. Cylindrical HA-coated implant—eight year observations. Compend Contin Educ Dent 1993; 15:526–532.
88. Cook SD, Salkeld SL, Gaisser DM, Wagner WR. The effect of surface macrotexture on the mechnical and histological characteristics of hydroxyapatite-coatd dental implants. J Oral Impl 1993; XIX: 288–294.
89. Kent JN, Block MS, Finger IM, Guerra L, Larsen H, Misiek DJ. Biointegrated hydroxyapatite-coatd dental implants: five-year clinical observations. J Am Dent Assoc 1990; 121:138–144.
90. Golec TS, Krauser JT. Long-term retrospective studies on hydroxyapatite-coated endosteal and subperiosteal implants. Dent Clin North Am 1992; 36:39–65.
91. Stultz ER, Lofland R, Sendax VI, Hornbuckle C. A multi-center 5 year retrospective survival analysis of 6200 integral implants. Compend Cont Educ Dent 1992; 14:428–486.
92. Buchs AU, Hahn J, Vassos DM. Interim clinical study report: threaded, HA coated implants—five-year post-restoration safety and efficacy. J Oral Impl 1995; XXI:266–274.
93. Saadoun AP, Le Gall MG. An 8-year compilation of clinical results obtained with Steri-Oss endosseous implants. Compend Cont Educ Dent 1996; 17:669–688.
94. Jones JD, Saigusa M, Sickels JE, Tiner BD, Gardner WA. Clinical evaluation of hydroxyapatite-coated titanium plasma-sprayed and titanium plasma-sprayed cylinder dental implants. A preliminary report. Oral Surg Oral Med Oral Pathol Oral Radiol Endod 1997; 84:137–141.
95. Roynesdal AK, Ambjornsen E, Stovne S, Haanas HR. A comparative clinical study of three different endosseous implants in edentulous mandibles. Int J Oral Maxillofac Impl 1998; 13:500–505.
96. Fettig RH, Kay JF. A seven-year clinical evaluation of soft-tissue effects of HA-coated vs. uncoated subperiosteal implants. J Oral Impl 1994; XX:42–48.
97. Fish JM, Misch CE. Mandibular bone growth induced by a hydroxyapatite-coated subperiosteal implant: a case report. J Oral Impl 2000; XXVI:267–275.
98. Toksvig-Larson S, Jorn LP, Ryd L, Lindstrand A. Hydroxyapatite-enhanced tibial prosthetic fixation. Clin Orthop 200; 370:192–200.
99. Nelissen RGHH, Valstar ER, Rozing PM. The effect of hydroxyapatite on the micromotion of total knee prostheses. J Bone Joint Surg 1998; 80A:1665–1672.

100. Onsten I, Nordqvist A, Carlsson AS, Besjakov J, Shott S. Hydroxyapatite augmentation of the porous coating improves fixation of tibial components. A randomised RSA study in 116 patients. J Bone Joint Surg 1998; 80B:417–425.
101. Engh CA, Culpepper WJ, Engh CA. Long-term results of use of the anatomic medullary locking prosthesis in total hip arthroplasty. J Bone Joint Surg 1997; 79A:177–184.
102. Xenos JS, Callaghan JJ, Heekin RD, Hopkinson WJ, Savory CG, Moore MS. The porous-coated anatomic total hip prosthesis, inserted without cement. J Bone Joint Surg 1999; 81A:74–82.
103. Callaghan JJ, Dysart SH, Savory CG. The uncemented porous-coated anatomic total hip prosthesis. J Bone Joint Surg 1988; 70A:337–346.
104. Kim Y, Kim J, Cho S. Primary total hip arthroplasty with a cementless porous-coated anatomic total hip prosthesis. J Arthroplasty 1999; 14:538–548.
105. Hellman EJ, Capello WN, Feinberg JR. Omnifit cementless total hip arthroplasty. Clin Orthop 1999; 364:164–174.
106. Purtill JJ, Rothman RH, Hozack WJ, Sharkey PF. Total hip arthroplasty using two different cementless tapered stems. Clin Orthop 2001; 393:121–127.
107. Bourne RB, Rorabeck CH, Patterson JJ, Guerin J. Tapered titanium cementless total hip replacement. A 10- to 13-year followup study. Clin Orthop 2001; 393:112–120.
108. Bourne RB, Rorabeck CH. A critical look at cementless stems. Clin Orthop 1998; 355:212–223.
109. McLaughlin JR, Lee KR. Total hip arthroplasty in young patients. 8- to 13-year results using an uncemented stem. Clin Orthop 2000; 373:153–163.
110. Donnelly WJ, Kobayashi A, Freeman MAR, Chin TW, Yeo H, West M, Scott G. Radiological and survival comparison of four methods of fixation of a proximal femoral stem. J Bone Joint Surg; 79B:351–360.
111. Kobayashi A, Donelly WJ, Scott G, Freeman MAR. Early radiological observations may predict the long-trem survival of femoral hip prostheses. 1997; 79B:583–589.
112. D'Antonio JA, Capello WN, Manley MT, Geesink R. Hydroxyapatite femoral stems for total hip arthroplasty. 2001; 393:101–111.
113. Capello WN, D'Antonio JA, Manley MT, Feinberg JR. Hydroxyapatite in total hip arthroplasty. Clin Orthop 1998; 355:200–211.
114. D'Antonio JA, Capello WN, Manley MT, Feinberg J. Hydroxyaptite coated implants. Clin Orthop 1997; 344:124–138.
115. MaNally SA, Shepperd HAN, Mann CV, Walczak JP. The results at nine to twelve years of the use of a hydroxyapatite-coated femoral stem. 2000; 82B:378–382.
116. D'Lima DD, Walker RH, Colwell CW. Omnifit-HA stem in total hip arthroplasty. Clin Orthop 1999; 363:163–169.
117. Geesink RGT, Hoefnagels NHM. Six-year results of hydroxyapatite-coated total hip replacement. J Bone Joint Surg 1995; 77B:534–547.
118. Geesink RGT. Hydroxyapatite-coated total hip prostheses. Two-year clinical and roentgenographic results of 100 cases. Clin Orthop 1990; 261:39–58.
119. Petersilge WJ, D'Lima DD, Walker RH, Colwell CW. Prospective study of 100 consecutive Harris-Galante porous total hip arthroplasty. 4- to 8-year follow-up study. J Arthroplasty 1997; 12:185–193.
120. Martell JM, Pierson RH, Jacobs JJ, Rosenberg AG, Maley M, Galante JO. Primary total hip reconstruction with a titanium fiber–coated prosthesis inserted without cement. J Bone Joint Surg 193; 75A:554–571.
121. Maloney WJ. National joint replacement registries: has the time come?. J Bone Joint Surg 2001; 83A:1582–5.
122. Mont MA, Hungerford DS. Proximally coated ingrowth prostheses. A review. Clin Orthop 1997; 244:139–149.
123. Emerson RH, Sanders SB, Head WC, Higgins L. Effect of circumferential plasma-spray porous coating on the rate of femoral osteolysis after total hip arthroplasty. J Bone Joint Surg 1999; 81A: 1291–1298.
124. Ichikawa T, Hirota K, Kanitani H, Wigianto R, Kawamoto N, Matsumoto N, Mitake Y. Rapid bone resorption adjacent to hydroxyapatite-coated implants. J Oral Impl 1996; XXII:232–235.

125. Johnson BW. HA-coated dental implants: long-term consequences. J Calif Dent Assoc 1992; 20: 33–41.
126. Piatelli A, Cosci F, Scarano A, Trisi P. Localized chronic suppurative bone infection as a sequel of peri-implants in a hydroxyapatite-coated dental implant. Biomaterials 1995; 16:917–920.
127. Weehler SL. Eight-year clinical retrospective study of titanium plasma-sprayed and hydroxyapatite-coated cylinder implants. Int J Oral Maxillofue Impl 1996; 11:340–350.
128. Radl R, Aigner C, Hungerford M, Pascher A, Windhager R. Proximal femoral bone loss and increased rate of fracture with a proximally hydroxyapatite-coated femoral component. J Bone Joint Surg 2000; 82B:1151–5.
129. Cheang P, Khor KA. Addressing processing problems associated with plasma spraying of hydroxyapatite coatings. Biomaterials 1996; 17:537–544.
130. Manley MT, Capello WN, D'Anotnio JA, Geesink RGT. Fixation of acetabular cups without cement in total hip arthroplasty. J Bone Joint Surg 1998; 80A:1175–1185.
131. Adell R, Lekholm V, Rockler B. A 15-year study of osseointegrated implants in the treatment of the edentulous jaw. Int J Oral Surg 1981; 10:387–416.
132. Furlong R. Proximal femoral bone loss and increased rate of fracture with a proximally hydroxyapatite-coated femoral component. J Bone Joint Surg 2001; 83B:461.
133. Baumgarten HS, Chiche GJ. Diagnosis and evaluation of complications and failures associated with osseointegrated implants. Compend Cont Educ Dent 1995; 16:814–823.
134. Cranin AN, Mehrali M, Baraoidan M. Hydroxylapatite-coated endosteal implants from the clinicians' perspective. J Oral Impl 1994; XX:235–239.
135. Steflik DE, Parr GR, Singh BB, Lake FT, Sisk AL, Howell FV, Shelton TW. Light microscopic and scanning electron microscopic analyses of dental implants retrieved from humans. J Oral Impl 1994; XX:8–24.
136. Corpe RS, Steflik DE, Young TR, Wilson MR, Jaramillo CA, Hipps M, Sisk A, Parr GR. Retrieval analyses of implanted biomaterials: light microscopic and scanning electron microscopic retrieval analyses of implants retrieved from humans. J Oral Impl 1999; XXV:162–178.
137. Steflik DE, Corpe RS, Young TR, Parr GR, Tucker M, Sims M, Tinley J, MaDaniel M. Light microscopic and scanning electron microscopic retrieval analyses of implanted biomaterials retrieved from humans and experimental animals. J Oral Implr 2001; XXVII:5–15.
138. Cranin AN, Baraoidan M, DeGrado J. A human clinical and histologic report of an osseointegrated titanium alloy root form implant. J Oral Impl 1997:21–24.
139. Rosenlicht JL, Tarnow DP. Human histologic evidence of integration of functionally loaded hydroxyapatite-coated implants placed simultaneously with sinus augmentation: a case report 21/2 years postplacement. J Oral Impl 1999; XXV:7–10.
140. Coathup MJ, Blunn GW, Flynn N, Williams C, Thomas NP. A comparison of bone remodelling around hydroxyapatite-coated, porous-coated and grit-blasted hip replacement retrieved at postmortem. J Bone Joint Surg 2000; 82B:118–123.
141. Tonino A, Oosterbos C, Rahmy A, Therin M, Doyle C. Hydroxyapatite-coated acetabular components. J Bone Joint Surg 2001; 83A:817–825.
142. Tonio AJ, Therin M, Doyle C. Hydroxyapatite-coated femoral stems. Histology and histomorphometry around five components retrieved at postmortem. J Bone Joint Surg 1999; 81B:148–154.
143. Bauer TW, Geesink PGT, Zimmerman R, McMahon JT. Hydroxyapatite-coated femoral stems. J Bone Joint Surg 1991; 73A:1439–1452.
144. Bauer TW, Stulberg BN, Jiang M, Geesink RGT. Uncemented acetabular components: Histologic analysis of retrieved components. J Arthroplasty 1993; 8:167–177.
145. Bloebaum RD, Bachus KN, Rubman MH, Dorr LD. Postmortem comparative analysis of titanium and hydroxyapatite porous-coated femoral implants retrieved from the same patient. J Arthroplasty 1993; 8:203–211.
146. Furlong RJ, Osborn JF. Fixation of hip prostheses by hydroxyapatite coatings. J Bone Joint Surg 1991; 73B:741–745.
147. Herman H. Plasma spray deposition processes. MRS Bulletin 1988; 12:60–67.
148. Filiaggi MJ, Coombs NA, Pilliar RM. Characterization of the interface in the plasma-sprayed HA coating/Ti-6Al-4V implant system. J Biomed Mater Res 1991; 25:1211–1229.

149. Palka V, Postrkova E, Koerten HK. Some characteristics of hydroxylapatite powders after plasma spraying. Biomaterials 1998; 19:1763–1772.
150. Nimb L, Gotfredsen K, Steen Jensen J. Mechanical failure of hydroxyapatite-coated titanium and cobalt–chromium–molybdenum alloy implants. An animal study. Acta Orthopaedica Belgica 1993; 59:333–338.
151. Dasarathy H, Riley C, Coble HD. Analysis of apatite deposits on substrates. J Biomed Mater Res 1993; 27:477–482.
152. Ong JL, Harris LA, Lucas LC, Lacefield WR, Rigney D. X-ray photoelectron spectroscopy characterization of ion beam sputter deposited calcium phosphate coatings. J Am Ceram Soc 1991; 74: 2301–2304.
153. Wolke JGC, van Dijk K, Schaeken HG, de Groot K, Jansen JA. Study of the surface characteristics of magnetron-sputter calcium phosphate coatings. J Biomed Mater Res 1994; 28:1477–1484.
154. Haman JD, Lucas LC, Crawmer D. Characterization of high velocity oxy-fuel combustion sprayed hydroxyapatite. Biomaterials 1995; 16:229–237.
155. Cooley DR, Van Dellen AF, Burgess JO, Windeler AS. The advantages of coated titanium implants prepared by radiofrequency sputtering from hydroxyapatite. J Prosthet Dent 1992; 67:93–100.
156. Lacefield WR, Lucas LC, Rigney ED, Ong JL, Chen TS, Gantenberg J. Sputter deposition of calcium phosphate coatings for biomedical applications. In: Sudarshan TS, Braza JF, eds. Surface Modification Technologies VI. Pennsylvania: Minerals, Metals and Materials Society, 1993: 591–604.
157. Lucas LC, Lacefield WR, Ong JL, Whitehead RY. Calcium-phosphate coatings for dental and medical implants. Coll Surf 1993; 77:141–147.
158. Barrer F, Layrolle F, van Blitterswijk CA, de Groot K. Biomimetic calcium phosphate coatings on Ti-6Al-4V. Bone 1999; 25:107S–111S.
159. Li P, Ducheyne P. Quasi-biological apatite film induced by titanium in a simulated body fluid. J Biomed Mater Res 1998; 41:341–348.

18
Calcium Phosphate Ceramics in Japan

Masataka Nishikawa
Osaka University Graduate School of Medicine
Osaka, Japan

Hajime Ohgushi
Tissue Engineering Research Center, National Institute of Advanced Industrial Science and Technology
Hyogo, Japan

I. INTRODUCTION

Recent technology has developed many types of clinically applicable biomaterials. In the field of bone reconstruction, osteoconductive ceramic materials have been used as artificial bone graft substitute [1–3]. Many different kinds of ceramics have been used. This chapter discusses the chemical and physical properties of the ceramics produced by Japanese companies and provides examples of their clinical applications in the field of orthopedics and craniofacial surgery.

Although the osteoconductive property resides in the ceramics, the ceramics alone do not have osteogenic abilities, i.e., the ceramics are not osteoinductive. If the ceramics themselves possessed osteoinductive properties, the expansion of clinical applications of the ceramics would be assured. Bone morphogenetic protein (BMP) is a protein that demonstrates osteogenesis when implanted subcutaneously or intramuscularly. Thus, BMP is osteoinductive [4]. Some authors have reported the usefulness of a BMP/ceramic composite to enhance the inherent osteogenic ability of BMP [5,6]. In contrast to the utilization of the osteoinductive activity of BMP, reports of others and ourselves found that composites of marrow cells and porous materials such as hydroxyapatite (HA) show an osteogenic response when implanted in vivo [7]. The osteogenic capability of the marrow cells resides not only in fresh marrow cells, but also in in vitro culture expanded mesenchymal stem cells (MSCs) derived from the marrow cells [8]. This chapter also focuses on the osteogenic capabilities of MSCs in combination with ceramics.

II. CHARACTERIZATIONS OF CALCIUM PHOSPHATE CERAMICS IN JAPAN

A. Hydroxyapatite

The crystalline phase of natural bone is basically hydroxyapatite (HA), and HA ceramics have been used extensively as a substitute in bone grafts. The ceramics are available as dense or porous types and the shape types are granular or blocklike. Different pore sizes, porosities, and strengths are available. Here we describe four types of hydroxyapatite ceramics that are used clinically:

1. Bonefil (Mitsubishi Materials). These ceramics are porous blocks and porous granules and are most often used in orthopedics. The sintering temperature is 900°C. The compressive strength of porous blocks and porous granules are 15 MPa and 2 to 3 MPa, respectively. The pore shape is spongiose, and the pore size is 200 to 300 μm. The degree of porosity is 60 to 70%.
2. Bonetite (Mitsubishi Materials). These ceramics are porous blocks and dense granules and are most often used in dental surgery. The sintering temperature is 1200°C. The pore shape is spongiose and the pore size is 200 μm. The degree of porosity is 70%.
3. Boneceram (Sumitomo Osaka Cement). Porous block and porous granular types of these ceramics are available as Boneceram-P. The sintering temperature is 1150°C. The compressive strength is 44.1 to 68.6 MPa, and the bending strength is 12.7 to 19.6 MPa. The pore shape is spherical, and the pore size is 50 to 300 μm. The degree of porosity is 35 to 48%. Dense block types of the ceramics having a high mechanical strength are available as Boneceram-K. The sintering temperature is 1150°C. The bending strength is over 58.8 MPa.
4. Apaceram (Pentax). These ceramics are both dense and porous. The porous ceramic has a degree of porosity of 15 to 60%. The sintering temperature is 1200°C. The compressive strength and bending strength vary from 16 to 250 MPa and 8 to 47 MPa, respectively. The higher mechanical properties are associated with a decrease in the degree of porosity (Table 1). The pore shape is spherical. The pore structure is an interconnected bimodal pore configuration consisting of a combination of 300-μm macropores and 2-μm micropores (Fig. 1). The dense hydroxyapatite has a degree of porosity of less than 0.8%. The sintering temperature is 1050°C. The compressive strength and the bending strength are 750 MPa and 210 MPa, respectively (Fig. 2). Clinical applications began in 1985, with about 5000 clinical uses of Apaceram ceramics (custom-designed porous type plate; Fig. 3) in cranioplasty since then. The numbers of clinical cases involving spinal surgery and ENT surgery with ear ossicle substitutes are 70,000 and 20,000, respectively.

All four of these manufactured hydroxyapatite ceramics are essentially nonresorbable.

Table I Mechanical Strength of Porous Type Apaceram

Porosity (%)	Compressive strength (MPa)	Bending strength (MPa)
15	250	47
30	180	25
40	66	16
50	30	11
55	21	9.5
60	16	8

Source: Courtesy of T. Ogawa, Pentax Corp., Tokyo, Japan.

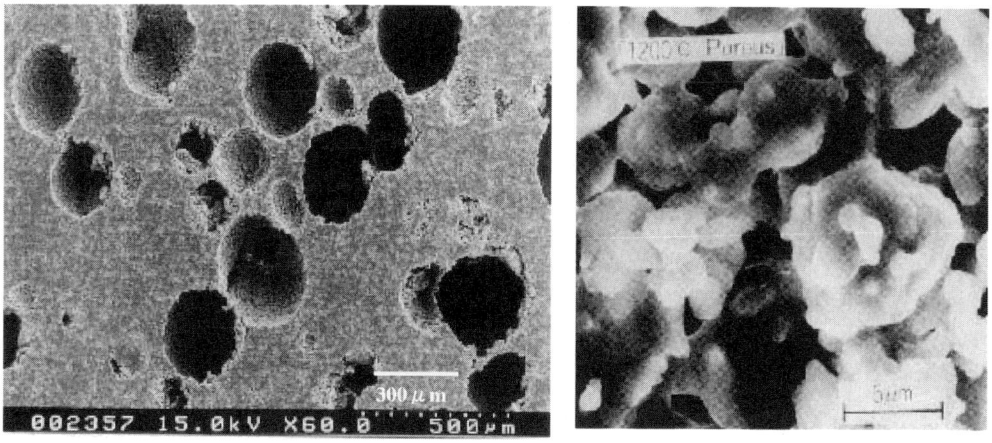

a b

Figure 1 SEM photographs of macro- and microstructures of Apaceram. (a) The pores (300 μm) of Apaceram are randomly distributed and interconnected. (b) The surface of Apaceram is rough, and the particles are loosely attached to one another. (Courtesy of T. Ogawa, Pentax Corp., Tokyo, Japan.)

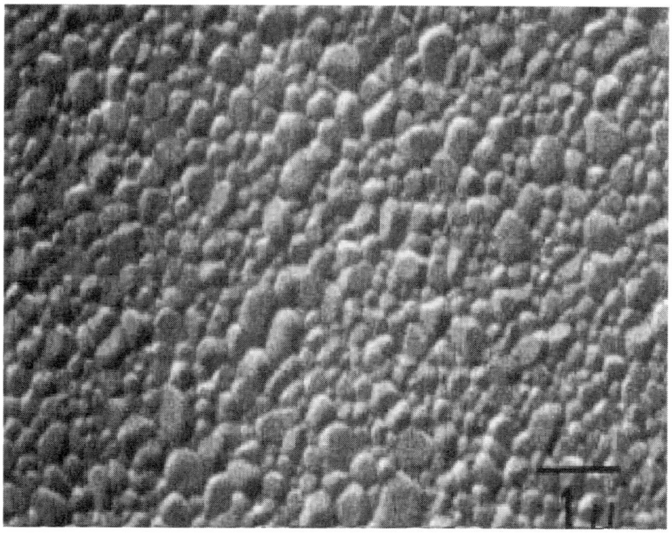

Figure 2 SEM photograph of dense type Apaceram. The surface of dense type Apaceram is very smooth, and the hydroxyapatite particles are tightly bound. (Courtesy of T. Ogawa, Pentax Corp., Tokyo, Japan.)

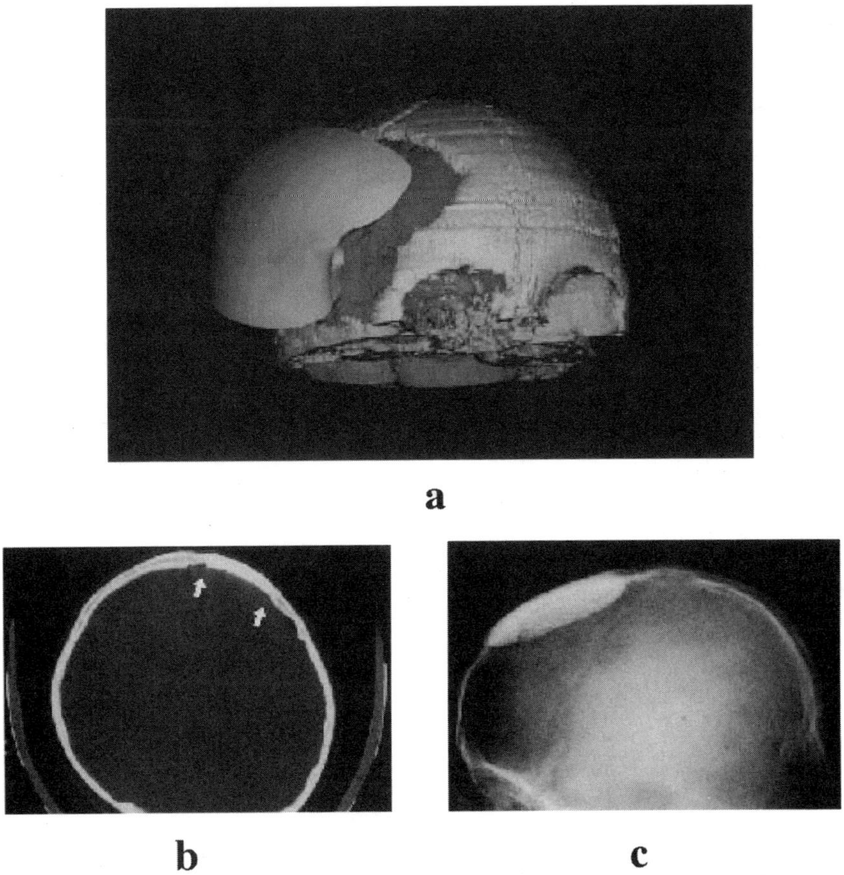

Figure 3 (a) Custom-made design of Apaceram using a three-dimensional computer-assisted design (3D CAD) model (skull). (b) Computed tomography (CT) of custom-made Apaceram (skull). The white arrows indicate the custom-made implants. (c) Radiograph of custom-made Apaceram. The Apaceram reconstructed the skull and bonded to the bone. (Courtesy of T. Ogawa, Pentax Corp., Tokyo, Japan.)

B. Tricalcium Phosphate

In contrast to HA ceramics, tricalcium phosphate (TCP) is a resorbable ceramic, and host-derived new bone can be expected relatively soon after replacement of the implanted ceramic. One Japanese company has produced TCP ceramics with a high degree of purity:

1. Osferion (Olympus Optical). Osferion is a porous type of β-TCP with high degree of purity. The porosity is 75% and the pore shape is sponge type with an interconnected bimodal pore structure. The macropore diameter is 100 to 400 μm and the micropore diameter is less than 1 μm. The sintering temperature is 950 to 1100°C. The compressive strength is very weak, at 2 MPa. Due to the low level of mechanical properties, the use of this product in high load-bearing areas, such as just beneath the knee joint, often calls for additional grafting of cancellous bone together with cortical bone and internal fixation. In addition, compared with the use of other types of ceramics, it is necessary to delay the start of load bearing. The ceramics are completely resorbed in

a relatively short time (6 months to a few years) and can be expected as new, regenerated bone tissue appears. If the bone tissue replacement proves promising, these ceramics might be able to be used in load-bearing areas and are expected to be safe for use in younger, growing patients.

C. HA/TCP Hybrids

As described above, after implantation TCP ceramics can be completely resorbed. However, because of the high rate of resorption, use of these ceramics is difficult in load-bearing areas. If we can control the rate of resorption, their range of application can be expanded. One method of controlling resorption is to make hybrid materials of nonresorbable HA and resorbable TCP in different ratios.

1. Ceratite (NGK Spark Plug). The composition of Ceratite is 70% HA and 30% TCP. Ceratite supports the osteogenic action of host bone tissue near the implantation, resulting in new bone formation around the implanted ceramic. The implant bonds directly with newly formed bone tissue to repair the site of a bone defect, while the TCP in the Ceratite gradually replaces the bone. Several types of Ceratite ceramics are available; there are dense and porous blocks as well as granular types with pores (Fig. 4). The porous ceramics have two different architectures. One type is a fine-pored (micropored) ceramic with a pore diameter of 1 to 10 μm. The other type is a large-pored ceramic with a pore diameter of 150 to 250 μm (Fig. 5). The fine-pored ceramic has a porosity ratio of 35 to 40%. The sintering temperature is 1100°C. The compressive and bending strengths are 120 to 160 MPa and 30 to 40 MPa, respectively. Compared with the large-pored type, the fine-pored type has a high degree of mechanical strength and can be manufactured into desired shapes. The large-pored type has a porosity ratio of 45 to 55%, and the sintering temperature is 1300°C. The compressive and bending strengths are 20 to 30 MPa and 7 to 10 MPa, respectively. The large-pored type has a bimodal pore size distribution. The finer porous architecture is in the ceramic matrix. The large pores have good porous connectivity (interconnection size of 60 to 80 μm), giving a three-dimensional pore structure. The sintering temperature of the dense type of Ceratite is 1300°C. The compressive and bending strengths are 600 to 700 MPa and 140 to 170 MPa, respectively. This data shows that the dense type of Ceratite has the highest degree of mechanical strength. Two clinical cases, one treated with the granular type and the other with the dense type, are discussed below.

   Case 1: Male, 26 years old (benign bone tumor in the side of the tibia). Ceratite granules were mixed with autogenous bone chips and implanted in the area of the excised bone tumor. Eight months after the operation, radiographs showed homogenous implants (Fig. 6).

   Case 2: Male, 53 years old (cerebral aneurysm). Ceratite buttons were implanted for a burr hole skull defect. The postoperative x-ray showed bone integration around the ceramic buttons (Fig. 7).

D. Calcium Phosphate Bone Cement

The several types of dense and porous ceramics described above show good osteoconductive properties. However, those ceramics cannot be molded during the operation. In this regard, calcium phosphate bone cement (CPC) is easy to handle because it can be formed into any shape during the operation.

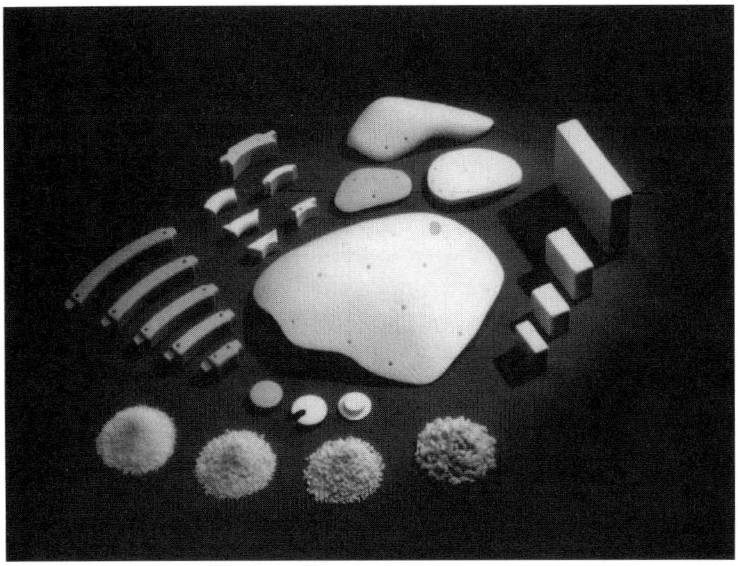

Figure 4 The different types of Ceratite. (Courtesy of K. Kondo and T. Okura, NGK Spark Plug Co., Ltd., Aichi, Japan.)

1. Biopex-R (Mitsubishi Materials). Powders and liquid solvents are mixed during the operation. The powder consists of α–tricalcium phosphate (α-TCP), tetracalcium phosphate, calcium hydrogen phosphate dihydrate, hydroxyapatite, and trimagnesium phosphate octahydrate. The liquid solvent consists of disodium succinate anhydrate, sodium chondroitin sulfate, and sodium hydrogen sulfide. After mixing, the powders and solvents become paste. After in vivo implantation of the paste, it sets and the α-TCP component is resorbed gradually, followed by formation of carbonate apatite with low crystallinity [9]. Although the set Biopex-R is of the porous type, it is not an open-pored type. The porosity ratio is 40 to 50%, depending on the ratio of powder to liquid. Because of the closed porous architecture, the absorbance and replacement by newly formed bone occurs only on the surface of the Biopex-R, and a long period is required for it to become completely absorbed in comparison with porous β-TCP ceramic. Biopex-R is an injectable bioactive bone-bonding cement. It is rapid setting and immediately load bearing. Two days after implantation by the injection method, the compressive strength is 63.8 MPa, and reaches a maximum of 77.8 MPa. Biopex-R can be used to repair ordinary bone defects as well as fractures. Due to its pastelike properties, which harden after implantation into bone defect sites, Biopex-R can be used to maintain normal bone architecture after reduction of fractures of bones such as the radius (metaphyseal areas) [10], tibia plateau, calcaneus [11], and vertebral body (compression types) [12]. Biopex-R also can also be used for unstable femoral intertrochanteric fractures with hip screw fixation devices [13].

IV. APPLICATION OF CALCIUM PHOSPHATE CERAMICS IN TISSUE ENGINEERING

It is well known that mesenchymal stem cells (MSCs) reside in marrow cells [14] and that the MSCs can differentiate into osteoblasts and produce mineralized bone matrices in in vitro cultures

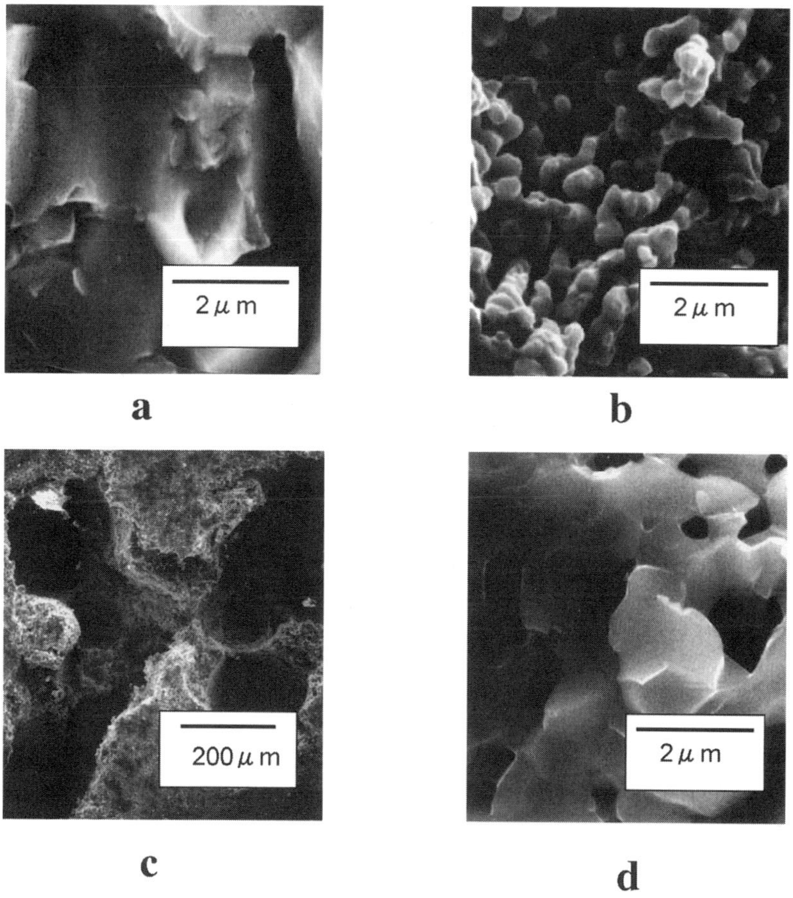

Figure 5 SEM photographs of Ceratite. (a) Dense type Ceratite; the hydroxyapatite particles are bound tightly. (b) Fine pore type Ceratite; the surface is rough and the particles are loosely attached to one another. (c) Macrostructure of large-pore type of Ceratite; the pores (150–250 μm) are interconnected with each other. (d) Microstructures of large-pore type of Ceratite; the surface is rough and the large particles are tightly bound to one another. (Courtesy of K. Kondo and T. Okura, NGK Spark Plug Co., Ltd., Aichi, Japan.)

in the presence of dexamethasone and β-glycerophosphate [15]. The number of MSCs in fresh marrow is extremely small, but their number can be expanded through the use of culture techniques without loss of their osteogenic potential [16]. When MSCs were cultured in a porous framework of HA under conditions where the stem cells are able to differentiate into osteoblasts and form a bone matrix, the porous areas of the HA were covered with a thin layer of in vitro bone and active osteoblasts [17]. These constructs can show the immediate capability of forming new bone after in vivo implantation (Fig. 8). Because constructs fabricated with human cultured MSCs can show promising osteogenic abilities, we initiated clinical applications of different types of ceramics loaded with in vitro bone tissue/osteoblasts derived from the marrow cells of patients [18]. These techniques of using a patient's cultured MSCs can convert nonosteogenic materials into osteogenic materials, i.e., osteoconductive ceramics could be converted into os-

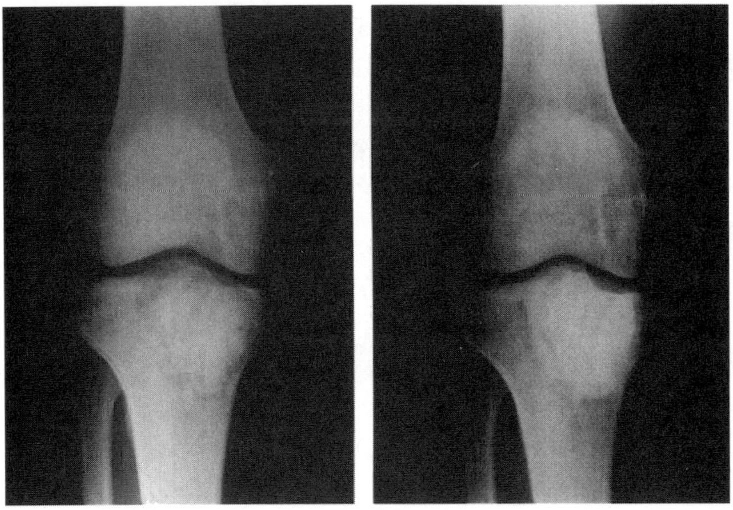

(a) (b)

Figure 6 X-ray photograph of bone tumor. Case 1: male, 26 years old, with a benign bone tumor in the side of the tibia. (a) Pre-operation and (b) eight months post-operation. (Courtesy of K. Kondo and T. Okura, NGK Spark Plug Co., Ltd., Aichi, Japan.)

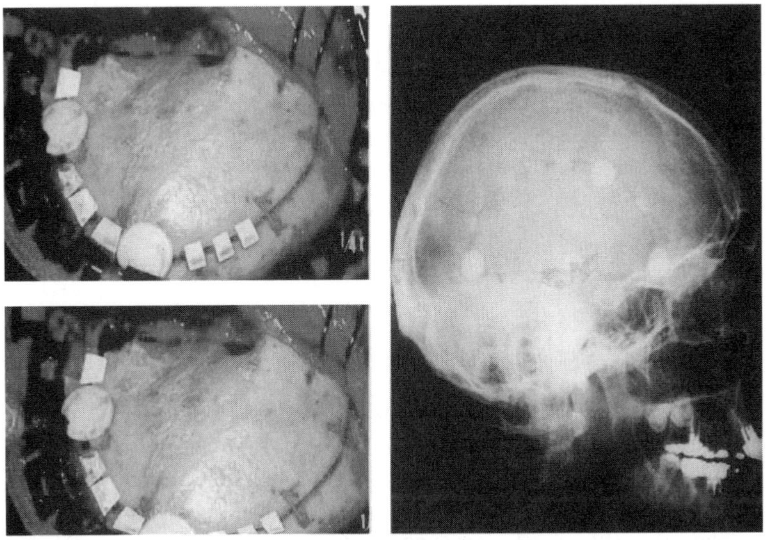

(a) (b)

Figure 7 Clinical application of Ceratite for cranial surgery. Case 2: male, 53 years old, with a cerebral aneurysm. (a) During operation and (b) post-operation. (Courtesy of K. Kondo and T. Okura, NGK Spark Plug Co., Ltd., Aichi, Japan.)

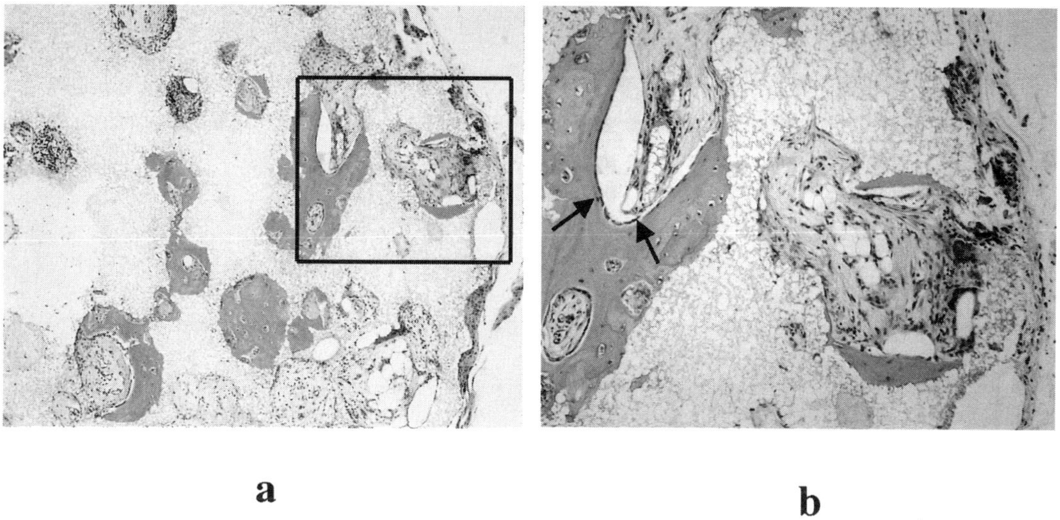

Figure 8 Histological section of composite of mesenchymal stem cells and ceramic material. Implants were harvested 4 weeks after implantation and stained with hematoxylin and eosin. (a) The white area indicates the shadow of the hydroxyapatite ceramic (Apaceram). Mature bone tissue is seen in the porous area. 40 ×. (b) Higher magnification of the rectangular area in (a). Many active osteoblasts (arrow) initiate bone formation. 100 ×.

teoinductive ceramics. These new approaches to tissue engineering [19,20] can therefore be applied to regenerate skeletal tissues in many patients with massive skeletal defects.

IV. CONCLUSIONS

We described the current status of ceramic bone graft substitutes in Japan, especially, in the field of orthopedics and craniofacial surgery. The ceramic materials available for clinical use include HA, TCP, HA/TCP hybrids, and CPC. All the materials can exhibit a biocompatible and an osteoconductive nature. However, these characteristics differ based on their degree of porosity, pore size, mechanical strength, ability to resorb, etc. We also addressed the approaches to tissue engineering that make use of composites of mesenchymal stem cells and ceramics. These approaches to tissue engineering can offer new methods of cell-based therapy for patients requiring skeletal reconstruction.

ACKNOWLEDGMENTS

We wish to express our appreciation to Mr. T. Ogawa (Pentax Corp.), Mr. Y. Takeuchi and Mr. N. Asaoka (Mitsubishi Materials Corp.), Dr. K. Kondo and Mr. T. Okura (NGK Spark Plug Co., Ltd.), Mr. Y. Hakamatsuka and Mr. H. Irie (Olympus Optical Co., Ltd.) and Mr. K. Oomamiuda (Sumitomo Osaka Cement Co., Ltd.) for their generous offering of the data and figures used in this chapter.

This work was done by the Three-Dimensional Tissue Module Project, METI (a Millenium Project) and was also in part supported by the R&D Projects in ''Advanced Support System for Endoscopic and Other Minimally Invasive Surgery'' entrusted from the New Energy and Industrial Technology Development Organization (NEDO) to the Japan Fine Ceramics Center.

REFERENCES

1. Bucholz RW, Carlton A, Holmes RE. Hydroxyapatite and tricalcium phosphate bone graft substitute. Orthop. Clin. North Am 1987; 18:323–334.
2. Holmes RE, Buchplz RW, Monney V. Porous hydroxyapatite as a bone-graft substitute in metaphyseal defect. J. Bone. Joint. Surg. Am 1986; 68:904–911.
3. Ishihara K, Arai H, Nakabayashi N, Morita S, Furuya KI. Adhesive bone cement containing hydroxyapatite particles as bone compatible filter. J. Biomed. Mater. Res 1992; 26:937–945.
4. Urist MR. Bone: formation by autoinduction. Science 1965; 150:893–899.
5. Noshi T, Yoshikawa T, Dohi Y, Ikeuchi M, Horiuchi K, Ichijima K, Sugimura M, Yonemasu K, Ohgushi H. Recombinant human bone morphogenetic protein-2 potentiates the in vivo osteogenic ability of marrow/hydroxyapatite composites. Artif. Organs 2001; 25:201–208.
6. Takaoka K, Nakahara H, Yoshikawa H, Masuhara K, Tsuda T, Ono K. Ectopic bone induction on and in porous hydroxyapatite combined with collagen and bone morphogenetic protein. Clin. Orthop 1988; 234:250–254.
7. Ohgushi H, Okumura M, Tamai S, Shors EC, Caplan AI. Marrow cell induced osteogenesis in porous hydroxyapatite and tricalcium phosphate: a comparative histomorphometric study of ectopic bone formation. J. Biomed. Mater. Res 1990; 24:1563–1570.
8. Ohgushi H, Dohi Y, Yoshikawa T, Tamai S, Tabata S, Okunaga K, Shibuya T. Osteogenic differentiation of cultured marrow stromal stem cells on the surface of bioactive glass ceramics. J. Biomed. Mater. Res 1996; 32:341–348.
9. Frankenburg EP, Goldstein SA, Bauer TW, Harris SA, Poser RD. Biomechanical and histological evaluation of a calcium phosphate cement. J. Bone Joint Surg. Am 1998; 80:1112–1124.
10. Jupiter JB, Winters S, Sigman S, Lowe C, Pappas C, Ladd AL, Van Wagoner M, Smith ST. Repair of five distal radius fractures with an investigational cancellous bone cement: a preliminary report. J. Orthop. Trauma 1997; 11:110–116.
11. Thordarson DB, Hedman TP, Yetkinler DN, Eskander E, Lawrence TN, Poser RD. Superior compressive strength of a calcaneal fracture construct augmented with remodelable cancellous bone cement. J. Bone Joint Surg. Am 1999; 81:239–246.
12. Cornell CN, Lane JM, Chapman M, Merkow R, Seligson D, Henry S, Gustilo R, Vincent K. Multicenter trial of Collagraft as bone graft substitute. J. Orthop. Trauma 1991; 5:1–8.
13. Elder S, Frankenburg E, Goulet J, Yetkinler D, Poser R, Goldstein S. Biomechanical evaluation of calcium phosphate cement–augmented fixation of unstable intertrochanteric fractures. J. Orthop. Trauma 2000; 14:386–393.
14. Caplan AI. Mesenchymal stem cells. J. Orthop. Res 1991; 9:641–650.
15. Maniatopoulos C, Sodek J, Melcher AH. Bone formation in vitro by stromal cells obtained from marrow of young adult rats. Cell. Tissue Res 1988; 254:317–330.
16. Yoshikawa T, Ohgushi H, Tamai S, Ichijima K. Self-setting hydroxyapatite cement as a career for bone-forming cells. Biomed. Mater. Eng 1996; 6:345–351.
17. Yoshikawa T, Ohgushi H, Akahane M, Tamai S, Ichijima K. Analysis of gene expression in osteogenic cultured marrow/hydroxyapatite construct implanted at ectopic sites: a comparison with the osteogenic ability of cancellous bone. J. Biomed. Mat. Res 1998; 41:568–573.
18. Yoshikawa T, Ohgushi H, Uemura T, Nakajima H, Ichijima K, Tamai S, Tateisi T. Human marrow cells–derived cultured bone in porous ceramics. Biomed. Mater. Eng 1998; 8:311–320.

19. Ohgushi H, Caplan AI. Stem cell technology and bioceramics: from cell to gene engineering. J. Biomed. Mater. Res 1999; 48:913–927.
20. Bruder SP, Kraus KH, Goldberg VM, Kadiyala S. The effect of implants loaded with autologous mesenchymal stem cells on the healing of canine segmental bone defects. J. Bone Joint Surg. Am 1998; 80:985–996.

19

Aspects of the Clinical Application of Ni-Ti and Ni-Ti-Cu Shape Memory Alloys

F. J. Gil and J. A. Planell
Universitat Politècnica de Catalunya
Barcelona, Spain

I. INTRODUCTION

The shape memory effect and superelasticity are related to a thermoelastic martensitic transformation (austenite → martensite) which can be produced either by cooling or be stress induced [1–5]. Martensite formation can be initiated by cooling the material below M_s, defined as the temperature at which the martensitic transformation begins. M_f is the temperature at which martensitic transformation finishes. The transformation is reversible, A_s being the temperature at which the reverse austenitic transformation (martensite → austenite) begins upon heating and A_f the temperature at the end of the reverse austenitic transformation. In Fig. 1 can be observed the microstructure of the martensite in a orthodontic archwire.

When a stress is applied to the parent phase material above its A_f temperature, a mechanically elastic martensite is stress induced in alloys which exhibit thermoelastic behavior. That is, the deformed material reverts to its original shape when the stress is released. This effect is called superelasticity or pseudoelasticity [6–8]. Figure 2 shows the stress–strain curve for a superelastic alloy.

The stress necesary to induce the formation of stress-induced martensite (SIM) is a linear function of temperature. The critical stress, $\sigma^{SIM \rightarrow AUS}$, increases with increasing temperature, while the yield stress of the parent phase decreases with increasing temperature. Hence, above a certain temperature plastic deformation of the austenite occurs before stress-induced martensite can be formed [9–12].

The size of the hysteresis (i.e., the difference between $\sigma^{AUS \rightarrow SIM}$ and $\sigma^{SIM \rightarrow AUS}$) reflects the amount of irreversible process taking place during the formation of stress-induced martensite [13–16].

When the applied stress is released below A_s, the produced shape change remains since reverse rearrangements of twins and martensite variants have not occurred. However, upon heating through the A_s to A_f temperature range, the material regains its original shape by a reverse transformation from martensite to austenite. Thus, the original shape of the parent phase is obtained and the ''deformation'' of the martensitic phase recovered; this is known as the shape memory effect [2,3].

It should be noticed that when the alloy is excessively deformed, the martensite is plastically deformed by slip processes and that plastic deformation is, of course, not recoverable on subsequent heating.

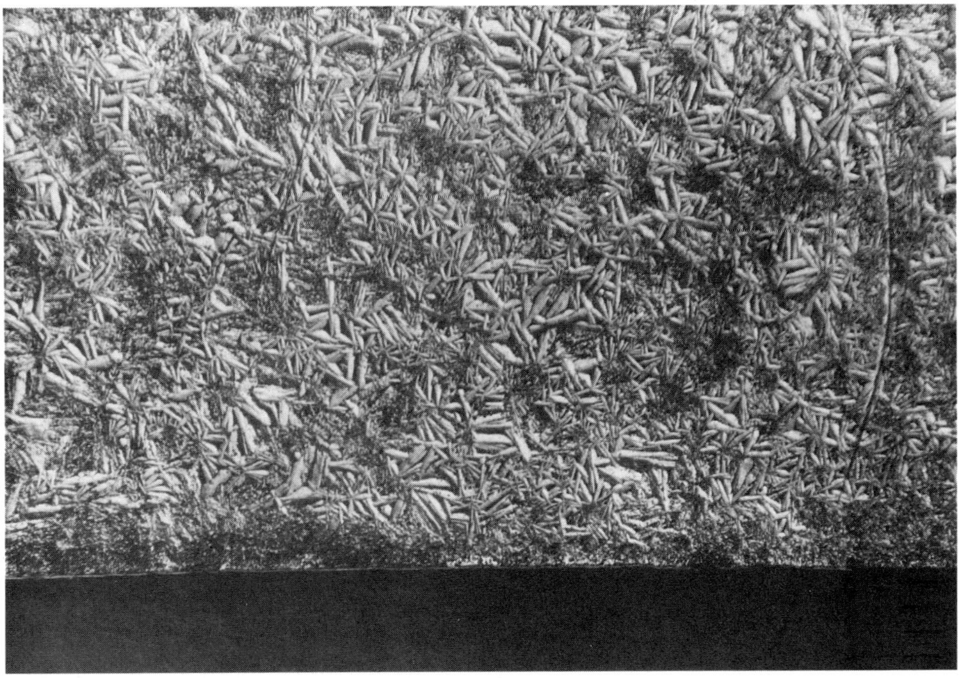

Figure 1 Martensitic microstructure of the Ni-Ti orthodontic archwire.

When shape memory alloys are thought of as candidates to be applied in medical devices, they must be able to fullfil functional requirements related not only to their mechanical reliability, but also to their chemical reliability (in vivo degradation, decomposition and dissolution, corrosion, etc.) and their biological reliability (biocompatibility, cytoxicity, carcinogenicity, antithrombogenicity, antigenicity, etc.). Nickel–titanium (Ni-Ti) is the most important shape memory

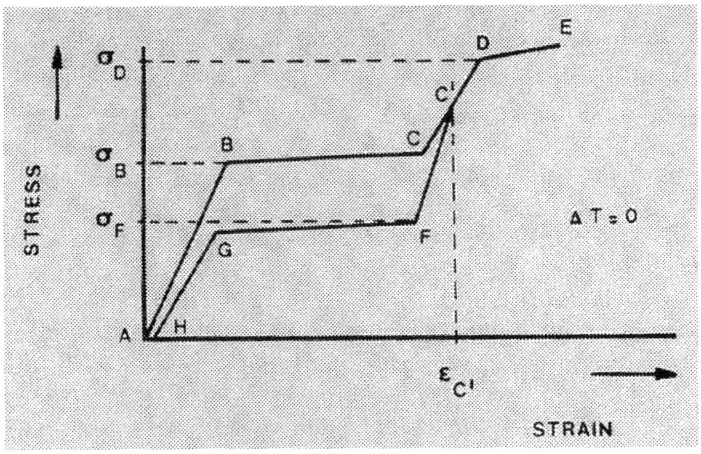

Figure 2 Stress–strain curve of superelastic alloy.

Table I Chemical Composition of Ni-Ti Wires

Alloy	Ni (wt%)	Ti (wt%)
1	55.8 ± 0.2	44.1 ± 0.3
2	69.3 ± 0.1	30.7 ± 0.2
3	69.4 ± 0.1	30.6 ± 0.2
4	69.6 ± 0.2	30.3 ± 0.2
5	64.6 ± 0.3	36.4 ± 0.2
6	62.4 ± 0.2	37.6 ± 0.1
7	63.0 ± 0.1	37.0 ± 0.1

alloy used as a biomaterial—that is to say, a material that comes into contact with and is tolerated by the living tissues, without any deterioration in its properties and in its specific function after implantation. However, besides internal applications, Cu-based shape memory alloys can be used in external biomedical devices where no biocompatibility requirements need to be fullfilled [17–21].

Nickel–Titanium alloy combines the characteristics of shape memory effect and superelasticity with excellent corrosion resistance, wear characteristics, and mechanical properties and good biocompatibility. These properties make it an ideal biological engineering material, especially in orthopedic surgery and orthodontics. Some possible applications of shape memory alloys are connectors, temperature control devices, orthodontic dental archwires, intracranial aneurism clips, vena cava filters, contractile artificial muscles for an artificial heart, orthopedic implants, and other medical devices [22–40].

In most clinical applications the device may be heat treated, and during its life in service it will be cyclically deformed and will be thermally cycled. It is therefore important to investigate the effect of thermal cycles and cyclic strainning and heat treatment upon transformation stresses and material temperatures.

II. EFFECT OF COPPER ADDITION ON Ni-Ti

Nickel–titanium and Ni-Ti-Cu alloys with different chemical compositions (Tables 1 and 2, respectively) were studied. The transformation temperatures were measured by means of a calorimeter, and the stress–strain curves were used to obtain the residual deformation values and the critical stress required to induce the martensitic transformation and their variations with the number of cycles. The transformation temperatures were determined after different numbers of cycles.

Table 2 Chemical Composition of Ni-Ti-Cu Alloys

Alloy	Ni (wt%)	Ti (wt%)	Cu (wt%)
1	49.0 ± 0.1	45.6 ± 0.2	5.4 ± 0.1
2	49.1 ± 0.1	45.2 ± 0.1	5.7 ± 0.2
3	49.5 ± 0.3	45.0 ± 0.1	5.5 ± 0.1
4	49.6 ± 0.1	45.0 ± 0.2	5.4 ± 0.2
5	49.9 ± 0.2	45.1 ± 0.2	5.0 ± 0.1

Table 3 Transformation Temperatures and Thermal Hysteresis (°C) for the Ni-Ti Alloys Studied

Alloy	M_s	M_f	A_s	A_f	ΔT_0
1	27.2 ± 0.3	16.1 ± 0.4	20.0 ± 0.1	32.3 ± 0.7	10.1
2	23.3 ± 0.2	1.2 ± 0.3	5.1 ± 0.4	28.4 ± 0.5	10.3
3	22.4 ± 0.4	14.2 ± 0.6	20.1 ± 0.3	26.5 ± 0.9	10.5
4	20.7 ± 0.1	−5.0 ± 0.4	−1.1 ± 0.2	26.1 ± 0.8	11.1
5	10.9 ± 0.2	−9.0 ± 0.3	−2.2 ± 0.2	15.2 ± 0.7	12.1
6	23.6 ± 0.3	−1.4 ± 0.3	5.1 ± 0.4	28.1 ± 0.6	10.9
7	12.4 ± 0.4	−13.4 ± 0.1	7.3 ± 0.5	16.2 ± 0.9	11.2

The addition of even small concentrations of many third elements to Ni-Ti results in a large change in the M_s temperature such that controllable adjustments of M_s are not easily achieved [13–16]. In contrast, substitution of even large concentrations of Cu do not change the M_s temperature significantly. Tables 3 and 4 show the transformation temperatures and thermal hysteresis for different Ni-Ti and Ni-Ti-Cu shape memory alloys of different chemical compositions [41,42].

From these results it can be noticed that small chemical composition changes produce large variations in the transformation temperatures for Ni-Ti alloys: a variation of 0.6% in Ni (alloy 6: 62.4% in Ni; alloy 7: 63.0% in Ni) produces a change in the M_s temperature of 11.2°C (alloy 6: M_s = 23.6°C; alloy 7: M_s = 12.4°C). However, for Ni-Ti-Cu alloys (Cu content ranges usually between 5 and 10%) the transformation temperatures are much more stable in relation to changes in the chemical composition. In this case, a variation of 0.6% in Ni (alloy 1: 49.0% in Ni; alloy 4: 49.6% in Ni) produces a change in the M_s temperature of 0.8°C (alloy 1: M_s; 18.7°C alloy 4: M_s = 17.9°C). Moreover, although M_s is insensitive to the substitution of Ni by Cu, M_s decreases as Cu substitutes Ti. The presence of Cu also makes the M_s temperature less sensitive to variations in the Ni-Ti ratio. A lower concentration dependent M_s allows for easier production of commercial quantities of materials having a controlled M_s for thermal sensor and actuator uses.

Calorimetric measurements indicate that alloys with copper have substantially narrower hysteresis than the binary alloy. Tables 3 and 4 list the hysteresis values (ΔT_0), measured as the temperature difference between the peaks on the calorimetric curves upon heating and cooling. It can be seen that this hysteresis is reduced from around 10°C for the binary alloy to less than 4.5°C for material with copper. Further Cu addition decreases the hysteresis only slightly, with a 5.7% Cu alloy having a width of only 3.5°C.

The transformation stresses (β↔stress induced martensite) are shown in Table 5. A stress hysteresis is associated with the transformation, as the diferrence between the critical stresses

Table 4 Transformation Temperatures and Thermal Hysteresis (°C) for the Ni-Ti-Cu Alloys Studied

Alloy	M_s	M_f	A_s	A_f	ΔT_o
1	18.7 ± 0.1	2.1 ± 0.2	14.0 ± 0.2	37.1 ± 0.6	4.5
2	17.8 ± 0.5	1.7 ± 0.3	13.8 ± 0.2	37.0 ± 0.8	3.5
3	16.2 ± 0.4	2.0 ± 0.4	13.2 ± 0.2	36.8 ± 0.8	3.7
4	17.9 ± 0.3	1.9 ± 0.6	14.3 ± 0.4	36.9 ± 0.9	3.9
5	17.6 ± 0.2	2.3 ± 0.7	14.2 ± 0.4	37.0 ± 1.0	4.4

Table 5 Critical Stresses at Different Test Temperatures

	NiTi				NiTiCu			
	$\sigma^{\beta\rightarrow SIM}$ (MPa)		$\sigma^{SIM\rightarrow\beta}$ (MPa)		$\sigma^{\beta\rightarrow SIM}$ (MPa)		$\sigma^{SIM\rightarrow\beta}$ (MPa)	
Alloy	20°C	37°C	20°C	37°C	20°C	37°C	20°C	37°C
1	230 (15)	320 (22)	65 (9)	151 (19)	150 (17)	287 (24)	55 (5)	211 (19)
2	235 (23)	331 (29)	70 (8)	155 (24)	155 (22)	291 (34)	60 (10)	215 (23)
3	247 (18)	333 (32)	77 (7)	156 (18)	160 (32)	299 (31)	65 (32)	216 (9)
4	260 (12)	350 (20)	79 (3)	166 (18)	154 (3)	290 (21)	59 (7)	216 (17)
5	280 (12)	354 (23)	80 (5)	160 (12)	159 (6)	287 (16)	61 (12)	214 (23)

Standard deviation in parentheses.

(stress for inducing martensitic transformation due to loading and the reverse transformation upon unloading). This stress hysteresis is much narrower for Ni-Ti-Cu alloys (≅70 MPa) than for the binary alloys (≅150 MPa). The narrower hysteresis of Ni-Ti-Cu alloys has practical importance in engineering applications requiring a fast response time on thermal cycling. Besides, the narrower stress hysteresis of Ni-Ti-Cu means that for a similar process condition, the unloading or reversion stress is higher. The stored energy density in the Ni-Ti-Cu superelastic material is correspondingly higher than that of the binary alloy.

The magnitude of the stress is very important when the bone must be remodeled, as for example in orthodontics: the elastic deformation of an orthodontic wire and the subsequent release of its elastic energy over a period of time gives rise to the correcting forces. It is generally assumed that optimal tooth movement is achieved by applying forces that are low in magnitude and continuous in nature. Such forces minimize tissue destruction and produce a relatively constant stress in the periodontal ligament during tooth movement. The superelasticity of Ni-Ti and Ni-Ti-Cu archwires allows the orthodontist to apply an almost continuous light force with larger activations, which results in the reduction of tissue trauma and the patient discomfort, thus facilitating enhanced tooth movement.

In contrast, forces that are high in magnitude encourage hyalinization of the periodontal ligament and may cause irreversible tissue damage such as root resorption. The Ni-Ti archwires produce teeth movement with greater efficiency and in a shorter time when compared to other orthodontic alloys, and it is especially appropriate in situations requiring large deflections of an orthodontic archwire such as a preliminary bracket alignment stage in the load deflection stage in orthodontic therapy [3–10].

III. EFFECT OF LOAD CYCLING

Different numbers of fatigue cycles were applied up to 3% of maximum strain controlled by means of extensometry with $R=0$ and 1 Hz. The stress–strain curves were used to obtain the residual deformation values and the critical stress required to induce the martensitic transformation and their variations with the number of cycles. The transformation temperatures were determined after different numbers of cycles.

The stress–strain curves at different numbers of cycles were used to obtain the variation of the critical stress necessary to produce the superelastic effect and to determine the optimal working stress for the alloy. A relevant result obtained is that in contrast to what happens in

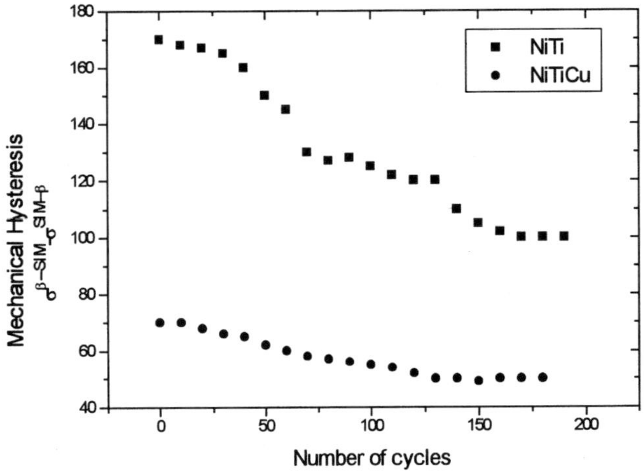

Figure 3 Mechanical hysteresis versus the number of load cycles.

Ni-Ti alloys, Ni-Ti-Cu alloys show stable superelasticity characteristics when cyclically loaded with small stress hysteresis. It is clear that the hysteresis is smaller and the change of stress–strain curve by cyclic deformation is more stable in the Ni-Ti-Cu alloy, as can be observed in Fig. 3.

Figure 4 shows the critical stress required to induce the martensitic transformation as a function of the number of cycles. From the results shown in Fig. 4 it can be noticed that the critical stress required to induce the martensitic transformation decreases with the number of cycles up to saturation for the Ni-Ti, while the Ni-Ti-Cu does not present any changes [42,43].

It should be emphasized that above a certain number of cycles, about 50 load/unload cycles in this case, the critical stress for Ni-Ti alloys reaches saturation. This suggests that the

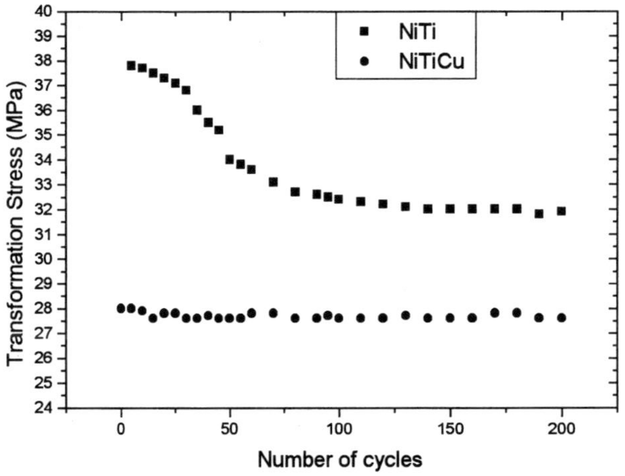

Figure 4 Transformation stress versus the number of cycles for Ni-Ti and Ni-Ti-Cu shape memory alloys.

transformation is stabilized around a practically constant stress and, even more important, this occurs at a relatively low number of cycles. The results of the transformation temperatures in relation to the load cycles are shown in Fig. 5. It was observed that the transformation temperature to martensite, increases and the transformation stress decreases with the number of cycles. Such decrease in the transformation stress may be understood by the arresting of martensitic plates by dislocations, inducing the nucleation of new plates at M_s and producing an increase in the transformation temperature.

The increase in the A_s temperature is also shown; the reason for this is that the sample contains a large amount of stabilized martensite when the load cycles increase. The M_f and A_f are practically constant in relation to the number of cycles. For Ni-Ti-Cu, the transformation

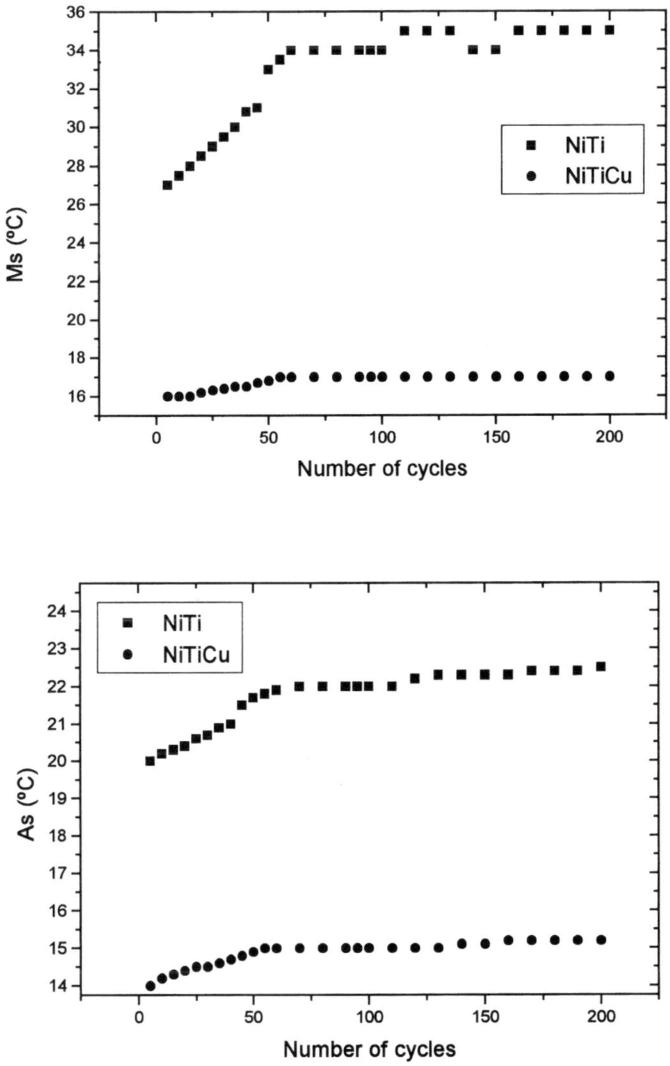

Figure 5 Transformation temperatures versus the number of cycles.

temperatures are kept constant with the load cycles. The above results of the small stress hysteresis and the stability against cyclic loading suggest that the interface between the parent and martensitic phases can move easily without producing a large amount of defects such as dislocations. This is reasonable when considering the fact that the habit plane of the Ni-Ti-Cu alloy is plain since the lattice invariant shear is not necessary in order to form the habit plane in this alloy, whose crystal structure is orthorhombic [5,7,41,45].

It is assumed that optimal movcmcnt is achieved when the applied force is low, stable, and continuous in order to minimize tissue destruction and to produce a constant stress on the tissues. The present results seem to provide a material where the control of the applied forces by the doctor can be successfully achieved.

IV. THERMAL CYCLING EFFECTS

The samples were thermally cycled at 300 and 37°C both for 10 min followed by cooling in water at 20°C. Transformation temperatures were determined at different numbers of thermal cycles.

For the samples cycled up to 100,000 cycles between 37 and 20°C, the transformation temperatures do not change with thermal cycling. The temperatures are low to produce aging effects. However, for the samples cycled between 300 and 20°C, small changes in the transformation temperatures are observed.

The thermal cycling effect on the transformation temperatures of the samples cycled at 300°C is shown in Fig. 6, where M_s and A_s are plotted against the number of cycles. All the samples studied show a light decrease of M_s during thermal cycling, while A_s increases with the number of cycles. Miyazaki et al. [50,51] studied the thermal cycling effects by means of transmission electron microscopy. The observation of the samples revealed fine Ti_3Ni_4 precipitates. These precipitates can be formed in Ni-Ti alloys with more than 50.5% at%Ni [46,47]. When these precipitates are formed the titanium content decreases into the parent phase, and

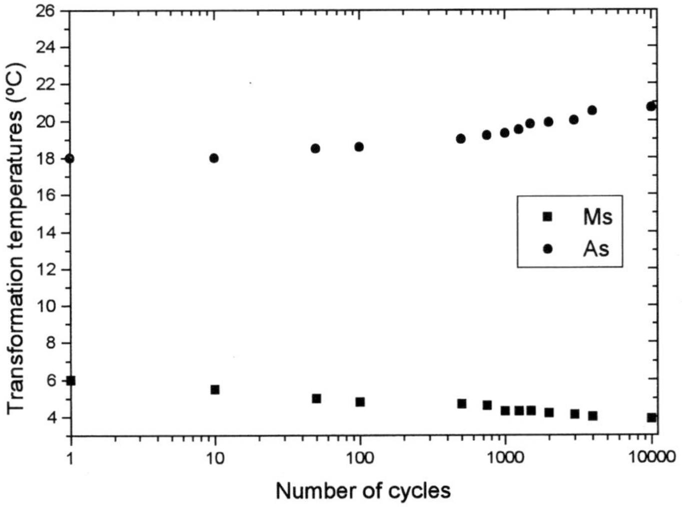

Figure 6 Variation of the transformation temperatures as a function of the thermal cycles.

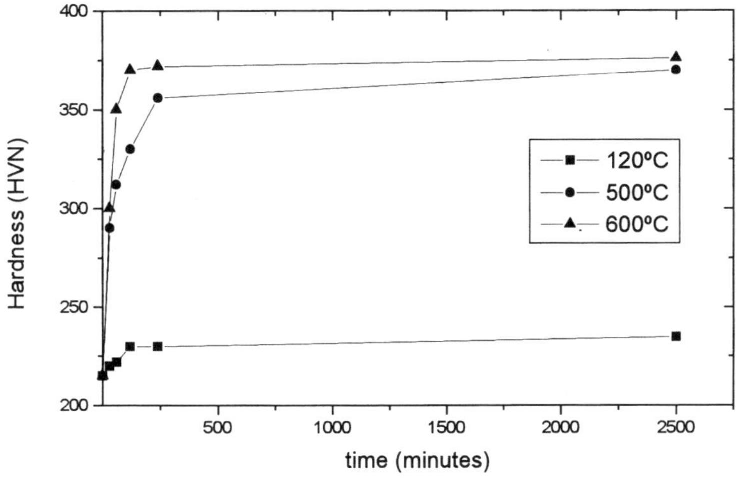

Figure 7 Evolution of the hardness with different heat treatment times and temperatures.

consequently the M_s temperature decreases as thermal cycling takes place. The increase of A_s can be explained by the obstruction of the precipitates to the martensitic plates' contraction when the sample is heated.

V. EFFECT OF AGING

The aging of these austenitic specimens produced different degrees of decomposition depending on the temperature and the heating time. The hardness evolution with the annealing time is shown in Fig. 7 [42,43]. The aging in the austenitic phase of the Ni-Ti alloy is observed at high temperatures (500 and 600°C), producing an increase in hardness of the samples. At each temperature the curve of hardness verses heating time shows a sigmoidal shape, indicating that the precipitation occurs by a nucleation and growth process controlled by atom diffusion. The heat treatment at 120°C shows a constant value of hardness due to the low temperature for starting the precipitation process. This hardening is produced by the precipitation of $Ti_{11}Ni_{14}$.

This precipitation produces an important variation in the chemical composition and consequently in the temperatures and the stresses of the transformation [42,43]. At 6 h of heat treatment at 500 and 600°C, the precipitation reaches completion as shown by the high hardness saturation value (370 HV). The microstructure obtained at high annealing temperatures is shown in Fig. 8. A relevant finding is that under these circumstances the alloy loses its shape memory and superelastic properties, the reason being the important chemical composition variation. However, these properties can be recovered by means of a heat treatment at 900°C for 30 min, which is able to dissolve the precipitates.

ACKNOWLEDGMENTS

The authors are grateful to CICYT for the financial support of this study with reference MAT 98–0415.

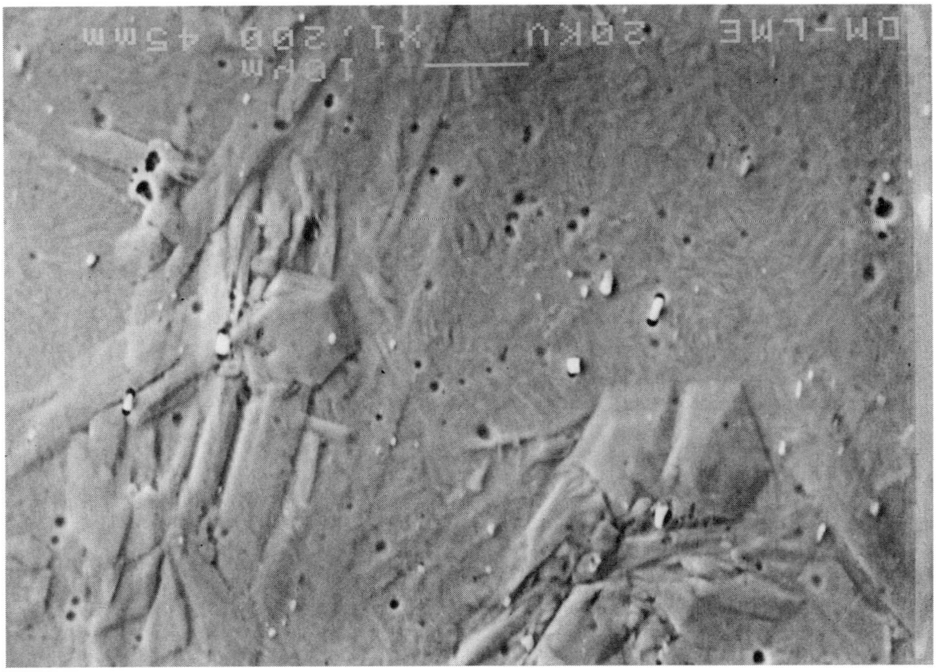

Figure 8 Microstructure of the Ni-Ti alloy, where the $Ti_{11}Ni_{14}$ precipitates are shown.

REFERENCES

1. Duerig TW, Zadno R. Engineering Aspects of Shape Memory Alloys. London: Butterworth-Heinemann, 1990:124–132.
2. Tosho S. Shape Memory Alloys Funakubo H, ed. Vol. 1. Tokyo: Gordon and Breach Science Publishers, 1984:23.
3. Delaey L, Krishnan RV, Tas H, Warlimont H. Review: thermoelasticity, pseudoelasticity and the shape memory effects associated with martensitic transformation. J.Mater.Sci 1974; 9:1521.
4. Delaey L, Deruyttere L, Aernoudt E, Roos J. Shape memory effect, superelasticity and damping in copper–zinc–aluminium alloys, Rapport 78R1, Department Metaalkunde. Belgium: Katholieke Universiteit Leuven, 1978.
5. Krousbrock R. Shape memory alloys. In Metal and Ceramic Biomaterials Ducheyne P, Hastings GW, eds. Vol. II. Boca Raton, Florida: CRC Press, 1984:82.
6. Wayman CM, Duering TW. An introduction to martensite. In Shape Memory in Engineering: Aspects of Shape Memory Alloys Duering TW, Melton KN, Stöckel D, Wayman CM, eds. London: Butterworth-Heinemann, 1990.
7. Wasilewski RJ. On the nature of the martensitic transformation. Metall. Trans 1975; 6A:1405.
8. Otsuka K, Shimizu K. Pseudoelasticity and shape memory effect. Int. Metals. Rev 1976; 31(3):93.
9. Delaey L. Diffusionless transformations. Materials Science and Technology. 5 Ed Haasen P, ed. Phase Transformations in Materials: VCH, 1987:339–404.
10. Perkins J. Shape Memory Effects in alloys. New York: Plenum Press, 1975:12.
11. Li L, Aernoudt E, Delaey L. Hysteresis effects of martensitic transformation during thermomechanical cycling. Scripta Met 1988; 22:1435–1440.
12. Zhongguo W, Dazhi Y. On the hysteresis loops and characteristic temperatures of thermoelastic martensitic transformations. Scripta Met 1988; 22:1245–1249.

13. Jardine AP. Calorimetric techniques for the evaluation of thermal efficiencies of shape memory alloys. J Mater. Sci 1989; 24:2587–2593.
14. Otsuka K, Shimizu K. Proc. Int. Conf. Solid–solid phase transformations. Pittsburgh, 1981:1267.
15. Golestaneh AA. Energetic shape recovery associated with martensitic transformation in shape memory alloys. Acta Met 1980; 28:1427–1436.
16. Mellor BG. Shape memory effects. In The Science and Technology of Shape Memory Alloys. COMETT Course. Barcelona: Ed. Univ. Illes Balears, 1980.
17. Melton KN. NiTi based shape memory alloys. In Shape Memory in Engineering; Aspects of Shape Memory Alloys Duering TW, Melton KN, Stöckel D, Wayman CM, eds. London: Butterworth-Heinemann, 1990:21–35.
18. Lu S. Medical applications of NiTi alloys in China. In Shape Memory in Engineering: Aspects of Shape Memory Alloys Duering TW, Melton KN, Stöckel D, Wayman CM, eds. London: Butterworth-Heinemann, 1990:445–446.
19. Castleman LS, Motzkin SM. Biocompatibility of clinical implant materials Williams DF, ed. in CRC Series in Biocompatibility. Boca Raton, FL: CRC Press, 1981.
20. Castleman LS, Motzkin SM, Alicandri FP, Bonawit VL. Biocompatibility of nitinol alloys as implant materials. J. Biomed. Mater. Res 1976; 10:695–731.
21. Cutright DE, Bhaskar SN, Pérez B, Johnson RM, Cowan GSM. Tissue reaction to nitinol wire alloy. Oral Surg 1973; 35:578.
22. Haarsters J, Salis-Solio G, Bensmann G. The use of NiTi as an implant material In orthopedics. In Shape Memory in Engineering: Aspects of Shape Memory Alloys Duering TW, Melton KN, Stöckel D, Wayman CM, eds. London: Butterworth-Heinemann, 1990:426–427.
23. Baumgart F, Bensmann G, Hartwig H. Tech. Mitt. Krupp, Forsch-Ber 1977; 35:157.
24. Sekiguci Y. Medical applications. In Shape Memory Alloys Funakubo H, ed: Gordon and Breach Science Publishers, 1984:10–23.
25. Hughes JL. Evaluation of nitinol for use as a material in the construction of orthopaedic implants, DAMD 17-74-C-4041. Fort Detrick, Frederick. MD: US Army Medical Research and Development Command, 1977.
26. Haasters J, Baumgart F, Bensmann G. Memory alloys—new material for implantation in orthopedic surgery, part 2. In Current Concepts of Internal Fixation of Fractures Uthoff HK, ed. New York: Springer-Verlag, 1980:128.
27. Schmerling MA, Wilkov MA, Sandres AE, Wooseley JE. Using the shape recovery of nitinol in the Harrington rod treatment of scoliosis. J.Biomed. Mater.Res 1976; 10:879–902.
28. Hamanaka H, Miura I, Dohi H, Suzuki Y. Proceedings of the 8th Annual Meeting of the Society of Biomaterials, 1982.
29. Hamada H, Miyagi M, Onishi O. Proceedings of the 4th Conference of the Japanese Society for Biomaterials.
30. Krousbroeck R, Van der Perre G, Aernoudt E, Mulier JC. Shape memory effect in biomedical devices. In Advances in Biomaterials Winter GD, Gibbons DF, Plenk H, eds. Vol. 3. New York: John Wiley & Sons, 1982:767.
31. Gil FJ, Fernández E, Manero JM, Planell JA, Sabrià J, Cortada M, Giner L. A study of load cycling in a NiTi alloy with superelastic behaviour used in dental prosthetic fixators. Biomed. Mater Eng 1996; 6(3):153–158.
32. Sabrià J, Cortada M, Giner L, Fernández E, Gil FJ, Manero JM, Planell JA. A study of load cycling in a NiTi alloy with pseudoelastic behaviour used in dental prosthetic fixators. Biomed. Mater Eng 1995; 5(3):161–167.
33. Andreasen GF, Morrow RE. Laboratory and clinical analysis of nitinol wire. Am. J. Orthod 1978; 73:142–149.
34. Andreasen GF. A clinical trial of alignment of teeth using a 0.019 inch thermal nitinol wire with a transition temperature range between 31°C and 45°C. Am. J. Orthod 1980; 78:528–536.
35. Libenson C, Gil FJ, Planell JA. Untersuchung der Nickel-Titandrähte auf ihre Verwendbarkeit in der Klinischen Kieferorthopädie. Orthodonthie & Kieferorthopädie 1996; 28(1):135–154.
36. Miura F, Mogi M, Ohura Y, Karibe M. The superelastic Japanese NiTi alloy wire for use in orthodontics. Am. J. Orthod. Dentofac. Orthop 1988; 94(2):89–96.

37. Burstone CJ, Qin B, Morton JY. NiTi archwire a new orthodontic alloy. Am J Orthod 1985; 87(6): 445–452.
38. Libenson C, Gil FJ, Planell JA. Etude des caractéristiques des fils NiTi à usage orthodontique. Revue Orthopedic Dento Faciale 1996; 30:129–141.
39. Libenson C, Gil FJ, Planell JA. Untersuchung der Nickel-Titandrähte auf ihre Verwendbarkeit in der Klinischen Kieferorthopädie. Orthodonthie & Kieferorthopädie 1996; 28(1):135–154. 18.
40. Marcotte MR. Biomechanics in Orthodontics. Ontario: Ed. MR. Masson, 1992:324.
41. Shugo Y, Shimizu T, Morii K, Yamada T, Kusaka K. Effects of heat treatment and thermal cycle on helical spring properties in NiTiCu alloy. Proceedings of the International Conference on Martensitic Trnaformation. 1993:1259.
42. Gil FJ, Planell JA. Effect of copper addition on the superelastic behaviour of NiTi shape memory alloys for orthodontic applications. J Biomed Mater Res Appl Biomater 1999; 48:682–688.
43. Gil FJ, Planell JA. In vitro thermomechanical ageing of NiTi alloys. J Biomater Appl 1998; 12: 237–248.
44. Furuya Y, Matsumoto M, Matsumoto T. Mechanical properties and microstructure of rapidly solidified NiTiCu alloy. Proceedings of the International Conference on Martensitic Transformation. 1993: 905.
45. Saburi T, Tatsumi T, Nenno S. Effects of heat treatment on mechanical behaviour of TiNi alloys. Journal de Physique, Colloque C4 1982; 12(43):261–266.
46. Miyazaki D, Otsuka K. Metallurgical Transactions. 1986; A 4:53.
47. Miyazaki D, Imai T, Otsuka K, Suzuki Y. Journal de Physique 1982; 43(Suppl. 12):C4–255.

Index